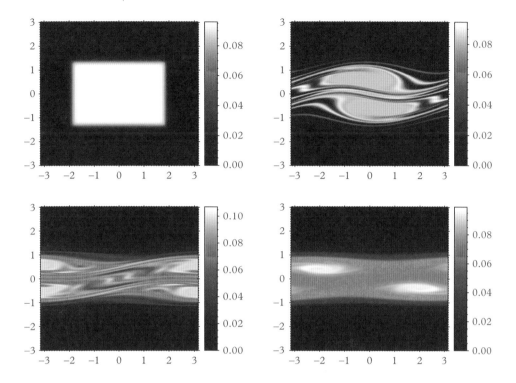

Figure 9.9 *Simulation of the Vlasov equation (9.47) that starts from a* water-bag *initial condition with* $\varepsilon = 0.69$ *and* $m_0 = 0.5$. *The final snapshot (lower right panel) is the quasi-stationary state. In the plots, the horizontal and vertical axes refer to* θ *and* p, *respectively. Reprinted from Campa* et al. *(2009),* © *2009, with permission from Elsevier.*

Figure 10.1 *A spectacular image of the large spiral galaxy NGC 1232. The image is based on three exposures in ultra-violet, blue and red light, respectively. The central core contains older stars of reddish colour, while the spiral arms are populated by young, blue stars and many star-forming regions. Courtesy of ESO.*

PHYSICS OF LONG-RANGE INTERACTING SYSTEMS

Physics of Long-Range Interacting Systems

A. Campa
Istituto Superiore di Sanità, Rome, Italy

T. Dauxois
CNRS & ENS de Lyon, France

D. Fanelli and S. Ruffo
Università degli Studi di Firenze, Italy

OXFORD
UNIVERSITY PRESS

OXFORD
UNIVERSITY PRESS

Great Clarendon Street, Oxford, OX2 6DP,
United Kingdom

Oxford University Press is a department of the University of Oxford.
It furthers the University's objective of excellence in research, scholarship,
and education by publishing worldwide. Oxford is a registered trade mark of
Oxford University Press in the UK and in certain other countries

Published in the United States of America by Oxford University Press
198 Madison Avenue, New York, NY 10016, United States of America

British Library Cataloguing in Publication Data
Data available

Library of Congress Control Number: 2014933928

ISBN 978-0-19-958193-1

Printed in Great Britain by
Clays Ltd, St Ives plc

Foreword

The statistical mechanics of systems with long-range interactions is a special branch of statistical mechanics aimed at providing a theoretical framework that accounts for the intriguing properties of these systems. In these systems particles interact via a potential which decays algebraically at large distance with a sufficiently small power law. They are rather common in nature; examples include self gravitating systems, dipolar magnets, plasmas and many others. Due to the long-range nature of the interactions, these systems are not additive in the sense that the energy, free energy and other thermodynamic functions of a system composed of two sub-systems are not necessarily equal to the sum of the corresponding quantities of the two sub-systems. As a result, many of the common properties of systems with short-range interactions are not shared by systems with long-range interactions. These include the convexity of the free energy, equivalence of the various statistical mechanical ensembles, the non-negativity of response functions such as specific heat and compressibility, and other properties. While some of the peculiar features of these systems have been recognized and understood long ago, extensive activity in this field during the last twenty years has led to much progress in the understanding of the thermodynamic and dynamical features of these systems. This book, the authors of which have been a driving force and major contributors to the progress made in this field in recent years, presents a clear, coherent and comprehensive summary of the state of the art in this field. It serves as an excellent introduction for those who wish to enter the field of long-range interacting systems, and at the same time it could be used as a reference book for researchers working in this field.

The book is divided into three parts each of which is to a large extent self-contained. Part I yields a detailed account of the static and equilibrium properties of systems with long-range interactions. It contains a general derivation of the main features of these systems as well as numerous examples of models in which these features can be clearly demonstrated. In fact, this part can serve as a very nice addition to any textbook on statistical mechanics. In deriving the general principle of statistical mechanics, typical introductory books on the subject assume right from the beginning that one is dealing with systems with short-range interactions, and hence the various thermodynamic potentials are additive. The discussion of the effect of long-range interactions is usually deferred to a later stage when, say, gravitational or other long-range systems are considered and is thus done with a specific system in mind. Here, on the other hand, a very clear and comprehensive discussion based on general principles is presented yielding an insight into the question of how the general principles of statistical mechanics are modified if one discards the assumption of additivity. This, by itself, clarifies some of the assumptions made in the general derivation of the principles of statistical mechanics and it would be very helpful to students taking an introductory course on statistical mechanics. The numerous examples of simple exactly soluble models given in this part of the book are rather easy to follow and are thus very helpful in demonstrating the general principles underlying the thermodynamics of systems with long-range interactions.

Part II is devoted to dynamical properties of systems with long-range interactions, a subject which has witnessed considerable progress in recent years. After a general discussion of kinetic theories this part of the book considers the dynamics and stability of quasistationary states which are typical to systems with long-range interactions. These are long-lived non-Gibbs–Boltzmann (GB) states which relax to the GB state over a time scale which diverges with the system's size. As a result these states become stable non-GB states in the thermodynamic limit. Various approaches for analysing such states and studying their stability are described including the Lynden–Bell maximum entropy approach and a more recent insightful approach based on a careful analysis of the Vlasov equation.

The third part of the book is devoted to applications to various physical systems. It provides a particularly useful account of the rather broad spectrum of systems and phenomena where long-range interactions play a role. These include self-gravitating systems, geophysical fluid mechanics, coulomb systems, plasmas, free electron lasers, where the coupling of the electromagnetic waves to the electrons is of long-ranged nature, and dipolar systems.

The book serves as a reference volume in the field of long-range interacting systems where the various features of these systems are presented in a coherent way. On the one hand it makes the field easily accessible to newcomers and on the other hand it provides a guide to the vast literature on the subject which will be useful to researchers active in the field. It is a welcomed timely contribution.

David Mukamel
The Weizmann Institute, Rehovot, May 2014

Preface

Physical systems for which the interaction potential decays as a power of the inverse interparticle distance with an exponent smaller than the dimension of the embedding space are called long-range interacting systems. Although different aspects of these systems have been tackled in the past in specific scientific communities, notably astrophysics and plasma physics, this has not constituted immediately a seed for more general theoretical studies. However, in the past 15 years, it has become progressively clear that the ubiquitous presence of long-range forces needs an approach that integrates different methodologies. This observation stimulated a renewed and widespread interest in long-range systems throughout numerous research groups, leading nowadays to a much better understanding of both their equilibrium and out-of-equilibrium properties.

In this book, we will focus on the statistical physics of such systems. For a long time, the application of methods and tools borrowed from statistical mechanics to long-range systems has been questioned, mainly due to the lack of additivity that descends from the specific nature of the interaction. In fact, although gravitation and electromagnetism are central topics, covered in Bachelor's and Master's courses in physics, they are not discussed in statistical mechanics classes and in the very large majority of textbooks devoted to this field. It is now well understood that non-additivity does not hinder a formal statistical mechanics treatment. However, it is at the root of basic facts and concepts that at first appeared odd, like ensemble inequivalence, negative specific heat, negative susceptibility and ergodicity breaking. These are all features that called for an extended statistical mechanics description, beyond its standard domain of applications. Moreover, the great richness of the out-of-equilibrium properties of systems with long-range interactions, and especially the presence of long relaxation times towards equilibrium and of anomalous diffusion, resonate with the current emphasis on non-equilibrium statistical mechanics in general.

Having moved onto a more mature period, this field, besides strengthening its foundations, has incorporated methods and tools originally developed in other disciplines. General perspectives of this evolution have been presented in several proceedings of workshops and schools (Dauxois *et al.*, 2002a, 2009; Campa *et al.*, 2008b). Moreover, several interesting reviews have been recently published. Some of them concentrate more on models (Campa *et al.*, 2009; Bouchet *et al.*, 2010; Levin *et al.*, 2014), while others deal with specific physical systems: gravitation (Padmanabhan, 1990; Chavanis, 2006a), Coulomb systems (Brydges and Martin, 1999) and hydrodynamics (Bouchet and Venaille, 2012). In this book, we present the material at a more introductory level than all these very useful references have done. Moreover, we emphasize the interdisciplinary aspects and the applications in several branches of physics.

The structure of the book is as follows.

In Part I, we present the statistical physics approach, which allows us to deal with non-additive systems. We discuss paradigmatic examples, which are simple and general enough to be useful also for specific applications. The main problems emerging from these examples, the novel technical tools that we apply and the main features of the solutions that we obtain are carefully examined and proposed using a language amenable to beginners in the field, although relying on some previous knowledge of basic thermodynamics and statistical mechanics. This part begins with a presentation of statistical mechanics for short-range systems, before emphasizing the novelties introduced by the long-range nature of the interaction. Equilibrium statistical mechanics solutions of simple mean-field toy models are extensively presented as a first step for understanding phase diagrams of more complex long-range systems. We then go beyond mean-field models, before a final excursus on related quantum aspects.

In Part II, we discuss the essential properties of out-of-equilibrium dynamics. Relaxation to equilibrium is a key topic in systems with long-range interactions, which can be historically traced back to pioneering studies on self-gravitating systems. This part of the book begins with an introductory presentation of the kinetic equations in the framework of short-range systems. Then, we proceed by emphasizing the differences introduced by long-range interactions, discussing several kinetic equations: Klimontovich, Vlasov, Lenard-Balescu and Fokker-Planck. By considering a simple model, that is nevertheless representative of long-range interacting systems, we discuss the complex process of relaxation to equilibrium.

Finally, Part III deals with the different physical systems in which long-range interactions are important. We begin of course with the gravitational interaction, but long-range couplings play also a crucial role in a large gallery of interesting problems in physics. Different chapters are dedicated to distinct applications for which long-range features are important. In particular, we discuss equilibrium and out-of-equilibrium aspects of two-dimensional hydrodynamics, cold Coulomb system, hot plasmas, wave–particles systems and, finally, dipolar systems.

While preparing this textbook, our ambition was threefold:

- Master's and PhD degree students were, of course, at the heart of our objectives when organizing the content and the style of the book. Only basic knowledge in physics and in particular statistical mechanics is required to understand it. Moreover, we have tried to explain pedagogically the different steps in all derivations, privileging simple models in the first two parts, while considering physical systems in the third.
- We also hope that this book can be useful to more experienced colleagues, in particular those interested in this domain of research, which has significantly expanded in the past 15 years, but also colleagues who have been teaching statistical mechanics for years and might appreciate the new perspectives that the fascinating field of long-range systems can give to their lectures. Mean-field methods constitute here excellent approximations, or are even exact, as opposed to their applications to short-range systems. Microcanonical ensemble becomes central. Several models

which could be useful for renovating tutorials or preparing original exams can also be found.

- Theoretical results obtained in the study of systems with long-range interactions have nowadays reached a level of understanding that might allow, in the near future, the experimental verifications of effects like negative specific heat, ergodicity breaking and quasi-stationary states. We hope that this book will enhance this possibility, by offering a general view of this domain to experimentalists. We look towards realizations of well-devised experiments that could lead to the observation of the effects predicted by the theory. This book has also been prepared in this spirit.

If all or part of these objectives will be attained, only the future will tell. Anyway, writing a book is always an adventure: starting to write a book is an easy task; finishing it is quite an achievement. The pleasure is even more vivid when the book is written with friends. Along the years, numerous meetings in our different host cities, Florence, Lyon and Roma, have allowed, of course, many scientific discussions, but also gastronomical discoveries, wine tasting, cultural visits, football discussions and even running marathons! It was really a pleasure, and we hope you will enjoy and taste the fruit of our labour.

January 2014
A. Campa, T. Dauxois, D. Fanelli, S. Ruffo

Acknowledgements

We would like to warmly thank our collaborators Angel Alastuey, Mickael Antoni, Andrea Antoniazzi, Malbor Asllani, Erik Aurell, Romain Bachelard, Julien Barré, Fausto Borgonovi, Freddy Bouchet, Francesco Califano, Timoteo Carletti, Giuseppe-Luca Celardo, Maxime Champion, Cristel Chandre, Pierre-Henri Chavanis, Giovanni De Ninno, Yves Elskens, Marie-Christine Firpo, Andrea Giansanti, Dieter Gross, Alessio Guarino, Shamik Gupta, Haye Hinrichsen, Peter Holdsworth, Ramaz Khomeriki, Tetsuro Konishi, Hiroko Koyama, Raman Johal, Vito Latora, Xavier Leoncini, Stefano Lepri, François Leyvraz, Leonardo Lori, George Miloshevich, Gianluca Morelli, Daniele Moroni, David Mukamel, Cesare Nardini, Jean-Pierre Nguenang, Francesco Piazza, Alessandro Pluchino, Andrea Rapisarda, Florence Raynal, Nir Schreiber, Luca Sguanci, Fabio Staniscia, Takayuki Tatekawa, Alessandro Torcini, Alessio Turchi, Hugo Touchette, Antoine Venaille and Yoshi Yamaguchi.

We thank for financial support Contract ANR-10-CEXC-010-01 (France), Prin-2009 (Italy), Istituto Nazionale di Fisica Nucleare (Italy) and the hospitality of ENS-Lyon, Università degli Studi di Firenze and Università 'La Sapienza' di Roma, which has facilitated several discussion meetings.

Contents

Part I

Static and Equilibrium Properties

1

Basics of Statistical Mechanics of Short-Range Interacting Systems

Statistical mechanics has had enormous success describing many-body systems. The fulfilment of its founding purpose, the derivation of the properties of macroscopic bodies, in particular of their thermodynamical behaviour, from the microscopic laws governing the dynamics, can be considered as one of the greatest achievements of physics.

Probabilistic concepts are at the heart of statistical mechanics. They are implemented through the introduction of statistical ensembles. Statistical ensembles are conceptual constructions, in which the system under study is thought to be replicated an *infinite* number of times. Each system in this collection is in one of its accessible microscopic states. This idealization aims at describing a system under given macroscopic conditions; note that microscopic dynamics is absent in this approach. These macroscopic conditions pose constraints on the possible microscopic states, but nevertheless the number of accessible states remains huge, typically exponential in the number of microscopic constituents. *A priori*, the system can be in any state compatible with conservation laws.

The interactions between the constituents of a macroscopic system usually decay very fast with the distance. In these *short-range* systems, the energy of interaction between two systems that are in contact is entirely accounted for by the superficial layers lying in the contact region whose thickness is of the order of the molecular length scale. The ratio of this interaction energy to the energy of any one of the two systems is very small, tending to 0 when the systems size increases. This is the *additivity* property of short-range systems.

Let us now suppose that at large distance r the decay of the two-body interaction potential $V(r)$ goes like

$$V(r) \propto \frac{1}{r^{\alpha}} \quad \text{for } r \gg \ell, \tag{1.1}$$

with $\alpha \geq 0$ and ℓ a typical microscopic length scale. The microscopic constituents may be atoms, molecules or whatever are the elementary units that define the inspected system. We will see in details in Chapter 2 that if $\alpha \leq d$ in d dimensions (i.e. $\alpha \leq 3$ in ordinary three-dimensional space), additivity does not hold, and the interaction energy

Physics of Long-Range Interacting Systems. First Edition. A. Campa *et al.*
© A. Campa, T. Dauxois, D. Fanelli, and S. Ruffo 2014. Published in 2014 by Oxford University Press.

cannot be neglected even for macroscopic systems. In this case we have a *long-range system*. The necessity to include the interaction energy has profound consequences on the equilibrium and dynamical properties, which are generally very different from those of short-range systems. In this book, we will analyse these peculiar properties, first from a general point of view, with the help of simple models with long-range interactions, and then studying the most relevant physical situations in which long-range forces appear.

Our readers are assumed to have already followed a basic course in statistical mechanics. Therefore, the purpose of this chapter is to provide a short review of the most important concepts, especially those concerning the physical and mathematical aspects related to the use of the statistical ensembles. This will prepare the ground for the analysis of long-range systems that will start in the next chapter.

1.1   The Microcanonical Ensemble

The microscopic state of a Hamiltonian system of N particles in three dimensions is uniquely represented by the $3N$ canonical coordinates $q_1, \ldots, q_{3N}$ and the $3N$ canonical momenta $p_1, \ldots, p_{3N}$, living in a $6N$-dimensional Γ space, the phase space of the system. Therefore each point of Γ represents a single state of the system. The Hamilton canonical equations of motion

$$\frac{dq_i}{dt} = +\frac{\partial H(\{q_i, p_i\})}{\partial p_i} \qquad (1.2)$$

$$\frac{dp_i}{dt} = -\frac{\partial H(\{q_i, p_i\})}{\partial q_i}, \qquad (1.3)$$

where $H(\{q_i, p_i\})$ is the Hamiltonian of the system, determine the evolution in time of the representative point in Γ space. In principle, all properties of a given system can be extracted from the solution of the equations of motion and from the possible symmetries of the Hamiltonian that give rise to additional integrals of motion besides the total energy. This is what is actually done in the study of few-body systems. We know that, apart from few completely integrable cases, the solution of Eqs. (1.2) and (1.3) can only be obtained numerically. Despite this limitation, the behaviour and the general properties of the examined system are found by the time evolution as determined by the Hamilton equations.

For a macroscopic system, the number N is enormously high. To give an idea of the largeness of this number, textbooks on statistical mechanics generally remind that we are now concerned with systems where N is not smaller than 10^{23}, which is the order of magnitude of the Avogadro's number. It is of the order of the number of molecules in a mole but also of the number of stars in the universe. Obviously it is not possible to numerically simulate the dynamics of such a large number of interacting particles, and therefore a different methodology must be adopted.

Actually, the necessity to study macroscopic systems with a different conceptual framework is learned by physics students quite early, with the study of thermo-dynamics, before approaching the methods and the tools of statistical mechanics. In thermodynamics, the state of a macroscopic system is specified by the value of very few macroscopic observables; the most important for a one-component fluid are the tem-perature, the pressure and the density. Through its basic laws, and with the use of the equations of state (often empirically derived), thermodynamics is able to describe the relation among the macroscopic observables and how they are modified in response to a change of the external parameters or more generally to a perturbation of the system. This is achieved completely neglecting any relation of the macroscopic system with its microscopic constituents.

Of course, in statistical mechanics, this relation is not neglected, since it is exactly on its basis that one wants to derive the macroscopic properties. The inaccessibility, and uselessness, of the detailed dynamics, embodied by the solution of the equations of motion (1.2) and (1.3), imposes that the microscopic laws are taken into account not through their integration, but with a different method.

We can start from the following considerations. Even if the integration of the equa-tions of motion were feasible, we could not exploit this possibility, since the initial conditions are impossible to determine. Indeed, our ignorance about the position on Γ space of the representative point of the system, at any given instant of time, is very large. For example, if we know the total energy E of an isolated system in thermodynamic equilibrium, we do not have any information about the region of the hypersurface of con-stant energy E in Γ where the representative point is located. However, the constancy in time of the macroscopic observables in equilibrium tells us something very import-ant, a property that emerges when the number of degrees of freedom becomes very large: during the ceaseless wandering of the representative point on the constant energy hypersurface, these observables (not only the total energy, which is obvious) are con-stant, at least within the tolerance of the measurements. Therefore, for the vast majority of the points on the constant energy hypersurface the macroscopic observables take an almost constant value. Then, we should be able to derive the equilibrium value of an observable by averaging, over the hypersurface, the function of $q_1, \ldots, q_{3N}, p_1, \ldots, p_{3N}$ that corresponds to it.

The concept of microcanonical ensemble can be thought to stem from the facts stated earlier. For an isolated system, we can specify the number N of particles, the volume V where it is enclosed and its total energy E. Apart from being on the hypersurface of energy E, we have a complete lack of knowledge of the microscopic state. We therefore represent our system with an ensemble of systems, which are uniformly distributed on the constant energy hypersurface. In the microcanonical ensemble, all the points of this hypersurface are assigned an equal probability, while all other points of Γ have zero probability.

Before proceeding, it is useful to derive the equation governing the evolution of a general density distribution of systems in Γ space. This distribution can be indicated with $\rho(\{q_i, p_i\}, t)$, such that $\rho(\{q_i, p_i\}, t) \mathrm{d}q^{3N} \mathrm{d}p^{3N}$ is the number of systems that are in the infinitesimal volume $dq^{3N} dp^{3N} = d\Gamma$ about the position $\{q_i, p_i\}$, at time t. Then, for an arbitrary volume A in Γ, the number $\mathcal{N}_A(t)$ of systems inside A at time t is given by

$$\mathcal{N}_A(t) = \int_A dq^{3N} dp^{3N} \rho(\{q_i, p_i\}, t). \tag{1.4}$$

The total number of points (i.e. of systems in Γ space) is conserved; therefore the rate of variation of $\mathcal{N}_A(t)$ is equal to the opposite of the flux of systems out of the volume A

$$\frac{d}{dt}\mathcal{N}_A(t) = \frac{d}{dt}\int_A dq^{3N} dp^{3N} \rho(\{q_i, p_i\}, t) = -\int_{\partial A} dS\,\hat{n} \cdot [\mathbf{v}\rho(\{q_i, p_i\}, t)], \tag{1.5}$$

where ∂A is the hypersurface in Γ space bounding A, $\hat{n}$ is the unit normal to ∂A pointing outwards A and $\mathbf{v} = (\dot{q}_1, \ldots, \dot{q}_{3N}, \dot{p}_1, \ldots, \dot{p}_{3N})$ is the vector of Γ (more precisely, of the tangent space of Γ at the point $(q_1, \ldots, q_{3N}, p_1, \ldots, p_{3N})$) giving the velocity of the representative point. Since the volume A is arbitrary, we get from the divergence theorem that

$$\frac{\partial}{\partial t}\rho(\{q_i, p_i\}, t) + \nabla \cdot [\mathbf{v}\rho(\{q_i, p_i\}, t)] = 0. \tag{1.6}$$

The equations of motion (1.2) and (1.3) imply that

$$\mathbf{v} = \left(\left\{ \frac{\partial H}{\partial p_i}, -\frac{\partial H}{\partial q_i} \right\} \right), \tag{1.7}$$

from which we immediately derive that $\nabla \cdot \mathbf{v} = 0$, so that Eq. (1.6) becomes

$$\frac{\partial}{\partial t}\rho(\{q_i, p_i\}, t) + \mathbf{v} \cdot \nabla\rho(\{q_i, p_i\}, t) \equiv \frac{d}{dt}\rho(\{q_i, p_i\}, t) = 0, \tag{1.8}$$

where we have explicitly shown that the left-hand side is the total time derivative of $\rho(\{q_i, p_i\}, t)$, i.e. its time derivative taken along the path of the representative point. The last equation is the celebrated Liouville theorem, which shows that the density of systems is constant in time if it moves together with the representative point.

Using Eq. (1.7), we see that the second term on the left-hand side of Eq. (1.8) is the Poisson bracket of the functions ρ and H:

$$\mathbf{v} \cdot \nabla\rho(\{q_i, p_i\}, t) = \sum_{i=1}^{3N} \left(\frac{\partial\rho}{\partial q_i}\frac{\partial H}{\partial p_i} - \frac{\partial\rho}{\partial p_i}\frac{\partial H}{\partial q_i} \right) \equiv [\rho, H]. \tag{1.9}$$

The Liouville theorem shows that if ρ is a function of the point in Γ space through the Hamiltonian H, i.e. if $\rho = \rho\,[H(\{q_i, p_i\})]$, then ρ is stationary. It is then natural to take such a ρ as the density function of ensembles representing systems in equilibrium.

Once we have the density distribution function, the values of the macroscopic observables are expressed through averages over this distribution. For example, if $O(\{q_i, p_i\})$ is the microscopic mechanical function corresponding to the observable O, its average value in the macroscopic system at time t will be given by

$$\langle O \rangle (t) = \frac{\int dq^{3N} \, dp^{3N} \, O(\{q_i, p_i\}) \rho(\{q_i, p_i\}, t)}{\int dq^{3N} \, dp^{3N} \, \rho(\{q_i, p_i\}, t)}, \qquad (1.10)$$

where we have introduced the bracket notation $\langle \cdot \rangle$ for ensemble averages.

Coming back to our isolated system with energy E, we see that a density function which is constant on the hypersurface of energy E and 0 elsewhere, is stationary. Consequently, as previously argued, it is the proper distribution function to represent the system. In conclusion, the *microcanonical ensemble* for an isolated system with energy E is expressed by

$$\rho_{MC}(\{q_i, p_i\}) = \delta \left[E - H(\{q_i, p_i\}) \right], \qquad (1.11)$$

in which $\delta(x)$ stands for the Dirac delta function. To avoid a δ-like singularity, another definition is sometimes used, in which ρ is constant on the portion of Γ space between the two hypersurfaces corresponding to energies E and $E + \Delta E$, with $\Delta E \ll E$, and 0 otherwise. A third definition uses the Heaviside step function $\theta(E - H)$, which corresponds to keep all accessible states with an energy lower than E. These three definitions lead to exactly the same results, for instance infinite time averages in (1.10), provided

$$N \to \infty \qquad \text{(thermodynamic limit)}. \qquad (1.12)$$

The relation to thermodynamics is obtained through the integral of the density function over Γ space, which gives the microcanonical partition function

$$\Omega(E, V, N) = \frac{1}{N!} \int_{\Gamma_V} dq^{3N} \, dp^{3N} \, \delta \left[E - H(\{q_i, p_i\}) \right], \qquad (1.13)$$

where we have explicitly indicated the quantities on which Ω depends, i.e. the energy E, the three-dimensional volume V in which the system is enclosed and the number of particles N. The subscript Γ_V in the integral indicates that the coordinates q_i cannot go outside this volume. The quantity $\Omega(E, V, N)$ is thus proportional to the number of accessible microstates; the factor $1/N!$ in front of the integral provides the 'correct Boltzmann counting' (Huang, 1987) for the number of microstates. Since the coordinates q_i and p_i are continuous, the meaning of 'number of microstates' needs some attention. Obviously the microstates would be a continuous infinity. They can become a finite number if, e.g., for one microstate we mean a small volume in Γ. This identification can be put on a firm basis in quantum mechanics, where the volume of Γ to be associated with one microstate is related to the Planck constant, being given by h^{3N}. For this reason, the definition of the microcanonical partition function $\Omega(E, V, N)$ is often found with an additional h^{3N} in the denominator in front of the integral in Eq. (1.13); this has also the advantage of making $\Omega(E, V, N)$ dimensionless.

From the partition function, we define the entropy of the system

$$S(E, V, N) = \ln \Omega(E, V, N), \tag{1.14}$$

where we adopt units for which the Boltzmann constant k_B is equal to 1. From the entropy, expressed as a function of the energy, the volume and the number of particles, all other thermodynamic functions are obtained.

The identification of the logarithm of the microcanonical partition function with the thermodynamical entropy is generally justified, in the textbooks of statistical mechanics, by showing that the former has the same properties of the latter. This chapter is only a remainder of some basic facts and we do not repeat the argument. We would just like to point out that, apart possibly from an additive constant, Eq. (1.14) equals the entropy to the logarithm of the number of accessible microstates: this is exactly the famous definition introduced by Ludwig Boltzmann.

To end this section, we write the normalized microcanonical probability density, which is obtained from Eqs. (1.11) and (1.13) as

$$p_{MC}(\{q_i, p_i\}) = \frac{1}{\Omega(E, V, N)} \frac{1}{N!} \delta \left[E - H(\{q_i, p_i\}) \right]. \tag{1.15}$$

1.2   The Canonical and the Grand-Canonical Ensembles

1.2.1   The canonical ensemble

As we have seen, the microcanonical ensemble is appropriate for an isolated system. Often, however, the macroscopic system under consideration is not isolated, and it can exchange energy with the surrounding environment. Therefore, the accessible region of Γ space is different, since the system is not constrained to a hypersurface of constant energy. The computation of the density distribution is obtained by thinking our system to be a part of a much larger system, with total energy, say, E_0, volume V_0 and number of particles N_0. This large system is supposed to be isolated, but the system under study (the 'small' system in the following), which is a small part of it, can exchange energy with the large remainder, which plays the role of a heat bath.

Let us denote with the subscript '1' and '2' the quantities related to the small system and to the heat bath, respectively. The small system and the heat bath are enclosed in the separate volumes V_1 and V_2, respectively, with $V_0 = V_1 + V_2$ and with $V_2 \gg V_1$. We have also $N_0 = N_1 + N_2$, with $N_2 \gg N_1$. We can write the Hamiltonian of the whole system as

$$H(\Gamma_1, \Gamma_2) = H_1(\Gamma_1) + H_2(\Gamma_2) + H_{int}(\Gamma_1, \Gamma_2), \tag{1.16}$$

where H_1 depends only on the canonical coordinates of the small system, denoted collectively with Γ_1, H_2 only on those of the heat bath, denoted collectively with Γ_2, while the interaction term H_{int} describes the interaction between the small system and the heat bath, and depends in principle on all the coordinates. When H has a given value E_0, for all the allowed values of the coordinates compatible with this constraint, we will have $E_0 = E_1 + E_2 + E_{int}$, where the various terms are obviously associated with the terms on the right-hand side of Eq. (1.16). It is at this point that we introduce the additivity property of macroscopic systems: the interaction energy E_{int} is neglected, so that $E_0 = E_1 + E_2$. This is equivalent to neglecting the interaction part of the Hamiltonian (1.16). The additivity property is generally taken for granted, and it is based on the observation that the interaction energy is due to the interaction between the superficial layers of the small system and those of the heat bath; it is then argued that for sufficiently large systems this interaction energy can safely be neglected with respect to E_1 and E_2. However, this argument is strongly based on the short-rangedness of the interaction. We therefore begin to appreciate the point where long-range systems present an important difference. In fact, later, we will see that the lack of additivity, i.e. the fact that the total energy E_0 is not equal to the sum of the energies E_1 and E_2 of its parts, is the main reason for the peculiar properties of systems with long-range interactions.

Using the definition of the previous section for the microcanonical distribution function, Eq. (1.11), the microcanonical distribution for the whole system is

$$\rho(\Gamma_1, \Gamma_2) = \delta[E - H(\Gamma)] = \delta[E - H_1(\Gamma_1) - H_2(\Gamma_2) - H_{int}(\Gamma_1, \Gamma_2)]. \tag{1.17}$$

The probability density that the small system is in the neighbourhood $d\Gamma_1$ of a point in Γ_1, regardless of the state of the heat bath, will be proportional to the integral of the previous expression in $d\Gamma_2$. Exploiting the additivity property, i.e. neglecting H_{int} in the right-hand side of (1.17), we obtain

$$\rho(\Gamma_1, \Gamma_2) \simeq \delta[E - H_1(\Gamma_1) - H_2(\Gamma_2)] \tag{1.18}$$

$$= \int dE_1 dE_2 \, \delta[E_1 - H_1(\Gamma_1)] \, \delta[E_2 - H_2(\Gamma_2)] \, \delta[E - E_1 - E_2]. \tag{1.19}$$

Using this equation together with (1.13) and (1.14) we therefore obtain

$$\rho(\Gamma_1) = \frac{1}{N_2!} \int_{\Gamma_{V_2}} d\Gamma_2 \, \delta[E - H(\Gamma)] \tag{1.20}$$

$$= \int dE_1 dE_2 \, \delta[E_1 - H_1(\Gamma_1)] \, \Omega_2(E_2, V_2, N_2) \delta[E - E_1 - E_2] \tag{1.21}$$

$$= \int dE_1 \, \delta[E_1 - H_1(\Gamma_1)] \, \Omega_2(E - E_1, V_2, N_2) \tag{1.22}$$

$$= \Omega_2[E - H_1(\Gamma_1), V_2, N_2]. \tag{1.23}$$

In the last expression, we use again Eq. (1.14) and we expand to first order to obtain

$$\rho(\Gamma_1) = \exp\{S_2[E - H_1(\Gamma_1), V_2, N_2]\} \tag{1.24}$$

$$\approx \exp\{S_2(E, V_2, N_2) - \beta H_1(\Gamma_1)\} \tag{1.25}$$

$$= \Omega_2(E, V_2, N_2) \exp[-\beta H_1(\Gamma_1)], \tag{1.26}$$

where $\beta = 1/T$ is the inverse of the temperature T of the heat bath, defined according to the thermodynamic relation by

$$\frac{1}{T} = \left.\frac{\partial S_2}{\partial E}\right|_{N_2, V_2}. \tag{1.27}$$

The truncation at first order in the expansion of S_2 in Eq. (1.24) can be easily justified. In fact, the term $S_2(E, V_2, N_2)$ is of the order N_2, the second term is of the order $N_1 \ll N_2$ and the first neglected term is of the order $N_1^2/N_2 \ll N_1$.

The first factor on the right-hand side of Eq. (1.26) is a constant in Γ_1. When the coordinates of the small system are inserted again explicitly, the canonical distribution function is defined by

$$\rho_C(\{q_i, p_i\}) = \exp[-\beta H(\{q_i, p_i\})]. \tag{1.28}$$

Therefore, the small system with N_1 particles and enclosed in the volume V_1 is characterized by the same temperature $T = 1/\beta$ of the heat bath. Forgetting the adjective 'small' and the subscript for the volume and the number of particles, the canonical partition function of a system at inverse temperature β, enclosed in a volume V and made of N particles, is therefore, analogously to Eq. (1.13), defined by

$$Z(\beta, V, N) = \frac{1}{N!} \int_{\Gamma_V} dq^{3N} dp^{3N} \exp[-\beta H(\{q_i, p_i\})], \tag{1.29}$$

where the usual notation Z has been adopted for this function. Again the factor $1/N!$ provides the 'correct Boltzmann counting'. As for the microcanonical partition function, an additional h^{3N} is often introduced in the denominator in front of the integral in Eq. (1.29), making the partition function dimensionless.

The normalized canonical probability density is obtained from Eqs. (1.28) and (1.29) as

$$p_C(\{q_i, p_i\}) = \frac{1}{Z(\beta, V, N)} \frac{1}{N!} \exp[-\beta H(\{q_i, p_i\})]. \tag{1.30}$$

The logarithm of the canonical partition function is related to a thermodynamic potential, as for the microcanonical case. The following procedure shows that now the potential is the Helmholtz free energy. Let us write $Z(\beta, V, N) = \exp[-\beta F(\beta, V, N)]$, which defines F. Then the probability density (1.30) can be written as

$$p_C(\{q_i, p_i\}) = \frac{1}{N!} \exp[\beta F(\beta, V, N) - \beta H(\{q_i, p_i\})]. \tag{1.31}$$

Since $p_C(\{q_i, p_i\})$ is normalized, we have

$$0 = \frac{\partial}{\partial \beta} \int_{\Gamma_V} dq^{3N} dp^{3N} p_C(\{q_i, p_i\}) = F(\beta, V, N) + \beta \frac{\partial}{\partial \beta} F(\beta, V, N) - \langle H \rangle, \quad (1.32)$$

where the last term, the average of the Hamiltonian on the canonical distribution function, gives the average energy E of the system. We therefore have

$$F(\beta, V, N) = E(\beta, V, N) - \beta \frac{\partial}{\partial \beta} F(\beta, V, N). \quad (1.33)$$

Passing from β to $T = 1/\beta$, the last equation becomes

$$F(T, V, N) = E(T, V, N) + T \frac{\partial}{\partial T} F(T, V, N). \quad (1.34)$$

We see that this is the equation satisfied by the Helmholtz free energy $E - TS$, expressed as a function of (T, V, N); in fact, we know from thermodynamics that the derivative with respect to the temperature of the Helmholtz free energy at constant volume and number of particles is the opposite of the entropy. In conclusion

$$F(T, V, N) = -T \ln Z(T, V, N) \quad (1.35)$$

will be simply called the free energy.

1.2.2   The grand-canonical ensemble

The grand-canonical ensemble represents a situation in which the system can exchange with the heat bath not only energy, but also particles. The N_0 particles are no more divided into two constant numbers, N_1 particles in the system of interest and N_2 particles in the heat bath, but in principle each particle can go from the bath to the system and vice versa. Therefore, there will be a distribution also for the number of particles. Starting again from Eq. (1.17), we can proceed as in the case of the canonical ensemble and obtain Eq. (1.20). However, the quantity in the last equation will now be proportional to the probability density that the system of interest is in the neighbourhood $d\Gamma_1$ of a point in Γ_1, regardless of the state of the heat bath, *provided* that there are N_1 particles in the system and $N_2 = N_0 - N_1$ particles in the heat bath. We can denote this quantity as

$$\rho_{N_1}(\Gamma_1) = \Omega_2 [E - H_1(\Gamma_1), V_2, N_0 - N_1]. \quad (1.36)$$

Performing again a first-order expansion as in Eq. (1.24), we arrive at

$$\rho(\Gamma_1; N_1) = \Omega_2 [E, V_2, N_0] \exp [-\beta H_1(\Gamma_1)] \exp (\beta \mu N_1), \quad (1.37)$$

where now

$$\frac{1}{T} = \frac{\partial S_2}{\partial E}\bigg|_{N_0, V_2} \qquad \text{and} \qquad \frac{\mu}{T} = -\frac{\partial S_2}{\partial N_0}\bigg|_{E, V_2}, \qquad (1.38)$$

the second thermodynamic relation defining the chemical potential. We therefore arrive at the grand-canonical distribution function

$$\rho_{GC}(\{q_i, p_i\}; N) = \exp(\beta\mu N) \exp[-\beta H_N(\{q_i, p_i\})], \qquad (1.39)$$

where we have dropped the subscript for the number of particles of the system, which we have explicitly indicated in the Hamiltonian. Then the grand-canonical partition function of a system at inverse temperature β, enclosed in a volume V and characterized by the chemical potential μ, is

$$\mathcal{Z}(\beta, V, \mu) = \sum_{N=0}^{\infty} \frac{\exp(\beta\mu N)}{N!} \int_{\Gamma_V} dq^{3N} dp^{3N} \exp[-\beta H(\{q_i, p_i\})] \qquad (1.40)$$

$$= \sum_{N=0}^{\infty} \exp(\beta\mu N) Z(\beta, V, N). \qquad (1.41)$$

From Eqs. (1.39) and (1.41), we obtain the normalized grand-canonical probability density

$$p_{GC}(\{q_i, p_i\}; N) = \frac{1}{\mathcal{Z}(\beta, V, \mu)} \frac{\exp(\beta\mu N)}{N!} \exp[-\beta H_N(\{q_i, p_i\})]. \qquad (1.42)$$

To find the thermodynamic potential associated with the grand-canonical partition function we proceed as earlier, and we write $\mathcal{Z}(\beta, V, \mu) = \exp[-\beta\Xi(\beta, V, \mu)]$. Then the grand-canonical probability density is written as

$$p_{GC}(\{q_i, p_i\}; N) = \frac{1}{N!} \exp[\beta\Xi(\beta, V, \mu) + \beta\mu N - \beta H_N(\{q_i, p_i\})]. \qquad (1.43)$$

From the normalization condition, we obtain

$$0 = \frac{\partial}{\partial\beta} \sum_{N=0}^{\infty} \int_{\Gamma_V} dq^{3N} dp^{3N} p_{GC}(\{q_i, p_i\}; N) \qquad (1.44)$$

$$= \Xi(\beta, V, \mu) + \beta\frac{\partial}{\partial\beta}\Xi(\beta, V, \mu) + \mu\langle N\rangle - \langle H\rangle, \qquad (1.45)$$

where $\langle N\rangle$ is the average number of particles. Passing from β to T, and writing simply N for the average number of particles, as we have replaced $\langle H\rangle$ for E from (1.32) to (1.33), we are led to

$$\Xi(T, V, \mu) = E(T, V, \mu) + T \frac{\partial}{\partial T} \Xi(T, V, \mu) - \mu N(T, V, \mu). \qquad (1.46)$$

We see that this is the equation satisfied by the grand potential $E - TS - \mu N$, expressed as a function of (T, V, μ); in fact, we know from thermodynamics that the derivative with respect to the temperature of the grand potential at constant volume and chemical potential is the opposite of the entropy.

However, we also know from thermodynamics that μN is equal to the Gibbs free energy $G = E + pV - TS$, where p is the pressure of the system. In conclusion, we get for the grand potential

$$\Xi(T, V, \mu) = -T \ln \mathcal{Z}(T, V, \mu) = -p(T, V, \mu)V. \qquad (1.47)$$

1.3   Equivalence of Ensembles for Short-Range Interactions

Let us summarize the connection between the different ensembles and the different physical situations.

- The microcanonical ensemble describes an isolated system, with a given energy E, volume V and number of particles N.

- The canonical ensemble is related to a system with a given volume V and number of particles N that can exchange energy with a heat bath at temperature T.

- The grand-canonical ensemble describes a system in a volume V that can exchange both energy and particles with a bath which is at temperature T and is characterized by a chemical potential μ.

In each case the logarithm of the partition function is proportional to the thermodynamic potential that is a function of the corresponding variables: the entropy $S(E, V, N)$, the free energy $F(T, V, N)$ and the grand potential $\Xi(T, V, \mu)$, respectively. Although much less used, it is possible to define also the isothermal-isobaric ensemble, which describes a system without a given fixed volume V. This system can exchange energy with the heat bath at temperature T and pressure p. The corresponding partition function is related to the Gibbs free energy $G(T, p, N)$.

As the various ensembles refer to different physical situations, in principle we could expect that the resulting equilibrium state is different in each case. However, we can make the following argument. In thermodynamics, the macroscopic systems are characterized by an equation of state that relates thermodynamic variables among them. For example, for a one-component system, like the one for which we have described the ensembles in the previous section, the knowledge of two thermodynamic quantities, e.g. the temperature T and the number density $n = N/V$, is sufficient to determine all the others through the equation of state. This fact tells us two important properties. The first is that all thermodynamic quantities have given values or, more precisely, that their relative fluctuations are negligible. The second is that this is true independently of the preparation of the system. In the example of the one-component system just discussed, if we fix temperature and density at given values T^* and n^*, the internal energy will have a value E^*. If, on the other hand, we fix the density and the energy at n^* and E^*, then the temperature will have the value T^*.

For the statistical mechanics description, these facts mean the following. As we already noted at the beginning of the previous section, the observables take practically the same value for the greatest part of the microscopic states accessible in each ensemble, so that the relative fluctuations are negligible; secondly, at least in the thermodynamic limit $N \to \infty$, each ensemble should give the same equation of state. More specifically, let us suppose that we study a system within the microcanonical ensemble, choosing the values of E, V and N. Using Eq. (1.10) we can compute, e.g. the expected value of the temperature T and of its fluctuation ΔT. The former argument tells us that $\Delta T/T \ll 1$, and, above all, that if we study the same system in the canonical ensemble, with the same V and N as before, and at T equal to the expected value obtained from the microcanonical ensemble, then the expected value of the energy will be equal to the value E used before in the microcanonical ensemble, and in addition $\Delta E/E \ll 1$. What we have described in this paragraph is the equivalence of the ensembles.

Discussions of ensemble equivalence can be found, e.g., in the books of Ruelle (1969) or Gallavotti (1999). Here we will just make a summary of the main results. To this purpose, we must introduce two simple mathematical concepts: the concavity and convexity of functions and the Legendre–Fenchel transform (LFT). They are the main ingredients on which the equivalence of ensembles is built, and thus they are useful in the remainder of this chapter; but above all they will be important throughout this book. Let us begin with the concept of concavity of functions.

1.3.1   Concave functions

A function $f(x)$ of one variable is said to be concave if, for any x_1 and x_2 in its domain of definition and for any c satisfying $0 \leq c \leq 1$, we have

$$f(cx_1 + (1-c)x_2) \geq cf(x_1) + (1-c)f(x_2). \tag{1.48}$$

In words, this means that the graph of the function is always above the straight line connecting any couple of its points. When the inequality is satisfied strictly, the function is said to be strictly concave. If there is an interval (x_1, x_2) where Eq. (1.48) is satisfied as an equality for all $0 \leq c \leq 1$, in that interval the graph of the function is a straight line. If f is twice differentiable, concavity is expressed by $d^2f/dx^2 \leq 0$ for all x. If Eq. (1.48) is satisfied only when x_1 and x_2 belong to a range of the domain of definition of f, then f is said to be concave in that range.

Convexity is defined in an analogous way, substituting '$\geq$' with '$\leq$' in the just-mentioned inequalities. The graph of a convex function is always below the straight line connecting any couple of its points.

The former definitions extend naturally to functions of more than one variable. In the case of a function of two variables $f(x, y)$, the concavity with respect to one of the variables, e.g. x, holds when

$$f(cx_1 + (1-c)x_2, y) \geq cf(x_1, y) + (1-c)f(x_2, y) \qquad (1.49)$$

for any y, x_1, x_2 in the domain of definition, and $0 \leq c \leq 1$. Concavity with respect to the other variable y is defined in an analogous way. The function is completely concave when

$$f(cx_1 + (1-c)x_2, cy_1 + (1-c)y_2) \geq cf(x_1, y_1) + (1-c)f(x_2, y_2) \qquad (1.50)$$

for any x_1, x_2, y_1, y_2 in the domain of definition, and for all $0 \leq c \leq 1$. Complete concavity implies separate concavity with respect to both x and y. On the other hand, separate concavity with respect to both x and y is not sufficient to guarantee complete concavity. The straight line connecting any two points of a completely concave function is entirely below the graph of the function. If $f(x, y)$ is twice differentiable, complete concavity occurs when the Hessian is negative definite at all points. If there is a region in the (x, y) plane where the completely concave function satisfies (1.50) as an equality for all $0 \leq c \leq 1$, then in that region the graph of the function is a portion of a plane. As stated previously, the concavity of a function $f(x, y)$, either complete or with respect to x or y, could hold for only a portion of its domain of definition.

Again, convexity is defined substituting '$\geq$' with '$\leq$'. The straight line connecting any two points of a completely convex function is entirely above the graph of the function.

1.3.2 The Legendre–Fenchel transform (LFT)

In thermodynamics, we learn that the different thermodynamic potentials are related through the Legendre transform. Here we introduce a generalization, the Legendre–Fenchel transform. This is another tool that will be very important. Given a function $f(x)$, we can define its Legendre–Fenchel transform, function of y, by

$$g(y) = \inf_x [yx - f(x)]. \qquad (1.51)$$

It is an easy exercise, which we leave to the reader, to show that $g(y)$ is a concave function for any $f(x)$. Let us now compute the Legendre–Fenchel transform of $g(y)$, which will give a function $\tilde{f}(x)$ that, in principle, is different from $f(x)$. We obtain

$$\tilde{f}(x) = \inf_y [xy - g(y)]. \qquad (1.52)$$

Since we know that $\tilde{f}(x)$ is concave, we are sure *a priori* that it is different from $f(x)$ if the latter function is not concave. However, it is possible to prove the following properties. In general $\tilde{f}(x) \geq f(x)$; if $f(x)$ is concave, then $\tilde{f}(x) = f(x)$; if $f(x)$ is not concave, then $\tilde{f}(x)$ is equal to the concave envelope of $f(x)$, which is defined as the smallest concave function $f_c(x)$ such that $f_c(x) \geq f(x)$ (i.e. for any concave function $h(x)$ such that $h(x) \geq f(x)$, then $h(x) \geq f_c(x)$).

Let us now suppose that $f(x)$ in Eq. (1.51) is twice differentiable. Then the infimum in x on the right-hand side will necessarily satisfy

$$y = \frac{df}{dx} \quad \text{with} \quad \frac{d^2f}{dx^2} < 0. \tag{1.53}$$

If in addition $f(x)$ is strictly concave (so that $d^2f/dx^2 < 0$ everywhere in the range of definition), the first expression is uniquely invertible to give $x = x(y)$ (for y in a given range, outside which there is no solution).

It is possible to define the other transform in which the infimum in Eq. (1.51) is substituted by a supremum

$$h(y) = \sup_x [yx - f(x)]. \tag{1.54}$$

In this case, $h(y)$ is a convex function for any $f(x)$. Now the inverse transform

$$\tilde{\tilde{f}}(x) = \sup_y [xy - h(y)] \tag{1.55}$$

gives a function $\leq f(x)$. Analogously to the previous case, if $f(x)$ is convex, then $\tilde{\tilde{f}}(x) = f(x)$; if $f(x)$ is not convex, then $\tilde{\tilde{f}}(x) = f(x)$ is equal to the convex envelope of $f(x)$, i.e. the largest convex function $f_x(x)$, such that $f_x(x) \leq f(x)$. If $f(x)$ is twice differentiable, the supremum in x on the right-hand side of Eq. (1.54) will necessarily satisfy

$$y = \frac{df}{dx} \quad \text{with} \quad \frac{d^2f}{dx^2} > 0. \tag{1.56}$$

The relation between concavity and convexity of functions, on the one hand, and the Legendre–Fenchel transform, on the other, will be exploited in the theorems concerning ensemble equivalence, which will be introduced shortly. The following property is also useful: if the function $f(x)$, defined for $x > 0$, is concave (convex), then the function $g(x) = \frac{1}{x}f(x)$, also defined for $x > 0$, is concave (convex) with respect to $1/x$. The easy proof is left as an exercise to the reader.

1.3.3   Stable and tempered potentials

We now study in some details the conditions normally satisfied by short-range potentials. The Hamiltonian of a generic N-body system can be written as

$$H = \sum_{i=1}^{N} \frac{\mathbf{p}_i^2}{2m} + U(\mathbf{q}_1, \ldots, \mathbf{q}_N) = \sum_{i=1}^{N} \frac{\mathbf{p}_i^2}{2m} + \sum_{1 \leq i < j \leq N} V(\mathbf{q}_i - \mathbf{q}_j), \tag{1.57}$$

where we have used bold letters for three-dimensional vectors. The total interaction potential U is given by the sum of two-body potentials V; the dependence only on the difference $\mathbf{q}_i - \mathbf{q}_j$ reflects the translational invariance of the Hamiltonian. Although it is possible, with simple extensions, to treat the more general case in which there are also n-body interactions, with $n > 2$ (still translationally invariant), for simplicity we consider here only $n = 2$, which is the most common case. We also consider the case without an external potential, which would break the translational invariance of the Hamiltonian and would give rise to an inhomogeneous equilibrium state. Note that the enclosure of the system in a given volume can obviously break translational invariance, unless we use periodic boundary conditions. However, we know that the interaction of the system with the volume borders is confined to a microscopic layer, and apart from this layer the equilibrium state is homogeneous. We also assume that the potential is isotropic, i.e. $V(\mathbf{q}) = V(|\mathbf{q}|)$.

There are two conditions on the interaction potential V that assure 'normal' thermodynamic behaviour and ensemble equivalence. We have already introduced the first one, the *additivity*, directly related to the short rangedness of the interaction: the interaction is short range when Eq. (1.1) holds at large r ($r \equiv |\mathbf{q}|$) for $\alpha > 3$. As we have noted, this condition is necessary in order to have the additivity property, i.e. the possibility to neglect the interaction energy between two macroscopic systems, which is very small with respect to the energy of each of the systems.

The second one is related to the other property that we expect in normal systems, i.e. *extensivity*. This is best described with the help of the thermodynamic potentials. For a macroscopic system the energy, the entropy, the free energy and the Gibbs free energy should be proportional to the number N of particles, for N sufficiently large, and the grand potential Ξ should be proportional to the volume V. Indicating with $\varepsilon = E/N$ the energy per particle, with $s = S/N$ the entropy per particle, with $f = F/N$ the free energy per particle, with $\mu = G/N$ the Gibbs free energy per particle (we have already mentioned before Eq. (1.47) this relation between G and the chemical potential μ), with $\omega = \Xi/V = -p$ the grand potential per unit volume and with $n = N/V$ the density of the system, extensivity means that the thermodynamic potentials have the form $S = Ns(\varepsilon, n)$, $F = Nf(T, n)$, $G = N\mu(T, p)$ and $\Xi = V\omega(T, \mu) = -Vp(T, \mu)$. As for the total energy, the inverse of the first of these relations is immediately seen to give $E = N\varepsilon(s, n)$, as expected. These relations show that the extensive quantities, E, S, F, G, Ξ, are equal to an extensive quantity, N or V, times a function of intensive quantities, i.e. quantities that do not depend on the size of the system. There are two kinds of intensive quantities: the specific values of extensive quantities, like $\varepsilon, s, f, n, \omega$, and the quantities governing the equilibrium between macroscopic systems, like T, p. Interestingly, μ, being the specific Gibbs free energy and chemical potential, belongs to both categories.

We have already used the additivity of the energy in the construction of the canonical ensemble. For a macroscopic system with N particles, thought to be divided into two subsystems with N_1 and N_2 particles, we can write the total energy, putting in evidence also the extensivity property, as $E = N_1 s(\varepsilon, n) + N_2 s(\varepsilon, n)$. This will hold for all thermodynamic potentials: in particular $S = N_1 s(\varepsilon, n) + N_2 s(\varepsilon, n)$ and $F = N_1 f(T, n) + N_2 f(T, n)$. However, these last expressions hold also when the two subsystems in contact are in different equilibrium states, provided that they are also in equilibrium among them. In particular, they must have the same T, p and μ, but they could have different values of the other thermodynamic variables. The reader surely recognizes in this description what happens when two phases coexist and are in contact. For example at a given T, a fluid could be divided into two phases, one in the condensed and the other in the vapour phase. In this case, we would have $F = N_1 f(T, n_1) + N_2 f(T, n_2)$: given T and the volume available to the system, this divides into two phases, one with the density n_1 and the other with density n_2, proper of the two phases at that temperature, with N_1 and N_2 such that the total occupied volume is V.

We know from thermodynamics that the free energy of a system is the smallest possible compatible with assigned values of T and V (or that the Gibbs free energy is the smallest possible compatible with assigned values of T and p), and that the entropy is the largest possible with assigned values of ε and V. Together with the considerations of the previous paragraph, we can infer the concavity properties of s and f. We first introduce the specific volume $v = V/N$, i.e. the inverse of the density n, and rewrite the above expression of F when the system is divided into two phases: $F = N_1 f(T, v_1) + N_2 f(T, v_2)$. Then, suppose that the specific free energy f is of the form represented by the dotted line in Fig. 1.1, and that v_1 and v_2 are the two values indicated in the figure, with the straight line connecting the two points of the dotted line with those abscissas. Dividing the last expression by $N = N_1 + N_2$, we get $F/N = cf(T, v_1) + (1 - c)f(T, v_2)$, where $c = N_1/N$. Varying c between 0 and 1, we see that F/N is given by the points of the full straight line of Fig. 1.1. For any v in the range (v_1, v_2), this line is below the dotted line, and it is therefore thermodynamically favoured. The substitution in that range of the dotted line with the full straight line realizes the convex envelope of the dotted line. In conclusion, for short-range systems we expect from thermodynamics that f, expressed as a function of T and $v = 1/n$, is convex in v for each T. When two phases coexist, f will not be strictly convex in v, but there will be a straight line for a given range. The reader might wonder why the convexity holds with respect to v and not with respect to n. This is because for any v between v_1 and v_2 the

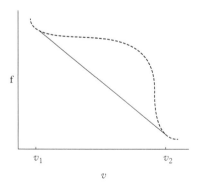

Figure 1.1 *A hypothetical specific free energy (dotted line) and its convex envelope, obtained by substituting the dotted curve between v_1 and v_2 with the full straight line.*

value of $c = N_1/N$ that gives v as the average specific volume is determined exactly by $cv_1 + (1 - c)v_2 = v$. However, this means that $c/n_1 + (1 - c)/n_2 = 1/n$, which is obviously different from $cn_1 + (1 - c)n_2 = n$.

An analogous argument can be made for the specific entropy function $s(\varepsilon, v)$, with the difference that s must be concave, since thermodynamics favours the largest entropy value. Another difference is that now the concavity property does not concern only the specific volume v for any ε, but $s(\varepsilon, v)$ is completely concave; in fact, two systems in equilibrium might be in equilibrium among them even with different densities and specific energies. Also in this case the presence of phase coexistence implies the presence of straight segments in the graph of s.

Let us now consider a potential $V(\mathbf{q})$ that has a negative absolute minimum at $|\mathbf{q}| = 0$, i.e. $V(0) = -a$, with $a > 0$. It is not difficult to see that with this potential we can choose, in the microcanonical ensemble, an energy $E = -N^2 a$, which is also obtained in the canonical ensemble in the limit $T \to 0$. This violation of extensivity is due to the large weight given to configurations in which an arbitrarily large number of particles is in a finite region of space, i.e. to configurations where the system is 'collapsed'. In this case, the configuration is the one with all the particles at rest at the same point in space. This violation happens even if this potential is short range (Compagner *et al.*, 1989; Posch *et al.*, 1990), e.g. if $V(|\mathbf{q}|) = -a$ for $|\mathbf{q}| \le r_0$ and $V(|\mathbf{q}|) = 0$ for $|\mathbf{q}| > r_0$. More generally, to avoid the violation of extensivity, we require that for any possible configuration the total potential is lower bounded by a constant times the number of particles. More precisely, we ask that there exists $A \ge 0$ such that

$$\sum_{1 \le i < j \le N} \sum V(\mathbf{q}_i - \mathbf{q}_j) \ge -NA, \tag{1.58}$$

uniformly for each N and each configuration $(\mathbf{q}_1, \ldots, \mathbf{q}_N)$. This is called stability condition (sometimes H-stability condition), and the potentials that satisfy it are called stable potential (Ruelle, 1969).

When the inequality (1.58) holds, this implies that there exist r_0, $B > 0$ and $\alpha > 3$ such that

$$V(\mathbf{q}) > -B|\mathbf{q}|^{-\alpha} \tag{1.59}$$

when $|\mathbf{q}| \ge r_0$. In fact, if (1.59) does not hold, then a homogeneous configuration violates (1.58). We see that the last inequality assures the sufficiently fast decay of the interaction potential from below. Of course, to have additivity the potential must decay in absolute value. Therefore, together with Eq. (1.58), we complete our requests on the potential with the temperedness condition, which holds when there exist r_0, $C > 0$ and $\alpha > 3$ such that

$$V(\mathbf{q}) < C|\mathbf{q}|^{-\alpha} \tag{1.60}$$

when $|\mathbf{q}| \ge r_0$. The stability condition (1.58) and the temperedness condition (1.60) are sufficient for ensemble equivalence. Without giving the proofs that the interested reader might find in Ruelle (1969), we now show how this equivalence is expressed.

1.3.4   Ensemble equivalence

We mentioned before that ensemble equivalence is expected to hold in the thermo-dynamic limit, $N \to \infty$. Let us express more precisely this limit. Together with N, also the volume V of the system and its energy E must go to infinite, so that the energy per particle E/N and the density N/V tend to given limits: $E/N \to \varepsilon$ and $N/V \to n > 0$. As mentioned already, the proofs of the theorems stated below can be found, e.g. in the book of Ruelle (1969).

We begin by considering Eq. (1.14), which defines the entropy of the system with the logarithm of the microcanonical partition function. For stable and tempered potentials the following limit exists:

$$s(\varepsilon, n) = \lim_{N \to \infty} \frac{1}{N} S(E, V, N) \tag{1.61}$$

for the specific entropy $s(\varepsilon, n)$. We therefore see that for large N we can write

$$S(E, V, N) = Ns(\varepsilon, n), \tag{1.62}$$

as requested by the extensivity property. The stability condition (1.58) implies that, for any n, the specific energy ε has a lower bound given by $-An$; there might also be an upper bound to n, e.g. in the presence of a hard core in the potential V. In addition, the continuous function $s(\varepsilon, n)$ has the following properties. For systems with a kinetic energy, $s(\varepsilon, n)$ is increasing in ε for any given n, and decreasing in n for any given ε; furthermore, as we argued previously from thermodynamic considerations, $s(\varepsilon, 1/n)$ is a completely concave function. We note that these properties assure the positivity of the temperature, of the pressure, of the specific heat at constant volume and of the isothermal compressibility.

We now write the canonical partition function (1.29) as

$$Z(\beta, V, N) = \frac{1}{N!} \int_{\Gamma_V} dq^{3N} \, dp^{3N} \, dE \, \delta \, [E - H(\{q_i, p_i\})] \exp [-\beta E] \, . \tag{1.63}$$

Putting aside mathematical rigour, we substitute $E = N\varepsilon$, valid in the thermodynamic limit, we invert the order of integration, we exploit the definitions (1.13) and (1.14), and in the resulting expression we introduce $S(E, V, N) = Ns(\varepsilon, n)$, also valid in the thermodynamic limit. We thus obtain

$$Z(\beta, V, N) = \int d(N\varepsilon) \, \exp \{N \, [s(\varepsilon, n) - \beta\varepsilon]\} \, . \tag{1.64}$$

For very large N, this integral can be performed using the saddle point method, which shows that the leading dependence on N is exponential

$$Z(\beta, V, N) \propto \exp\left\{N \sup_{\varepsilon} [s(\varepsilon, n) - \beta\varepsilon]\right\} \tag{1.65}$$

$$= \exp\left\{-N \inf_{\varepsilon} [\beta\varepsilon - s(\varepsilon, n)]\right\}. \tag{1.66}$$

The factor of proportionality has an algebraic dependence on N; therefore in the thermodynamic limit, using (1.35), the specific free energy is given by

$$f(\beta, n) = \lim_{N \to \infty} \frac{1}{N} F(\beta, V, N) = -\lim_{N \to \infty} \frac{1}{\beta N} \ln Z(\beta, V, N) \tag{1.67}$$

$$= \frac{1}{\beta} \inf_{\varepsilon} [\beta\varepsilon - s(\varepsilon, n)]. \tag{1.68}$$

This relation, which can be proven rigorously, shows that the function

$$\phi(\beta, n) \equiv \beta f(\beta, n), \tag{1.69}$$

which we will call the rescaled free energy (it is sometimes called a Massieu potential), is the Legendre–Fenchel transform, with respect to ε, of the specific entropy $s(\varepsilon, n)$. Accordingly, $\phi(\beta, n)$ is concave with respect to β. From the property mentioned at the end of Section 1.3.2, we obtain that $f(\beta, n)$ is concave with respect to $1/\beta$. Concerning the other variable, because of the minus sign in front of s in the definition of the transform, $\phi(\beta, n)$ and $f(\beta, n)$ are decreasing and convex with respect to $1/n$.

Since $s(\varepsilon, n)$ is concave, the Legendre–Fenchel transform is invertible, and we have

$$s(\varepsilon, n) = \inf_{\beta} [\beta\varepsilon - \phi(\beta, n)]. \tag{1.70}$$

Equations (1.68) and (1.70) express the equivalence between microcanonical and canonical ensembles. For each positive value of β, there will be a value of ε (depending on n) satisfying (1.68).[1] For that value of ε, Eq. (1.70) will be satisfied by the previous value of β. Figure 1.2 provides a visual explanation of the relation between s and ϕ and of the correspondence between ε and β.

In the presence of phase transitions there will be points where s and f are non-analytic, e.g. with points of discontinuity in derivatives of a given order or ranges where the functions are not strictly concave or convex. In this case, this correspondence between ε and β is not one-to-one. This fact, however, does not spoil ensemble equivalence (MacDowell *et al.*, 2004).

---

[1] We expect that the system can have all possible temperatures from 0 to ∞, and all specific energies from the lowest allowed values to ∞; accordingly the derivative of $s(\varepsilon, n)$ with respect to ε will decrease from ∞, at the lowest value of ε to 0, for $\varepsilon \to \infty$. In lattice systems without a kinetic energy, the specific energy is generally upper bounded, and this can result in the presence of an energy range where the entropy decreases when ε increases, and then the presence of negative temperatures (Onsager, 1949; Purcell and Pound, 1951).

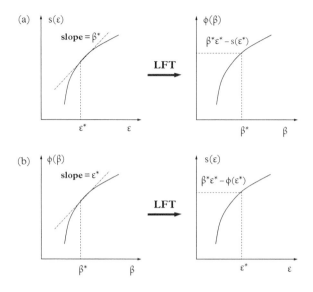

Figure 1.2 *Relation between the entropy per particle $s(\varepsilon, n)$ and $\phi(\beta, n) = \beta f(\beta, n)$ by the Legendre–Fenchel transform (LFT). $f(\beta, n)$ stands for the free energy per particle and n is kept fixed. Reprinted from Campa et al. (2009), © 2009, with permission from Elsevier.*

Let us now pass to the grand-canonical ensemble. Introducing the fugacity $z = e^{\beta\mu}$, we can write the grand partition function, Eq. (1.41), as

$$\mathcal{Z}(\beta, V, z) = \sum_{N=0}^{\infty} z^N Z(\beta, V, N). \tag{1.71}$$

Again, we do not worry about mathematical rigour and we follow a procedure similar to that used for the canonical ensemble. We substitute for the canonical partition function $Z(\beta, V, N)$ its expression valid in the thermodynamic limit, to obtain

$$\mathcal{Z}(\beta, V, z) = \sum_{N=0}^{\infty} \exp\{N[\ln z - \phi(\beta, N/V)]\}. \tag{1.72}$$

From Eq. (1.47), we get that the pressure of the system is

$$p(\beta, z) = \frac{1}{\beta} \lim_{V \to \infty} \frac{1}{V} \ln \mathcal{Z}(\beta, V, z). \tag{1.73}$$

In the thermodynamic limit, there will be a dominant term in the sum on the right-hand side of Eq. (1.72). The value of N of this dominant term depends on V, and we can write

$$\mathcal{Z}(\beta, V, z) \propto \sup_{N} \left[\exp\left\{ N \left[\ln z - \phi(\beta, N/V) \right] \right\} \right] \tag{1.74}$$

$$= \exp\left\{ N(V) \left[\ln z - \phi(\beta, N(V)/V) \right] \right\} . \tag{1.75}$$

Substituting in Eq. (1.73), we obtain

$$p(\beta, z) = \frac{1}{\beta} \sup_{n} \left[n \ln z - n\phi(\beta, n) \right] . \tag{1.76}$$

The function $\beta p(\beta, z)$ is therefore the Legendre–Fenchel transform, with respect to n, of the function $n\phi(\beta, n)$. Since $\phi(\beta, n)$ is convex with respect to $1/n$, $n\phi(\beta, n)$ is convex with respect to n. This assures that the transformation (1.76) is invertible, and we have

$$\phi(\beta, n) = \frac{1}{n} \sup_{z > 0} \left[n \ln z - \beta p(\beta, z) \right] . \tag{1.77}$$

Equations (1.76) and (1.77) show the equivalence between the canonical and the grand-canonical ensembles, giving the correspondence between the values of the density n and of the fugacity z. As previously, when there are phase transitions and nonanaliticities in ϕ and p, this correspondence is not one-to-one.

1.3.5   Equivalence in presence of phase transition: the Maxwell construction

Phase transitions are associated with singularities of thermodynamic functions (Huang, 1987). In the microcanonical and canonical ensemble, they will be hence signalled by discontinuities in the derivatives (at some specific order) of the entropy s and/or rescaled free energy ϕ.

To clarify this point we imagine to work at constant n and solely focus on the dependence of s (resp. ϕ) on ε (resp. β). Assume $s(\varepsilon)$ has a zero curvature in some energy range $[\varepsilon_1, \varepsilon_2]$: The second derivative of $s(\varepsilon)$ is hence 0 in that interval, which in turn amounts to assuming a linear dependence of $s(\varepsilon)$ versus ε, as depicted in Fig. 1.3. Clearly, at the edges of the interval, the second derivative of s is discontinuous. Moreover for $\varepsilon \in [\varepsilon_1, \varepsilon_2]$, the function is not strictly concave and relation (1.48), with $f = s$ and $x = \varepsilon$, is satisfied with an equality sign. If the energy of the system falls in this range, the system itself can hence separate into two phases, respectively characterized by the energy values ε_1 and ε_2. Because of additivity, the following condition must be met,

$$\varepsilon = c\varepsilon_1 + (1 - c)\varepsilon_2, \tag{1.78}$$

where c quantifies the fraction in phase 1 and $1-c$ refers to the remaining fraction in phase 2. The Legendre–Fenchel transform of $s(\varepsilon)$ makes it possible to calculate the canonical (rescaled) free energy: see Fig. 1.3. Note that all energies in the interval $[\varepsilon_1, \varepsilon_2]$ yield the same $\beta = \beta_t$, i.e. the slope of the straight segment of $s(\varepsilon)$. As depicted in Fig. 1.3a, the

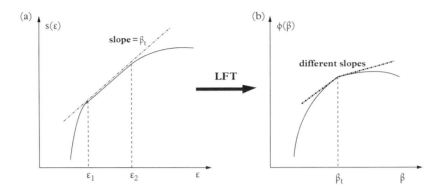

Figure 1.3 *(a) Entropy $s(\varepsilon)$ vs ε. (b) $\phi(\beta)$ vs β. The situation corresponding to a first-order phase transition is here schematized. The inverse transition temperature is β_t and $[\varepsilon_1, \varepsilon_2]$ is the energy range of phase coexistence. Reprinted from Campa et al. (2009), © 2009, with permission from Elsevier.*

entropy function $s(\varepsilon)$ changes its slope, when approaching the interval $[\varepsilon_1, \varepsilon_2]$ from the left or right. The rescaled free energy (Fig. 1.3b) presents therefore a discontinuity in the first derivative with respect to β at β_t $((d\phi/d\beta)_{-,+} = \varepsilon_{1,2})$. Following Ehrenfest's classification, this is a first-order phase transition.

Note that no one-to-one correspondence is found here between ε and β: Several microcanonical macroscopic states collapse, in fact, into a single canonical state. In the presence of first-order phase transitions the relation between the ensembles must be, hence, handled with extreme care. Indeed, as we shall see, this is a marginal case of a more extended inequivalence class. The one-to-one correspondence between the values of ε and β is formally restored when adjusting the parameters of the model so to shrink the straight segment of Fig. 1.3 to one single point. This is the case of a second-order phase transition, for which the discontinuity in the first derivative of the rescaled free energy is removed.

As a further important point, let us consider the case where the entropy function shows a convex region, see Fig. 1.4a. For short-range interactions, this scenario is, for instance, found for a finite system near a phase transition. When increasing the system size the entropy approaches the so-called *concave envelope*, represented with a tick dashed line in Fig. 1.4a. This fact is deduced on the basis of numerical (Labastie and Whetten, 1990; Jellinek *et al.*, 1998; Gross, 2001; Chomaz and Gulminelli, 2002; Chomaz, 2008) and analytical (Lynden–Bell, 1995; Lynden-Bell, 1996) investigations.

From the definition of the inverse temperature $\beta = ds/d\varepsilon$, we get

$$s(\varepsilon_2) = s(\varepsilon_1) + \int_{\varepsilon_1}^{\varepsilon_2} d\varepsilon \, \beta(\varepsilon). \tag{1.79}$$

The function $\beta(\varepsilon)$, corresponding to the entropy profile of Fig. 1.4a, is reported in Fig. 1.4b. The value $s(\varepsilon_2)$ is alternatively obtained by integrating along the thick dashed line in Fig. 1.4a which yields

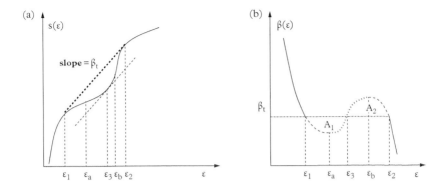

Figure 1.4 *(a) Schematic shape of the entropy s as a function of the energy ε (solid line) showing a 'globally' convex in region in the range $[\varepsilon_1, \varepsilon_2]$, the thick dashed line realizes the 'concave envelope'. (b) Inverse temperature β as a function of energy ε. According to Maxwell construction $A_1 = A_2$. The curve β(ε) represents states that are stable (solid line), unstable (dotted line) and metastable (dashed lines). Reprinted from Campa et al. (2009), © 2009, with permission from Elsevier.*

$$s(\varepsilon_2) = s(\varepsilon_1) + (\varepsilon_2 - \varepsilon_1)\beta_t, \tag{1.80}$$

where $\beta_t = \beta(\varepsilon_1) = \beta(\varepsilon_2)$. This implies that

$$\int_{\varepsilon_1}^{\varepsilon_2} d\varepsilon\, \beta(\varepsilon) = (\varepsilon_2 - \varepsilon_1)\beta_t. \tag{1.81}$$

Splitting the integral into two intervals, $[\varepsilon_1, \varepsilon_3)$ and $[\varepsilon_3, \varepsilon_2]$, we get

$$(\varepsilon_1 - \varepsilon_3)\beta_t + \int_{\varepsilon_1}^{\varepsilon_3} d\beta\, \beta(\varepsilon) = (\varepsilon_2 - \varepsilon_3)\beta_t + \int_{\varepsilon_2}^{\varepsilon_3} d\beta\, \beta(\varepsilon), \tag{1.82}$$

where ε_3 is the energy value (within the convex region) where the entropy has slope β_t. Condition (1.82) amounts to requiring equal areas A_1 and A_2 in Fig. 1.4. Introducing the *generalized free energy*, which is a function of both energy and inverse temperature,

$$\widehat{f}(\beta, \varepsilon) = \varepsilon - \frac{1}{\beta}s(\varepsilon), \tag{1.83}$$

we obtain from Eq. (1.82) that

$$\widehat{f}(\beta_t, \varepsilon_1) = \widehat{f}(\beta_t, \varepsilon_2). \tag{1.84}$$

In other words, and as it is visually clear from Fig. 1.4, the requirement that the entropy is concave is equivalent to Maxwell equal areas construction. Moreover, the generalized free energies, computed at the transition inverse temperature β_t and at the two energies ε_1 and ε_2 which delimit the coexistence region, are equal (and looking at Fig. 1.4a, also equal to $f(\beta_t)$).

The Maxwell construction is related to the application of a maximum entropy principle for additive systems. Indeed, for all energies in the range $(\varepsilon_1, \varepsilon_2)$, the entropy corresponding to the full line in Fig. 1.4a is smaller than the entropy corresponding to the dashed line at the same energy. This latter entropy is related to a system which has performed *phase separation* and is therefore obtained as a mixture composed of a certain fraction of a state with energy ε_1 and the remaining fraction with energy ε_2, as in formula (1.78). With this latter system having a larger entropy, the natural tendency will be to phase separate. Hence the 'concave envelope' recovers maximum entropy states.

It should be remarked that the truly 'locally' convex part of the entropy is the one in the range $[\varepsilon_a, \varepsilon_b]$, while the range $[\varepsilon_1, \varepsilon_2]$ is 'globally' convex. We should therefore expect a difference in the properties of the physical states in the various ranges. Indeed, states in the range $[\varepsilon_a, \varepsilon_b]$ are *unstable* (dotted line in Fig. 1.4b), while states in the ranges $[\varepsilon_1, \varepsilon_a]$ and $[\varepsilon_b, \varepsilon_2]$ are *metastable* (dashed lines in Fig. 1.4b): at a solid–liquid phase transition, they would correspond to superheated solids and supercooled liquids, respectively. While the unstable states cannot be observed, the metastable states are observable but are not true equilibrium states, because higher entropy phase separated states are accessible.

In the next chapter, we will see that when the interactions are long range, an entropy function with a convex 'intruder' (Gross, 2001), like the one depicted with a solid line in Fig. 1.4a, can represent the truly stable equilibrium. This fact has important consequences for the general concept of ensemble inequivalence, as we shall substantiate in the following chapter.

1.4   Lattice Systems

Up to now, we have considered continuous systems, in which the dynamical variables are the coordinates and the momenta of the particles. Another important class is that of lattice systems: in each site of a given lattice there is an elementary constituent (that we can call particle) of the system that has one or more internal degrees of freedom. Let us denote by s_i the degrees of freedom of the particle at the ith site. The total interaction potential of a translationally invariant lattice system with N sites is taken as

$$U(s_1, \ldots, s_N) = \sum_{1 \le i < j \le N} \mathcal{J}(|\mathbf{r}_i - \mathbf{r}_j|) V(s_i, s_j) + h \sum_{i=1}^{N} V_h(s_i). \qquad (1.85)$$

The coupling constant $\mathcal{J}$ between the particles at the ith and at the jth sites depends on the distance between the sites. Obviously the $\{\mathbf{r}_i\}$ are not dynamical variables, but they just denote the position of the lattice sites. There is also an external potential V_h, with the same coupling constant h with all the particles. The dynamical variables $\{s_i\}$ may take continuous or discrete values; in the former case the Hamiltonian may include, together with U, also the kinetic energy term. The typical case with discrete variables is that of a spin system. Generally, the $\{s_i\}$ take values in a compact set, and the potentials V and V_h are bounded.

A lattice system is short range when the coupling constant $\mathcal{J}(r)$ decays at large distance faster than $1/r^3$ in three dimensions. As in continuous systems, this implies the additivity property and the existence of analogous theorems concerning ensemble equivalence. Because of the nature of a lattice system, we are restricted to consider only the microcanonical and the canonical ensembles. The microcanonical partition function depends on the total energy E, the coupling constant h and the number of sites N.[2] It is given by

$$\Omega(E, h, N) = \int ds^N dp^N \, \delta\left[E - H(\{s_i, p_i\}, h)\right], \tag{1.86}$$

where p_i denotes the momenta associated with the ith site, if there is a kinetic energy. For discrete variables $\{s_i\}$, the momenta are not present, and the integral is replaced by a sum. The distinguishability of different sites implies that the factor $1/N!$ is not necessary. The thermodynamic limit is defined by $N \to \infty$ and $E \to \infty$, with $E/N \to \varepsilon$. We then define the specific entropy by

$$s(\varepsilon, h) = \lim_{N \to \infty} \frac{1}{N} \ln \Omega(E, h, N). \tag{1.87}$$

It can be proven that this limit exists, and that $s(\varepsilon, h)$ is concave in ε.
   From the canonical partition function

$$Z(\beta, h, N) = \int ds^N dp^N \exp\left[-\beta H(\{s_i, p_i\}, h)\right], \tag{1.88}$$

we define the specific free energy by the following limit, which can be proven to exist

$$f(\beta, h) = -\frac{1}{\beta} \lim_{N \to \infty} \ln Z(\beta, h, N). \tag{1.89}$$

With the same procedure used for continuous systems, we can see that

$$f(\beta, h) = \frac{1}{\beta} \inf_{\varepsilon} \left[\beta \varepsilon - s(\varepsilon, h)\right]. \tag{1.90}$$

Therefore the function $\phi(\beta, h) = \beta f(\beta, h)$ is the Legendre–Fenchel transform of $s(\varepsilon, h)$, and as such it is concave in β. Since $s(\varepsilon, h)$ is concave in ε, the last relation is invertible. We thus have

$$s(\varepsilon, h) = \inf_{\beta} \left[\beta \varepsilon - \phi(\beta, h)\right]. \tag{1.91}$$

Equations (1.90) and (1.91) show the equivalence between microcanonical and canonical ensembles, with a correspondence between the values of ε and of β. Again, phase transitions give rise to points or ranges where s and f are non-analytic.

---

[2] The coupling constant h is a thermodynamical variable since it is not a parameter characterizing the system like the coupling constants $\mathcal{J}$, but can be varied, as in the case, e.g., of an external magnetic field.

Treating continuous systems in the previous section, we have considered for simplicity the case where only translational degrees of freedom are present. Of course, the particles could have also internal degrees of freedom, e.g. like the vibrational and rotational degrees of freedom of a molecular fluid. With only some technical complications, the same theorems on ensemble equivalence hold also in this more general case.

1.5   Microstates and Macrostates

We introduce here an important concept that will be relevant in the following chapters. We have already referred to the *microscopic states*, or *microstates*, as those represented by each point in the Γ space of a system. We have also explained that a macroscopic system for which we specify the value of few macroscopic observables can be in any one of a huge number of microstates, and that the accessible microstates depend on the nature and particular values of the macroscopic observables, to which we can associate a function of the phase space Γ (i.e. function of the microscopic state). Furthermore, the equilibrium values of each observable can be computed by performing an average over the appropriate ensemble of the corresponding function of Γ. This leads to the following concept. Let us consider a small number of independent functions $g_i(\Gamma)$, $i = 1, \ldots, n$, each one corresponding to a given observable; one of these could be the Hamiltonian itself, obviously associated with the total energy of the system. The restriction to a particular set of values g_i^* for these functions defines a *macrostate* of the system. Clearly, each macrostate is associated with a very large number of microstates, i.e. all those satisfying $g_i(\Gamma) = g_i^*$, and thus the macrostate description of a system is much coarser than the microstate one.

The physical relevance of the macrostate description can be inferred from the fact that we have emphasized previously: the vast majority of the accessible microstates of a system will practically have the same value for the usual macroscopic observables. In other words, if e.g. $g_1(\Gamma)$ is the Hamiltonian, and therefore g_1^* is the particular value of the total energy, then for the large majority part of the points in Γ for which $g_1(\Gamma) = g_1^*$ the other functions $g_i(\Gamma)$ ($i > 1$) take practically the same constant values g_i^*. Therefore, in the following, we will sometimes refer, with an abuse of language from the mathematical point of view, to a macrostate of a system as that represented by the equilibrium values of the observables.

However, there is another important point concerning the macrostate description. In this chapter, we have shown that for short-range systems ensembles are equivalent, and thus it is not relevant which of the observables are kept fixed (e.g. the energy in the microcanonical ensemble or the temperature in the canonical ensemble). We will see that in long-range systems, although this equivalence does not hold in general, microscopic functions corresponding to the macroscopic observables are still almost constant, within each ensemble, and the macrostate description is useful for studying ensemble equivalence or inequivalence. We will provide a demonstration of this feature within the large deviations method, treated in Chapter 3.

1.6   A Summary of the Most Relevant Points

In this chapter, we have offered a brief overview of the statistical mechanics description of short-range systems and of the theorems that assure the equivalence of the different ensembles. We want to summarize here the most relevant points, emphasizing the physical properties on the basis of these theorems.

- Without any doubt, the most important physical property of macroscopic short-range systems, as far as their statistical mechanics treatment is concerned, is that of **additivity**. We have seen that, for this to hold, it is necessary that the two-body potential decays sufficiently fast at large distance, more precisely, that it decays faster than $1/r^3$. It is easy to understand the reason for this requirement. An interaction potential $V(r)$ such that $|V(r)|$ decays as $1/r^\alpha$ with $\alpha > 3$ is integrable in three dimensions; this means that (i) the interaction of a given particle with all the others of the system is bounded even in the thermodynamic limit; (ii) the interaction of two macroscopic systems is confined to the boundary layers in contact. Therefore this is what we call a short-range potential.

- The additivity property implies that the energy and the thermodynamic potentials of two macroscopic systems in equilibrium that are in contact (and that are also in equilibrium among them) is the sum of the energies and of the potentials of the two systems. In particular, from the thermodynamic properties of the entropy and the free energy, we have inferred the **concavity** properties of these two functions, which in turn imply that the system divides in different phases for some values of the thermodynamic variables. Mathematical theorems confirm these expectations (Ruelle, 1969).

- We have seen that, besides the sufficiently fast decay at large distance, also the stability property is required for standard thermodynamic behaviour. Together, these two properties hold when the inequalities (1.58) and (1.60) with $\alpha > 3$ are satisfied. The **stability** property avoids that a large weight is given to configurations with a diverging number of particles in a finite region of space, and so it avoids the collapse of the system. We have emphasized that there might be potentials that are short range but that nevertheless do not satisfy the stability property, because the repulsion for $r \to 0$ is not sufficiently strong. The reader will have no difficulty to see that a very important potential, i.e. the gravitational one, is long range and at the same time does not avoid the collapse. In the study of concrete systems, it is therefore important to consider the behaviour of the potential both at large and at small distance.

We will see, in the remainder of this book, that the violation of the additivity property in long-range systems is the seed from which many peculiar and interesting properties arise.

2

Equilibrium Statistical Mechanics of Long-Range Interactions

2.1 Non-additivity

2.1.1 Definition of long-range interactions

To provide an operative definition of long-range interactions we consider a two-body potential $V(r)$ in d dimensions that decays as a power law at large distances. Specifically, we assume that

$$V(r) \propto \frac{\mathcal{J}}{r^\alpha},\tag{2.1}$$

where r stands for the modulus of the interparticle distance and $\mathcal{J}$ represents the coupling constant. Let us calculate the interaction energy ε of a particle placed at the centre of a d-dimensional sphere of radius R. In doing so, we will assume the other particles to be homogeneously distributed within the considered volume. As we are interested only in the interaction with the distant particles, we exclude the particles located in a small neighbourhood of radius δ (see Fig. 2.1).

The calculation yields

$$\varepsilon = \int_\delta^R d^d r \, \rho \frac{\mathcal{J}}{r^\alpha} = \rho \mathcal{J} \Omega_d \int_\delta^R r^{d-1-\alpha} dr \tag{2.2}$$

$$= \frac{\rho \mathcal{J} \Omega_d}{d - \alpha} \left[R^{d-\alpha} - \delta^{d-\alpha} \right], \qquad \text{if } \alpha \neq d, \tag{2.3}$$

where ρ is the generic (e.g. mass, charge) density and Ω_d is the angular volume in dimension d (2π in $d = 2$, 4π in $d = 3$, etc.).

When the radius R is made to increase, the energy ε stays finite provided $\alpha > d$. The total energy $E = V\varepsilon$ increases hence linearly with the volume size $V \propto R^d$, or in other words, the system is extensive. This is the case of short-range interactions.

Physics of Long-Range Interacting Systems. First Edition. A. Campa *et al.*
© A. Campa, T. Dauxois, D. Fanelli, and S. Ruffo 2014. Published in 2014 by Oxford University Press.

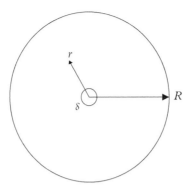

Figure 2.1 *Schematic picture of the domain considered for the evaluation of the energy ε of a particle. It is a spherical shell of outer radius R and inner radius δ. Reprinted from Campa* et al. *(2009), © 2009, with permission from Elsevier.*

In contrast, if the potential does not decay sufficiently fast, the total energy grows superlinearly with the volume at constant density, which violates extensivity. For α smaller than the embedding dimension d, in fact, the energy ε diverges with the volume, obeying the scaling $V^{1-\alpha/d}$ (logarithmically in the marginal case $\alpha = d$). Following the customary paradigms (Dauxois *et al.*, 2002a), we will define an interaction to be long range if $\alpha \leq d$ (i.e. $\alpha \leq 3$ in ordinary three-dimensional space). This definition is given here for systems with continuous translational degrees of freedom, which do not possess internal degrees of freedom. However, it can be generalized to the lattice systems, defined in Section 1.4. In that case the decay property determining the short or long rangedness of the system is that of the coupling constant associated with the interaction, i.e. $\mathcal{J}(r)$ of Eq. (1.85).

Remarkably, and according to the definition just given, the surface contribution to the energy cannot be neglected when long-range couplings are at play. Each particle is hence influenced by all other constituents, not just local neighbours, as it happens instead for short-range interactions.

Extensivity can be formally recovered by redefining the coupling constant $\mathcal{J}$ as $\mathcal{J}V^{\alpha/d-1}$. A special class of transformations, Kac's rescaling (Van Hove *et al.*, 1949; Lebowitz and Penrose, 1966), applies to the so-called mean-field models: These latter correspond to having $\alpha = 0$, since the interaction does not depend on the distance, and therefore $\mathcal{J} \to \mathcal{J}/V$. However, as we shall be commenting in the following, the fact that energy can be made extensive does not imply that the system becomes additive. It is indeed the lack of additivity which eventually materializes in the intriguing equilibrium and dynamical properties of long-range systems as compared to short-range ones.

2.1.2 Extensivity vs additivity

To clarify the crucial concepts of *extensivity* and *additivity*, we shall here make explicit reference to a concrete example, namely the mean field Curie–Weiss Hamiltonian. This latter is defined by

$$H_{CW} = -\frac{\mathcal{J}}{2N}\sum_{i,j=1}^{N} S_i S_j = -\frac{\mathcal{J}}{2N}\left(\sum_{i=1}^{N} S_i\right)^2, \tag{2.4}$$

where the spin variables $S_i = \pm 1$ are linked to specific sites labelled by the discrete index $i = 1, \ldots, N$. In this case, the interaction is not sensitive to the distance, each spin being equally influenced by all other spins belonging to the system under scrutiny. These are the so-called mean-field models. The regularizing $1/N$ prefactor in (2.4) is introduced following the aforementioned Kac's prescription to guarantee the convergence of the energy per particle to a finite value in the thermodynamic limit. Model (2.4) is *extensive*. In fact, for any given intensive magnetization

$$m = \frac{\sum_i S_i}{N} = \frac{M}{N}, \tag{2.5}$$

when doubling the number of spins, the energy of the system gets magnified by a factor of 2. The quantity M in Eq. (2.5) is the extensive magnetization. Hamiltonian (2.4) is, however, *not additive*, as it can be proven on the basis of a simple reasoning. Imagine dividing the system into equal portions, as depicted in Fig. 2.2. Furthermore, consider the particular case where all spins belonging to the left part are set to +1, whereas the others are −1. The energy of the two parts is thus straightforwardly evaluated as $E_1 = E_2 = -\mathcal{J}N/4$. However, the total energy of the full system is clearly $E = 0$. Since $E \neq E_1 + E_2$, at least for this specific configuration, the system is not additive.

Summing up, *extensivity* in the Curie–Weiss model is provided by the $1/N$ factor introduced in Hamiltonian (2.4), that makes the energy proportional to N. Nonetheless, the model is *non-additive*, as shown earlier. *Extensivity* and *additivity* are hence two distinct concepts that are to be handled with caution. As we shall be discussing in the following, the peculiar properties of long-range interacting systems ultimately stem from

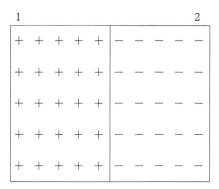

Figure 2.2 *Schematic picture of a system separated into two equal parts with N/2 spins up in domain 1 and N/2 spins down in domain 2. Reprinted from Campa et al. (2009), © 2009, with permission from Elsevier.*

the lack of additivity, as depicted earlier. More specifically, additivity implies extensivity (thus non-extensivity implies non-additivity), but not the reverse.

The reader might wonder about the physical plausibility of the Kac's prescription. Obviously in a real system the interaction potential between two elementary constituents of the system does not change with the number N. The prescription mathematically recovers the extensivity of the system, but this by itself does not justify, from the physical point of view, its use. However, we will see that the prescription, besides being a convenient mathematical procedure, gives meaningful thermodynamic and statistical properties, see Chapter 8, after the description of the tools employed in the study of the dynamics of long-range systems. In fact, at that point we will have more arguments to explain the consequences of the Kac's prescription. We will see that once the equilibrium and dynamical properties of a system with the Kac's scaling have been determined, the corresponding properties of the system without such scaling can be derived for any large finite N.

2.1.3 Non-additivity and the lack of convexity in thermodynamic parameters

Let us start by recalling the basic definition of *convex* set in a d-dimensional space. A set is said to be convex if it contains *all* the line segments connecting any pairs of its points.

Consider the space of the intensive thermodynamic parameters and focus in particular on the $2D$ projection (ε, m). Here m stands for the magnetization per particle (see Eq. (2.5)), while ε labels the energy per particle of the system. In general, a limited portion of the available space can be visited by the system, due to the specific peculiarities of the model. The allowed region is, however, always convex when short-range interactions are at play. This latter property descends directly from additivity, as it is argued in the following.

Imagine dividing a short-range system into two subsystems with different energies ε_1 and ε_2, and different magnetizations, m_1 and m_2. Introduce then a parameter λ which takes values between 0 and 1, depending on the relative size of the subsystems. As a consequence of additivity, the energy per particle obtained by composing together the contributions associated with each subsystem reads $\varepsilon = \lambda\varepsilon_1 + (1 - \lambda)\varepsilon_2$, and the corresponding magnetization $m = \lambda m_1 + (1 - \lambda)m_2$. Any value of λ in the interval $[0, 1]$ corresponds to a specific configuration of the system that can be thermodynamically realized, by properly tuning the relative size of the subsystems. Mathematically, this amounts to requiring a *convexity property* of the attainable region in the thermodynamic parameters space. Moreover, convexity implies that the space of thermodynamic parameters is connected.

As opposed to this view, systems with long-range interactions are not additive, and the associated parameters space may be not convex. Intermediate values of the extensive parameters are hence not necessarily accessible, as exemplified in Fig. 2.3. This is a rather profound observation, which has important consequences on the dynamics of long-range interacting systems. Since the space of thermodynamic parameters

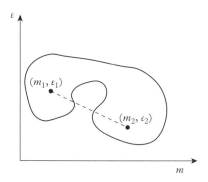

Figure 2.3 *The set of accessible macrostates in the (m, ε) space can be concave for systems subject to long-range interactions. Even if the macroscopic configurations (m_1, ε_1) and (m_2, ε_2) can be realized, this is not necessarily true for all the states connecting these two along the straight dashed line. Reprinted from Campa et al. (2009), © 2009, with permission from Elsevier.*

may be not connected, *ergodicity breaking* can materialize when focusing on a continuous microcanonical description (Borgonovi *et al.*, 2004; Mukamel *et al.*, 2005).

2.1.4 Non-additivity and the canonical ensemble

As discussed earlier, the lack of additivity impacts crucially the topological properties of the space of thermodynamical parameters. Concave domains are often found when the additivity is lost, due to the long-range nature of the couplings, consequently yielding peculiar dynamical behaviours. However, in this first part of the book we are concerned with the equilibrium properties, and we now show how the non-additivity reflects on the construction of the canonical ensemble.

As it should be clear from the previous chapter, additivity is crucial in order to justify the factorization hypothesis employed in (1.18). It is, hence, because of the underlying additivity property that we eventually recover the canonical distribution. This is not true for long-range interactions which are non-additive. The failure of the usual derivation of the canonical ensemble suggests that non-additive systems might have a very peculiar behaviour if they are in contact with a thermal reservoir (see Posch and Thirring, 2006; Lynden–Bell, 2008; Ramirez-Hernandez *et al.*, 2008a,2008b; Velazquez and Curilef, 2009 for recent literature on this topic).

Different attitudes have then catalysed a vigorous discussion, when it comes to characterizing the equilibrium statistical mechanics of long-range systems:

- On the one hand, it is claimed that the canonical ensemble *cannot* be defined for long-range systems. This implies that all analyses should be carried out within the microcanonical ensemble (Gross, 2001, 2002).

- On the other hand, researchers are reluctant to abandon the canonical viewpoint and suggest that it could be still formally defined and used (Kiessling, 1989; Padmanabhan, 1990, 2009; Chavanis, 2002b).
- Alternative approaches are finalized to developing operative construction of the heat bath (Baldovin and Orlandini, 2006a,b; De Buyl *et al.*, 2013). More recently, it has been argued (Bouchet and Barré, 2005; Villain *et al.*, 2008) that the canonical ensemble of long-range systems does not describe the fluctuations of a small part of the system, but rather the fluctuations of the whole system when coupled to a thermostat via a negligibly small coupling. This interesting line of investigation deserves to be scrutinized further.

Summarizing, there are difficulties to justify physically the canonical ensemble, although it is possible to envisage a consistent and plausible physical picture represented by it. Furthermore, it is clear that it is always possible to formally define the canonical distribution function as in Eq. (1.28), and that, if the system is made extensive (although not additive), the canonical partition function (1.29) will provide an extensive free energy.

2.2   Ensemble Inequivalence and Negative Specific Heat

At the end of the previous section, we saw the conceptual problems related to the canonical ensemble. It is then clear that for the elaboration of a thermodynamic picture of long-range systems, it is better to start by calculating the microcanonical entropy $s(\varepsilon)$ associated with a given energy ε. Once the microcanonical entropy is made available, we can obtain the (canonical) free energy $f(\beta)$ by applying the Legendre–Fenchel transform, as in Eq. (1.68) (here we consider only the variables, ε and β, interested in the transform). The rescaled free energy $\phi(\beta) = \beta f(\beta)$ is thus concave also for long-range systems, because of the properties of the Legendre–Frenchel transform that we have described. However, for long-range systems there is no theorem guaranteeing that the microcanonical entropy $s(\varepsilon)$ is concave. When this function is not concave, we cannot obtain it by the inverse Legendre–Frenchel transform, as in Eq. (1.70); this operation does not yield the correct microcanonical entropy, but rather its 'concave envelope'. For short-range systems we have shown in Section 1.3.5 that, if the computation of the microcanonical entropy for a homogeneous system leads to a function with a convex 'intruder', as in Fig. 1.4a, then, in the thermodynamic limit, there will be a phase separation, based on Eq. (1.78), so that the real entropy will be the concave envelope of the computed one; this restores the possibility of obtaining it as the inverse Legendre–Fenchel transform of $\phi(\beta)$. In long-range systems, the presence of a convex 'intruder' is possible even in the thermodynamic limit, and it cannot be avoided by invoking the phase separation of the system and the maximum entropy principle, since that argument was based on the additivity property. Therefore, the real entropy will be that with the convex intruder, and not the one represented by the concave envelope. In the next section we will give a concrete example of a simple long-range system where the microcanonical entropy has such a convex region. In conclusion, while in short-range systems the microcanonical entropy and the canonical free energy are obtainable, one by the other, with a Legendre–Fenchel transform, this does not happen in general for long-range systems, and the entropy cannot be recovered by the Legendre–Fenchel transform of the free energy.

This fact constitutes the thermodynamic ensemble inequivalence. However, this phenomenon of inequivalence is best described using the concept of *macrostate*, introduced in Section 1.5. Recall that a *microstate* is identified via the phase-space variables of the system, and so it refers to a precise microscopic configuration. Conversely, a *macrostate* is described in terms of few coarse-grained variables. It can indeed correspond to a large variety of distinct microscopic states, all yielding the same values of the macroscopic variables (Touchette, 2008). From this point of view, specifying the energy ε, as in a microcanonical description, or the inverse temperature β, as in the canonical description, means defining a macrostate. We thus see that ensemble inequivalence at the level of macrostates is expressed by the fact that all the microcanonical macrostates whose energies, in Fig. 1.4a, lay in the interval from ε_1 to ε_2, do not have a corresponding macrostate in the canonical ensemble (Ellis *et al.*, 2003).

The physical consequences of ensemble inequivalence are many-fold. First, we note that the existence of a convex 'intruder' in the entropy–energy curve, as in Fig. 1.4a, is associated with the presence of negative specific heat. Indeed,

$$\frac{\partial^2 S}{\partial E^2} = -\frac{1}{C_V T^2}, \tag{2.6}$$

where $C_V = \partial E/\partial T$ is the heat capacity at fixed volume. Hence, in the energy range $[\varepsilon_a, \varepsilon_b]$, the convexity of the entropy, $\partial^2 S/\partial E^2 > 0$, implies that the heat capacity is negative. The same conclusion applies to the normalized specific heat, namely $c_V = C_V/N$. The inequivalence stems from the observation that in the canonical ensemble, the specific heat ought to be positive. Because of the concavity of the function ϕ, we have in fact,

$$\frac{\partial^2 \phi}{\partial \beta^2} = -\frac{c_V}{T^2} < 0, \tag{2.7}$$

implying that $c_V > 0$.

Moreover, we realize that, while in the microcanonical ensemble there is no phase transition, varying the energy ε, the canonical ensemble presents a first-order phase transition. In fact, as we proved in Section 1.3.5, for a system where the microcanonical entropy has a convex intruder, as in Fig. 1.4a (full line), the canonical free energy, obtainable by the Legendre–Fenchel transform, is the same as if the entropy would be the concave envelope (dashed line in the figure). The corresponding rescaled free energy $\phi(\beta)$ is depicted in Fig. 1.3b, while the inverse temperature $\beta(\varepsilon)$ is plotted in Fig. 1.4b, with the Maxwell construction between ε_1 and ε_2. As we emphasized earlier, there is no canonical macrostate with energy between these two values: when β is decreased from values larger than $\beta(\varepsilon_1)$, the energy increases continuously. At $\beta = \beta(\varepsilon_1) = \beta(\varepsilon_2) = \beta_t$ the energy will jump from ε_1 to ε_2; afterwards, the energy again increases continuously with decreasing β. This energy jump in the canonical ensemble, i.e. the first-order phase transition with the associated latent heat is related to a discontinuity in the derivative of the function $\phi(\beta)$, visible in Fig. 1.3b.

It turns out that the presence of a first-order phase transition in the canonical ensemble is a necessary condition to obtain ensemble inequivalence. This statement was first conjectured in Barré *et al.* (2001), and then rigorously substantiated in Ellis *et al.* (2003) and Bouchet and Barré (2005), analysing the convexity properties of the entropy $s(\varepsilon)$.

In fact, it has been shown (Ellis *et al.*, 2003) that if the rescaled free energy $\phi(\beta)$ is differentiable, then the entropy $s(\varepsilon)$ can be obtained by its Legendre–Fenchel transform. This applies also for second-order phase transitions, when the second derivative of $\phi(\beta)$ is discontinuous, while the first derivative is continuous. Hence, in presence of a second-order phase transition in the canonical ensemble, the microcanonical and canonical ensembles are equivalent.

We have here considered the case in which the microcanonical entropy has a convex intruder but no singularities, since this case is sufficient to describe the features related to ensemble inequivalence. However, in general there will be singularities also in the microcanonical ensemble. In the example treated in the next section we will see that there can be ensemble inequivalence with a first-order phase transition in the canonical ensemble and a second-order phase transition in the microcanonical ensemble.

As a final point, we shortly comment about another intriguing physical phenomenon, which is related to the existence of regions with negative specific heat. As we have remarked, the energies between ε_1 and ε_2 in Fig. 1.4a correspond to the same value of β in the canonical ensemble. We can then try to speculate on what may happen to an initially isolated system with negative specific heat and energy between ε_1 and ε_2, when put in contact with a heat bath with inverse temperature β_{bath}. Refer to Fig. 2.4 and assume the system has energy U, in the interval $[\varepsilon_a, \varepsilon_b]$ where the specific heat is negative. The system becomes unstable to small perturbations: In fact, if it gets a small amount of energy from the bath, its temperature lowers, since the specific heat is negative. Consequently, further energy will flow from the bath to the system, inducing a further lowering of system's temperature. In contrast, when the initial energy fluctuation decreases the system's energy, its temperature rises, inducing a further energy outflow from the system towards the bath, and, hence, a further increase of system's temperature.

Thus, in contact with a heat bath, the system does not maintain energies in which its microcanonical specific heat is negative. The flow of energy initiated by tiny energy fluctuations ends when the system reaches again the same temperature of the bath, but at an energy for which its specific heat is positive. Looking at Fig. 2.4, this can either occur outside the range $[\varepsilon_1, \varepsilon_2]$, i.e. at point S, or inside this interval, namely in M. The same conclusion applies to all points U inside $[\varepsilon_a, \varepsilon_b]$. Once in M, the system will be in a thermodynamically metastable state. A sufficiently large fluctuation in the energy will force it to abandon this metastable configuration. The system will migrate towards a state with energy falling outside the interval $[\varepsilon_1, \varepsilon_2]$, i.e. point S, which has the same

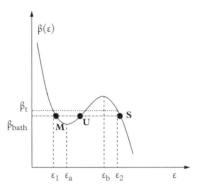

Figure 2.4 *Inverse temperature β as a function of energy ε. The Maxwell construction is shown by the dotted line, while the dash-dotted line stands for the inverse temperature of the bath β_{bath}. U denotes an unstable macroscopic state with negative specific heat, while M and S are metastable and stable configurations, respectively. Reprinted from Campa et al. (2009), © 2009, with permission from Elsevier.*

inverse temperature of the bath β_{bath}. At variance, if the system reaches directly S from U, it will stay there indefinitely because such a point belongs to a thermodynamically stable branch.

2.3   An Analytical Solvable Example: The Mean-Field Blume–Emery–Griffiths (BEG) Model

In the previous section, we presented the main physical and mathematical aspects related to ensemble equivalence or inequivalence in the study of long-range systems. Other mathematical approaches and tools will be presented in connection with concrete examples. Actually, this section is dedicated to a toy model that exhibits all the features that have been discussed so far, in particular ensemble inequivalence and negative specific heat in the microcanonical ensemble. Historically, the relation between first-order phase transition and negative specific heat for long-range systems in the thermodynamic limit was first pointed out by Antoni and Torcini (1998, 1999). The phenomenology we are going to discuss in this section has been heuristically described in Antoni *et al.* (2002).

2.3.1   Qualitative remarks

We consider the Blume–Emery–Griffiths (BEG) lattice–spin model with infinite range, mean-field-like interactions. Its phase diagram can be obtained analytically within both the canonical and the microcanonical ensembles. This study enables us to compare the two resulting phase diagrams and get a better understanding of the effect of the non-additivity on the thermodynamic behaviour of the model.

The model we consider is a simplified version of the BEG model (Blume *et al.*, 1971), known as the Blume–Capel model, where the quadrupole-quadrupole interaction is absent. The model is intended to reproduce the relevant features of superfluidity in He^3–He^4 mixtures. Recently, it has also been proposed as a realistic model for metallic ferromagnetism (Ayuela and March, 2008). It is a lattice system, and each lattice point i is occupied by a spin-1 variable, i.e. a variable S_i assuming the values $S_i = 0, \pm 1$. We will consider the mean-field version of this model, for which all lattice points are coupled with the same strength.

The Hamiltonian is given by

$$H = \Delta \sum_{i=1}^{N} S_i^2 - \frac{\mathcal{J}}{2N} \left(\sum_{i=1}^{N} S_i \right)^2,$$ (2.8)

where $\mathcal{J} > 0$ is a ferromagnetic coupling constant and $\Delta > 0$ controls the energy difference between the ferromagnetic $S_i = 1, \forall i$, or $S_i = -1, \forall i$, and the paramagnetic, $S_i = 0, \forall i$, states. In the following, we will set $\mathcal{J} = 1$, without loss of generality since we consider only ferromagnetic couplings. The paramagnetic configuration has zero energy, while the uniform ferromagnetic configurations have an energy $(\Delta - 1/2)N$. In the canonical ensemble, the minimization of the free energy $F = E - TS$ at zero temperature is equivalent to the minimization of the energy. We thus find that the paramagnetic state is the most favourable from the thermodynamic point of view if $E(\{\pm 1\}) > E(\{0\})$, which corresponds to $\Delta > 1/2$. At the point $\Delta = 1/2$, there is therefore a phase transition; it is a *first*-order phase transition since it corresponds to a sudden jump of magnetization from the ferromagnetic state to the paramagnetic state.

For vanishingly small Δ, the first term of Hamiltonian (2.8) can be safely neglected so that we recover the Curie–Weiss Hamiltonian (2.4) with spin 1, usually introduced to solve the Ising model within the mean-field approximation. It is well known that such a system has a *second*-order phase transition when $T = 2/3$ (we remind that we are adopting units for which $\mathcal{J} = 1, k_B = 1$). Since we have phase transitions of different orders on the T and Δ axis (see Fig. 2.5), we expect that the (T, Δ) phase diagram displays a *transition line* separating the low-temperature ferromagnetic phase from the high-temperature paramagnetic phase. The transition line is indeed found to be first order at large Δ values, while it is second order at small Δ values.

2.3.2   The solution in the canonical ensemble

The canonical phase diagram of this model in the (T, Δ) has been known since a long time (Blume, 1966; Capel, 1966; Blume *et al.*, 1971). The partition function reads

$$Z(\beta, N) = \sum_{\{S_1, \dots, S_N\}} \exp \left(-\beta \Delta \sum_{i=1}^{N} S_i^2 + \frac{\beta}{2N} \left(\sum_{i=1}^{N} S_i \right)^2 \right).$$ (2.9)

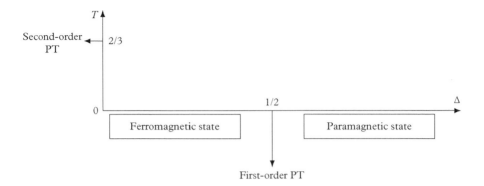

Figure 2.5 *Elementary features of the phase diagram of the Blume–Emery–Griffiths model, showing the phase transitions on the temperature T and local coupling Δ axis, respectively.*

Using the Gaussian identity (often called the Hubbard–Stratonovich transformation)

$$\exp(bm^2) = \sqrt{\frac{b}{\pi}} \int_{-\infty}^{+\infty} dx \exp(-bx^2 + 2mbx), \qquad (2.10)$$

with $m = \sum_i S_i/N$ and $b = N\beta/2$, we obtain

$$Z(\beta, N) = \sum_{\{S_1,...,S_N\}} \exp\left(-\beta\Delta \sum_{i=1}^{N} S_i^2\right) \sqrt{\frac{N\beta}{2\pi}} \int_{-\infty}^{+\infty} dx \exp\left(-\frac{N\beta}{2}x^2 + mN\beta x\right). \qquad (2.11)$$

We then easily get

$$Z(\beta, N) = \sqrt{\frac{N\beta}{2\pi}} \int_{-\infty}^{+\infty} dx \exp(-N\beta\tilde{f}(\beta, x)), \qquad (2.12)$$

where

$$\tilde{f}(\beta, x) = \frac{1}{2}x^2 - \frac{1}{\beta}\ln[1 + e^{-\beta\Delta}(e^{\beta x} + e^{-\beta x})]. \qquad (2.13)$$

The integral in (2.12) can be computed using the saddle point method where N is the large parameter. The free energy is thus

$$f(\beta) = \inf_x \tilde{f}(\beta, x). \qquad (2.14)$$

It is not difficult to see that the spontaneous magnetization m is equal to the value of x at the extremum computed in Eq. (2.14). We should also note that $\tilde{f}(\beta, x)$ is even in

x; therefore, if there is a value of x different from 0 realizing the extremum, also the opposite value realizes it. This means that if the minimum $\bar{x}$ is equal to 0 the system is in the paramagnetic phase, while if $\bar{x} \neq 0$ the system is in the ferromagnetic phase, where it can assume a positive or a negative magnetization. The phase diagram, in the (T, Δ) plane, is then divided into a paramagnetic region ($\bar{x} = 0$) and a ferromagnetic one ($\bar{x} \neq 0$).

Let us now show that the two regions are divided by a second-order phase transition line and a first-order phase transition line, which meet at a tricritical point. As in the Landau theory of phase transitions, we find a second-order transition line by a power series expansion in x of the function $\tilde{f}(\beta, x)$ in Eq. (2.13). The second-order line is obtained by equating to 0 the coefficient A_c of x^2, i.e. by the relation

$$A_c \equiv \beta - \frac{1}{2}e^{\beta\Delta} - 1 = 0, \tag{2.15}$$

provided that the coefficient B_c of x^4 is positive, i.e. provided that

$$B_c \equiv 4 - e^{\beta\Delta} > 0. \tag{2.16}$$

The tricritical point is obtained when $A_c = B_c = 0$. This gives $\Delta = \ln(4)/3 \simeq 0.4621$ and $\beta = 3$. The continuation of the critical line after the tricritical point is the first-order phase transition line, which can be obtained by finding numerically the local maximum value $\bar{x} \neq 0$ (magnetic phase) for which $\tilde{f}(\beta, x)$ is equal to $\tilde{f}(\beta, 0)$ (paramagnetic phase), i.e. by equating the free energies of the ferromagnetic and the paramagnetic phases. The behaviour of the function $\tilde{f}(\beta, x)$ as β varies is shown in Fig. 2.6: (a) represents the case

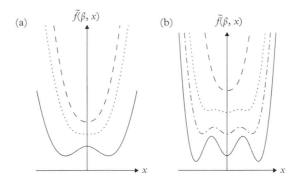

Figure 2.6 *The function $\tilde{f}(\beta, x)$ vs x for different values of the inverse temperature $\beta = 1/T$. (a) The case of a second-order phase transition, temperature values $T = 0.8$ (dashed line), 0.63 (dotted), 0.4 (solid) when $\Delta = 0.1$ are displayed. (b) The case of a first-order phase transition with $\Delta = 0.485$ when $T = 0.5$ (dashed), 0.24 (dotted), 0.21 (dash-dotted), 0.18 (solid). Reprinted from Campa et al. (2009), © 2009, with permission from Elsevier.*

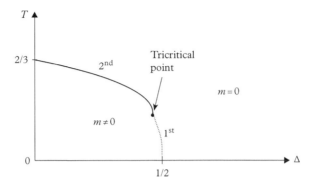

Figure 2.7 *Phase diagram of the Blume–Emery–Griffiths model in the canonical ensemble. The second-order phase transition line (solid) ends at the tricritical point (●), where the transition becomes first order (dotted). Reprinted from Campa* et al. *(2009), © 2009, with permission from Elsevier.*

of a second-order phase transition (Δ = 0.1) and (b) the case of a first-order phase transition (Δ = 0.485).

A picture of the complete phase diagram is shown in Fig. 2.7.

2.3.3   The solution in the microcanonical ensemble

The derivation of the phase diagram of the BEG model (2.8) in the *microcanonical ensemble* relies on a simple counting problem (Barré *et al.*, 2001), since all spins interact with equal strength, independently of their mutual distance. A given macroscopic configuration is characterized by the numbers N_+, N_-, N_0 of up, down and zero spins, with $N_+ + N_- + N_0 = N$. The energy E of this configuration is only a function of N_+, N_- and N_0 and is given by

$$E = \Delta Q - \frac{1}{2N} M^2, \tag{2.17}$$

where the quadrupole moment

$$Q = \sum_{i=1}^{N} S_i^2 = N_+ + N_- \tag{2.18}$$

and the magnetization

$$M = \sum_{i=1}^{N} S_i = N_+ - N_- \tag{2.19}$$

are the two order parameters. The number of microscopic configurations Ω compatible with the macroscopic occupation numbers N_+, N_- and N_0 is

$$\Omega = \frac{N!}{N_+!N_-!N_0!}. \tag{2.20}$$

Using Stirling's approximation in the large N limit, the entropy, $S = \ln \Omega$, is given by

$$S = -N\Big[(1-q)\ln(1-q) + \frac{1}{2}(q+m)\ln(q+m)+$$

$$+ \frac{1}{2}(q-m)\ln(q-m) - q\ln 2\Big], \tag{2.21}$$

where

$$q = Q/N \quad \text{and} \quad m = M/N \tag{2.22}$$

are the quadrupole moment and the magnetization per site, respectively. Equation (2.17) may be written as

$$q = 2K\varepsilon + Km^2, \tag{2.23}$$

where $K = 1/(2\Delta)$. Using this relation, the entropy per site $\tilde{s} = S/N$ can be expressed in terms of m and ε, as follows:

$$\tilde{s}(\varepsilon, m) = -(1 - 2K\varepsilon - Km^2)\ln(1 - 2K\varepsilon - Km^2)$$

$$- \frac{1}{2}(2K\varepsilon + Km^2 + m)\ln(2K\varepsilon + Km^2 + m)$$

$$- \frac{1}{2}(2K\varepsilon + Km^2 - m)\ln(2K\varepsilon + Km^2 - m)$$

$$+ (2K\varepsilon + Km^2)\ln 2. \tag{2.24}$$

In the last expression, the use of the tilde for the entropy per site is due to the fact that this entropy is defined as the logarithm of the number of states at fixed energy ε *and* fixed magnetization m. Therefore, it is not the standard entropy that is defined as the logarithm of the number of all the states at a given energy, independently from the value of the magnetization. Then, $\tilde{s}(\varepsilon, m)$ can be interpreted as a constrained entropy, where the constraint is given by the value of the order parameter m. In Section 2.4.1 we will give a simple argument to show that the entropy $s(\varepsilon)$ is given by the maximum in m of $\tilde{s}(\varepsilon, m)$, and the corresponding value of m is the equilibrium value of the magnetization. We already met something similar in Eqs. (2.13) and (2.14), although in that case x was not the order parameter, but the auxiliary variable for the Hubbard–Stratonovich transformation (however, its extremizing value in Eq. (2.14) is indeed the equilibrium magnetization). Then, at fixed ε, the value of m which maximizes the entropy corresponds to the equilibrium magnetization. The corresponding equilibrium entropy

$$s(\varepsilon) = \sup_m \tilde{s}(\varepsilon, m) \tag{2.25}$$

contains all the relevant information about the thermodynamics of the system in the microcanonical ensemble. As usual in systems where the energy per particle is bounded from above, the model has both a positive and a negative temperature region: entropy is a one-humped function of the energy. In order to locate the continuous transition line, we develop $\tilde{s}(\varepsilon, m)$ in powers of m, in analogy with what has been done earlier for the canonical free energy

$$\tilde{s} = \tilde{s}_0 + A_{mc}\, m^2 + B_{mc}\, m^4 + O(m^6), \tag{2.26}$$

where

$$\tilde{s}_0 = \tilde{s}(\varepsilon, m = 0) = -(1 - 2K\varepsilon)\ln(1 - 2K\varepsilon) - 2K\varepsilon \ln(K\varepsilon), \tag{2.27}$$

and

$$A_{mc} = -K \ln \frac{K\varepsilon}{(1 - 2K\varepsilon)} - \frac{1}{4K\varepsilon}, \tag{2.28}$$

$$B_{mc} = -\frac{K}{4\varepsilon(1 - 2K\varepsilon)} + \frac{1}{8K\varepsilon^2} - \frac{1}{96K^3\varepsilon^3}. \tag{2.29}$$

In the paramagnetic phase, both A_{mc} and B_{mc} are negative, and the entropy is maximized by $m = 0$. The continuous transition to the ferromagnetic phase takes place at $A_{mc} = 0$ for $B_{mc} < 0$. In order to obtain the critical line in the (T, Δ) plane, we first observe that the temperature $T = (\partial s/\partial \varepsilon)^{-1}$ is calculable on the critical line ($m = 0$) using (2.27). We get

$$\frac{1}{T} = 2K \ln \frac{1 - 2K\varepsilon}{K\varepsilon}. \tag{2.30}$$

Requiring now that $A_{mc} = 0$, we get the following expression for the critical line:

$$\beta = \frac{1}{2K\varepsilon}. \tag{2.31}$$

Equivalently, this expression may be written as

$$\beta = \frac{1}{2} \exp\left[\frac{\beta}{2K}\right] + 1. \tag{2.32}$$

The microcanonical critical line thus coincides with the critical line (2.15) obtained for the canonical ensemble. The tricritical point of the microcanonical ensemble is obtained at $A_{mc} = B_{mc} = 0$. Combining these equations with Eq. (2.30), we find that, at the tricritical point, β satisfies the equation

$$\frac{K^2}{2\beta^2}\left[1 + 2\exp\left(-\frac{\beta}{2K}\right)\right] - \frac{K}{2\beta} + \frac{1}{12} = 0. \tag{2.33}$$

Equations (2.32) and (2.33) yield a tricritical point at $K \simeq 1.0813$, $\beta = 3.0272$. This must be compared with the canonical tricritical point located at $K = 1/(2\Delta) = 3/\ln(16) \simeq 1.0820$, $\beta = 3$. The two points, although very close to each other, do not coincide. The microcanonical critical line extends beyond the canonical one. This feature, which is a clear indication of ensemble inequivalence, was first found analytically for the BEG model (Barré *et al.*, 2001) and later confirmed for gravitational models (Chavanis, 2002b, 2006a). The non-coincidence of microcanonical and canonical tricritical points is a generic feature, as proven in Bouchet and Barré (2005).

2.3.4   Inequivalence of ensembles

We have already discussed in general terms the question of ensemble equivalence or inequivalence in Sections 2.1 and 2.2. Inequivalence is associated with the existence of a convex region of the entropy as a function of energy. This is exactly what happens for the BEG model in the region of parameters $1 < K < 3/\ln(16)$. Since the interesting region is extremely narrow for this model (Barré *et al.*, 2001), it is more convenient to plot a schematic representation of the entropy and of the free energy (see Fig. 2.8). We show what happens in a region of K where both a *negative specific heat* and a *temperature jump* are present. The entropy curve consists of two branches: the high-energy branch is obtained for $m = 0$ (dotted line), while the low-energy one corresponds to $m \neq 0$ (full line). The $m = 0$ branch has been extended also in a region where it corresponds to metastable states, just to emphasize that these correspond to a smaller entropy and that it remains a concave function over all the energy range. We have not extended the $m \neq 0$ branch in the high-energy region not to make the plot confusing: also this region corresponds to metastable states. The two branches merge at an energy value ε_t where the left and right derivatives do not coincide; hence, microcanonical temperature is different on the two sides, leading to a *temperature jump*. It has been proven in Bouchet and Barré (2005) that for all types of bifurcations the temperature jump is always negative. In the low-energy branch, there is a region where entropy is locally convex (thick line in Fig. 2.8), giving a *negative specific heat* according to formula (2.6). The convex envelope, with constant slope β_t, is also indicated by the dash-dotted line. In the same figure (right panel), we plot the rescaled free energy $\phi(\beta)$, which is a concave function, with a point β_t where left and right derivatives (given by ε_1 and ε_2, respectively) are different. This is the first-order phase transition point in the canonical ensemble.

A schematic phase diagram near the canonical tricritical point (CTP) and the microcanonical one (MTP) is given in Fig. 2.9. In the region between the two tricritical

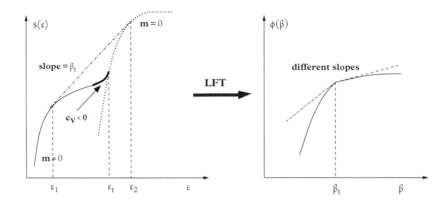

Figure 2.8 *(Left) Schematic plot of the entropy $s(\varepsilon)$ as a function of energy density ε for the BEG model in a case where negative specific heat coexists with a temperature jump. The dash-dotted line is the concave envelope of $s(\varepsilon)$ and the region with negative specific heat $c_V < 0$ is explicitly indicated by the thick line. (Right) Rescaled free energy $\phi(\beta)$: the first-order phase transition point β_t is shown. Reprinted from Campa et al. (2009), © 2009, with permission from Elsevier.*

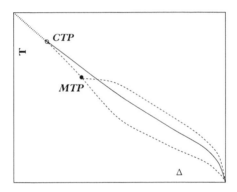

Figure 2.9 *A schematic representation of the phase diagram of the BEG model, where we expand the region around the canonical (CTP) and the microcanonical (MTP) tricritical point. The second-order line, common to both ensembles, is dotted, the first-order canonical transition line is solid and the microcanonical transition lines are dashed (with the bold dashed line representing a continuous transition). Reprinted from Campa et al. (2009), © 2009, with permission from Elsevier.*

points, the canonical ensemble yields a first-order phase transition at a higher temperature, while in the microcanonical ensemble the transition is still continuous. It is in this region that negative specific heat appears. Beyond the microcanonical tricritical point, the temperature has a jump at the transition energy in the microcanonical ensemble. The two lines emerging on the right side from the MTP correspond to the two limiting temperatures which are reached when approaching the transition energy from below and from above (see Figs. 2.10c and 2.10d). The two microcanonical temperature lines and the canonical first-order phase transition line all merge on the $T = 0$ line at $\Delta = 1/2$.

To get a better understanding of the microcanonical phase diagram and also in order to compare our results with those obtained for self-gravitating systems (Chavanis, 2002b, 2006a) and for finite systems (Gross 2001 2002; Chomaz and Gulminelli, 2002; Chomaz, 2008) we consider the temperature-energy relation $T(\varepsilon)$ (also called in the literature 'caloric curve'). Also this curve has two branches: a high-energy branch (2.30) corresponding to $m = 0$, and a low-energy branch obtained from the definition of the microcanonical temperature by using the spontaneous magnetization $m_s(\varepsilon) \neq 0$. At the intersection point of the two branches, the two entropies become equal. However, their first derivatives at the crossing point can be different, resulting in a jump in the temperature, i.e. *a microcanonical first-order transition*. When the transition is continuous in the microcanonical ensemble, i.e. the first derivative of the two entropy branches at the crossing point are equal, the BEG model always displays a discontinuity in the second derivative of the entropy. This is due to the fact that here we have a true symmetry breaking transition (Bouchet and Barré, 2005). Figure 2.10 displays the $T(\varepsilon)$ curves for decreasing values of K. For $K = 3/\ln(16)$, corresponding to the canonical tricritical point, the lower branch of the curve has a zero slope at the intersection point (Fig. 2.10a). Thus, the specific heat of the ordered phase diverges at this point. This effect signals the canonical tricritical point through a property of the microcanonical ensemble. Decreasing K down to the region between the two tricritical points, a *negative*

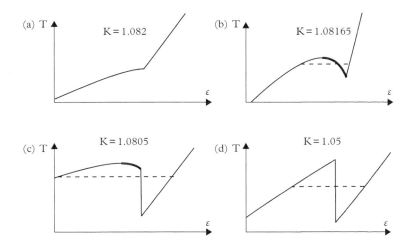

Figure 2.10 *The temperature–energy relation in the microcanonical ensemble for different values of K. The dashed horizontal line is the Maxwell construction in the canonical ensemble and identifies the canonical first-order transition temperature at the point where two minima of the free energy coexist. Thick lines identify negative specific heat in the microcanonical ensemble. We don't report the numerical values on the axes in order to improve the readability of the figure. Reprinted from Campa et al. (2009), © 2009, with permission from Elsevier.*

specific heat in the microcanonical ensemble first arises ($\partial T/\partial \varepsilon < 0$); see Fig. 2.10b. At the microcanonical tricritical point, the derivative $\partial T/\partial \varepsilon$ of the lower branch diverges at the transition point, yielding a vanishing specific heat. For smaller values of K, a jump in the temperature appears at the transition energy (Fig. 2.10c). The lower temperature corresponds to the $m = 0$ solution (2.30) and the upper one is given by $\exp(\beta/2K) = 2(1 - q^*)/\sqrt{(q^*)^2 - (m^*)^2}$, where m^*, q^* are the values of the order parameters of the ferromagnetic state at the transition energy. The negative specific heat branch disappears at even smaller values of K, leaving just a temperature jump (see Fig. 2.10d). In the $K \to 1$ limit, the low-temperature branch, corresponding to $q = m = 1$ in the limit, shrinks to 0 and the $m = 0$ branch (2.30) occupies the full energy range.

2.4 Entropy and Free Energy Dependence on the Order Parameter

In this section, we will discuss in detail the dependence of both the canonical free energy and the microcanonical entropy on the order parameter. This will enable us to understand more deeply the relation between the two ensembles by revisiting Maxwell construction. Besides that, we will also discover an interesting physical effect, *negative susceptibility*.

2.4.1 Basic definitions

Let us start from Eq. (2.24), which gives the entropy per site $\tilde{s}(\varepsilon, m)$, for the BEG model, as a function of the energy per site ε *and* the magnetization m. This entropy is proportional to the logarithm of the number of configurations which have a given energy *and* a given magnetization. In the general expression Eq. (1.86), which gives the microcanonical partition function for lattice systems, these configurations can be obtained by adding a further Dirac delta function in the integrand, so that only the configurations with a given m would be counted. Thus, we get

$$\tilde{s}(\varepsilon, m) = \lim_{N \to \infty} \frac{1}{N} \ln \int \prod_i^N ds_i \, \delta \left(E - U(\{s_i\}) \right) \delta \left(Nm - M(\{s_i\}) \right), \tag{2.34}$$

where M is the total magnetization corresponding to configuration $\{s_i\}$. For simplicity, we have considered the case without kinetic energy, so that the Hamiltonian is given by potential energy, $H = U$. We know that for spin models, where $\{s_i\}$ take discrete values, this is always true, while with continuous internal coordinates (e.g. angles) a kinetic energy term might appear. However, our discussion here does not depend on the presence of such a term. For spin models the integral in Eq. (2.34) is replaced by a discrete sum. The calculation of entropy (2.34) is often an intermediate step in the calculation of $s(\varepsilon)$. In order to get $s(\varepsilon)$, we compute the global maximum of the constrained entropy (2.34), as we did in the previous section for the BEG model, in Eq. (2.25). In fact, in the thermodynamic limit, this procedure is fully justified, since the relative contribution of all configurations corresponding to values of the order parameter that are different from the one realizing the global maximum vanishes. This can be seen by the following argument, which is not rigorous, but that should grasp the essential features. From the definition of $s(\varepsilon)$, which uses the thermodynamic limit $N \to \infty$, we know that for finite but large N we can write $\Omega(N\varepsilon) \simeq \exp[Ns(\varepsilon)]$, where Ω is the microcanonical partition function. Similarly, calling $\tilde{\Omega}(N\varepsilon, Nm)$ the integral in Eq. (2.34), we can write $\tilde{\Omega}(N\varepsilon, Nm) \simeq \exp[N\tilde{s}(\varepsilon, m)]$. Since to obtain $\Omega(N\varepsilon)$ we must integrate $\tilde{\Omega}(N\varepsilon, Nm)$ over m, the application of the saddle point method shows that $s(\varepsilon)$ is the maximum of $\tilde{s}(\varepsilon, m)$ with respect to m. As we will see in Chapter 3, this type of scaling with N can be fully justified within large deviations theory.

In the canonical ensemble, the computation of the partition function for a given value of the order parameter, i.e. for the system at a given temperature *and* at a given magnetization m, can be obtained by adding a Dirac delta function to the integrand in Eq. (1.29):

$$\tilde{f}(\beta, m) = -\frac{1}{\beta} \lim_{N \to \infty} \frac{1}{N} \ln \sum_{\{s_1,\dots,s_N\}} ds_i \exp\left[-\beta H(\{s_i\})\right] \delta\left(Nm - M(\{s_i\})\right). \tag{2.35}$$

Thus, the free energy depends on both β *and* the magnetization m. Finally, a relation analogous to Eq. (1.68) holds between entropy and free energy

$$\tilde{f}(\beta, m) = \inf_{\varepsilon}\left[\varepsilon - \frac{1}{\beta}\tilde{s}(\varepsilon, m)\right], \tag{2.36}$$

valid, as before, for all systems, independently of the range of the interactions.

We are therefore led to the introduction of the *generalized free energy* (see Eq. (1.83))

$$\widehat{f}(\beta, \varepsilon, m) = \varepsilon - \frac{1}{\beta}\tilde{s}(\varepsilon, m), \tag{2.37}$$

which will be used, as previously, to study the relation between microcanonical and canonical equilibrium states. Needless to say, it is not at all guaranteed that the function $\tilde{s}(\varepsilon, m)$ can be easily derived, in general, as we have done for the BEG model; nevertheless this general discussion is useful to show how the properties of this function explain the occurrence, or not, of ensemble equivalence.

In the microcanonical ensemble, the entropy $s(\varepsilon)$ of the system at a given energy ε is given by formula (2.25). In the canonical ensemble, the free energy $f(\beta)$ of the system at a given inverse temperature β will be given by

$$f(\beta) = \inf_{\varepsilon, m}\widehat{f}(\beta, \varepsilon, m) = \inf_{m}\tilde{f}(\beta, m) = \inf_{\varepsilon, m}\left[\varepsilon - \frac{1}{\beta}\tilde{s}(\varepsilon, m)\right], \tag{2.38}$$

as can be easily deduced from Eqs. (2.36) and (2.37). The two extremal problems (2.25) and (2.38), which basically contain the single function $\tilde{s}(\varepsilon, m)$, can be employed to study ensemble equivalence. Suppose we fix β and solve the extremal problem (2.38), finding the values of ε and m that realize an extremum. Then, we will have ensemble equivalence if the same values will realize the extremum in formula (2.25) while at the same time the derivative $\partial s/\partial \varepsilon$ will be equal to the fixed value of β. In conclusion, we seek in both extremal problems the solution of the first-order conditions

$$\frac{\partial \tilde{s}}{\partial m} = 0 \tag{2.39}$$

$$\frac{\partial \tilde{s}}{\partial \varepsilon} = \beta. \tag{2.40}$$

We denote by $\varepsilon^*(\beta), m^*(\beta)$ the solution of the variational problem (2.39) and (2.40). Using (2.39) and (2.40), it is straightforward to verify that

$$\frac{d(\beta f)}{d\beta} = \varepsilon^*(\beta), \tag{2.41}$$

meaning that the value of ε at the extremum is indeed the canonical mean energy.

However, we must consider also the stability of these extrema. We denote derivatives by subscripts; e.g. $\tilde{s}_m$ is the first derivative of $\tilde{s}$ with respect to m. The only condition required by (2.25) is that $\tilde{s}_{mm} < 0$. In order to discuss the stability of the canonical solution, we must determine the sign of the eigenvalues of the Hessian of the function to be minimized in (2.38). The Hessian is

$$\mathcal{H} = -\frac{1}{\beta} \begin{pmatrix} \tilde{s}_{mm} & \tilde{s}_{m\varepsilon} \\ \tilde{s}_{\varepsilon m} & \tilde{s}_{\varepsilon\varepsilon} \end{pmatrix}. \tag{2.42}$$

The extremum is a minimum if and only if both the trace and the determinant of the Hessian are positive:

$$-\tilde{s}_{\varepsilon\varepsilon} - \tilde{s}_{mm} > 0 \tag{2.43}$$

$$\tilde{s}_{\varepsilon\varepsilon}\tilde{s}_{mm} - \tilde{s}_{m\varepsilon}^2 > 0. \tag{2.44}$$

This implies that $\tilde{s}_{\varepsilon\varepsilon}$ and $\tilde{s}_{mm}$ must be negative, and moreover $\tilde{s}_{\varepsilon\varepsilon} < -\tilde{s}_{m\varepsilon}^2/|\tilde{s}_{mm}|$. This has strong implications on the canonical specific heat, which must be positive, as it has been shown on general grounds in Section 2.2; see Eq. (2.7). Indeed, taking the derivatives of Eqs. (2.39) and (2.40) with respect to β, after having substituted into them $\varepsilon^*(\beta), m^*(\beta)$, we get

$$\tilde{s}_{\varepsilon\varepsilon}\frac{d\varepsilon^*}{d\beta} + \tilde{s}_{\varepsilon m}\frac{dm^*}{d\beta} = 1 \tag{2.45}$$

$$\tilde{s}_{m\varepsilon}\frac{d\varepsilon^*}{d\beta} + \tilde{s}_{mm}\frac{dm^*}{d\beta} = 0, \tag{2.46}$$

where all second derivatives are computed at $\varepsilon^*(\beta)$ and $m^*(\beta)$. Recalling now that the specific heat per particle at constant volume is

$$c_V = \frac{\mathrm{d}\varepsilon^*}{\mathrm{d}T} = -\beta^2 \frac{\mathrm{d}\varepsilon^*}{\mathrm{d}\beta}, \tag{2.47}$$

we get

$$c_V = \beta^2 \frac{\tilde{s}_{mm}}{\tilde{s}_{\varepsilon m}^2 - \tilde{s}_{\varepsilon\varepsilon}\tilde{s}_{mm}}, \tag{2.48}$$

which is always positive if the stability conditions (2.43) and (2.44) are satisfied. Since the stability condition in the microcanonical ensemble only requires that $\tilde{s}_{mm} < 0$, a canonically stable solution is also microcanonically stable. The converse is not true: we may well have an entropy maximum, $\tilde{s}_{mm} < 0$, which is a free energy saddle point with $\tilde{s}_{\varepsilon\varepsilon} > 0$. This implies that the specific heat (2.48) can be negative in the microcanonical ensemble.

These results are actually quite general, provided the canonical and microcanonical solutions are expressed through variational problems of the type (2.25) and (2.38). The extrema, and thus the caloric curves $T(\varepsilon)$, are the same in the two ensembles, but the stability of the different branches is different. This aspect was first discussed by Katz (1978) in connection with self-gravitating systems (see also Chavanis, 2002b).

2.4.2 Maxwell construction in the microcanonical ensemble

We have already discussed Maxwell construction in Section 1.3.5. We have shown that, for short-range systems, where microcanonical and canonical ensembles are always equivalent, Maxwell construction derives from the concave envelope construction for microcanonical entropy. This, in turn, is a consequence of *additivity* and of the presence of a first-order phase transition in the canonical ensemble. Here, we discuss Maxwell construction for long-range systems, where the microcanonical entropy can have a stable convex intruder, leading to ensemble inequivalence.

The study of the BEG model, in Section 2.3, emphasized the presence of an extremely rich phenomenology. In particular, in a specific region of the control parameter K, the canonical and microcanonical ensembles show a first-order phase transition, with a forbidden energy range in the former ensemble and a temperature jump in the latter (see Fig. 2.10c). Both the $\varepsilon(\beta)$ and the $\beta(\varepsilon)$ curves become multiply valued if we include metastable and unstable states. Since we know that the Maxwell construction leads to an equal area condition for the $\beta(\varepsilon)$ curve, which defines the phase transition inverse temperature β_t in the canonical ensemble, we wonder here whether a similar construction exists for the $\varepsilon(\beta)$ relation which would lead to the determination of the transition energy ε_t in the microcanonical ensemble.

In the following discussion of Maxwell construction, it is crucial to understand the mechanism that generates multiple branches of the $\beta(\varepsilon)$ curve. Let us define

$$\tilde{\beta}(\varepsilon, m) = \frac{\partial \tilde{s}(\varepsilon, m)}{\partial \varepsilon}. \tag{2.49}$$

We have seen that the equilibrium magnetization m^*, at any energy ε, is the global maximum of the entropy per site (2.24). Once m^* has been computed, the equilibrium inverse temperature is given by $\beta(\varepsilon) = \tilde{\beta}(\varepsilon, m^*)$. However, also local maxima, local minima and saddles of the entropy exist, corresponding to different values of m. Following such critical points as a function of ε, we determine the different branches of $\beta(\varepsilon)$. In particular, we have the continuation at energies lower than the microcanonical transition energy ε_t of the high energy $m = 0$ branch (dashed part of $\beta_H(\varepsilon)$ in Fig. 2.11) and the continuation at higher energies of the magnetized branch $\beta_L(\varepsilon)$. It's interesting to remark that the $m = 0$ point remains an extremum for all values of ε since $\tilde{s}(\varepsilon, m)$ is even in m.

An example of inverse temperature $\beta(\varepsilon)$ relation is plotted in Fig. 2.11. The lower branch β_L starts at low energy and ends at the energy ε_H, where its derivative becomes infinite. The upper branch β_H starts at high energy and ends at the energy ε_L, where again its derivative is infinite. These two branches are connected by the vertical line at energy ε_t, and by the intermediate branch β_I that goes from ε_L to ε_H. The equilibrium state is given by the lower branch for $\varepsilon < \varepsilon_t$ and by the upper branch for $\varepsilon > \varepsilon_t$. The lower branch for $\varepsilon_t < \varepsilon < \varepsilon_H$ (dashed) and the upper branch for $\varepsilon_L < \varepsilon < \varepsilon_c$ (dashed) represent metastable states, while the intermediate branch represents unstable states (dotted). Therefore, when the energy increases, the equilibrium value of β jumps from the lower to the upper branch (thus following the vertical line) at the transition energy ε_t. It is easy to show that the vertical line realizes a Maxwell construction, i.e. the two areas A_1 and A_2 are equal. The curve $\beta(\varepsilon)$ has therefore three branches, which we denote by $\beta_L(\varepsilon)$ (the low-energy magnetized branch), $\beta_I(\varepsilon)$ (the intermediate branch of unstable states) and $\beta_H(\varepsilon)$ (the high-energy paramagnetic branch). Then, we have

$$A_2 - A_1 = \int_{\varepsilon_t}^{\varepsilon_H} \beta_L(\varepsilon) d\varepsilon + \int_{\varepsilon_H}^{\varepsilon_L} \beta_I(\varepsilon) d\varepsilon + \int_{\varepsilon_L}^{\varepsilon_t} \beta_H(\varepsilon) d\varepsilon \tag{2.50}$$

$$= (s_L(\varepsilon_H) - s_L(\varepsilon_t)) + (s_I(\varepsilon_L) - s_I(\varepsilon_H)) + (s_H(\varepsilon_t) - s_H(\varepsilon_L)), \tag{2.51}$$

where on the right-hand side $s_i(\varepsilon)$ is the function whose derivative gives the branch $\beta_i(\varepsilon)$, with $i = H, I, L$. We use now the continuity property of the entropy, imposing that $s_L(\varepsilon_H) = s_I(\varepsilon_H)$ and $s_I(\varepsilon_L) = s_H(\varepsilon_L)$. Moreover, the transition occurs at the energy where the entropies of the low-energy branch and of the high-energy branch are equal, i.e. $s_L(\varepsilon_t) = s_H(\varepsilon_t)$. We then obtain that $A_1 = A_2$. It should be remarked that the values of the three branches of $\beta(\varepsilon)$ at ε_t determine the size of the temperature jump. Indeed, $\beta_L(\varepsilon_t) = \beta_L^*$, $\beta_I(\varepsilon_t) = \beta_I^*$ and $\beta_H(\varepsilon_t) = \beta_H^*$ (see Fig. 2.11).

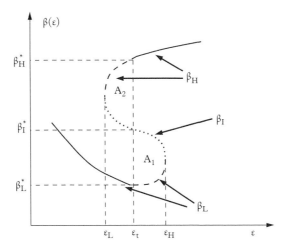

Figure 2.11 *Typical shape of the $\beta(\varepsilon)$ curve at a microcanonical first-order phase transition. The transition energy ε_t is determined by an equal area $A_1 = A_2$ Maxwell construction. All states are represented: stable (solid line), metastable (dashed lines) and unstable (dotted line). The inverse temperature jump is given by $\beta_H^* - \beta_L^*$. Reprinted from Campa et al. (2009), © 2009, with permission from Elsevier.*

The equal area condition implies that

$$\int_{\beta_L}^{\beta_H} d\beta \, [\varepsilon(\beta) - \varepsilon_t] = 0. \tag{2.52}$$

Using $\varepsilon = d\phi/d\beta$, we get

$$\phi(\beta_H) - \phi(\beta_L) - \varepsilon_t(\beta_H - \beta_L) = 0, \tag{2.53}$$

which, after defining the *generalized entropy*

$$\widehat{s}(\beta, \varepsilon) = \beta\varepsilon - \phi(\beta), \tag{2.54}$$

leads to

$$\widehat{s}(\beta_L, \varepsilon_t) = \widehat{s}(\beta_H, \varepsilon_t), \tag{2.55}$$

which is the condition analogous to (1.84).

Maxwell construction first appeared for self-gravitating systems in Aronson and Hansen (1972), and was later extended to microcanonical phase transitions in Chavanis (2002a) and Barré *et al.* (2002). On the other hand, canonical and microcanonical caloric curves were studied in Stahl *et al.* (1995). If the caloric curve is made of several disconnected branches the evaluation of Maxwell areas must be done cautiously (Chavanis, 2002b, 2006a).

2.4.3 Negative susceptibility

We have seen that ensemble inequivalence can give rise to negative specific heat in the microcanonical ensemble. We show here that another consequence of ensemble inequivalence is the existence of equilibrium microcanonical states with a negative magnetic susceptibility, a non-negative quantity in the canonical ensemble. We will follow a treatment close to that of Section 2.4.1.

Fixing the energy and the magnetization, the entropy is given by Eq. (2.34). From the first principle of thermodynamics, that for magnetic systems reads $T dS = dE - h dM$, with E and M the internal energy and the total magnetization of the system, respectively, it is straightforward to prove the following formula for the average effective magnetic field in the microcanonical ensemble (see also Campa *et al.* (2007c)):

$$h(\varepsilon, m) = -\frac{\dfrac{\partial \tilde{s}}{\partial m}}{\dfrac{\partial \tilde{s}}{\partial \varepsilon}} = -\frac{1}{\beta(\varepsilon, m)} \frac{\partial \tilde{s}}{\partial m}. \tag{2.56}$$

Taking into account that, like β is canonically conjugated to ε, βh is canonically conjugated to $-m$, it is natural to define the partition function

$$Z(\beta, h, N) = \sum_{\{S_1, \dots, S_N\}} \exp\{-\beta \, [H(\{S_i\}) - hM(\{S_i\})]\}, \tag{2.57}$$

where, as in Eq. (2.34), $M(\{S_i\})$ is the total magnetization corresponding to configuration $\{S_i\}$. Analogously to Eq. (2.38), the free energy is

$$f(\beta, h) = -\frac{1}{\beta} \lim_{N \to \infty} \frac{1}{N} \ln Z(\beta, h, N) = \inf_{\varepsilon, m} \left[\varepsilon - hm - \frac{1}{\beta} \tilde{s}(\varepsilon, m) \right]. \tag{2.58}$$

For $h = 0$, we obviously recover Eq. (2.38). As in Section 2.4.1, we see that the relation between the two ensembles can be studied by analysing the single function $\tilde{s}(\varepsilon, m)$. Taking into account Eq. (2.56), the variational problem in Eq. (2.58), together with the variational problem that defines $s(\varepsilon)$, can be solved by imposing that

$$\frac{\partial \tilde{s}}{\partial m} = -h\beta \tag{2.59}$$

$$\frac{\partial \tilde{s}}{\partial \varepsilon} = +\beta, \tag{2.60}$$

that generalize the conditions (2.39) and (2.40) to $h \neq 0$, providing the functions $\varepsilon(\beta, h)$ and $m(\beta, h)$. Independently of the value of h, the stability conditions in the canonical ensemble are the same as those given in Eqs. (2.43) and (2.44). In the microcanonical ensemble, since we are not maximizing with respect to m, we have no condition on the second derivative of $\tilde{s}(\varepsilon, m)$ with respect to m.

Magnetic susceptibility is defined as

$$\chi = \frac{\partial m}{\partial h}. \tag{2.61}$$

Deriving (2.59) and (2.60) with respect to h, we obtain

$$\tilde{s}_{m\varepsilon} \frac{\partial \varepsilon}{\partial h} + \tilde{s}_{mm} \frac{\partial m}{\partial h} = -\beta, \tag{2.62}$$

$$\tilde{s}_{\varepsilon\varepsilon} \frac{\partial \varepsilon}{\partial h} + \tilde{s}_{\varepsilon m} \frac{\partial m}{\partial h} = 0, \tag{2.63}$$

from which we get

$$\chi = -\beta \frac{\tilde{s}_{\varepsilon\varepsilon}}{\tilde{s}_{\varepsilon\varepsilon}\tilde{s}_{mm} - \tilde{s}_{\varepsilon m}^2}. \tag{2.64}$$

This formula is valid in both the canonical and microcanonical ensemble. However, the results can differ in the two ensembles because the quantities in this formula are computed at different stationary points in the two ensembles. In the canonical ensemble, χ is positive definite for all h because of the stability conditions (2.43) and (2.44).

As for the specific heat, the positivity of magnetic susceptibility in the canonical ensemble can be derived on general grounds from Eq. (2.57), since it is easily shown that susceptibility is proportional to the canonical expectation value $\langle (M - \langle M \rangle)^2 \rangle$.

On the other hand, in the microcanonical ensemble, as already remarked, no condition on the second derivatives of $\tilde{s}$ with respect to m is required, and therefore susceptibility can have either sign. Indeed, in Campa *et al.* (2007c), it is shown that a simple ϕ^4 model can display a negative microcanonical susceptibility.

With ensemble equivalence, as it always happens in short-range systems, magnetic susceptibility is positive also in the microcanonical ensemble. We note that this result is also a byproduct of the convexity property discussed in Section 2.1.3, since the attainable region in the (ε, m) plane is necessarily convex for short-range systems. In fact, this implies that Eqs. (2.43) and (2.44) are satisfied for all equilibrium values (ε, m).

2.5 The Min–Max Procedure

We have already emphasized that the microcanonical partition function is generally more difficult to obtain than the canonical one. In this section we will show how the microcanonical entropy can be obtained when the canonical free energy has been derived through an optimization procedure. At the same time, this discussion will allow us to understand how ensemble inequivalence can arise. However, it should be remarked that the derivation presented here is not rigorous and relies on strong assumptions. We will follow the argument presented in Leyvraz and Ruffo (2002), and later generalized in Campa and Giansanti (2004), although we will give here a somewhat different proof of the result.

It is crucial to assume that the canonical partition function is given by the integral

$$Z(\beta, N) = \int_{-\infty}^{+\infty} dx \, \exp\left[-N\tilde{\phi}(\beta, x)\right], \tag{2.65}$$

with $\tilde{\phi}(\beta, x)$ a differentiable function of β for $\beta \geq 0$ and of a dummy variable x. This is a crucial assumption, which is at the core of the applicability of the method. We do not explicitly give any clue to what is the nature and physical meaning of the variable x, but, of course, we have in mind the integration variable which appears in the Hubbard–Stratonovich transformation. In principle the range of x is the full real axis and the integral in Eq. (2.65) is required to converge; for example, in the Hubbard–Stratonovich transformation there is a dependence of the integrand of the kind $\exp(-x^2)$.

In the thermodynamic limit, we therefore have

$$\phi(\beta) = \inf_x \tilde{\phi}(\beta, x). \tag{2.66}$$

The *canonical entropy* is defined by the variational problem

$$s_{\mathrm{can}}(\varepsilon) = \inf_{\beta \geq 0} [\hat{s}(\beta, \varepsilon)], \tag{2.67}$$

where $\hat{s}$ is defined in formula (2.54). This is nothing but the Legendre–Fenchel transform of $\phi(\beta)$, and it comes from the thermodynamic expression relating entropy to free energy. We note that the concavity of $\phi(\beta)$ assures that there can be at most one solution to the minimization problem (2.67). We observe that this expression is valid also in the case in which ε is such that there is no β value for which the derivative of $\phi(\beta)$ is equal to ε. In our case, we can insert Eq. (2.66) to obtain

$$s_{\mathrm{can}}(\varepsilon) = \inf_{\beta \geq 0} \left\{ \sup_x \left[\beta\varepsilon - \tilde{\phi}(\beta, x) \right] \right\}. \tag{2.68}$$

We should remark that this relation is valid only if the assumption made for the partition function in Eq. (2.65) is fulfilled. In particular the x-dependence of $\tilde{\phi}$ cannot be neglected in order for the integral in Eq. (2.65) to converge. The relation between $s(\varepsilon)$, $\phi(\beta)$ and $s_{can}(\varepsilon)$ is summarized in Fig. 2.12. From the properties of the Legendre–Fenchel transform we know that $s_{can}(\varepsilon) \geq s(\varepsilon)$. When ensembles are equivalent $s_{can}(\varepsilon) = s(\varepsilon)$.

$$s(\varepsilon) \xrightarrow{\text{LFT}} \phi(\beta) \xrightarrow{\text{LFT}} s_{can}(\varepsilon)$$

Figure 2.12 *Relation between microcanonical entropy $s(\varepsilon)$, rescaled free energy $\phi(\beta)$ and canonical entropy $s_{can}(\varepsilon)$. LFT indicates the Legendre–Fenchel transform.*

We will prove here that, once the function $\tilde{\phi}(\beta, x)$ is known, we can introduce its Legendre–Fenchel transform

$$\tilde{s}(\varepsilon, x) = \inf_{\beta \geq 0} \left[\beta \varepsilon - \tilde{\phi}(\beta, x) \right] \tag{2.69}$$

and obtain the microcanonical entropy by the formula

$$s(\varepsilon) = \sup_x [\tilde{s}(\varepsilon, x)] = \sup_x \left\{ \inf_{\beta \geq 0} \left[\beta \varepsilon - \tilde{\phi}(\beta, x) \right] \right\}. \tag{2.70}$$

The only difference in expressions (2.68) and (2.70) is just the order in which the minimum with respect to β and the maximum with respect to x are taken. Although this might seem a detail, it determines a different result.

We can first check that Eq. (2.70) gives indeed a function which is $\leq s_{can}(\varepsilon)$. In fact, let us consider a function of two variables $f(x, y)$. Under quite general conditions on f, the following inequality holds:

$$\sup_x \inf_y f(x, y) \leq \inf_y \sup_x f(x, y). \tag{2.71}$$

This can be proven as follows. Let us denote by (x_1, y_1) the extremum that satisfies the left-hand side of (2.71), then

$$f(x_1, y_1) \leq f(x_1, y), \qquad \forall y. \tag{2.72}$$

Similarly, the extremum that satisfies the right-hand side, denoted (x_2, y_2), satisfies

$$f(x_2, y_2) \geq f(x, y_2), \qquad \forall x. \tag{2.73}$$

Then,

$$f(x_1, y_1) \leq f(x_1, y_2) \leq f(x_2, y_2), \tag{2.74}$$

which proves inequality (2.71).

Applying this inequality to the microcanonical entropy (2.70) and the canonical entropy (2.68) leads therefore to

$$s(\varepsilon) \leq s_{can}(\varepsilon). \tag{2.75}$$

Let us remark that the entropy defined in formula (2.69) does not coincide with $\tilde{s}(\varepsilon, m)$. In particular, we can easily verify that $\tilde{s}(\varepsilon, x)$ is always concave in ε for all x values, since it is obtained from a Legendre–Fenchel transform. In contrast $\tilde{s}(\varepsilon, m)$ can be non-concave in ε.

It is interesting to propose a sketchy proof of formula (2.70), which is based on the analysis of the microcanonical partition function. Using the Laplace representation of the Dirac delta function, this latter can be expressed as

$$\Omega(E, N) = \sum_{\{S_1, \dots, S_N\}} \delta(E - H(\{S_i\})) \tag{2.76}$$

$$= \frac{1}{2\pi i} \int_{\beta - i\infty}^{\beta + i\infty} d\lambda \sum_{\{S_1, \dots, S_N\}} \exp[\lambda(E - H(\{S_i\}))] \tag{2.77}$$

$$= \frac{1}{2\pi i} \int_{\beta - i\infty}^{\beta + i\infty} d\lambda \, Z(\lambda, N) \exp(\lambda E), \tag{2.78}$$

where $\beta = \mathrm{Re}(\lambda) > 0$ is the inverse temperature. We use λ, instead of β, as an integration variable, because we are considering the analytical continuation of $Z(\lambda, N)$ to the complex plane. The last integral cannot be solved, in the thermodynamic limit, using the saddle point method because, after expressing $Z(\beta, N) \sim \exp[-N\beta f(\beta)]$, the function $\beta f(\beta)$ is not in general differentiable for all β. Despite this, we can heuristically argue that the integral will be dominated by the value of the integrand at a real value of λ ($\lambda = \beta + i\lambda_I$): otherwise, we would obtain an oscillatory behaviour of $\Omega(E, N)$, giving rise to negative values, which are impossible for the density of states. To have a proof of this we can proceed as follows.

We are assuming to be in cases where $Z(\beta, N)$, for real β, can be expressed as in (2.65), with $\tilde{\phi}$ analytic. Then, this integral representation will be valid, in the complex λ plane, for at least a strip that includes the real axis: let's say for $|\lambda_I| < \Delta$, with $\Delta > 0$.

We now divide the integral in (2.78) in three intervals, defined by $\lambda_I < -\delta, -\delta < \lambda_I < \delta$ and $\lambda_I > \delta$, respectively, with $0 < \delta < \Delta$. In Appendix B, we show that the contribution to the integral in λ coming from values of λ_I outside the strip, i.e. for values of λ_I with $|\lambda_I| > \Delta$, is exponentially small in N. Therefore, the calculation of the microcanonical partition function reduces to performing the integral

$$\Omega(E, N) = \frac{1}{2\pi i} \int_{\beta - i\delta}^{\beta + i\delta} d\lambda \, e^{N\lambda\varepsilon} Z(\lambda, N) \tag{2.79}$$

$$= \frac{1}{2\pi i} \int_{-\infty}^{+\infty} dx \int_{\beta - i\delta}^{\beta + i\delta} d\lambda \, \exp\left(N[\lambda\varepsilon - \tilde{\phi}(\lambda, x)] \right), \tag{2.80}$$

where $0 < \delta < \Delta$, and where, in the second equality, we have used the fact that inside the strip $|\lambda_I| < \Delta$ we can represent $Z(\beta, N)$ as in (2.65). Since $\tilde{\phi}$ is analytic in all the domain of integration, we can perform the integral in λ in the large N limit using the saddle point method. It can be shown that for each value of x the real part of the argument of the exponential in (2.79) is larger if computed on the real axis $\lambda_I = 0$. Indeed, we have that

$$\int_{-\infty}^{+\infty} dx \exp\{N[\lambda\varepsilon - \tilde{\phi}(\lambda, x)]\} = \exp(N\lambda\varepsilon)Z(\lambda, N) \tag{2.81}$$

$$= \exp(N\lambda\varepsilon) \sum_{\{S_1, \dots, S_N\}} \exp\{-\beta H(\{S_i\}) - i\lambda_I H(\{S_i\})\} \tag{2.82}$$

$$= e^{N\beta\varepsilon} \sum_{\{S_1,...,S_N\}} \exp\{-\beta H(\{S_i\}) - i\lambda_I H(\{S_i\}) + i\lambda_I N\varepsilon\} \tag{2.83}$$

$$= e^{N\beta\varepsilon} \langle \exp(i\lambda_I[N\varepsilon - H(\{S_i\})]) \rangle Z(\beta, N), \tag{2.84}$$

where in the last expression the average $\langle\cdot\rangle$ is performed with Boltzmann weight $\exp[\beta H(\{S_i\})]$. Using now the definition $Z(\beta, N) = \int dx \exp[-N\tilde{\phi}(\beta, x)]$, we can rewrite $\Omega(E, N)$ as

$$\Omega(E, N) = \frac{1}{2\pi i} \int_{-\infty}^{+\infty} dx \int_{\beta-i\delta}^{\beta+i\delta} d\lambda \exp\left(-N\tilde{\phi}(\beta, x) + N\beta\varepsilon + \ln\langle\exp(i\lambda_I[N\varepsilon - H(\{S_i\})])\rangle\right). \tag{2.85}$$

The real part of the exponent in the integral is obtained by replacing $\ln(\langle\cdot\rangle)$ with $\ln(|\langle\cdot\rangle|)$. The maximum of this logarithm is obtained for $\lambda_I = 0$, since for all other non-zero values of λ_I, when performing the average in $\langle\cdot\rangle$, we would sum unit vectors $\exp(i\lambda_I\alpha)$ with different values of the phase α, depending on the specific configuration $\{S_i\}$, obtaining a result which will have certainly a smaller modulus than summing all the vectors in phase with $\lambda_I = 0$. Therefore, if there is a saddle point on the real axis, it will certainly give a larger real part for the argument of the exponential than other saddle points eventually present outside the real axis. Moreover, we have just proved that the saddle is necessarily a maximum along the imaginary λ direction. Therefore, from the general properties of holomorphic functions, it will be a minimum along the $\beta = \text{Re}(\lambda)$ axis. Once the integral over λ is performed, the remaining integral over x can be computed using again the saddle point method, now for a real function, which gives a maximum over x. Combining all this, we get the formula for the microcanonical entropy (2.70).

Another way of arguing (Leyvraz and Ruffo, 2002) is to remark that, since $\tilde{\phi}(\lambda, x)$ is obtained by analytically continuing a real function, if saddle points are present in the complex λ plane, they necessarily appear in complex conjugate pairs. This would induce oscillations in the values of $\Omega(E, N)$ as a function of N, which would imply that $\Omega(E, N)$ could even take negative values. This is absurd and leads to excluding the presence of saddle points out of the real λ axis.

We have already remarked that ensemble inequivalence is a consequence of inequality (2.75), which in turn derives from the different order of the minimum in β with respect to the maximum in x in expressions (2.68) and (2.70). Let us now study in more detail the extrema defined by the two different variational problems. The first-order stationarity conditions are

$$\frac{\partial\tilde{\phi}}{\partial\beta} = \varepsilon \tag{2.86}$$

$$\frac{\partial\tilde{\phi}}{\partial x} = 0, \tag{2.87}$$

which are, of course, the same for the two ensembles. However, the stability conditions deriving from the two problems are different. For what concerns the canonical entropy $s_{can}(\varepsilon)$, we have the conditions

$$\frac{\partial^2 \tilde{\phi}}{\partial x^2} > 0 \tag{2.88}$$

$$\frac{\partial^2 \tilde{\phi}}{\partial \beta^2} \frac{\partial^2 \tilde{\phi}}{\partial x^2} - \left(\frac{\partial^2 \tilde{\phi}}{\partial \beta \partial x} \right)^2 < 0, \tag{2.89}$$

while for the microcanonical entropy $s(\varepsilon)$ we have

$$\frac{\partial^2 \tilde{\phi}}{\partial \beta^2} < 0 \tag{2.90}$$

$$\frac{\partial^2 \tilde{\phi}}{\partial \beta^2} \frac{\partial^2 \tilde{\phi}}{\partial x^2} - \left(\frac{\partial^2 \tilde{\phi}}{\partial \beta \partial x} \right)^2 < 0. \tag{2.91}$$

A necessary and sufficient condition to satisfy all four conditions (2.88), (2.89), (2.90) and (2.91) is that $\partial^2 \tilde{\phi}/\partial \beta^2 < 0$ and $\partial^2 \tilde{\phi}/\partial x^2 > 0$. However, since the conditions are different in the two ensembles, we can find values of β and x that correspond to stable states in one ensemble but are unstable in the other. It should be noted that we could find more than one stationary point in a given ensemble, satisfying the corresponding stability condition. Obviously, in this case we must choose the global extremum. If the global stable extrema are different in the two ensembles, then we have ensemble inequivalence. Tightly linked to stability is the sign of specific heat. Indeed, using the expression

$$c_V = -\beta^2 \frac{\partial \tilde{\phi}(\beta, x(\beta))}{\partial \beta^2}, \tag{2.92}$$

where $x(\beta)$ is obtained by solving (2.86) and (2.87), we can obtain a formula for the specific heat which is valid in both the ensembles:

$$c_V = -\beta^2 \frac{\dfrac{\partial^2 \tilde{\phi}}{\partial \beta^2} \dfrac{\partial^2 \tilde{\phi}}{\partial x^2} - \left(\dfrac{\partial^2 \tilde{\phi}}{\partial \beta \partial x} \right)^2}{\dfrac{\partial^2 \tilde{\phi}}{\partial x^2}}. \tag{2.93}$$

We see from (2.88) and (2.89) that this expression is positive in the canonical ensemble. However, in the microcanonical ensemble the conditions (2.90) and (2.91) do not determine the sign of $\partial^2 \tilde{\phi}/\partial x^2$, and thus the specific heat can have either sign in the microcanonical ensemble.

3

The Large Deviations Method and Its Applications

3.1 Introduction

The mathematical theory of large deviations is a field in its own. It is obviously outside the scope of this book to provide a detailed account on the theory and treat it with mathematical rigour. However, the interested reader should know that the methods based on large deviations theory and their applications to problems in statistical mechanics have been popularized among theoretical physicists in several books and review papers (Ellis, 1985; Dembo and Zitouni, 1998; Touchette, 2008).

Many particle systems with long-range interactions often offer a relatively simple field of application of the theory of large deviations. This is very interesting, especially if one considers that the outcome of the calculation is the entropy function. In this chapter, we will discuss a few representative examples, which have been selected for their pedagogical importance. In the next chapters, we will then turn to consider more complex applications.

Let us, however, emphasize a few interesting works in which physical systems with long-range interactions have been studied using this method. For example, Michel and Robert (1994) successfully used large deviation techniques to rigorously prove the applicability of statistical mechanics to two-dimensional fluid mechanics, proposed earlier (Miller, 1990; Robert, 1991; Robert and Sommeria, 1991). Ellis (1999) pursued the approach of Robert and Sommeria (1991) to solve two-dimensional geophysical systems.

The structure of this chapter is the following. In the next section, we introduce the method: we will show, in particular, how to obtain the entropy for a general class of systems of broad interest for what concerns applications to long-range applications. In Section 3.3, we will consider the case of a 3-states Potts model. Working in this context it is possible to operate a direct comparison between the predictions of the theory and a numerical estimate of the entropy. Let us emphasize that such a simple discrete system displays ensemble inequivalence. Then, we will present a different solution of the Blume–Emery–Griffiths (BEG) model, alternative to that derived by means of usual techniques in the preceding chapter.

Physics of Long-Range Interacting Systems. First Edition. A. Campa *et al.*
© A. Campa, T. Dauxois, D. Fanelli, and S. Ruffo 2014. Published in 2014 by Oxford University Press.

In the next chapter, we will introduce the application of the method to systems with continuous variables by considering a simple but very representative model of systems with long-range interactions: the Hamiltonian Mean-Field (HMF) model. Interesting applications will be treated in Sections 4.2 and 4.3, both devoted to systems presenting ensemble inequivalence: a generalized HMF model and the so-called ϕ^4 model. Finally, in Section 14.5, we will present the equilibrium solution of the Colson–Bonifacio model of a linear free electron laser.

3.2   The Computation of the Entropy for Long-Range Interacting Systems

3.2.1   The method in three steps

Denoting again collectively with $x \equiv (\{p_i\}, \{q_i\})$ the phase-space variables of a Hamiltonian system, let us suppose that the energy per particle $\varepsilon(x) = H(x)/N$ can be expressed as a function of (few) global 'mean fields' $\mu_1(x), \ldots, \mu_n(x)$. In other words, let us suppose that it is possible to write

$$\varepsilon(x) = \bar{\varepsilon}(\mu_1(x), \ldots, \mu_n(x)). \tag{3.1}$$

This is a situation often realized in long-range interacting systems. Actually, we could require that formula (3.1) be valid only asymptotically for $N \to \infty$, while for large but finite N the right-hand side could contain a remainder $R(x)$ that can be neglected in the thermodynamic limit. However, in mean-field systems, representation (3.1) is exact for all N. According to the definitions given previously, specifying the microscopic configuration x corresponds to defining what is generally referred to as a microstate of the system. In contrast, by specifying that the system is in a state in which the global variables have given values $\mu_1, \ldots, \mu_n$, we are defining what is called a macrostate of the system. Once the macrostate is chosen, the microscopic configuration is not determined, since all x that satisfy $\mu_k(x) = \mu_k$ for $k = 1, \ldots, n$ belong to the same macrostate.

The application of the large deviations method to long-range interacting systems consists of three different steps.

The **first step** is the identification, in a concrete system, of the global variables. We will present a few examples in the remainder of the chapter.

The **second step** is the computation of the entropy of the different macrostates, i.e. the calculation of the function

$$\bar{s}(\mu_1, \ldots, \mu_n) = \lim_{N \to \infty} \frac{1}{N} \ln \int dx \, \delta(\mu_1(x) - \mu_1) \ldots \delta(\mu_n(x) - \mu_n). \tag{3.2}$$

Leaving aside for the moment the problem of computing $\bar{s}(\mu_1, \ldots, \mu_n)$, which at first sight does not seem to be any simpler than computing the entropy function $s(\varepsilon)$, it is easy to see how this last function can be obtained from $\bar{s}(\mu_1, \ldots, \mu_n)$. In fact, we have

$$\int dx\, \delta\, [N(\varepsilon(x) - \varepsilon)] \;=\; \int dx\, \delta\, [N(\bar{\varepsilon}(\mu_1(x), \ldots, \mu_n(x)) - \varepsilon)] \tag{3.3}$$

$$=\; \int dxd\mu_1 \ldots d\mu_n\, \delta(\mu_1(x) - \mu_1) \ldots \delta(\mu_n(x) - \mu_n)$$
$$\times \delta\, [N(\bar{\varepsilon}(\mu_1, \ldots, \mu_n) - \varepsilon)] \tag{3.4}$$

$$\overset{N \to +\infty}{\sim}\; \int d\mu_1 \ldots d\mu_n\, \exp\, [N\bar{s}(\mu_1, \ldots, \mu_n)]$$
$$\times \delta\, [N(\bar{\varepsilon}(\mu_1, \ldots, \mu_n) - \varepsilon)] . \tag{3.5}$$

This therefore leads to

$$s(\varepsilon) = \sup_{\mu_1, \ldots, \mu_n}\, \left[\bar{s}(\mu_1, \ldots, \mu_n)\,|\,\bar{\varepsilon}(\mu_1, \ldots, \mu_n) = \varepsilon\right]. \tag{3.6}$$

Finally, the **third step** in the application of the large deviations method requires calculating the solution of the above extremal problem.

3.2.2 The computation of the entropy of the different macrostates

We are therefore left with the problem of computing $\bar{s}(\mu_1, \ldots, \mu_n)$, which corresponds to the actual implementation of the second step of the method. Let us introduce the generating function

$$\psi\, (\lambda_1, \ldots, \lambda_n) = \int dx\, \exp\, [-N\, (\lambda_1\mu_1(x) + \cdots + \lambda_n\mu_n(x))]. \tag{3.7}$$

Besides being a useful mathematical trick, we can give a physical interpretation of this function on thinking of a set of N uncoupled systems, each of them taking values according to a given probability distribution. Few steps completely analogous to those relating $\phi(\beta) = \beta f(\beta)$ (see Eq. (1.68) and surrounding ones) to $s(\varepsilon)$ show that the free energy associated with $\psi\, (\lambda_1, \ldots, \lambda_n)$ is given by the (multidimensional) Legendre–Fenchel transform of $\bar{s}(\mu_1, \ldots, \mu_n)$

$$\bar{\phi}(\lambda_1, \ldots, \lambda_n) \equiv -\lim_{N \to \infty}\, \frac{1}{N}\, \ln \psi\, (\lambda_1, \ldots, \lambda_n) \tag{3.8}$$

$$=\; \inf_{\mu_1, \ldots, \mu_n}\, [\lambda_1\mu_1 + \cdots + \lambda_n\mu_n - \bar{s}(\mu_1, \ldots, \mu_n)]. \tag{3.9}$$

We know that in general $s(\varepsilon)$ is not concave and cannot be obtained by the Legendre–Fenchel transform of $\phi(\beta)$. We would expect the same difficulty in the inversion of (3.8). However, if it happens that $\bar{\phi}(\lambda_1, \ldots, \lambda_n)$ is differentiable for real values of the λ_i, then we are guaranteed that $\bar{s}(\mu_1, \ldots, \mu_n)$ can be obtained by the inversion of (3.9) and that it is therefore concave (Ellis *et al.*, 2003)

$$\bar{s}(\mu_1, \ldots, \mu_n) = \inf_{\lambda_1, \ldots, \lambda_n} \left[\lambda_1 \mu_1 + \cdots + \lambda_n \mu_n - \bar{\phi}(\lambda_1, \ldots, \lambda_n) \right]. \tag{3.10}$$

Obviously the practical usefulness of this method resides in the fact that generally, *even in the presence of phase transitions*, the function $\bar{\phi}(\lambda_1, \ldots, \lambda_n)$ for real values of the λ_i happens to be differentiable. Indeed, for the cases in which the global variables $\mu_k(x)$ are given by sums of one-particle functions

$$\mu_k(x) = \frac{1}{N} \sum_{i=1}^{N} g_k(q_i, p_i) \qquad k = 1, \ldots, n, \tag{3.11}$$

the generating function is given by

$$\psi(\lambda_1, \ldots, \lambda_n) = \left(\int dp\, dq \, \exp\left[-\sum_{k=1}^{n} \lambda_k g_k(q, p) \right] \right)^N, \tag{3.12}$$

which leads to

$$\bar{\phi}(\lambda_1, \ldots, \lambda_n) = -\ln\left(\int dp\, dq \, \exp\left[-\sum_{k=1}^{n} \lambda_k g_k(q, p) \right] \right). \tag{3.13}$$

In the range of λ_i where the integral in Eq. (3.13) is defined, $\bar{\phi}$ is continuous. Moreover, differentiating with respect to λ_i, we have

$$\frac{\partial \bar{\phi}}{\partial \lambda_i}(\lambda_1, \ldots, \lambda_n) = \frac{\int dp\, dq \, g_i(q, p) \exp\left[-\sum_{k=1}^{n} \lambda_k g_k(q, p) \right]}{\int dp\, dq \, \exp\left[-\sum_{k=1}^{n} \lambda_k g_k(q, p) \right]}, \tag{3.14}$$

which is also a continuous function, under the only hypothesis that the expectation value of the observable $g_i(q, p)$ is finite for all the allowed values of the λ_i.

Finally, it is not difficult to see how $\phi(\beta)$ is related to $\bar{s}(\mu_1, \ldots, \mu_n)$. Again, with steps analogous to those linking $\phi(\beta)$ to $s(\varepsilon)$, we find that $\phi(\beta)$ is given by the extremal problem

$$\phi(\beta) = \beta f(\beta) = \inf_{\mu_1, \ldots, \mu_n} \left[\beta \bar{\varepsilon}(\mu_1, \ldots, \mu_n) - \bar{s}(\mu_1, \ldots, \mu_n) \right]. \tag{3.15}$$

The two variational problems (3.6) and (3.15) express the microcanonical entropy and the canonical free energy as a function of $\bar{s}(\mu_1, \ldots, \mu_n)$ and of the energy function $\bar{\varepsilon}(\mu_1, \ldots, \mu_n)$, and offer a tool to study ensemble equivalence in concrete systems.

It is important to emphasize that the large deviations method makes it possible to reduce the statistical mechanics study of a model to an optimization problem. The method of global variables reduces exactly (or sometimes approximately) the search of the equilibrium solution to a variational problem. This approach often drastically simplifies the derivation of the statistical mechanics properties and is not very well known among physicists.

However, such a procedure does not apply to all long-range interacting systems. In particular, those for which statistical mechanics cannot be reduced (even approximately) to a mean-field variational problem are excluded. As shown, the method is strongly dependent on the possibility to introduce global or coarse-grained variables (examples are the averaged magnetization, the total kinetic energy, *etc.*) such that the Hamiltonian can be expressed as a function of these variables, as in Eq. (3.1), modulo a remaining term $R(x)$ whose relative contribution vanishes in the thermodynamic limit. If also short-range interactions are present, the procedure is not applicable. We might still be able to express the Hamiltonian as a function of coarse-grained variables plus a rest, but the rest will not vanish in the thermodynamic limit.

The global variables $\{\mu_i\}$ could be given by fields. For instance, they could correspond to a local mass density in a gravitational system, or a coarse-grained vorticity density in 2D turbulence. In these cases, they would be mathematically infinite dimensional variables. However, with natural extensions, the steps that we have shown can be repeated (Michel and Robert, 1994; Ellis *et al.*, 2000, 2002). Other mathematical proofs can be found in Caglioti *et al.* (1992) and Kiessling and Neukirch (2003).

Finally, we note that we could be interested in entropy functions, depending not only on the energy density ε, but also on other quantities, see e.g. Eq. (2.34) for the dependence on magnetization. In this case, also these additional quantities should be expressed as functions of the global variables $\mu_1, \ldots, \mu_n$ (e.g. $m(x) = \bar{m}(\mu_1, \ldots, \mu_n)$), while the variational problem (3.6) will be constrained to fixed values of $\bar{\varepsilon}(\mu_1, \ldots, \mu_n) = \varepsilon$ and $\bar{m}(\mu_1, \ldots, \mu_n) = m$. Examples of this sort will be discussed in the following.

3.3 The Three-States Potts Model: An Illustration of the Method

Here, we apply the large deviations method to the three-state Potts model with infinite range interactions (Barré *et al.*, 2005). This simple example has been used as a toy model to illustrate peculiar thermodynamic properties of long-range systems (Ispolatov and Cohen, 2001). The diluted three-state Potts model with short-range interactions has also been studied by Gross and Votyakov (2000).

The Hamiltonian of the three-state Potts model is

$$H_N = -\frac{\mathcal{J}}{2N} \sum_{i,j=1}^{N} \delta_{S_i, S_j}. \tag{3.16}$$

Each lattice site i is occupied by a spin variable S_i, which assumes three different states a, b, or c. A pair of spins gives a ferromagnetic contribution $-\mathcal{J}$ ($\mathcal{J} > 0$) to the total energy if they are in the same state, and no contribution otherwise. It is important to stress that the energy sum is extended over *all* pairs (i,j): the interaction is infinite range.

Let us apply the method, following the three steps described earlier. The first step consists in associating, to every microscopic configuration x, global (coarse-grained) variables, such that the Hamiltonian can be expressed as a function of them. For Hamiltonian (3.16), the appropriate global variables are

$$\mu \equiv (\mu_a, \mu_b) = (n_a, n_b), \tag{3.17}$$

where (n_a, n_b) are the fractions of spins in the two different states a, b. As a function of the microscopic configuration, we have

$$\mu_k = \frac{1}{N} \sum_{i=1}^{N} \delta_{S_i,k}, \tag{3.18}$$

with $k = a, b$. The energy per particle is expressed in terms of the global variables as

$$\bar{\varepsilon}(n_a, n_b) = -\frac{\mathcal{J}}{2} \left[n_a^2 + n_b^2 + (1 - n_a - n_b)^2 \right], \tag{3.19}$$

which is exact for any N.

The second step consists the computation of $\bar{s}(\mu_1, \mu_2)$. According to what we have shown in the previous section, we must first compute the generating function, where in this case the integral over the configurations is a discrete sum. We get

$$\psi(\lambda_a, \lambda_b) = \sum_{\{S_i = a,b,c\}} \exp\left(-N[\lambda_a \mu_a + \lambda_b \mu_b]\right) \tag{3.20}$$

$$= \sum_{\{S_i = a,b,c\}} \exp\left(-N\left[\lambda_a \frac{1}{N} \sum_{i=1}^{N} \delta_{S_i,a} + \lambda_b \frac{1}{N} \sum_{i=1}^{N} \delta_{S_i,b}\right]\right) \tag{3.21}$$

$$= \sum_{\{S_i = a,b,c\}} \exp\left(-\lambda_a \sum_{i=1}^{N} \delta_{S_i,a} - \lambda_b \sum_{i=1}^{N} \delta_{S_i,b}\right). \tag{3.22}$$

The above sum must be interpreted with each S_i ranging independently over the three values a, b, c. Following Huang (1987) (see Chapter 14) for the Ising canonical partition function, we can rewrite the summation above as

$$\psi\left(\lambda_a, \lambda_b\right) = \sum_{S_1} \sum_{S_2} \cdots \sum_{S_N} \exp\left(-\lambda_a \sum_{i=1}^{N} \delta_{S_i,a} - \lambda_b \sum_{i=1}^{N} \delta_{S_i,b}\right) \tag{3.23}$$

$$= \prod_{i=1}^{N} \left[\sum_{S_i=a,b,c} \exp\left(-\lambda_a \delta_{S_i,a} - \lambda_b \delta_{S_i,b}\right)\right], \tag{3.24}$$

$$= \prod_{i=1}^{N} \left(e^{-\lambda_a} + e^{-\lambda_b} + 1\right) \tag{3.25}$$

$$= \left(e^{-\lambda_a} + e^{-\lambda_b} + 1\right)^N. \tag{3.26}$$

We therefore obtain

$$\bar{\phi}(\lambda_a, \lambda_b) = -\ln\left(e^{-\lambda_a} + e^{-\lambda_b} + 1\right). \tag{3.27}$$

As this function is evidently analytic, we can compute the function $\bar{s}$ by a Legendre–Fenchel transform. The entropy function $\bar{s}(n_a, n_b)$ will then be

$$\bar{s}(n_a, n_b) = \inf_{\lambda_a, \lambda_b}\left[\lambda_a n_a + \lambda_b n_b + \ln\left(e^{-\lambda_a} + e^{-\lambda_b} + 1\right)\right], \tag{3.28}$$

which is easily solved to get

$$\bar{s}(n_a, n_b) = -n_a \ln n_a - n_b \ln n_b - (1 - n_a - n_b) \ln(1 - n_a - n_b). \tag{3.29}$$

We now proceed to the third and final step of the calculation of the entropy $s(\varepsilon)$. The variational problem (3.6) becomes

$$s(\varepsilon) = \sup_{n_a, n_b}\left(-n_a \ln n_a - n_b \ln n_b - (1 - n_a - n_b) \ln(1 - n_a - n_b)\right.$$
$$\left.\left| -\frac{\mathcal{J}}{2}\left(n_a^2 + n_b^2 + (1 - n_a - n_b)^2\right) = \varepsilon\right). \tag{3.30}$$

As we have anticipated, this expression could have been obtained by direct counting. The variational problem (3.30) can be solved numerically. Microcanonical entropy has a non-concave region as it can be shown by drawing the microcanonical inverse temperature (see Fig. 3.1) in the allowed energy range $[-\mathcal{J}/2, -\mathcal{J}/6]$. Ispolatov and Cohen (2001) have obtained the same result by determining the density of states. A negative specific heat region appears in the energy range $[-0.215\,\mathcal{J}, -\mathcal{J}/6]$.

Let us now consider the canonical ensemble. Applying Eq. (3.15) to Hamiltonian (3.16), we have

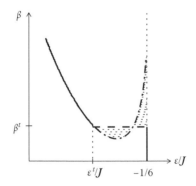

Figure 3.1 *Caloric curve (inverse temperature vs energy density) for the three-states infinite range Potts model. The canonical solution is represented by a solid line. The microcanonical solution coincides with the canonical one for $\varepsilon \leq \varepsilon^t$ and is instead indicated by the dash-dotted line for $\varepsilon^t \leq \varepsilon < -\mathcal{J}/6$. The increasing part of the microcanonical dash-dotted line corresponds to a negative specific heat region. In the canonical ensemble, the model displays a first-order phase transition at β^t. The two dotted regions bounded by the dashed line and by the microcanonical dash-dotted line have the same area (Maxwell construction). Reprinted from Campa et al. (2009), © 2009, with permission from Elsevier.*

$$\phi(\beta) = \inf_{n_a, n_b} \left(n_a \ln n_a + n_b \ln n_b + (1 - n_a - n_b) \ln(1 - n_a - n_b) \right.$$
$$\left. - \frac{\beta \mathcal{J}}{2} \left(n_a^2 + n_b^2 + (1 - n_a - n_b)^2 \right) \right). \qquad (3.31)$$

To obtain the caloric curve, we must compute $\varepsilon = d\phi/d\beta$. Figure 3.1 shows that at the canonical transition inverse temperature $\beta^t \simeq 2.75$, corresponding to the energy $\varepsilon^t/\mathcal{J} \simeq -0.255$, a first-order phase transition appears, with an associated latent heat. The low energy 'magnetized' phase becomes unstable, while the high-energy 'homogeneous' phase, which has the constant energy density, $\varepsilon/\mathcal{J} = -1/6$, is stabilized. In Fig. 3.1, the two dotted regions have the same area, respecting Maxwell construction. At the inverse transition temperature β, there is also a jump in the global variables $(n_a, n_b, 1 - n_a - n_b)$, which are the order parameters of the model.

This extremely simple example shows already ensemble inequivalence. In the microcanonical ensemble, there is no phase transition and the specific heat becomes negative. On the other hand, in the canonical ensemble, there is a first-order phase transition with a latent heat. The caloric curves do not coincide. We observe that in the energy range of ensemble inequivalence, microcanonical temperatures do not coincide with any canonical one.

3.4 The Solution of the BEG Model Using Large Deviations

An interesting application to consider is the Blume–Emery–Griffiths model introduced and discussed extensively in Section 2.3. The large deviations method leads indeed straightforwardly to the solution in that case, also as shown originally in Barré *et al.* (2002).

The first step corresponding to the identification of the global variables for the Hamiltonian (2.8) is here again direct for this mean-field model. The appropriate global variables are $\mu \equiv (q = Q/N, m = M/N)$, the quadrupole and magnetization per site defined in (2.22).

If we set $\mathcal{J} = 1$ as in Chapter 2, the Hamiltonian (2.8) can thus be rewritten as

$$H = N \left(\Delta q - \frac{m^2}{2} \right) = N \, \bar{\varepsilon}(q, m), \qquad (3.32)$$

an expression exact for any value N. This mean-field model is well suited for the use of the large deviations method with $\bar{\varepsilon}(q, m) = \Delta q - m^2/2$.

The second step is performed by computing the entropy associated with μ, i.e. $\bar{s}(q, m)$. According to what has been explained earlier, we must compute the generating function whose calculation is very similar to the different steps performed for the Potts model (see Eqs. (3.20) to (3.26)):

$$\psi(\lambda, \rho) = \sum_{\{S_i = -1, 0, 1\}} \exp\left[-N(\lambda q + \rho m)\right] \qquad (3.33)$$

$$= \sum_{S_1} \sum_{S_2} \cdots \sum_{S_N} \exp\left[-\sum_{i=1}^{N} (\lambda S_i^2 + \rho S_i)\right] \qquad (3.34)$$

$$= \prod_{i=1}^{N} \left[\sum_{S_i = -1, 0, 1} \exp-(\lambda S_i^2 + \rho S_i) \right] \qquad (3.35)$$

$$= \left(e^{(-\lambda + \rho)} + 1 + e^{(-\lambda - \rho)} \right)^N \qquad (3.36)$$

$$= \left(1 + 2e^{-\lambda} \cosh \rho \right)^N. \qquad (3.37)$$

We therefore obtain

$$\bar{\phi}(\lambda, \rho) = -\ln\left(1 + 2e^{-\lambda} \cosh \rho\right). \qquad (3.38)$$

The entropy

$$\bar{s}(q, m) = \inf_{\lambda, \rho} \left[\lambda q + \rho m - \bar{\phi}(\lambda, \rho)\right], \qquad (3.39)$$

is derived by computing $\partial \bar{\phi}/\partial \lambda = q$ and $\partial \bar{\phi}/\partial \rho = m$, which leads to the extrema

$$\bar{\lambda} = -\ln \frac{\sqrt{q^2 - m^2}}{2(1 - q)} \quad \text{and} \quad \bar{\rho} = -\ln \sqrt{\frac{q + m}{q - m}}. \qquad (3.40)$$

We therefore get

$$\bar{s}(q, m) = -\frac{q+m}{2} \ln(q+m) - \frac{q-m}{2} \ln(q-m) - (1-q) \ln(1-q) + q \ln 2, \qquad (3.41)$$

which corresponds to the expression (2.21).

Finally the third step is achieved by computing the microcanonical solution for an energy per spin ε as

$$s(\varepsilon) = \sup_{q,m}(\bar{s}(q, m) \mid \Delta q - m^2/2 = \varepsilon), \qquad (3.42)$$

while the rescaled free energy per particle can be written

$$\phi(\beta) = \inf_{q,m}\left(\frac{q+m}{2} \ln(q+m) + \frac{q-m}{2} \ln(q-m) + (1-q)\ln(1-q) \right.$$
$$\left. -q \ln 2 + \beta \Delta q - \beta \frac{m^2}{2} \right). \qquad (3.43)$$

The fact that the canonical solution (3.43) coincides with (2.14) is less transparent, but it is indeed the case.

Summing up, we have shown in this chapter that the large deviations method can prove adequate to derive the thermodynamic properties of long-range interacting systems. For this reason, it will be occasionally employed in the following. Already in the next chapter, we will apply the large deviations method to systems with continuous variables. More specifically, after considering a simple but very representative model of systems with long-range interactions, the so-called Hamiltonian Mean-Field (HMF) model, interesting applications will be discussed in Sections 4.2 and 4.3, both devoted to systems that display ensemble inequivalence: a generalized version of the HMF model and the ϕ^4 model. Finally, in Section 14.5, we will present the equilibrium solution of the Colson–Bonifacio model of the linear free electron laser.

4

Solutions of Mean Field Models

This chapter is devoted to discussing a selection of problems that can be studied via the techniques discussed in the previous chapter. The purpose of this chapter is to introduce the reader to the peculiar traits of systems subject to long-range forces and so open up the perspective for the systematic treatment of the applications, to which the third part of the book is entirely devoted.

4.1 The Hamiltonian Mean-Field (HMF) Model

We shall here begin by studying the Hamiltonian Mean-Field (HMF) model. More precisely we will derive the equilibrium solution of the model by operating in both the canonical and microcanonical ensemble. Comparing the solutions we will conclude that the statistical ensembles are equivalent. In the subsequent section we shall in turn discuss a generalized version of the HMF for which ensembles are instead non-equivalent.

The Hamiltonian Mean-Field model (Antoni and Ruffo, 1995; Dauxois *et al.*, 2002b; Chavanis *et al.*, 2005) is defined by the Hamiltonian

$$
H_N = \sum_{i=1}^{N} \frac{p_i^2}{2} + \frac{\varepsilon_{\mathcal{J}}}{2N} \sum_{i,j} \left[1 - \cos(\theta_i - \theta_j)\right],
\tag{4.1}
$$

where $\theta_i \in [0, 2\pi[$ is the position (angle) of the ith unit mass particle on a circle and p_i the corresponding conjugated momentum. The interacting potential can be seen as the first mode of the Fourier expansion of the potential of one-dimensional gravitational and charged sheet models, to which we shall make extensive reference in the third part of the book.

The infinite range cosine interaction can be of the attractive ($\varepsilon_{\mathcal{J}} > 0$) or repulsive ($\varepsilon_{\mathcal{J}} < 0$) type. The system can also be seen as classical XY-rotators with infinite range ferromagnetic ($\varepsilon_{\mathcal{J}} > 0$) or antiferromagnetic ($\varepsilon_{\mathcal{J}} < 0$) couplings. The renormalization factor N of the potential energy allows for a correct definition of the energy per particle and temperature in the $N \to \infty$ limit. This is the Kac's prescription that we have already met.

Physics of Long-Range Interacting Systems. First Edition. A. Campa *et al.*
© A. Campa, T. Dauxois, D. Fanelli, and S. Ruffo 2014. Published in 2014 by Oxford University Press.

Historically, this model has been independently introduced in the continuous time version in Ruffo (1994), Inagaki (1993), Inagaki and Konishi (1993), and Del Castillo-Negrete (1998a,b), and had been previously considered in its time discrete representation (Konishi and Kaneko, 1992).

For practical reason, it is instructive to rewrite the Hamiltonian (4.1) in a different form, by making use of the definition of the x and y components of the microscopic magnetization $\mathbf{m} = (m_x, m_y)$

$$m_x = \frac{1}{N} \sum_{i=1}^{N} \cos \theta_i \qquad \text{and} \qquad m_y = \frac{1}{N} \sum_{i=1}^{N} \sin \theta_i. \qquad (4.2)$$

The modulus of the magnetization $m = (m_x^2 + m_y^2)^{1/2}$ measures the degree of bunching of the particles on the circles and results in a global parameter that can be employed to monitor the equilibrium as well as the out-of-equilibrium dynamics of the HMF model.

Making use of the definition (4.2) we can then recast the Hamiltonian (4.1) in the form

$$H_N = \sum_{i=1}^{N} \frac{p_i^2}{2} + \frac{\varepsilon_J N}{2} \left(1 - m^2\right). \qquad (4.3)$$

As we shall comment in the following, the HMF model displays a rather rich out-of equilibrium dynamics and, for this reason, it has been extensively studied as a paradigmatic representative of long-range interacting systems. We shall return to discussing these peculiarities when studying the dynamics of long-range systems and their slow relaxation to equilibrium. When it comes to the equilibrium properties of the HMF model, one can derive both the canonical and the microcanonical solutions. The microcanonical solution can in particular be obtained by several different methods, which we shall hereafter review. As anticipated, the two ensembles, canonical and microcanonical, are equivalent. The system displays in fact a second-order phase transition, in the canonical ensemble. In the following we will start by deriving the equilibrium solution in the canonical ensemble, and limit our presentation to the ferromagnetic case by setting $\varepsilon_J = 1$.

4.1.1   The canonical solution

The exact canonical solution of the HMF is obtained by applying the Hubbard–Stratonovich transformation. The partition function for this model is given by

$$Z = \int \prod_{i=1}^{N} \mathrm{d}p_i \mathrm{d}\theta_i \exp(-\beta H), \qquad (4.4)$$

where the integration extends over the phase-space variables and β stands for the inverse temperature, namely $\beta = 1/T$ (we remind that we are using $k_B = 1$). Integrating the partition function (4.4) over the momenta yields

$$Z = \left(\frac{2\pi}{\beta}\right)^{N/2} \exp\left(-\frac{\beta N}{2}\right) Z_{\text{conf}}, \tag{4.5}$$

with

$$Z_{\text{conf}} = \int_0^{2\pi} \prod_{i=1}^N d\theta_i \exp\left[\frac{\beta}{2N} \sum_{i,j=1}^N \cos(\theta_i - \theta_j)\right] \tag{4.6}$$

$$= \int_0^{2\pi} \prod_{i=1}^N d\theta_i \exp\left[\frac{\beta}{2N} \left(\sum_i^N \mathbf{m}_i\right)^2\right], \tag{4.7}$$

where $\mathbf{m}_i = (\cos\theta_i, \sin\theta_i)$. In order to evaluate this integral, we can use the Hubbard–Stratonovich transformation which can be cast in the useful form

$$\exp\left[\frac{\mu}{2}\mathbf{x}^2\right] = \frac{1}{\pi} \int_{-\infty}^{+\infty} \int_{-\infty}^{+\infty} d\mathbf{y} \exp[-\mathbf{y}^2 + \sqrt{2\mu}\mathbf{x}\cdot\mathbf{y}], \tag{4.8}$$

where $\mu > 0$ and where $\mathbf{x}$ and $\mathbf{y}$ are two-dimensional vectors. Thus Eq. (4.7) becomes

$$Z_{\text{conf}} = \frac{1}{\pi} \int_0^{2\pi} \prod_{i=1}^N d\theta_i \int_{-\infty}^{+\infty} \int_{-\infty}^{+\infty} d\mathbf{y} \exp[-\mathbf{y}^2 + \sqrt{2\mu} \sum_{i=1}^N \mathbf{m}_i \cdot \mathbf{y}], \tag{4.9}$$

with $\mu = \beta/N$. We can now exchange the order of the integrals in (4.9) and factorize the integration over the coordinates of the N particles. Introducing then the rescaled variable $\mathbf{y} \to \mathbf{y}\sqrt{N/(2\beta)}$, we obtain

$$Z_{\text{conf}} = \frac{N}{2\pi\beta} \int_{-\infty}^{+\infty} \int_{-\infty}^{+\infty} d\mathbf{y} \exp\left[-N\left(\frac{y^2}{2\beta} - \ln(2\pi I_0(y))\right)\right], \tag{4.10}$$

where y is the modulus of $\mathbf{y}$ and I_0 is the modified Bessel function of order 0. Let us recall that for integer indices the modified Bessel functions of order n can be defined by

$$I_n(z) = \frac{1}{2\pi} \int_0^{2\pi} d\theta\, e^{z\cos\theta} \cos(n\theta). \tag{4.11}$$

The integral in Eq. (4.10) can be evaluated by employing the saddle point technique in the mean-field limit $N \to \infty$. By operating in this limit, the rescaled Helmholtz free energy $\phi = \beta f$ reads

$$\phi(\beta) = \beta f(\beta) = -\lim_{N\to\infty} \frac{1}{N} \ln Z = -\frac{1}{2} \ln\left(\frac{2\pi}{\beta}\right) + \frac{\beta}{2} + \min_{y\geq 0}\left(\frac{y^2}{2\beta} - \ln(2\pi I_0(y))\right). \tag{4.12}$$

The minimization of the last term in (4.12) results in the consistency equation

$$\frac{y}{\beta} = \frac{I_1(y)}{I_0(y)}.$$

(4.13)

For $\beta < 2$, Eq. (4.13) admits a minimal free energy solution for $y = 0$, which corresponds to a homogeneous equilibrium distribution with zero magnetization. In contrast, for $\beta > 2$ the minimal free energy solution is a non-vanishing, β-dependent value of y, denoted $\bar{y}$, that can be calculated numerically. We hence obtain a magnetized equilibrium solution. The value of m for such non-homogeneous states is given by the ratio of the Bessel functions

$$m = \frac{\bar{y}}{\beta} = \frac{I_1(\bar{y})}{I_0(\bar{y})}.$$

(4.14)

As a consequence, the system displays a phase transition at the critical temperature $\beta_c = 2$, which can be easily proven to be of the second-order type with the classical exponent $1/2$ (Antoni and Ruffo, 1995). The magnetization passes continuously from zero to a finite value, when decreasing the temperature (or increasing β).

This prediction is fully confirmed by numerical simulations, as displayed in Fig. 4.1, where the equilibrium magnetization is plotted against the internal energy per particle ε of the system. This latter can be easily obtained from the free energy by

$$\varepsilon = \frac{\partial(\beta f)}{\partial \beta},$$

(4.15)

which yields

$$\varepsilon = \frac{1}{2\beta} + \frac{1 - m^2}{2}.$$

(4.16)

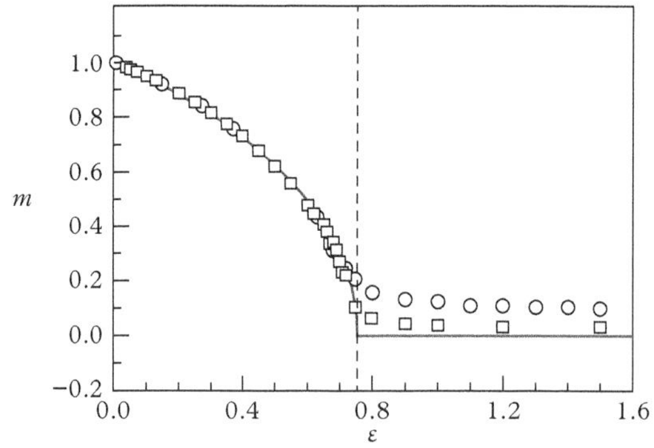

Figure 4.1 *Equilibrium magnetization m as a function of the energy per particle ε. Symbols refer to molecular dynamics data for $N = 10^2$ and 10^3, while the solid lines refer to the canonical prediction. The vertical dashed line indicates the critical energy, located at $\varepsilon_c = 0.75$, $\beta_c = 2$.*

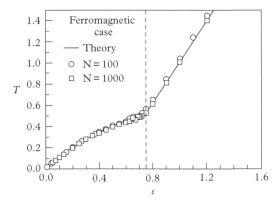

Figure 4.2 *Equilibrium temperature T vs energy per particle ε. Symbols refer to molecular dynamics data for N = 10² and 10³, while the solid lines refer to the canonical prediction.*

The temperature $T = 1/\beta$ as a function of the internal energy ε, at equilibrium, is reported in Fig. 4.2. In the plot comparisons with numerical simulations are also displayed.

As remarked earlier, the HMF model can be also solved in the microcanonical ensemble, an interesting exercise that allows us to prove the equivalence between the canonical and microcanonical statistical ensemble for this specific case. This equivalence was discussed in a heuristic way by Antoni *et al.* (2002). The next section is entirely devoted to discussing in a rigorous manner the microcanonical solution of the HMF model.

4.1.2   The microcanonical solution

Since we have shown that the HMF model has a second-order phase transition in the canonical ensemble, we could immediately conclude that ensembles are equivalent for this model. It is then straightforward to derive the entropy by performing the inverse Legendre–Fenchel transform of the free energy formula. We shall here follow, pedagogically, an alternative approach by applying a method which is frequently employed in the study of gravitational systems (Binney and Tremaine, 1987) and hereafter revisited.

Let us denote by K and U the values of the kinetic energy and the potential energy, respectively, and by $U(\{\theta_i\})$ the potential energy as a function of the microscopic configuration. The microcanonical partition function corresponding to the energy E is given by

$$\Omega(E, N) = \int \prod_i dp_i d\theta_i \, \delta(E - H_N) \tag{4.17}$$

$$= \int \prod_i dp_i d\theta_i \underbrace{\int dK \, \delta\left(K - \sum_i \frac{p_i^2}{2}\right)}_{=1} \delta(E - K - U(\{\theta_i\})) \tag{4.18}$$

$$= \int dK \int \underbrace{\prod_i dp_i \, \delta \left(K - \sum_i \frac{p_i^2}{2} \right)}_{\Omega_{kin}(K)} \underbrace{\int \prod_i d\theta_i \delta \left(E - K - U(\{\theta_i\}) \right)}_{\Omega_{conf}(E-K)}. \quad (4.19)$$

The factor Ω_{kin} is related to the surface of the hypersphere with radius $R = \sqrt{2K}$ in N dimensions and can be computed by using the properties of the Dirac δ function. We obtain the expression

$$\Omega_{kin} = 2\pi^{N/2} R^{N-2}/\Gamma(N/2). \quad (4.20)$$

Using for large N the asymptotic expression of the Γ function, $\ln \Gamma(N) \simeq (N-1/2) \ln N - N + (1/2) \ln(2\pi)$, and keeping only the terms that contribute to the entropy per particle in the thermodynamic limit yields

$$\Omega_{kin}(K) \overset{N \to +\infty}{\sim} \exp \left(\frac{N}{2} \left[1 + \ln \pi - \ln \frac{N}{2} + \ln(2K) \right] \right) \quad (4.21)$$

$$= \exp \left(\frac{N}{2} \left[1 + \ln(2\pi) + \ln u \right] \right), \quad (4.22)$$

where $u = 2K/N$. We now define the configurational entropy per particle as

$$s_{conf}(\tilde{u}) = (\ln \Omega_{conf}(N\tilde{u}))/N, \quad (4.23)$$

where $\tilde{u} = U/N = (E - K)/N = \varepsilon - u/2$. Equation (4.19) can be rewritten as

$$\Omega(N\varepsilon, N) \overset{N \to +\infty}{\sim} \frac{N}{2} \int du \exp \left[N \left(\frac{1}{2} + \frac{\ln(2\pi)}{2} + \frac{1}{2} \ln u + s_{conf}(\tilde{u}) \right) \right]. \quad (4.24)$$

By solving the integral in the saddle point approximation, we end up with the expression

$$s(\varepsilon) = \lim_{N \to +\infty} \frac{1}{N} \ln \Omega_N(\varepsilon N) \quad (4.25)$$

$$= \frac{1}{2} + \frac{1}{2} \ln(2\pi) + \sup_u \left[\frac{1}{2} \ln u + s_{conf}(\tilde{u}) \right]. \quad (4.26)$$

It is worth emphasizing that the expression (4.26) is quite general. It is, in fact, valid for any system in which the kinetic energy takes the canonical quadratic form. To progress in the analysis, an explicit expression for the configurational entropy s_{conf} is needed.

To this end we make use of the fact that the potential energy of the HMF model is a simple function of the microscopic magnetization $\mathbf{m} = (m_x, m_y)$. There is in particular a one-to-one correspondence between the value of the potential energy U and the modulus of the microscopic magnetization $m^2 = m_x^2 + m_y^2$. By defining

$$\Omega_m(m) = \int \prod_i d\theta_i \, \delta \left(\sum_i \cos\theta_i - Nm \right) \delta \left(\sum_i \sin\theta_i \right), \tag{4.27}$$

we have that this function will be proportional to $\Omega_{\text{conf}}(U)$ for $\tilde{u} = U/N = (1/2 - m^2/2)$. The coefficient of proportionality will give a vanishing contribution to $s_{\text{conf}}(\tilde{u})$ in the thermodynamic limit. We note that, as in the canonical case, there is a continuous degeneracy on the direction of the spontaneous magnetization; therefore, we do not lose generality by choosing the spontaneous magnetization in the direction of the x axis. The integral in (4.27) can be computed using the Fourier representation of the δ function. We therefore have

$$\Omega_m(m) = \left(\frac{1}{2\pi} \right)^2 \int_{-\infty}^{+\infty} dq_1 \int_{-\infty}^{+\infty} dq_2 \int \prod_i d\theta_i$$

$$\times \exp\left[iq_1 \left(\sum_i \cos\theta_i - Nm \right) \right] \exp\left[iq_2 \left(\sum_i \sin\theta_i \right) \right] \tag{4.28}$$

$$= \left(\frac{1}{2\pi} \right)^2 \int_{-\infty}^{+\infty} dq_1 \int_{-\infty}^{+\infty} dq_2 \, \exp\left\{ N \left[-iq_1 m + \ln\left(2\pi \mathcal{J}_0((q_1^2 + q_2^2)^{\frac{1}{2}}) \right) \right] \right\}, \tag{4.29}$$

where $\mathcal{J}_0(z)$ is the Bessel function of order 0. For integer indices the Bessel functions of order n can be defined by

$$\mathcal{J}_n(z) = \frac{(-i)^n}{2\pi} \int_0^{2\pi} d\theta \, e^{iz\cos\theta} \cos(n\theta). \tag{4.30}$$

To solve the integral in (4.29) with the saddle point method, we must consider q_1 and q_2 as complex variables. Using that the derivative of $\mathcal{J}_0$ is $-\mathcal{J}_1$, the opposite of the Bessel function of order 1, the saddle point must satisfy the equations

$$-im - \frac{\mathcal{J}_1}{\mathcal{J}_0}((q_1^2 + q_2^2)^{\frac{1}{2}}) \frac{q_1}{(q_1^2 + q_2^2)^{\frac{1}{2}}} = 0 \tag{4.31}$$

$$-\frac{\mathcal{J}_1}{\mathcal{J}_0}((q_1^2 + q_2^2)^{\frac{1}{2}}) \frac{q_2}{(q_1^2 + q_2^2)^{\frac{1}{2}}} = 0. \tag{4.32}$$

The solution of these equations is $q_2 = 0$ and $q_1 = -i\gamma$, where γ solves the equation

$$\frac{I_1(\gamma)}{I_0(\gamma)} = m. \tag{4.33}$$

Here, use has been made of the properties $\mathcal{J}_0(iz) = I_0(z)$ and $\mathcal{J}_1(iz) = iI_1(z)$. Let us label with B_{inv} the inverse function of I_1/I_0. In the thermodynamic limit, we can then write

$$s_{\text{conf}}\left(\frac{1}{2} - \frac{1}{2}m^2\right) = \lim_{N \to +\infty} \frac{1}{N} \ln \Omega_m(m) = -mB_{\text{inv}}(m) + \ln\left(2\pi I_0(B_{\text{inv}}(m))\right). \qquad (4.34)$$

We now substitute relation (4.34) into Eq. (4.26). By using $u = 2(\varepsilon - 1/2 + m^2/2)$, and performing a maximization over m instead of that over u, we eventually obtain

$$s(\varepsilon) = \sup_{m \geq m_0}\left[\frac{1}{2}\ln\left(\varepsilon - \frac{1}{2} + \frac{1}{2}m^2\right) - mB_{\text{inv}}(m) + \ln\left(2\pi I_0(B_{\text{inv}}(m))\right)\right]$$
$$+ \frac{1}{2} + \frac{1}{2}\ln(2\pi) + \frac{1}{2}\ln 2, \qquad (4.35)$$

with $m_0^2 = \sup[0, 1 - 2\varepsilon]$. For $\varepsilon < 0.5$ the lowest possible value of m is $1 - 2\varepsilon$. The maximization problem over m is solved graphically, looking for the solutions of the equation

$$\frac{m}{2\varepsilon - 1 + m^2} - B_{\text{inv}}(m) = 0. \qquad (4.36)$$

The graphical solution is depicted in Fig. 4.3, and returns the following results. For $0 \leq \varepsilon \leq 3/4$, the magnetization $m(\varepsilon)$ decreases monotonically from 1 to 0, while for $\varepsilon > 3/4$ the solution is always $m = 0$. At $\varepsilon = 3/4$, a second-order phase transition takes place, the two ensembles giving equivalent predictions.

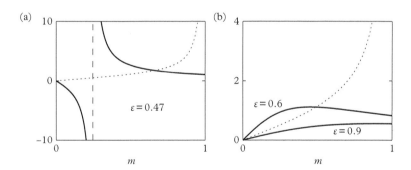

Figure 4.3 *Graphical solution of Eq. (4.36) for three choices of the energy ε, one above and two below the critical value $\varepsilon_c = 3/4$. In both plots the dotted curve is the function $B_{\text{inv}}(m)$, while the solid curve is the function of the first term in Eq. (4.36). This last function diverges for $m^2 = 1 - 2\varepsilon$, which is between 0 and 1 for $0 \leq \varepsilon \leq 1/2$. The corresponding asymptote is the vertical dashed line in (a). (a) $\varepsilon = 0.47$: the relevant solution is the one with $m > 0$. (b) $\varepsilon = 0.6$ and $\varepsilon = 0.9$: in the first case the relevant solution is the one with $m > 0$, while in the second case the only solution is $m = 0$. Reprinted from Campa et al. (2009), © 2009, with permission from Elsevier.*

Indeed, it is easy to prove that $s(\varepsilon)$ is concave. To this end, we first derive from (4.35) that the inverse temperature $\beta(\varepsilon)$ is

$$\beta(\varepsilon) = \frac{ds}{d\varepsilon} = \frac{1}{2\varepsilon - 1 + m^2(\varepsilon)}. \tag{4.37}$$

If $\beta(\varepsilon)$ has a negative derivative, then $s(\varepsilon)$ is concave. The negativity of this derivative is trivial for $\varepsilon > 3/4$, when $m(\varepsilon) = 0$. For $\varepsilon \leq 3/4$ we can proceed as follows. We note that, using the last equation, we can write (4.36) also as

$$m = \frac{I_1(m\beta(\varepsilon))}{I_0(m\beta(\varepsilon))}. \tag{4.38}$$

We have just proved graphically that (4.36), and thus (4.38), have a unique solution for each ε, which decreases if ε increases. Studying the canonical HMF solution we have found that m decreases when β decreases. Therefore, in the present case an increase in ε results in a decrease in $\beta(\varepsilon)$, also when $m(\varepsilon) > 0$. This proves the concavity of the entropy $s(\varepsilon)$ and therefore ensemble equivalence. The concavity property of the microcanonical entropy (4.35) ensures that it could also be computed by the Legendre–Fenchel transform of ϕ, the rescaled free energy computed in the canonical ensemble.

In conclusion of this section we plot the relevant thermodynamic variables of the model. In Fig. 4.4 we display the entropy as a function of the energy, as follows Eq. (4.35), and the rescaled free energy versus the inverse temperature, as dictated by Eq. (4.12). In Fig. 4.5 the caloric curve, see Eq. (4.37), is plotted together with the dependence of the order parameter on the energy, as prescribed by Eq. (4.36).

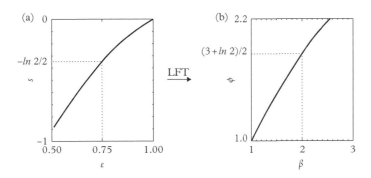

Figure 4.4 *Entropy versus energy (a) and rescaled free energy versus inverse temperature (b) for the HMF model (4.1) with $\varepsilon_{\overline{J}} = 1$. The dotted lines are traced at the phase transition point. Reprinted from Campa et al. (2009), © 2009, with permission from Elsevier.*

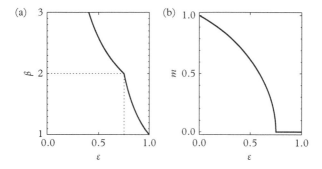

Figure 4.5 *Inverse temperature versus energy (a) and magnetization versus energy (b) for the HMF model (4.1) with $\varepsilon_J = 1$. The dotted lines are traced at the phase transition point. Reprinted from Campa et al. (2009), © 2009, with permission from Elsevier.*

4.1.3   The min–max solution

Let us check that, using the method described in Section 2.5, we can rederive the microcanonical entropy of the HMF model, Eq. (4.35). For the HMF model, the function $\tilde{\phi}(\beta, x)$ reads

$$\tilde{\phi}(\beta, x) = \frac{\beta}{2} - \frac{1}{2}\ln 2\pi + \frac{1}{2}\ln \beta + \frac{\beta x^2}{2} - \ln\left(2\pi I_0(\beta x)\right). \tag{4.39}$$

The stationarity conditions (2.86) and (2.87) read in this case

$$\frac{\partial \tilde{\phi}}{\partial \beta} = \frac{1}{2} + \frac{1}{2\beta} + \frac{1}{2}x^2 - x\frac{I_1(\beta x)}{I_0(\beta x)} = \varepsilon \tag{4.40}$$

$$\frac{\partial \tilde{\phi}}{\partial x} = \beta x - \beta\frac{I_1(\beta x)}{I_0(\beta x)} = 0. \tag{4.41}$$

Inserting the second equation in the first, we find that $\beta^{-1} = 2\varepsilon - 1 + x^2$. Substituting back in the second equation, and using the function B_{inv}, the inverse function of I_1/I_0 previously defined, we have

$$B_{\text{inv}}(x) - \frac{x}{2\varepsilon - 1 + x^2} = 0, \tag{4.42}$$

which is identical to Eq. (4.36). The computation of the second derivatives gives

$$\frac{\partial^2 \tilde{\phi}}{\partial \beta^2} = -\frac{1}{2\beta^2} - x\frac{\partial}{\partial \beta}\left(\frac{I_1(\beta x)}{I_0(\beta x)}\right) < 0 \tag{4.43}$$

$$\frac{\partial^2 \tilde{\phi}}{\partial x^2} = \beta - \beta\frac{\partial}{\partial x}\left(\frac{I_1(\beta x)}{I_0(\beta x)}\right) > 0. \tag{4.44}$$

These inequalities are both satisfied at the stationary points determined by (4.40) and (4.41). Actually, Eq. (4.43) is identically satisfied, since the derivative of I_1/I_0 is positive definite. This confirms ensemble equivalence for the HMF model.

Finally, Eqs. (2.68) and (2.70) give

$$s(\varepsilon) = s_{can}(\varepsilon) = \frac{1}{2} + \frac{1}{2}\ln(2\pi) + \frac{1}{2}\ln 2 + \frac{1}{2}\ln\left(\varepsilon - \frac{1}{2} + \frac{1}{2}x^2\right)$$

$$-\frac{x^2}{2\varepsilon - 1 + x^2} + \ln\left[2\pi I_0\left(\frac{x}{2\varepsilon - 1 + x^2}\right)\right], \tag{4.45}$$

with x satisfying Eq. (4.42). We thus obtain an expression identical to (4.35), taking into account Eq. (4.36).

It is also possible to solve the model by adapting the large deviations method to systems made of continuous variables. This is, however, a particular case of a calculation that we shall discuss in the next section when dealing with a generalized version of the HMF model.

4.2 The Generalized XY Model

4.2.1 Statistical mechanics via large deviations method

A generalized HMF model which yields ensemble inequivalence exists. The model is defined by the Hamiltonian (de Buyl *et al.*, 2005; Bouchet *et al.*, 2008)

$$H_N = \sum_i^N \frac{p_i^2}{2} + U, \tag{4.46}$$

where

$$U = N\,W(m) = N\left(-\frac{\mathcal{J}}{2}m^2 - K\frac{m^4}{4}\right), \tag{4.47}$$

with $m = (m_x^2 + m_y^2)^{1/2}$. The quantities m_x, m_y are defined according to formula (4.2). Models of this type have been invoked to describe the physics of nematic liquid crystals (Lebwohl and Lasher, 1972). In the following we shall assume $\mathcal{J}, K > 0$ as control parameters and discuss the statistical mechanics of model (4.46). We shall begin by deriving the microcanonical solution using the large deviations method; then, we will obtain the canonical solution by applying the Legendre–Fenchel transform. Note that the Hamiltonian (4.46) reduces, when $K = 0$ and $\mathcal{J} = 1$, to that of the HMF model (4.3) with $\varepsilon_{\mathcal{J}} = 1$. Consequently, using these change of variables in the derivation presented later will lead to the statistical mechanics solution of the HMF model presented in the previous section.

Besides m_x and m_y, two additional variables are used in the following, namely

$$u = \frac{1}{N}\sum_i p_i^2 \tag{4.48}$$

and

$$v = \frac{1}{N} \sum_i p_i. \tag{4.49}$$

The last variable, the average momentum, does not appear in the Hamiltonian, and it is a conserved quantity. We include it as a useful exercise. We will find, as it could be argued on physical basis, that for a given energy ε the entropy is maximum when the average momentum v is equal to 0. Hamiltonian (4.46) can be expressed in terms of the three global variables u, m_x and m_y, and the energy per particle reads

$$\bar{\varepsilon}(u, m_x, m_y) = \frac{1}{2} \left(u - \mathcal{J}(m_x^2 + m_y^2) - \frac{K}{2}(m_x^2 + m_y^2)^2 \right). \tag{4.50}$$

This relation is exact for any N. The second step in the large deviations procedure has to do with the computation of $\bar{s}(u, v, m_x, m_y)$. To this end we start by computing the generating function ψ

$$\psi(\lambda_u, \lambda_v, \lambda_x, \lambda_y) = \int \left(\prod_i d\theta_i dp_i \right) \tag{4.51}$$

$$\times \exp \left(-\lambda_u \sum_{i=1}^{N} p_i^2 - \lambda_v \sum_{i=1}^{N} p_i - \lambda_x \sum_{i=1}^{N} \cos\theta_i - \lambda_y \sum_{i=1}^{N} \sin\theta_i \right),$$

which results in

$$\psi(\lambda_u, \lambda_v, \lambda_x, \lambda_y) = \left[e^{\lambda_v^2/4\lambda_u} \sqrt{\frac{\pi}{\lambda_u}} \, 2\pi \, I_0 \left(\sqrt{\lambda_x^2 + \lambda_y^2} \right) \right]^N. \tag{4.52}$$

Here I_0, as previously, is the modified Bessel function of order 0. Note that the existence of the integral in (4.51) necessarily implies that $\lambda_u > 0$. The free energy associated with ψ reads

$$\bar{\phi}(\lambda_u, \lambda_v, \lambda_x, \lambda_y) = -\frac{\lambda_v^2}{4\lambda_u} - \frac{1}{2}\ln\pi + \frac{1}{2}\ln\lambda_u - \ln(2\pi) - \ln I_0 \left(\sqrt{\lambda_x^2 + \lambda_y^2} \right). \tag{4.53}$$

We can then calculate the function $\bar{s}$ as

$$\bar{s}(u, v, m_x, m_y) = \inf_{\lambda_u, \lambda_v, \lambda_x, \lambda_y} \left[\lambda_u u + \lambda_v v + \lambda_x m_x + \lambda_y m_y + \frac{\lambda_v^2}{4\lambda_u} + \frac{1}{2}\ln\pi \right.$$
$$\left. - \frac{1}{2}\ln\lambda_u + \ln(2\pi) + \ln I_0 \left(\sqrt{\lambda_x^2 + \lambda_y^2} \right) \right]. \tag{4.54}$$

This variational problem can be solved explicitly by formally separating the 'kinetic' (λ_u, λ_v) and 'configurational' (λ_x, λ_y) subspaces as

$$\bar{s}(u, v, m_x, m_y) = \bar{s}_{\text{kin}}(u, v) + \bar{s}_{\text{conf}}(m_x, m_y). \tag{4.55}$$

Call, as previously, B_{inv} the inverse function of I_1/I_0, where $I_0(z)$ and $I_1(z)$ are the modified Bessel function of order 0 and 1, respectively. Then, we obtain

$$\bar{s}_{\text{kin}}(u, v) = \frac{1}{2} + \frac{1}{2}\ln \pi + \ln(2\pi) + \frac{1}{2}\ln 2\,(u - v^2) \tag{4.56}$$

$$\bar{s}_{\text{conf}}(m) = -mB_{\text{inv}}(m) + \ln I_0(B_{\text{inv}}(m)). \tag{4.57}$$

The next step in the procedure concerns the calculation of the entropy function. Maximizing only with respect to u and m we obtain the function $s(\varepsilon, v)$ as

$$s(\varepsilon, v) = \sup_{u,m}\left[\bar{s}(u, v, m)\Bigg|\frac{u}{2} - \mathcal{J}\frac{m^2}{2} - K\frac{m^4}{4} = \varepsilon\right] \tag{4.58}$$

$$= \sup_{u,m}\left[\bar{s}_{\text{kin}}(u, v) + \bar{s}_{\text{conf}}(m)\Bigg|\frac{u}{2} - \mathcal{J}\frac{m^2}{2} - K\frac{m^4}{4} = \varepsilon\right] \tag{4.59}$$

$$= \frac{1}{2} + \frac{1}{2}\ln 2 + \frac{3}{2}\ln(2\pi) + \frac{1}{2}\ln\left(\varepsilon + \mathcal{J}\frac{m^2}{2} + K\frac{m^4}{4} - \frac{1}{2}v^2\right)$$
$$- mB_{\text{inv}}(m) + \ln I_0(B_{\text{inv}}(m)), \tag{4.60}$$

where m satisfies the self-consistency equation

$$B_{\text{inv}}(m) = \frac{\mathcal{J}m + Km^3}{2\varepsilon + \mathcal{J}m^2 + Km^4/2 - v^2}. \tag{4.61}$$

The entropy $s(\varepsilon)$ is obtained by maximizing with respect to v. It is immediate to see that this is obtained by putting $v = 0$ in Eqs. (4.60) and (4.61). Namely

$$s(\varepsilon) = \frac{1 + \ln 2 + 3\ln(2\pi)}{2} + \frac{1}{2}\ln\left(\varepsilon + \mathcal{J}\frac{m^2}{2} + K\frac{m^4}{4}\right) - mB_{\text{inv}}(m) + \ln I_0(B_{\text{inv}}(m)), \tag{4.62}$$

with m satisfying

$$B_{\text{inv}}(m) = \frac{\mathcal{J}m + Km^3}{2\varepsilon + \mathcal{J}m^2 + Km^4/2}. \tag{4.63}$$

By solving the last equation, we find the phase diagram and caloric curves, showing in particular that there are both first- and second-order phase transitions. Aiming at comparing the canonical and microcanonical viewpoints, we turn to calculate the rescaled canonical free energy.

Applying Eq. (3.15), we find that

$$\phi(\beta) = \frac{\beta \mathcal{J} m^2}{2} + \frac{3\beta K m^4}{4} - \ln(I_0[\beta(\mathcal{J}m + Km^3)]) + \frac{1}{2}\ln\beta - \frac{3}{2}\ln(2\pi), \qquad (4.64)$$

with m satisfying

$$B_{\mathrm{inv}}(m) = \beta(\mathcal{J}m + Km^3). \qquad (4.65)$$

We are now in a position to compare the microcanonical and canonical solutions and look at possible signatures of inequivalence. In Fig. 4.6, the phase diagram of model (4.46) is depicted in the parameter space $(T/\mathcal{J}, K/\mathcal{J})$, in both the canonical and the microcanonical ensemble. As expected (Barré *et al.*, 2001), the two phase diagrams differ in the region where the canonical transition line is first order. The second-order phase transition lines at $T/\mathcal{J} = 1/2$ coincide up to the canonical tricritical point, located at $K/\mathcal{J} = 1/2$. Then the predictions of the two ensembles differ. The canonical ensemble predicts a first-order phase transition, while in the microcanonical ensemble the critical line at $T/\mathcal{J} = 1/2$ extends up to the microcanonical tricritical point at $K/\mathcal{J} = 5/2$.

For $K/\mathcal{J} > 5/2$, two different temperatures coexist at the transition energy in the microcanonical ensemble, resulting in a temperature jump. This is a microcanonical first order phase transition: the associated 'phase coexistence' region is highlighted in grey in Fig. 4.6. As for the order parameter m, the most striking difference between the two

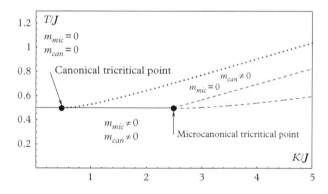

Figure 4.6 *Phase diagram of model (4.46). The canonical second-order transition line (solid line at $T/\mathcal{J} = 1/2$) becomes first order (dotted line, determined numerically) at the canonical tricritical point. The microcanonical second-order transition line coincides with the canonical one below $K/\mathcal{J} = 1/2$ but extends further right to the microcanonical tricritical point at $K/\mathcal{J} = 5/2$. At this latter point, the transition line bifurcates in two first-order microcanonical lines, corresponding to a temperature jump. The behaviour of the order parameter in the two ensembles is also shown in the figure, to highlight the striking difference in the predictions of the two ensembles. Reprinted from Campa et al. (2009), © 2009, with permission from Elsevier.*

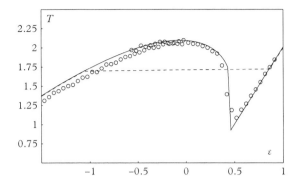

Figure 4.7 *Caloric curve $T(\varepsilon)$ at $K/\mathcal{J} = 10$. The microcanonical ensemble (solid line) predicts a region of negative specific heat, where temperature T decreases as the energy is increased. Moreover, a temperature jump is present at the transition energy. In the canonical ensemble, we have a first-order phase transition (dashed line). The points are the result of a molecular dynamics simulation performed by solving numerically the equations of motion given by Hamiltonian (4.46) with $N = 100$. Reprinted from Campa et al. (2009), © 2009, with permission from Elsevier.*

ensembles is detected in correspondence of the region delimited by the canonical first-order phase transition line and the upper microcanonical first-order line. The canonical ensemble predicts a non-vanishing order parameter $m_{can} \neq 0$, while the microcanonical ensemble returns $m_{mic} = 0$.

In Figure 4.7 we report the caloric curve as computed from direct simulations (symbols) as well as resulting from the theoretical predictions (lines). The comparison is drawn in the region of parameters where both canonical and microcanonical calculations predict a first-order phase transition. Interestingly, negative specific heat and temperature jumps are seen in the microcanonical ensemble, and are confirmed by the simulations. We thus see the inequivalence with the canonical ensemble, where the specific heat is instead positive.

In conclusion, the generalized XY model can be solved analytically in both the microcanonical and canonical ensembles, enabling us to gain quantitative insight into the important issue of ensemble inequivalence. Direct molecular dynamics simulations confirm the adequacy of the theory predictions.

4.2.2   Parameter space convexity

We here discuss a concrete example where we can show that the space of thermodynamic parameters is not convex, as already discussed in Section 2.1.3. Historically, this phenomenon was first observed for a spin chain with asymmetric coupling (Borgonovi *et al.*, 2004, 2006; Celardo *et al.*, 2006), and for an Ising model with both nearest neighbour and mean-field interactions (Mukamel *et al.*, 2005), see Section 5.1. However, it was later realized that this phenomenon occurs more generally, and even in the simple model studied here, with Hamiltonian given by Eq. (4.46). We will consider the case $\mathcal{J} = -1$

and $K > 0$ (Bouchet *et al.*, 2008) for which the m^2 term in the Hamiltonian (4.46) is antiferromagnetic. By operating in this context, we will in particular study the structure of the set of accessible states in the space of thermodynamic parameters in the microcanonical ensemble.

Intuitively, we expect that, for large values of K, the system is ferromagnetic while, for small values of K, the antiferromagnetic coupling will dominate and make the system paramagnetic. As we shall demonstrate later there exists a range of values of the parameter K for which the model exhibits a first-order microcanonical phase transitions between a paramagnetic phase at high energies and a ferromagnetic phase at low energies. In both phases there are regions in the (ε, K) plane in which the accessible magnetization interval exhibits a gap, resulting in breaking of ergodicity.

The specific kinetic energy $u/2 = \varepsilon - W(m)$ is by definition a non-negative quantity, which implies that

$$\varepsilon \geq W(m) = m^2/2 - Km^4/4. \tag{4.66}$$

We show that, as a result of this condition, not all the values of the magnetization m are attainable in a certain region in the (ε, K) plane; a disconnected magnetization domain is indeed typically found. As explained later, this situation is the one of interest. Let us characterize the accessible domains in the (ε, K) plane more precisely by analysing the different values of K (all this will become clear after looking at Fig. 4.8).

For $K < 1$, the local maximum of the potential energy W is not located inside the physical magnetization interval $[0, 1]$ (see Fig. 4.8a). The potential being a strictly increasing function of the magnetization in that interval, the maximum is reached at the extremum $m = 1$. The complete interval $[0, 1]$ is thus accessible when the energy ε is larger than $W(1)$: the corresponding domain is in $R1$ defined and illustrated in Figs. 4.8 and 4.9. In the latter figure the horizontally shaded region is forbidden, since the energy is lower than the minimum of the potential energy $W(0) = 0$. Finally, the intermediate region $0 < \varepsilon < W(1)$ defines region $R2$: it is important to emphasize that only the interval $[0, m_-(\varepsilon, K)]$, where $m_\pm(\varepsilon, K) = [(1 \pm \sqrt{1 - 4\varepsilon K})/K]^{1/2}$, is accessible ($m_-(\varepsilon, K) < 1$ for $0 < \varepsilon < W(1)$). Larger magnetization values correspond to a potential energy $W(m)$ larger than the energy density ε, which is impossible. Figure 4.8a also displays the value m_- corresponding to an energy in the intermediate region $R2$.

For $K \geq 1$, the maximum of the specific potential energy W, $W_{\max} = 1/(4K)$, is reached at $0 < m_{\max} = 1/\sqrt{K} \leq 1$. Figures 4.8b and 4.8c, where the potential energy per particle defined in Eq. (4.47) is plotted versus magnetization m, refer to such cases. For an energy ε larger than the critical value $W_{\max}$, condition (4.66) is satisfied for any value of the magnetization m. The complete interval $[0, 1]$ is thus accessible for the magnetization m. This region is $R1$ represented in Fig. 4.9.

Let us now consider the cases for which $\varepsilon \leq W_{\max}$. The minimum $W_{\min}$ of the potential energy for m in $[0, 1]$ is also important to distinguish between the different regions. For $1 \leq K \leq 2$ (see Fig. 4.8b), the minimum of $W(m)$ corresponds to the paramagnetic phase $m = 0$ where $W(0) = 0$. The quadrilled region shown in Fig. 4.9, which corresponds to negative energy values, is thus not accessible. In contrast, positive energy values are possible and correspond to very interesting cases, since only subintervals

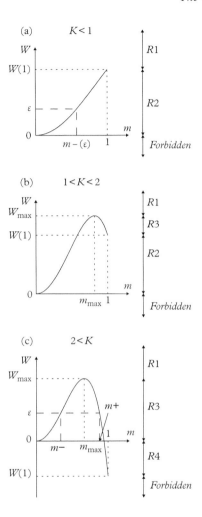

Figure 4.8 *Specific potential energy W vs magnetization m for three different cases: $K < 1$ (a), $1 < K < 2$ (b) and $2 < K$ (c). The location of the maximal magnetization $m_{\max}$ and the corresponding potential energy $W_{\max}$ are shown (see text). In (a) and (c), two examples of the location of the critical magnetization $m_{\pm}(\varepsilon, K)$ are indicated for energy density values ε in the intermediate regions.*

of the complete magnetization interval $[0, 1]$ are accessible. There are, however, two different cases:

- For $0 < W(1) < \varepsilon < W_{\max}$, the domain of possible magnetizations is $[0, m_{-}(\varepsilon, K)]$ $\cup [m_{+}(\varepsilon, K), 1]$ (now $m_{\pm}(\varepsilon, K) < 1$). These conditions are satisfied in $R3$ of Fig. 4.9.

- For $0 < \varepsilon < W(1)$, only the interval $[0, m_{-}(\varepsilon, K)]$ satisfies condition (4.66). This takes place within $R2$ of Fig. 4.9.

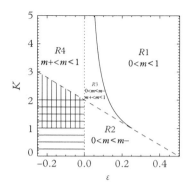

Figure 4.9 *The* (ε, K) *plane is divided into several regions. The solid curve corresponds to* $K = 1/(4\varepsilon)$, *the oblique dashed line to* $K = 2 - 4\varepsilon$, *while the dotted one to* $\varepsilon = 0$. *The vertically shaded, quadrilled and horizontally shaded regions are forbidden. The accessible magnetization interval in each of the four regions is indicated (see text for details). Reprinted from Campa* et al. *(2009),* © *2009, with permission from Elsevier.*

For the domain $2 \leq K$, the minimum of the potential energy for m in $[0, 1]$ is attained at the extremum, $m = 1$, implying that $\varepsilon > W(1) = 1/2 - K/4$. The vertically shaded region is thus forbidden. In the accessible region, two cases can be identified:

- For $W(1) < \varepsilon < 0$, only the interval $[m_+ (\varepsilon, K), 1]$ satisfies condition (4.66). It is important to note that $m_+ (\varepsilon, K) \leq 1$ provided $\varepsilon \geq 1/2 - K/4$. These cases correspond to region *R4*.
- For $0 \leq \varepsilon \leq W_{\max}$, the two intervals $[0, m_- (\varepsilon, K)]$ and $[m_+ (\varepsilon, K), 1]$ satisfy condition (4.66), corresponding to *R3* of Fig. 4.9.

In summary, the complete magnetization interval $[0, 1]$ is accessible only in the region *R1*. In *R2*, only $[0, m_-]$ is accessible, while only $[m_+, 1]$ is accessible in *R4*. Finally, we note that the phase space of the system is not connected in the region *R3*. Indeed, the magnetization cannot vary continuously from the first interval $[0, m_-]$ to the second one $[m_+, 1]$, although both are accessible. These restrictions yield the accessible magnetization domain shown in Fig. 4.10. The fact that for a given energy the space of the thermodynamic parameter m is disconnected implies ergodicity breaking for the Hamiltonian dynamics. It is important to emphasize that the discussion above is independent of the number of particles and ergodicity is expected to be broken even for a finite N.

4.2.3   Phase diagram in the microcanonical ensemble

We have thus found that, in certain regions in the (ε, K) plane, the magnetization cannot assume any value in the interval $[0, 1]$: for a given energy there exists a gap in this interval

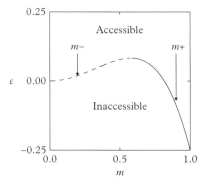

Figure 4.10 *Accessible region in the (m, ε) plane for $K = 3$. For energies in a certain range, a gap in the accessible magnetization values is present and defined by the two boundaries $m_{\pm}(\varepsilon, K)$. Reprinted from Campa et al. (2009), © 2009, with permission from Elsevier.*

to which no microscopic configuration can be associated. Let us now study the phase diagram in the microcanonical ensemble, i.e. in the parameter space (ε, K).

First, by comparing the low- and high-energy regimes, it is possible to show that a phase transition is present between the two regimes. Let us first write the entropy $\tilde{s}(\varepsilon, m)$ as a function of the energy and magnetization. Neglecting additive constants, it is obtained from Eq. (4.62) with $\tilde{J} = -1$, i.e.

$$\tilde{s}(\varepsilon, m) = \frac{1}{2} \ln \left(\varepsilon - \frac{m^2}{2} + K \frac{m^4}{4} \right) - m B_{\text{inv}}(m) + \ln I_0(B_{\text{inv}}(m)), \qquad (4.67)$$

with m free, instead of satisfying Eq. (4.63). The first term on the right-hand side is the kinetic entropy $\tilde{s}_{\text{kin}}(\varepsilon, m)$, while the other two terms constitute the configurational entropy $\tilde{s}_{\text{conf}}(\varepsilon, m)$.

In the domain *R4* of Fig. 4.9, for very low energy ε (close to the limiting value $1/2 - K/4$), the accessible range for m is a small interval located close to $m = 1$ (see Fig. 4.8c). The maximum in m of the entropy $\tilde{s}(\varepsilon, m)$ corresponds therefore to a magnetized state located very close to $m = 1$ (see the top-left inset in Fig. 4.11). In contrast, in the very large energy domain ($\varepsilon \gg m \simeq 1$), in the domain *R1*, the variations of the entropy with respect to m are dominated by the variations of the entropy $\tilde{s}_{\text{conf}}$, since the kinetic entropy

$$\tilde{s}_{\text{kin}}(\varepsilon, m) = \frac{1}{2} \ln \left(\varepsilon - \frac{m^2}{2} + K \frac{m^4}{4} \right) \qquad (4.68)$$

is roughly a constant $\tilde{s} \simeq (\ln \varepsilon)/2$ when m is of order 1. As expected, the configurational entropy is a decreasing function of the magnetization: the number of microstates corresponding to a paramagnetic macrostate being much larger than the same number for a ferromagnetic state. The configurational entropy has therefore a single maximum located at $m^* = 0$. A phase transition takes place between the paramagnetic state at large

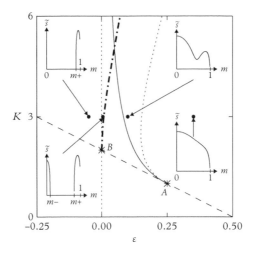

Figure 4.11 *Phase diagram of the mean field model (4.46) with $\tilde{J} = -1$. The dash-dotted curve corresponds to the first-order phase transition line, issued from the point $B(0, 2)$. As in Fig. 4.9, the solid curve indicates the right border of the region R3, where the space of thermodynamic parameters is disconnected. The dashed line corresponds to $K = 2 - 4\varepsilon$. The dotted line issued from the point $A(1/4, 1)$ represents the metastability line for the magnetized state, while the $\varepsilon = 0$ vertical dotted line is also the metastability line for the paramagnetic state. The four insets represent the entropy $\tilde{s}$ versus the magnetization m for the four energies: $\varepsilon = -0.05, 0.005, 0.1$ and 0.35, when $K = 3$. Reprinted from Campa et al. (2009), © 2009, with permission from Elsevier.*

energy and a magnetic state at small energy. Moreover, as the paramagnetic state is possible only for positive energies ε (see Fig. 4.8), the transition line is located in the domain $\varepsilon \geq 0$. In this region, the quantity $\partial^2\tilde{s}(\varepsilon, 0)/\partial m^2 = -(2 + 1/(2\varepsilon))$ is negative, which ensures that, for any value of ε and K, the paramagnetic state $m^* = 0$ is a local entropy maximum. The latter argument allows us to exclude a second-order phase transition at a positive critical energy, since the second derivative $\partial^2\tilde{s}(\varepsilon, 0)/\partial m^2$ would have to vanish, which is impossible. This argument leads to the conclusion that the phase transition must be *first order*.

Let us now focus on the behaviour of the entropy in the region $R3$, where the accessible range for m is the union of two disconnected intervals $[0, m_-] \cup [m_+, 1]$. For $\varepsilon > 0$ the entropy $\tilde{s}(\varepsilon, m)$ has a local maximum in the first interval $[0, m_-]$ located at $m^* = 0$ and associated with the entropy $\tilde{s}^1_{max}(\varepsilon) = \tilde{s}(\varepsilon, 0) = \log(\varepsilon)/2$. In the second interval $[m_+, 1]$, a maximum is also present with $\tilde{s}^2_{max}(\varepsilon) = \tilde{s}(\varepsilon, m^*)$, where $m^* \geq m_+ > 0$. As $\tilde{s}^1_{max}(\varepsilon)$ diverges to $-\infty$ when ε tends to 0, a magnetized state is expected on the line $\varepsilon = 0$, as long as $\tilde{s}^2_{max}$ remains finite. Since $K = 2$ is the only value for which $\tilde{s}^2_{max}(\varepsilon)$ diverges to $-\infty$, the first-order transition line originates at the point $B(0, 2)$ in Fig. 4.11. Although it is possible to study analytically the asymptotic behaviour of the transition line near this point, we can rather easily compute numerically the location of the first-order transition line, represented by the dash-dotted line in Fig. 4.11.

Figure 4.11 also shows the metastability line (the dotted line starting at point $A(1/4, 1)$), for the magnetized state $m^* \neq 0$. To the right of this metastability line, there is no metastable state, i.e. a local entropy maximum for a value $m > 0$ (see bottom-right inset in Fig. 4.11) while a metastable state (local maximum) exists at some non-vanishing magnetization on the other side (see top-right inset in Fig. 4.11). Finally, the vertical dotted line $\varepsilon = 0$ corresponds to the metastability line of the paramagnetic state $m^* = 0$.

One of the key issues we would like to address is the possible links between the breakdown of connectivity, and thus ergodicity breaking, on the one hand, and the phase transition, on the other. Obvious general properties can be identified: a region of parameters where the space is disconnected corresponds to a region where metastable states do exist. Let us justify this statement. At the boundary of any connected domain, when the order parameter is close to its boundary value m_b, there is a single accessible state. In a model with continuous variables, like the one we discuss here, this leads to a divergence of the entropy. In this case the singularity of the entropy is proportional to $\ln(m - m_b)$ (see, for instance, Eq. (4.68)). For a model with discrete variables, like an Ising model, the entropy would no more reach $-\infty$ as m tends to m_b, but would rather take a finite value. However, the singularity would still exist and would then be proportional to $(m - m_b) \ln(m - m_b)$. In both cases, of discrete and continuous variables, at the boundary of any connected domain, the derivative of the entropy as a function of the order parameter tends generally to $\pm\infty$ (with the entropy increasing from the boundary towards the inner part of the domain). As a consequence, entropy extrema cannot be located at the boundary. Thus, a local entropy maximum (metastable or stable) does exist in a region of parameters where the space is disconnected.

Hence, there is an entropy maximum (either local or global) in any connected domain of the space. For instance, considering the present model, in Fig. 4.11, the area *R3* is included in the area where metastable states exist (bounded by the two dotted lines and the dashed line). In such areas where metastable states exist, we generically expect first-order phase transitions. Thus the breakdown of phase-space connectivity is generically associated with first-order phase transition, as exemplified by the present study. However, this is not necessary; we can observe metastable states without first-order phase transitions, or first-order phase transitions without connectivity breaking.

A very interesting question is related to the critical points A and B shown in Fig. 4.11. As observed in the phase diagram, the end point for the line of first-order phase transition (point B) corresponds also to a point where the boundary of the region where the space is disconnected is not smooth. Similarly, the end point for the line of appearance of metastable states (point A) is also a singular point for the boundary of the area where the space is disconnected. It is thus possible to propose the conjecture that such a relation is generic, and that it should be observed in other systems where both first-order phase transitions and phase-space ergodicity breaking do occur.

4.2.4   Equilibrium dynamics

The feature of disconnected accessible magnetization intervals, which is typical of systems with long-range interactions, has profound implications on the dynamics. In particular, starting from an initial condition which lies within one of these intervals, the local dynamics is unable to move the system to a different accessible interval. A macroscopic change would be necessary to carry the system from one interval to the other. Thus the ergodicity is broken in the microcanonical dynamics even at finite N.

In Mukamel *et al.* (2005), this point has been demonstrated using the microcanonical Monte-Carlo dynamics suggested by Creutz (1983), see Section 5.1. Here, we use the Hamiltonian dynamics given by the equation of motions

$$\dot{\theta}_n = +\frac{\partial H_N}{\partial p_n} = p_n \tag{4.69}$$

$$\dot{p}_n = -\frac{\partial H_N}{\partial \theta_n} = \left(1 - Km^2\right)\left(\sin\theta_n m_x - \cos\theta_n m_y\right). \tag{4.70}$$

We display in Fig. 4.12 the evolution of the magnetization for two cases, since we have shown earlier that the gap opens up when ε decreases. The first case corresponds to the domain *R1*, in which the accessible magnetization domain is the full interval $[0, 1]$. Figure 4.12a presents the time evolution of m. The magnetization switches between the paramagnetic metastable state $m^* = 0$ and the ferromagnetic stable one $m^* > 0$. This is possible because the number of particles is small ($N = 20$) and, as a consequence,

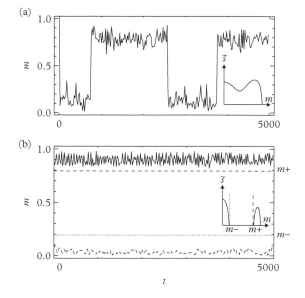

Figure 4.12 *Time evolution of the magnetization m (the entropy of the corresponding cases is plotted as an inset). (a) The case $\varepsilon = 0.1$ and $K = 8$; (b) $\varepsilon = 0.0177$ and $K = 3$. In (b), two different initial conditions are plotted simultaneously: the solid line corresponds to $m(t = 0) = 0.1$ while the dashed line to $m(t = 0) = 0.98$. The dashed (resp. dotted) straight line in (b) corresponds to the line $m = m_+ \simeq 0.794$ (resp. $m = m_- \simeq 0.192$). Reprinted from Campa et al. (2009), © 2009, with permission from Elsevier.*

the entropy barrier (see the inset) can be overcome. Considering a system with a small number of particles makes it possible to observe flips between local maxima, while such flips would be less frequent for larger N values.

In the other case, we consider a stable $m^* = 0$ state which is disconnected from the metastable one. This makes the system unable to switch from one state to the other. Note that this feature is characteristic of the microcanonical dynamics, since an algorithm reproducing canonical dynamics would allow the crossing of the forbidden region (by moving to higher energy states, which is impossible in the microcanonical ensemble). The result of two different numerical simulations is reported in Fig. 4.12b. One is initialized with a magnetization within $[0, m_-]$, while the other corresponds to an initial magnetization close to $m(0) = 1$ (i.e. within $[m_+, 1]$). We clearly see that the dynamics is blocked in one of the two possible regions, and not a single jump is visible over a long time span. This is a clear evidence of ergodicity breaking.

4.3   The phi-4 Model

Let us now turn to consider another interesting application, the so-called mean-field ϕ^4 spin model. The Hamiltonian of the model reads

$$H_N = \sum_{i=1}^{N} \left(\frac{p_i^2}{2} - \frac{1}{4} q_i^2 + \frac{1}{4} q_i^4 \right) - \frac{1}{4N} \sum_{i,j=1}^{N} q_i q_j. \tag{4.71}$$

The ϕ^4 model is a system of unit mass particles moving on a line. The particles are subject to a local double-well potential, and they are all coupled via an infinite range, mean-field interaction.

The model was introduced by Desai and Zwanzig (1978). More recently, its canonical solution was obtained by Dauxois *et al.* (2003), showing that the system exhibits a second-order ferromagnetic phase transition at a critical temperature $T_c \simeq 0.264$, corresponding to a critical energy $\varepsilon_c = T_c/2 \simeq 0.132$. Later, the entropy of the ϕ^4 model perturbed by an external magnetic field was explicitly computed by Campa and Ruffo (2007). In Hahn and Kastner (2005, 2006), Campa *et al.* (2007c), the entropy was characterized as a function of both the energy and the magnetization

$$m = \frac{1}{N} \sum_{i=1}^{N} q_i. \tag{4.72}$$

It is worth stressing that for the ϕ^4 model the magnetization m is not bounded inside the interval $[-1, 1]$.

In the following we shall review the solution of the ϕ^4 model by using the large deviations theory. As usual, the first step of the calculation consists in identifying the global variables, in terms of which we can express the energy ε. In addition to the magnetization m, we shall make use of u, twice the average kinetic energy, as well as the quantity z defined as

$$z = \frac{1}{4N} \sum_{i=1}^{N} \left(q_i^4 - q_i^2 \right), \tag{4.73}$$

related to the local potential. It is straightforward to see that

$$\bar{\varepsilon}(u, z, m) = \frac{1}{2} u + z - \frac{1}{4} m^2. \tag{4.74}$$

Then, as a second step in the calculation, we need to estimate the generating function

$$\psi(\lambda_u, \lambda_z, \lambda_m) = \int \left(\prod_i dq_i dp_i \right) \exp \left[-\lambda_u \sum_{i=1}^{N} p_i^2 - \lambda_m \sum_{i=1}^{N} q_i - \lambda_z \sum_{i=1}^{N} \left(q_i^4 - q_i^2 \right) \right]. \tag{4.75}$$

Also in this case, we can split the calculation by focusing on either the kinetic or the potential part. In formulae:

$$\psi(\lambda_u, \lambda_z, \lambda_m) = \psi_u(\lambda_u) \psi_{z,m}(\lambda_z, \lambda_m) \tag{4.76}$$

$$= \left[\sqrt{\frac{\pi}{\lambda_u}} \right]^{\frac{N}{2}} \left[\int_{-\infty}^{+\infty} dq \, e^{-\lambda_m q - \lambda_z \left(q^4 - q^2 \right)} \right]^N, \tag{4.77}$$

where the two terms on the right-hand side define $\bar{Z}_u(\lambda_u)$ and $\bar{Z}_{z,m}(\lambda_z, \lambda_m)$. For the integral to exist we need to require that $\lambda_u > 0$ and $\lambda_z > 0$. From the previous expression, we find the following contributions to the free energy associated with ψ,

$$\bar{\phi}(\lambda_u, \lambda_z, \lambda_m) = \bar{\phi}_u(\lambda_u) + \bar{\phi}_{z,m}(\lambda_z, \lambda_m), \tag{4.78}$$

with

$$\bar{\phi}_u(\lambda_u) = -\frac{1}{2} \ln \pi + \frac{1}{2} \ln \lambda_u \tag{4.79}$$

$$\bar{\phi}_{z,m}(\lambda_z, \lambda_m) = -\ln \left[\int_{-\infty}^{+\infty} dq \, e^{-\lambda_m q - \lambda_z \left(q^4 - q^2 \right)} \right]. \tag{4.80}$$

Both these functions are analytic, and we can therefore write

$$\bar{s}(u, z, m) = \bar{s}_{kin}(u) + \bar{s}_{conf}(z, m), \tag{4.81}$$

where

$$\bar{s}_{kin}(u) = \inf_{\lambda_u} \left[\lambda_u u - \bar{\phi}_u(\lambda_u) \right] = \frac{1}{2} + \frac{1}{2} \ln(2\pi) + \frac{1}{2} \ln u \tag{4.82}$$

$$\bar{s}_{conf}(z, m) = \inf_{\lambda_z, \lambda_m} \left[\lambda_z z + \lambda_m m - \bar{\phi}_{z,m}(\lambda_z, \lambda_m) \right]. \tag{4.83}$$

The third and final step in the calculation returns us with the sought entropy function

$$\tilde{s}(\varepsilon, m) = \sup_{u,z} \left[\tilde{s}_{kin}(u) + \tilde{s}_{conf}(z, m) \Big|_{\frac{u}{2} + z - \frac{m^2}{4}} = \varepsilon \right], \tag{4.84}$$

that can be cast in the form

$$\tilde{s}(\varepsilon, m) = \frac{1}{2} + \frac{1}{2} \ln(4\pi) + \sup_z \left[\frac{1}{2} \ln\left(\varepsilon - z + \frac{m^2}{4} \right) + \tilde{s}_{conf}(z, m) \right]. \tag{4.85}$$

This variational problem can be solved numerically and the result is depicted in Fig. 4.13 for different values of ε.

The study of the ϕ^4 spin model provides us with the interesting possibility of elucidating yet another important consequence of the ensemble inequivalence. As we will argue in the following, in fact, the magnetic susceptibility at constant temperature can be negative in the microcanonical ensemble, while it is bound to be positive in the canonical ensemble. Such a feature appears also for other long-range systems whose entropy depends on two macroscopic variables, e.g. models of two-dimensional geophysical flows (Venaille and Bouchet, 2009), for which entropy depends on total circulation and energy.

To clarify this important issue, and elaborate on the concept of inequivalence as applied to the ϕ^4 model, we shall derive its canonical solution. To this end we start by considering the canonical partition function

$$Z(\beta, h) = \int \left(\prod_i dq_i dp_i \right) \exp\left(-\beta \left[\sum_{i=1}^N \left(\frac{p_i^2}{2} - \frac{q_i^2}{4} + \frac{q_i^4}{4} \right) - \frac{1}{4N} \sum_{i,j=1}^N q_i q_j - h \sum_{i=1}^N q_i \right] \right) \tag{4.86}$$

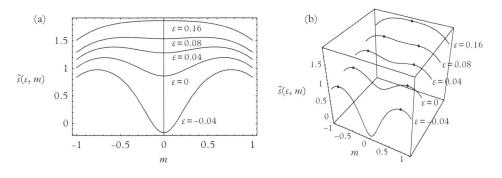

Figure 4.13 *(a) Entropy $\tilde{s}$ as a function of magnetization m for different values of the energy ε. (b) 3D plot of $\tilde{s}(\varepsilon, m)$. The black dots show the location of the equilibrium values m^* in the microcanonical ensemble. $\tilde{s}(\varepsilon, m)$ has a bimodal profile as a function of m and a region of convexity, for $\varepsilon < \varepsilon_c$ Reprinted from Campa et al. (2009), © 2009, with permission from Elsevier.*

from which we can derive the rescaled free energy

$$\phi(\beta, h) = \beta f(\beta, h) = - \lim_{N\to\infty} \frac{1}{N} \ln Z(\beta, h) = \inf_{\varepsilon, m} [\beta\varepsilon - \beta hm - \tilde{s}(\varepsilon, m)]. \qquad (4.87)$$

Note that for the computation of the canonical partition function we have introduced in the Hamiltonian the interaction with an external magnetic field h. This is necessary if we want to compare the susceptibilities in both ensembles. In fact, the magnetization and the magnetic field are conjugated variables (similarly to energy and temperature). In the microcanonical ensemble the magnetic field is related to the derivative of the entropy $\tilde{s}(\varepsilon, m)$ with respect to the magnetization m (similarly to the temperature as being related to the derivative of $\tilde{s}(\varepsilon, m)$ with respect to the energy ε). In the canonical ensemble the magnetic field is an external parameter (similar to the temperature).

The above variational problem yields

$$\frac{\partial \tilde{s}}{\partial m} = -h\beta \qquad (4.88)$$

$$\frac{\partial \tilde{s}}{\partial \varepsilon} = \beta. \qquad (4.89)$$

By deriving Eq. (4.88) and (4.89) with respect to h, we eventually obtain

$$\tilde{s}_{m\varepsilon} \frac{\partial \varepsilon}{\partial h} + \tilde{s}_{mm} \frac{\partial m}{\partial h} = -\beta \qquad (4.90)$$

$$\tilde{s}_{\varepsilon\varepsilon} \frac{\partial \varepsilon}{\partial h} + \tilde{s}_{\varepsilon m} \frac{\partial m}{\partial h} = 0, \qquad (4.91)$$

where $\tilde{s}_{kj} = (\partial^2 \tilde{s})/(\partial k \partial j)$ and $j, k = \varepsilon, m$. By introducing the magnetic susceptibility at constant temperature χ as

$$\chi = \frac{\partial m}{\partial h}, \qquad (4.92)$$

we immediately get

$$\chi = \left(\frac{\partial m}{\partial h}\right) = -\beta \frac{\tilde{s}_{\varepsilon\varepsilon}}{\tilde{s}_{\varepsilon\varepsilon}\tilde{s}_{mm} - \tilde{s}_{\varepsilon m}^2}, \qquad (4.93)$$

by solving Eqs. (4.90) and (4.91). This relation holds for both the canonical and the microcanonical ensembles. The stability of the canonical solution requires that both $\tilde{s}_{\varepsilon\varepsilon}$ and $\tilde{s}_{mm}$ be negative, with $\tilde{s}_{\varepsilon\varepsilon}\tilde{s}_{mm} - \tilde{s}_{\varepsilon m}^2 > 0$. This shows that χ is always positive in that ensemble. On the other hand, there is no constraint on the second derivatives of $\tilde{s}(\varepsilon, m)$ in the microcanonical ensemble, and thus χ can be negative. In our case, it turns out that $\tilde{s}_{\varepsilon\varepsilon}$ is always negative. Hence, the sign of the susceptibility is related to the sign of $\tilde{s}_{mm}$: when this latter quantity is positive, a possibility which is ruled out in the canonical ensemble, the susceptibility is negative.

In Fig. 4.13a, $\tilde{s}(\varepsilon, m)$ is plotted as a function of m for different values of ε. Regions exist where the function $\tilde{s}$ is convex. Figure 4.13b provides a 3D representation of $\tilde{s}$ as function of both (ε, m). The profile of $\tilde{s}(\varepsilon, m)$ versus m has a bimodal shape, and a region of convexity, for $\varepsilon < \varepsilon_c$.

A region of convexity in the microcanonical entropy $\tilde{s}(\varepsilon, m)$, as function of the global variable m, is associated with a first-order phase transition in the canonical ensemble. This provides a magnetization analogue of negative heat capacities, returning an alternative indicator of inequivalence of microcanonical and canonical ensembles.

The (effective) magnetic field of the microcanonical ensemble can have a sign which is opposite to that of the magnetization. In the canonical ensemble, this scenario is forbidden, since the equilibrium magnetization is always in the direction of the applied magnetic field.

4.4 The Self-Gravitating Ring (SGR) Model

4.4.1 Introduction of the model

As another interesting application, we here consider the self-gravitating ring (SGR) model introduced in Sota *et al.* (2001) to mimic, to some extent, the features of three-dimensional systems subject to gravitational interactions. As we shall discuss in the third part of this book, in Chapter 10, which is devoted to gravity, the main difficulty in handling self-gravitating systems stems from the observation that they cannot approach statistical equilibrium, because of the short-distance divergence of the potential and of the evaporation at the boundaries. By confining the system within a finite box, with adiabatic walls, we can eliminate the complications caused by the evaporation process. The *gravothermal catastrophe* (Antonov, 1962; Padmanabhan, 1990), which instead reflects the singularity, can be effectively prevented upon introduction of a small-scale softening in the interaction potential.

Direct studies of the full three-dimensional N-body gravitational dynamics are particularly demanding from the computational viewpoint (Heggie and Hut, 2003). For this reason, a plethora of simplified models that makes it possible to progress in the analytical, as well as numerical, investigations has been proposed.

In the SGR model, for example, point-like particles are confined on a ring and interact via the true Newtonian 3D potential; see Fig. 4.14. Furthermore, at short distances, an *ad hoc* regularization of the potential is imposed. The Hamiltonian of the SGR model reads

$$H = \frac{1}{2} \sum_{i=1}^{N} p_i^2 + \frac{1}{2N} \sum_{i,j} V_\varepsilon(\theta_i - \theta_j), \qquad (4.94)$$

in which

$$V_\epsilon(\theta_i - \theta_j) = -\frac{1}{\sqrt{2}} \frac{1}{\sqrt{1 - \cos(\theta_i - \theta_j) + \epsilon}}, \qquad (4.95)$$

where ϵ stands for the softening parameter at small scales (not to be confused with the energy density U defined later).

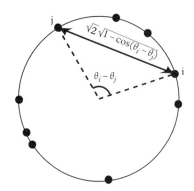

Figure 4.14 *Self-gravitating ring model with a fixed unitary radius. Particles are constrained to move on a ring and therefore their location is specified by the angles measured with respect to a fixed direction. Each pair of particles at θ_i and θ_j interacts through the inverse-square three-dimensional gravitational force. The distance is measured by the chord, as shown in the figure. Reprinted with permission from Tatekawa et al. (2005), © 2005 by the American Physical Society.*

Performing the large ϵ limit, the potential converges to

$$V_\epsilon = \frac{1}{\sqrt{2\epsilon}} \left[\frac{1 - \cos\left(\theta_i - \theta_j\right)}{2\epsilon} - 1 \right] + O(\epsilon^{-2}), \tag{4.96}$$

and we recover the Hamiltonian Mean-Field (HMF) model. The introduction of the softening parameter ϵ is not an innocent one and impacts dramatically on the system behaviour. To illustrate this fact, let us refer to Figs. 4.15a and 4.15b, where the caloric curve is reported, as determined from microcanonical simulations. More specifically, we are plotting the kinetic temperature of the system, estimated as twice the averaged kinetic energy per particle $T \equiv \beta^{-1} = 2 \langle K \rangle / N$, as a function of the total energy per particles, $U \equiv H/N$. When the SGR is sufficiently close to the HMF model ($\epsilon = 10$ in Fig 4.15a) the caloric curve is almost linear in the homogeneous phase, $U > U_c(\epsilon)$, while it bends downward in the clustered phase $U < U_c(\epsilon)$. However, the temperature grows with the energy: the specific heat is positive and no sign of inequivalence can be detected in agreement with the analysis reported earlier for the HMF case study.

However, when the softening gets reduced, a negative specific heat develops, in analogy with what happens in 3D. In Fig. 4.15b, the results of the simulations by Tatekawa *et al.* (2005) are displayed for two distinct choices of ϵ. Three phases can be identified (Sota *et al.*, 2001), as follows:

- A low-energy clustered phase for $U < U_{\text{top}}(\epsilon)$, U_{top} being defined as the energy for which $\partial T/\partial U = 0$;
- An intermediate-energy phase, $U_{\text{top}}(\epsilon) < U < U_c(\epsilon)$, with negative specific heat;
- A high-energy gaseous phase for $U_c(\epsilon) < U$.

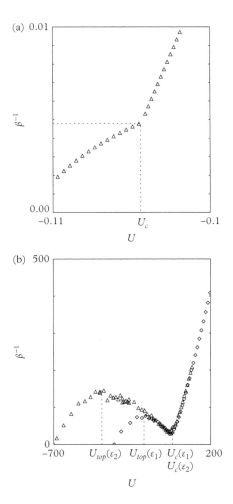

Figure 4.15 *Caloric curves of the self-gravitating ring (SGR) model as obtained from numerical simulations of Hamiltonian (4.94). (a) $\epsilon = 10$. A second-order phase transition appears at U_c. No backbending of the caloric curve, indicating a negative specific heat, is present. Simulations were presented in Tatekawa et al. (2005) for $N = 100$. (b) Caloric curves for $\epsilon_1 = 1.0 \times 10^{-6}$ and $\epsilon_2 = 2.5 \times 10^{-7}$, and $N = 100$. The transition is here first order in the microcanonical ensemble. The two transition energies $U_c(\epsilon_1)$ and $U_c(\epsilon_2)$ are close, suggesting a slow variation of the critical energy with the softening parameter ϵ. In contrast $U_{top}(\epsilon_1)$ is significantly larger than $U_{top}(\epsilon_2)$: this characteristic energy diminishes with ϵ. A negative specific heat phase appears for $U_{top} < U < U_c$, and expands as the softening parameter ϵ is reduced. All quantities are plotted in arbitrary units. Reprinted with permission from Tatekawa et al. (2005), © 2005 by the American Physical Society.*

The clustered phase is an effect induced by the presence of the softening ϵ. In absence of ϵ the particles would be attracted towards the zero distance singularity. Conversely, in the gas phase, the particles are hardly affected by the potential and hence behave as almost free particles. The intermediate phase bears the distinctive traits of gravity, persisting and widening in the relevant $\epsilon \to 0$ limit.

In Tatekawa *et al.* (2005), the SGR model has been studied analytically to shed light on the emergence of ensemble inequivalence. In the following we will revisit their main derivations and conclusions.

4.4.2   Inequivalence of ensemble

In the mean-field limit $N \to \infty$ (Messer and Spohn, 1982), we can introduce (see Chapters 7 and 8) the single particle distribution function $f(p, \theta)$ such that $f(p, \theta)\mathrm{d}p\mathrm{d}\theta$ quantifies the fraction of particles in $[\theta, \theta + \mathrm{d}\theta] \times [p, p + \mathrm{d}p]$. In terms of f, the potential energy takes the form

$$E_P[f] = \frac{1}{2} \int \mathrm{d}\theta \, \mathrm{d}\phi \, \mathrm{d}p \, \mathrm{d}p' \, f(p, \theta) V_\epsilon(\theta - \phi) f(p', \phi) \tag{4.97}$$

$$= \frac{1}{2} \int \mathrm{d}\phi \, \mathrm{d}\theta \, \rho(\theta)\rho(\phi) V_\epsilon(\theta - \phi), \tag{4.98}$$

where

$$\rho(\theta) = \int \mathrm{d}p \, f(p, \theta) \tag{4.99}$$

stands for the mass density (we remind that the particles have unit mass). The kinetic energy is

$$E_K[f] = \frac{1}{2} \int \mathrm{d}\theta \, \mathrm{d}p \, p^2 f(p, \theta) \tag{4.100}$$

and the total energy

$$E[f] = E_K[f] + E_P[f]. \tag{4.101}$$

The equilibrium distribution in the microcanonical ensemble is obtained by maximizing the entropy functional

$$S[f] = -\int \mathrm{d}\theta \, \mathrm{d}p \, f \log f \tag{4.102}$$

while imposing the constraints of fixed total energy $E[f] = U$, momentum

$$p[f] = \int p f(p, \theta)\mathrm{d}\theta \, \mathrm{d}p = 0, \tag{4.103}$$

and mass

$$M[f] = \int \rho \; d\theta = 1. \tag{4.104}$$

A necessary condition to get an entropy maximum is to require that the functional

$$F[f] \equiv S[f] - \beta E[f] - \alpha \int f \; dp \; d\theta - \gamma p[f], \tag{4.105}$$

is stationary, where α, β and γ are Lagrange multipliers. In formulae

$$\frac{\delta F[f]}{\delta f} = -\log f - 1 - \beta \left(\frac{p^2}{2} + W(\theta) \right) - \alpha - \gamma p = 0, \tag{4.106}$$

where $W(\theta)$ is defined as

$$W(\theta) \equiv \int_{-\pi}^{+\pi} \rho(\phi) V_\epsilon(\theta - \phi) d\phi. \tag{4.107}$$

For a symmetric state $p[f] = 0$, and the Lagrange multiplier γ is identically equal to 0. From Eq. (4.106), the normalized equilibrium distribution function takes the form

$$f(p, \theta) = A \exp \left[-\beta \left(\frac{p^2}{2} + W(\theta) \right) \right], \tag{4.108}$$

where $A = \exp(-1 - \alpha)$ stands for the normalization constant. The mass density is given by

$$\rho(\theta) = \tilde{A} \, e^{-\beta W(\theta)}, \tag{4.109}$$

with $\tilde{A} = A\sqrt{2\pi/\beta}$. When Eqs. (4.107) and (4.109) are combined, we end up with the consistency equation

$$W(\theta) = \tilde{A} \int_{-\pi}^{+\pi} e^{-\beta W(\phi)} V_\epsilon(\theta - \phi) d\phi, \tag{4.110}$$

and the equilibrium density expression

$$\rho(\theta) = \tilde{A} \exp \left[-\beta \tilde{A} \int_{-\pi}^{+\pi} \rho(\phi) V_\epsilon(\theta - \phi) d\phi \right], \tag{4.111}$$

which can be solved numerically. Once the equilibrium mass distributions ρ and the function W are obtained for each value of ϵ, the full single particle distribution function $f(\theta, p)$ can be estimated from Eq. (4.108). In addition, the potential energy and the

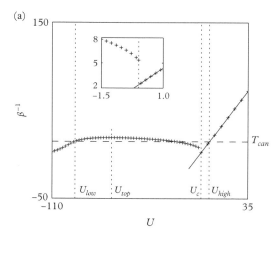

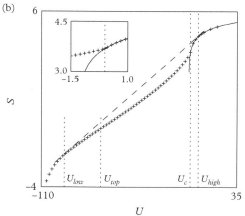

Figure 4.16 *Temperature (a) and entropy (b) versus energy U for $\epsilon = 10^{-5}$. Four values of the energy, indicated by the short-dashed vertical lines, can be identified: (i) $U_{low} \simeq -93$; (ii) $U_{high} \simeq 6$ bound from below and above the region of inequivalence of ensembles; (iii) $U_c \simeq 0$ is the transition energy in the microcanonical ensemble; (iv) $U_{top} \simeq -66$ limits from below the negative specific heat region, where temperature decreases as energy increases. $T_{can} \simeq 15$ is represented by a dashed line in (a). It stands for the canonical transition temperature and corresponds to the inverse slope of the entropy, at both U_{low} and U_{high}, as represented by the straight dashed line in (b). The full lines represent the analytical solutions of both temperature and entropy in the uniform case. They are extended slightly below U_c, in the metastable phase, for visualization purposes. The insets in (a) and (b) show a zoom of the temperature and of the entropy in the vicinity of $U = U_c$. A temperature jump is revealed at U_c and different slopes of the entropy above and below U_c are found, which points to the first-order nature of the phase transition. All quantities are plotted in arbitrary units. Reprinted with permission from Tatekawa et al. (2005), © 2005 by the American Physical Society.*

kinetic energy are determined by Eq. (4.98) and Eq. (4.100), respectively, allowing us to draw the caloric curve, i.e., $T \equiv \beta^{-1} = 2E_K$ against $U = E_K + E_P$.

As anticipated earlier, the problem can be handled numerically by implementing an iterative method, which ensures entropy increase and so global convergence of the procedure. The method is discussed in Tatekawa *et al.* (2005) and inspired to the algorithm used by Turkington *et al.* (2001) to compute the entropy maxima in the framework of two-dimensional turbulence studies. For technical details concerning the implementation the reader can refer to Tatekawa *et al.* (2005). In the following we will solely discuss the results of the calculations, with emphasis to the inequivalence issue.

In Fig. 4.16, both entropy and temperature $T = \beta^{-1}$ are reported as a function of energy U. A region of negative specific heat is found for $U_{\text{top}} \leq U \leq U_c$. For $U_{\text{low}} \leq U \leq U_{\text{high}}$, the entropy does not coincide with its convex envelope. Hence, microcanonical and canonical ensembles return different predictions. In practical terms, the macroscopic states in the microcanonical ensemble are stable, while they would be either metastable or unstable in the corresponding canonical picture. The mass density is uniform above U_c, while, it turns out to be localized for $U < U_c$. To characterize the effect of localization, we can employ an appropriate bunching parameter B, similar to the magnetization of the HMF model, defined as

$$B = \int_{-\pi}^{+\pi} d\theta \, e^{i\theta} \, \rho(\theta). \tag{4.112}$$

The modulus $|B|$ of this latter quantity vanishes if the mass distribution is uniform, while it takes the value $|B| = 1$, if the mass is concentrated in just one point. $|B|$ is plotted in Fig. 4.17 as a function of U. In $U = U_c$, a first-order microcanonical transition occurs: $|B|$ jumps from a finite values to 0. The first-order nature of the phase transition can be also appreciated by zooming the entropy S around U_c, see inset in Fig. 4.16b. The

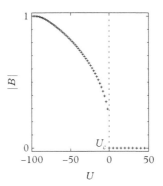

Figure 4.17 *The modulus of the bunching parameter $|B|$ versus energy U for $\epsilon = 10^{-5}$, which emphasizes the microcanonical first-order phase transition at $U_c \simeq 0$ by showing a jump in the order parameter. All quantities are plotted in arbitrary units. Reprinted with permission from Tatekawa et al. (2005), © 2005 by the American Physical Society.*

first-order phase transition is of the convex–concave (Bouchet and Barré, 2005) type and the canonical ensemble is obtained by taking the convex envelope of the microcanonical entropy.

In Tatekawa *et al.* (2005), the full phase diagram of the model is also derived in both the microcanonical and the canonical ensembles. To summarize the main conclusion, when the softening parameter is sufficiently small, a negative specific heat region appears in the microcanonical ensemble, in coincidence with the phase transition becoming first order in the canonical ensemble. Lowering further the softening parameter, the transition becomes first order in the microcanonical ensemble and a temperature jump appears at the transition energy. Also, it is analytically shown that the microcanonical and canonical tricritical points do not coincide.

5

Beyond Mean-Field Models

It is important to determine whether the interesting properties and behaviours described in the previous chapter are present only in mean-field models, or whether they are found also in other long-range systems. We have seen that the relevant factor responsible for the difference with respect to short-range systems is the absence of additivity, and therefore we are led to expect that the peculiar long-range features are found whenever additivity does not hold, thus not only in purely mean-field systems. The models described in this chapter will indeed confirm this expectation.

We will consider two non-mean-field types of long-range models:

i. Systems in which the interaction is given by a mean-field term plus a short-range one;

ii. Lattice systems of the type introduced in Section 1.4 with slowly decaying coupling, e.g. with a coupling that decays as in Eq. (2.1) with α smaller than the space dimension d.

Both types will present the properties already found in mean-field models. In particular, we will meet the typical characteristics of many mean-field systems, like ensemble inequivalence, negative specific heat, temperature jumps and ergodicity breaking.

In the first two sections, we will study a spin system defined on a one-dimensional lattice. In the first one, the spins interact with a mean-field potential augmented with a nearest neighbour term; the lattice will actually be a ring, due to the periodic boundary conditions. In the second section, the interaction is slowly decaying, and both free and periodic boundary conditions will be considered.

In the next two sections, we will focus on generalizations of the HMF model. First, we will look at a system of XY rotators residing on a one-dimensional lattice with periodic boundary conditions (a ring), which interact with a mean-field term identical to that of the HMF model plus a nearest neighbour term. Then, we will introduce the so-called α-HMF model, where the XY rotators, residing on a lattice of arbitrary dimension d with periodic boundary conditions, interact with long-range coupling constants that decay as $r^{-\alpha}$ with $\alpha < d$.

In the final part of the chapter, we will consider dipolar interactions, which will be treated in greater detail in Chapter 15. The dipolar interaction is only marginally long

Physics of Long-Range Interacting Systems. First Edition. A. Campa *et al.*
© A. Campa, T. Dauxois, D. Fanelli, and S. Ruffo 2014. Published in 2014 by Oxford University Press.

range, since $\alpha = d = 3$; however, its peculiar anisotropies give rise to very interesting behaviours.

We want to underline an important methodological aspect in the treatment of one-dimensional systems with both a mean field and a short-range interaction: the use of the transfer integral method combined with the Hubbard–Stratonovich transformations and the min–max procedure allows us to obtain, for both the discrete and the continuous variable cases, the canonical and the microcanonical solutions.

5.1 Ising Model

5.1.1 Introduction

An interesting Ising model combining long- with short-range interactions was originally introduced by Nagle (1970), while Kardar (1983a, b) later proposed further elaborations of the model.

The Hamiltonian is

$$H_N = -\frac{1}{2N}\left(\sum_{i=1}^{N} S_i\right)^2 - \frac{K}{2}\sum_{i=1}^{N}(S_i S_{i+1} - 1), \tag{5.1}$$

where $S_i = \pm 1$. In this one-dimensional spin chains, the first term has an infinite range and corresponds to the Curie–Weiss Hamiltonian (2.4). This term is responsible for the non-additive properties of the model, as shown in Section 2.1.2. Note that the pre-factor $\mathcal{J}$ present in Hamiltonian (2.4) has been set to 1 by an appropriate renormalization of the energy. This choice amounts to focus on the ferromagnetic coupling, with no further loss of generality. In contrast, the second term of (5.1) corresponds to an interaction between nearest neighbours along a one-dimensional lattice with periodic boundary conditions. The coupling constant K might be either positive or negative, and both cases will be hereafter discussed.

At $T = 0$, we can determine whether the model has a phase transition at some value of K by comparing the energy of the ferromagnetic state with that of the antiferromagnetic one, since only the energy term of the free energy matters (see Fig. 5.1).

The ferromagnetic state with all spins up ($S_i = 1, \forall i$) or down ($S_i = -1, \forall i$) has a negative energy $E_F = -N/2$. For the antiferromagnetic state with alternate signs of nearest neighbour spins, the first term of (5.1) gives a vanishing contribution to the energy. The total energy reads, therefore, $E_A = KN$. By imposing $E_A = E_F$ we get the

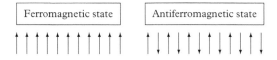

Figure 5.1 *Ferromagnetic and antiferromagnetic configurations.*

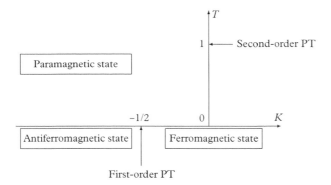

Figure 5.2 *Elementary features of the phase diagram of the short- plus long-range Ising model. Simple considerations show that one gets a* first-order *phase transition on the local coupling K axis, and a* second-order *one on the temperature T axis.*

phase transition value $K_t = -1/2$ at which a discontinuity of the order parameter is found, from $m = 0$ to $m = 1$. Therefore we have a *first-order* phase transition.

For non-zero temperature, we must take into account the entropic term of the free energy, which measures disorder. When the coupling constant K vanishes, we fully recover the Curie–Weiss Hamiltonian (2.4), which exhibits a *second-order* phase transition at $T = 1$. We therefore, expect that the (T, K) phase diagram displays a transition line which is first order at low T and second order at high T, dividing the (T, K) plane in different regions, as sketched in Fig. 5.2.

The transition line separates a ferromagnetic from a paramagnetic state, although exactly at $T = 0$ and $K < -1/2$ the state is antiferromagnetic. Let us now determine analytically this transition line in both the canonical and the microcanonical ensemble.

5.1.2   The solution in the canonical ensemble

The phase diagram has been studied in the canonical ensemble by Nagle (1970) and Kardar (1983a, b). The partition function can be written as

$$Z(\beta, N) = \sum_{\{S_1,...,S_N\}} e^{-\beta H} \tag{5.2}$$

$$= \sum_{\{S_1,...,S_N\}} \exp\left[\frac{\beta}{2N}\left(\sum_{i=1}^{N} S_i\right)^2 + \frac{\beta K}{2}\sum_{i=1}^{N}(S_i S_{i+1} - 1)\right]. \tag{5.3}$$

To get rid of the quadratic term, we use the Hubbard–Stratonovich transformation

$$\exp\left[\frac{\beta}{2N}\left(\sum_{i=1}^{N} S_i\right)^2\right] = \sqrt{\frac{\beta N}{2\pi}}\int_{-\infty}^{+\infty} dx\, \exp\left[-\frac{\beta N}{2}x^2 + \beta x \sum_{i=1}^{N} S_i\right], \tag{5.4}$$

so that the partition function (5.3) can be rewritten as

$$
Z(\beta, N) = \sqrt{\frac{\beta N}{2\pi}} \int_{-\infty}^{+\infty} dx \ e^{-\frac{\beta N}{2}x^2} \sum_{\{S_1,...,S_N\}} \left[e^{\beta x \sum_{i=1}^{N} S_i + \frac{\beta K}{2} \sum_{i=1}^{N} (S_i S_{i+1} - 1)} \right] \tag{5.5}
$$

$$
= \sqrt{\frac{\beta N}{2\pi}} \int_{-\infty}^{+\infty} dx \ e^{-N\beta \tilde{f}(\beta, x)}, \tag{5.6}
$$

if we introduce the free energy-like function $\tilde{f}(\beta, x)$ as

$$
\tilde{f}(\beta, x) = \frac{1}{2}x^2 + f_0(\beta, x), \tag{5.7}
$$

where $f_0(\beta, x)$ is the free energy of the nearest neighbour Ising model with an external field x. Such an expression can be easily derived using the transfer matrix (Kramers and Wannier, 1941; Kardar, 1983a; Huang, 1987). By an easy calculation, we find that

$$
f_0(\beta, x) = -\frac{1}{\beta N} \ln(\lambda_+^N + \lambda_-^N), \tag{5.8}
$$

where the two eigenvalues of the transfer matrix are given by the expressions

$$
\lambda_\pm = e^{\beta K/2} \cosh(\beta x) \pm \sqrt{e^{\beta K} \sinh^2(\beta x) + e^{-\beta K}}. \tag{5.9}
$$

As λ_+ is greater than λ_- for all values of x, only the larger eigenvalue λ_+ is relevant in the limit $N \to \infty$. We thus finally get the expression

$$
\tilde{\phi}(\beta, x) = \beta \tilde{f}(\beta, x) = \frac{\beta}{2}x^2 - \ln \left[e^{\beta K/2} \cosh(\beta x) + \sqrt{e^{\beta K} \sinh^2(\beta x) + e^{-\beta K}} \right], \tag{5.10}
$$

which is plotted in Figs. 5.3 and 5.4 for different values of the nearest neighbour coupling K.

In the large N-limit, the application of the saddle point method to Eq. (5.6) finally leads to the free energy $f(\beta)$, which is obtained by taking the value of x which minimizes $\tilde{\phi}(\beta, x)$ in formula (5.10).

From the knowledge of the free energy, as anticipated, we get either a second- (see Fig. 5.3) or a first-order (see Fig. 5.4) phase transition depending on the value of the coupling constant K.

As usual, the expansion of $\tilde{f}(\beta, x)$ in power of x is the appropriate procedure to define the critical lines and points. We get here

$$
\tilde{f}(\beta, x) = -\ln 2 \cosh \frac{\beta K}{2} + \frac{\beta}{2}x^2 \left(1 - \beta e^{\beta K}\right) + \frac{\beta^4}{24} e^{\beta K} \left(3e^{2\beta K} - 1\right) x^4 + \mathcal{O}(x^6). \tag{5.11}
$$

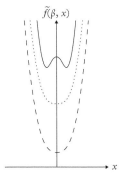

Figure 5.3 *Second-order phase transition: Evolution of the function $\tilde{f}(\beta, x)$ for different values of the inverse temperature $\beta = 1.1$ (dashed line), $\beta_c \simeq 1.4$ (dotted), $\beta = 2.5$ (solid) when $K = -0.25$. Note that the different curves have been vertically shifted for readability purposes. Reprinted from Campa et al. (2009), © 2009, with permission from Elsevier.*

The critical point of the second-order transition is obtained for each K by computing the value β_c at which the quadratic term of the expansion (5.11) vanishes, provided the coefficient of the fourth-order term is positive: we get, therefore, $\beta_c = \exp(-\beta_c K)$.

When also the fourth-order coefficient vanishes, which means when $3\exp(2\beta K) = 1$, we get the canonical tricritical point (CTP). This condition corresponds to particular values of the local coupling constant K and of the inverse temperature β

$$K_{\mathrm{CTP}} = -\frac{\ln 3}{2\sqrt{3}} \simeq -0.317 \qquad \text{with} \qquad \beta_{\mathrm{CTP}} = \sqrt{3}, \qquad (5.12)$$

in which β_{CTP} corresponds to the canonical tricritical inverse temperature.

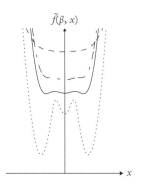

Figure 5.4 *First-order phase transition: Evolution of the function $\tilde{f}(\beta, x)$ for different values of the inverse temperature $\beta = 10$ (dotted), $\beta_t \simeq 2.4$ (solid), $\beta = 2.35$ (dash-triple dot), $\beta = 2$ (dashed) when $K = -0.4$. Note that the different curves have been vertically shifted for readability purposes. Reprinted from Campa et al. (2009), © 2009, with permission from Elsevier.*

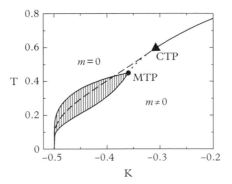

Figure 5.5 *The canonical and microcanonical (K, T) phase diagram. In the canonical ensemble, the large K transition is continuous (bold solid line) down to the tricritical point CTP, where it becomes first order (dashed line). In the microcanonical ensemble the continuous transition coincides with the canonical one at large K (bold line). It persists at lower K (dotted line) down to the tricritical point MTP, where it turns first order, with a branching of the transition line (solid lines). The shaded area is not accessible in the microcanonical ensemble. Reprinted from Campa* et al. *(2009), © 2009, with permission from Elsevier.*

We get finally the first-order line, numerically, by requiring that $f(\beta, 0) = f(\beta, x^*)$, where x^* is one of the two symmetric local minima. Figure 5.5 represents the phase diagram in both the canonical and the microcanonical ensemble. The features of this phase diagram are very close to those of the BEG model (see Fig. 2.9) and of the generalized HMF (see Fig. 4.6). We will come back to discussing these results, after having obtained the solution in the microcanonical ensemble (Mukamel *et al.*, 2005). The next subsection is devoted entirely to carrying out the calculation.

5.1.3 The solution in the microcanonical ensemble

In the microcanonical ensemble, we can proceed with a simple counting method as proposed in Mukamel *et al.* (2005). The first term of the Hamiltonian (5.1) can be straightforwardly rewritten as $-M^2/(2N)$, where $M = \sum_{i=1}^{N} S_i$ is the total magnetization. Concerning the second term of the Hamiltonian, since two identical neighbouring spins do not contribute to it ($S_i S_{i+1} - 1$ being equal to 0), while two different adjacent spins give a contribution equal to K, the total contribution associated with the second term is $K\mathcal{U}$, where $\mathcal{U}$ is the number of 'kinks' in the chain, i.e., links between two neighbouring spins of opposite sign. Therefore, the total energy can be written as $E = -M^2/(2N) + K\mathcal{U}$.

Introducing the number of spins up, N_+, and of spins down, N_-, with $N_+ + N_- = N$, the number of microstates in a chain of N spins corresponding to an energy E *and* a total magnetization M can be written as

$$\Omega(E, M; N) = \Omega(N_+, N_-, \mathcal{U}) \simeq \binom{N_+}{\mathcal{U}/2}\binom{N_-}{\mathcal{U}/2}, \tag{5.13}$$

where N_+, N_- and $\mathcal{U}$ are related to N, E and M by the above relations. The formula is derived by taking into account that we must distribute N_+ spins among $\mathcal{U}/2$ groups and N_- among the remaining $\mathcal{U}/2$. Each of these distributions gives a binomial term, and, since they are independent, the total number of states is the product of the two binomials. The expression is not exact because we are on a ring, but corrections are, however, of order N and do not affect the entropy. A slight correction to formula (5.13) is present for small N_+, N_- and $\mathcal{U}/2$, and all these numbers should be indeed reduced by a unity.

Introducing $m = M/N$, $u = \mathcal{U}/N$ and $\varepsilon = E/N = -m^2/2 + Ku$, we thus finally get the entropy

$$\tilde{s}(\varepsilon, m) = \frac{1}{N} \ln \Omega \tag{5.14}$$

$$= \frac{1}{2}(1+m)\ln(1+m) + \frac{1}{2}(1-m)\ln(1-m) - u\ln u$$

$$- \frac{1}{2}(1+m-u)\ln(1+m-u) - \frac{1}{2}(1-m-u)\ln(1-m-u), \tag{5.15}$$

where u can be expressed as a function of ε and m. As shown in Figs. 5.6 and 5.7 for different values of the energy ε, the shape of this function strongly depends on the value of the nearest neighbour coupling K.

In the large N-limit, the last step is to maximize entropy $\tilde{s}(\varepsilon, m)$ with respect to magnetization m, leading to the entropy $s(\varepsilon) = \tilde{s}(\varepsilon, m^*)$, where m^* is the equilibrium value. As anticipated, we get either a second (see Fig. 5.6) or a first-order (see Fig. 5.7) phase transition, depending on the value of the coupling constant K.

As usual, and analogously to what has been done in the canonical ensemble, the expansion of $\tilde{s}(\varepsilon, m)$ in powers of m is the appropriate procedure to define the critical lines and points. We get here

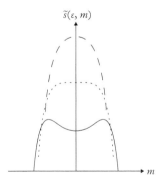

Figure 5.6 *Second-order phase transition. Evolution of the entropy $\tilde{s}(\varepsilon, m)$ for different values of the energy $\varepsilon = -0.1$ (dashed line), $\varepsilon_c \simeq -0.15$ (dotted), $\varepsilon = -0.2$ (solid) when $K = -0.25$. Note that the different curves have been vertically shifted for readability purposes. Reprinted from Campa et al. (2009), © 2009, with permission from Elsevier.*

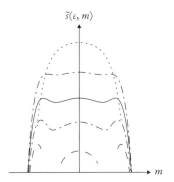

Figure 5.7 *First-order phase transition. Evolution of the entropy $\tilde{s}(\varepsilon, m)$ for different values of the energy $\varepsilon = -0.25$ (dotted), $\varepsilon = -0.305$ (dash-dotted), $\varepsilon_t = -0.3138$ (solid), $\varepsilon = -0.32$ (dash-triple dot), $\varepsilon = -0.33$ (dashed) when $K = -0.4$. The gaps present in the lower dashed curve are related to ergodicity breaking (see Section 5.1.4). Note that the different curves have been vertically shifted for readability purposes. Reprinted from Campa et al. (2009), © 2009, with permission from Elsevier.*

$$\tilde{s}(\varepsilon, m) = s_0(\varepsilon) + A_{mc}\, m^2 + B_{mc}\, m^4 + \mathcal{O}(m^4), \tag{5.16}$$

with the paramagnetic zero magnetization entropy

$$s_0(\varepsilon) = -\frac{\varepsilon}{K} \ln \frac{\varepsilon}{K} - \left(1 - \frac{\varepsilon}{K}\right) \ln \left(1 - \frac{\varepsilon}{K}\right) \tag{5.17}$$

and the expansion coefficients

$$A_{mc} = \frac{1}{2}\left[\frac{1}{K} \ln \frac{K-\varepsilon}{\varepsilon} - \frac{\varepsilon}{K-\varepsilon}\right] \tag{5.18}$$

$$B_{mc} = \frac{\varepsilon^3}{12(\varepsilon - K)^3} - \frac{K^2 + K}{4(\varepsilon - K)^2} + \frac{1}{8K\varepsilon}. \tag{5.19}$$

Using these expressions, it is straightforward to find the second-order phase transition line by requiring that $A_{mc} = 0$ ($B_{mc} < 0$), finding $\beta_c = \exp(-\beta_c K)$, which is the same equation found for the canonical ensemble. Again, as far as second-order phase transitions are concerned, the two ensembles are equivalent. The tricritical point is obtained by the condition $A_{mc} = B_{mc} = 0$, which leads to

$$K_{\mathrm{MTP}} \simeq -0.359 \qquad \text{and} \qquad \beta_{\mathrm{MTP}} \simeq 2.21, \tag{5.20}$$

which are close, but definitely different, from $K_{\mathrm{CTP}} \simeq -0.317$ and $\beta_{\mathrm{CTP}} = \sqrt{3}$ found in Eq. (5.12).

The microcanonical first-order phase transition line is obtained numerically by equating the entropies of the ferromagnetic and paramagnetic phases. At a given transition energy, there are two temperatures, thus giving a *temperature jump* as for the BEG and the generalized HMF model. In conclusion, similarly to the mean-field models discussed earlier, this model, that combines short- and long-range interactions, exhibits a region of negative specific heat when the phase transition is first order in the canonical ensemble. This striking feature is, therefore, not restricted to mean-field models.

5.1.4   Equilibrium dynamics: breaking of ergodicity

Now that we have shown in detail that the inequivalence of ensembles is a feature that might be preserved when adding a short-range component to a mean-field model, it is interesting to check whether other characteristic features of long-range interacting systems are preserved. We will show in this section that, as discussed in Mukamel *et al.* (2005), model (5.1) exhibits in particular a *breaking of ergodicity*. This phenomenon was first recognized in a different context in Borgonovi *et al.* (2004).

In order to reveal the dynamical consequences of this effect, we must define a *microcanonical dynamics*. An appropriate one is given by the Creutz algorithm (Creutz, 1983), which probes the microstates of the system with energy lower than or equal to the energy E. Indeed, there are two definitions of the microcanonical ensemble, which become equivalent in the thermodynamic limit. In the first one, only states contained on an energy shell $(E, E + dE)$ are counted and this is the definition we use throughout this book. In the second one, all the phase-space volume contained within the hypersurface of energy E is considered (Huang, 1987). Creutz algorithm is based on this second definition.

The algorithm is implemented by adding an auxiliary variable, called a *demon*, which has the following properties. We initiate the procedure with the demon at zero energy, while the system has an energy E. We then attempt to flip a spin. The move is accepted if it corresponds to an energy decrease and the excess energy is given to the demon. We then attempt to flip another spin. If the flip decreases the energy, it is accepted, whereas if it increases the total energy, it is accepted only if the needed energy can be withdrawn from the demon energy. Demon serves really as a bank with deposit and withdraw; however, the total energy of the 'bank' is always non-negative, a property which is clearly different from typical banks in the twenty-first century!

Creutz dynamics can be used to test the predictions on ergodicity breaking discussed in Section 4.2. Indeed, not all magnetization values are accessible in certain regions of the (K, ε) plane, which manifests itself in the gaps of the entropy curves shown in Fig. 5.7. This property is, in turn, a consequence of the non-convexity of the space of thermodynamic parameters (see Section 2.1.3)

To demonstrate ergodicity breaking, we present in Fig. 5.8 the time evolution of the magnetization in two cases: in the first case, where the whole magnetization interval is accessible, we clearly see switches between the metastable state $m^* = 0$ and two stable symmetric magnetized states. In contrast, in the second example, the metastable state belongs to an interval which is disconnected from the stable one (see the insets for the corresponding entropy curves). The dynamics maintains the system either in the stable or in the metastable interval, depending only on the initial condition. The system is unable to jump to the other interval even when the latter is more stable, having a larger entropy.

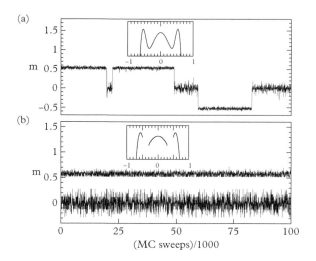

Figure 5.8 *Time evolution of the magnetization for $K = -0.4$ (a) in the ergodic region ($\varepsilon = -0.318$) and (b) in the non-ergodic region ($\varepsilon = -0.325$). Two different initial conditions are plotted simultaneously. The corresponding entropy curves are shown in the insets. Reprinted from Campa et al. (2009), © 2009, with permission from Elsevier.*

Similar results had been obtained using the Hamiltonian dynamics of the generalized HMF model; see Fig. 4.12.

5.2 α-Ising Model

Up to now, we have considered models with an infinite-range term in the Hamiltonian. This undoubtedly poses limits to the applicability of our analysis to realistic physical systems, where interactions decay with the distance. We begin in this section the study of several models of this kind which are direct generalizations of models presented earlier and represent a step forward in the study of long-range weakly decaying interactions. Some of the models are exactly solvable and reproduce features of the phenomenology observed for infinite-range systems (ensemble inequivalence, negative specific heat, ergodicity breaking, etc.).

Let us first consider the one-dimensional α-Ising Hamiltonian

$$H_N = \frac{\mathcal{J}}{N^{1-\alpha}} \sum_{i>j=1}^{N} \frac{1 - S_i S_j}{|i-j|^{\alpha}}, \tag{5.21}$$

where $\mathcal{J} > 0$ and the spins $S_i = \pm 1$ sit on a one-dimensional lattice with either free or periodic boundary conditions (in the latter case, $|i-j|$ is the minimal distance along the ring). The $N^{\alpha-1}$ prefactor is introduced in order to have an extensive energy as explained in Section 2.1.1. This model was first introduced by Dyson 1969 and studied for the 'integrable' case, $\alpha > 1$, in the canonical ensemble without the $N^{\alpha-1}$ prefactor. We show here that it is possible to obtain an exact microcanonical solution using large deviations theory when $0 \leq \alpha < 1$ (Barré, 2002; Barré *et al.*, 2002, 2005). Moreover, the study of this model gives also the opportunity to emphasize the important role played by boundary conditions when the interactions are long range.

We adopt the same scheme described in Chapter 3 to obtain the solution of model (5.21). The method can be generalized to lattices of higher dimension.

In the first step, the Hamiltonian H_N is rewritten in terms of global variables by introducing a *coarse-graining*. Let us divide the lattice in K boxes, each with $n = N/K$ sites, and let us introduce the average magnetization in each box m_k, $k = 1, \ldots, K$. In the limit $N \rightarrow \infty$, $K \rightarrow \infty$, $K/N \rightarrow 0$, magnetization becomes a continuous function $m(x)$ of the $[0, 1]$ interval. After a long but straightforward calculation, described in Barré *et al.* (2005), it is possible to express H_N as a functional of $m(x)$

$$H_N = NH[m(x)] + o(N), \tag{5.22}$$

where

$$H[m(x)] = \frac{\mathcal{J}}{2} \int_0^1 \mathrm{d}x \int_0^1 \mathrm{d}y \frac{1 - m(x)m(y)}{|x-y|^\alpha}. \tag{5.23}$$

In the second step, we evaluate the probability to get a given magnetization m_k in the kth box from all *a priori* equiprobable microscopic configurations. This probability obeys a local large deviation principle $P(m_k) \propto \exp[n\tilde{s}(m_k)]$, with

$$\tilde{s}(m_k) = -\frac{1 + m_k}{2} \ln \frac{1 + m_k}{2} - \frac{1 - m_k}{2} \ln \frac{1 - m_k}{2}. \tag{5.24}$$

Since microscopic random variables in different boxes are independent and no global constraint has been imposed, the probability of the full global variable $(m_1, \ldots, m_K)$ can be expressed in a factorized form as

$$P(m_1, m_2, \ldots, m_K) = \prod_{i=1}^{K} P(m_i) \simeq \prod_{i=1}^{K} e^{n\tilde{s}(m_i)} \tag{5.25}$$

$$= \exp\left[nK \sum_{i=1}^{K} \frac{\tilde{s}(m_i)}{K} \right] \tag{5.26}$$

$$\simeq \exp(N\bar{s}[m(x)]), \tag{5.27}$$

where $\bar{s}[m(x)] = \int_0^1 \tilde{s}(m(x)) \, \mathrm{d}x$ is the entropy functional associated with the global variable $m(x)$. Large deviation techniques rigorously justify these calculations (Boucher *et al.*, 1999), proving that entropy is proportional to N, also in the presence of long-range interactions. This result is independent of the specific model considered.

In the third step, we formulate the variational problem in the microcanonical ensemble to get the entropy

$$s(\varepsilon) = \sup_{m(x)} \left\{ \bar{s}[m(x)] \mid \varepsilon = H[m(x)] \right\}. \tag{5.28}$$

Let us remark that this optimization problem must be solved in a functional space. In general, this must be done numerically, taking into account boundary conditions. In the case of free boundary conditions, the only available solutions are numerical. An example of a maximal entropy magnetization profile obtained for free boundary conditions is shown in Fig. 5.9 for different values of α. The profile becomes more inhomogeneous when increasing α (for $\alpha = 0$, we recover the mean-field result with a homogeneous profile).

In the following, we will treat the periodic boundary conditions case, for which analytical results can be obtained. Both entropy and free energy can be obtained in analytical form for homogeneous magnetization profiles. As a matter of fact, in Campa *et al.* (2003), it was proven that for an entire class of systems, which includes the α-Ising model, the equilibrium states for periodic boundary conditions are states with homogeneous magnetization profiles. For an arbitrary magnetization profile, the variational problem (5.28), where $\bar{s}$ is defined in Eqs. (5.24), (5.26) and (5.27), leads to the self-consistency equation

$$\tanh^{-1}(m(x)) = \beta \mathcal{J} \int_0^1 \frac{m(y)}{|x-y|^\alpha} \, dy, \tag{5.29}$$

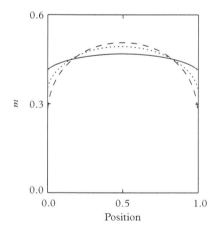

Figure 5.9 *Equilibrium magnetization profile for the α-Ising model with free boundary conditions at an energy density $\varepsilon = 0.1$ for $\alpha = 0.2$ (solid line), $\alpha = 0.5$ (dotted line) and $\alpha = 0.8$ (dashed line). Reprinted from Campa et al. (2009), © 2009, with permission from Elsevier.*

where β is a Lagrange multiplier, to be determined also from the value of the energy ε. Its value will then be the inverse temperature of the system. From Eq. (5.23), we see that fixing the magnetization profile implies fixing the energy; in turn, for homogeneous magnetization profiles, fixing the magnetization value implies fixing the energy and, consequently, the Lagrange multiplier β in Eq. (5.29). The relation between the energy and the homogeneous magnetization is obtained from Eq. (5.23), using the relations $\int_0^1 dx\,|x-y|^{-\alpha} = 2^\alpha/(1-\alpha)$, as $m = \pm\sqrt{1-\varepsilon/\varepsilon_{\max}}$, where $\varepsilon_{\max} = 2^\alpha \mathcal{J}/[2(1-\alpha)]$ is the upper bound for the energy of the system. The corresponding value of β is $\beta_c = 1/(2\varepsilon_c)$. Expressing the magnetization in terms of the energy allows us to derive the caloric curve, which is plotted in Fig. 5.10b (solid line). The limit temperature β_c (dotted line) is attained at zero magnetization, which is a boundary point.

In the canonical ensemble, we must solve the variational problem (3.15). This leads to exactly the same consistency equation (5.29), where the Lagrange multiplier is replaced by the fixed inverse temperature β. Solving this consistency equation for $\beta > 0$, one finds a zero magnetization for $\beta < \beta_c$ and a non vanishing one for $\beta > \beta_c$. We can also derive the canonical caloric curve, which is reported in Fig. 5.10b and superposes to the microcanonical caloric curve from $\beta = \infty$ down to β_c while it is represented by the dashed line for $\beta < \beta_c$. It follows that in the region $[0, \beta_c]$, the two ensembles are not equivalent. In this case, a single microcanonical state at $\varepsilon_{\max}$ corresponds to many canonical states with canonical inverse temperatures in the range $[0, \beta_c]$. Thus, in such a case, the canonical inverse temperature is not equal to the microcanonical one. In the microcanonical ensemble, the full high temperature region is absent and, therefore, no phase transition is present or, in other terms, the phase transition is at the boundary

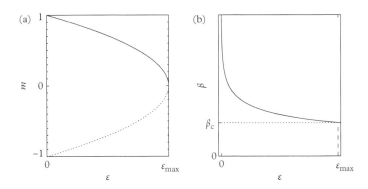

Figure 5.10 *(a) Equilibrium magnetization in the allowed energy range in the microcanonical ensemble for the α-Ising model with $\alpha = 0.5$; the negative branch is also reported with a dotted line. (b) Inverse temperature versus energy in the microcanonical ensemble (solid line). The canonical ensemble result superposes to the microcanonical one in the interval $[\beta_c, \infty)$ and is represented by a dashed line for $\beta \in [0, \beta_c]$. β_c is then the inverse critical temperature in the canonical ensemble. In the microcanonical ensemble, no phase transition is present. Reprinted from Campa et al. (2009), © 2009, with permission from Elsevier.*

of the accessible energy values. The entropy is always concave; hence no inequivalence can be present in the allowed energy range, apart from the boundaries. This situation is often called partial equivalence (Ellis, 1999; Touchette, 2008; Casetti and Kastner, 2007). Partial equivalence persists for all α values below 1, and is removed only for $\alpha = 1$ when $\varepsilon_{max} \to \infty$ and $\beta_c \to 0$: the phase transition is not present in both ensembles and the system is always in its magnetized phase. The main drawback of this analysis is the difficulty to obtain analytical solutions of Eq. (5.29) for non-constant magnetization profiles, which is the typical situation when boundary conditions are not periodic.

In Section 5.4, we will meet another model with weakly decaying long-range interactions.

5.3   XY Model with Long- and Short-Range Couplings

5.3.1   Introduction

In this section, we study yet another generalization of the HMF model, in which the Hamiltonian, besides the mean-field interaction, has a nearest neighbour interaction between rotators. The presence of such a term requires that we specify the properties of the lattice where the model is defined. As in Section 5.1, we study a one-dimensional lattice with a ring geometry; the first property allows the use of the transfer integral method, while the second property, equivalent to the introduction of periodic boundary conditions, is convenient for the calculations, but is irrelevant in the thermodynamic limit.

The model was introduced in Campa *et al.* (2006), and its Hamiltonian is given by

$$H_N = \sum_{i=1}^{N} \frac{p_i^2}{2} + \frac{1}{2N} \sum_{i,j=1}^{N} \left[1 - \cos\left(\theta_i - \theta_j\right)\right] - K \sum_{i=1}^{N} \cos\left(\theta_{i+1} - \theta_i\right), \qquad (5.30)$$

with $\theta_{N+1} = \theta_1$. The parameter K is the coupling constant of the nearest neighbour interaction. The similarities of this model with the Ising model studied earlier are clear: there is a mean-field interaction and a nearest neighbour interaction. However, there is also a kinetic energy term, which makes it possible to define a Hamiltonian dynamics.

As for the Ising model with short- and long-range interactions, it is possible to draw the main transition lines of the phase diagram of this model in the (K, T) plane. For $K = 0$, Hamiltonian (5.30) reduces to the HMF model, which has a second-order phase transition at $T_c = 0.5$ as shown in Chapter 4. Here also, comparing the energy per particle of the fully magnetized $m = 1$ state with that of the staggered antiferromagnetic $m = 0$ state makes it possible to locate the phase transition at $T = 0$. The energy density of the former is $\varepsilon = -K$, while that of the latter is $\varepsilon = 1/2 + K$. Consequently, also in this case the most interesting properties are found for $K < 0$ and we derive that the magnetized state is favoured for $K \geq -1/4$. We get, therefore, a phase diagram, which is very similar to that obtained for the Ising model with long-range and short-range interactions, and corresponds to the schematic picture sketched in Fig. 5.11.

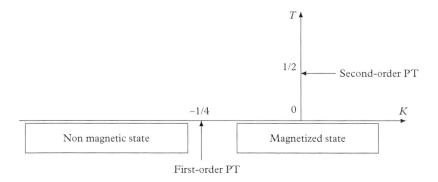

Figure 5.11 *Elementary features of the phase diagram of the short- plus long-range XY model.*

5.3.2 Solutions in the canonical and microcanonical ensembles

To solve the model, we follow the min–max procedure, described in Section 2.5. Once an expression for the canonical partition function has been obtained in the form of Eq. (2.65), we can not only compute the canonical free energy $\phi(\beta)$ by the minimization procedure defined in Eq. (2.66), but also obtain the microcanonical entropy $s(\varepsilon)$ by Eq. (2.70).

Let us first derive the canonical partition function. As for the HMF model, we use the Hubbard–Stratonovich transformation (4.8), in which the two-dimensional variable $\mathbf{y} = (y_1, y_2)$ is used. The model shares with the HMF model invariance under global rotations; therefore spontaneous magnetization is defined only in modulus, while there is a degeneracy with respect to its direction. We can exploit this fact to slightly simplify the computation; namely, we assume from the beginning that the saddle point value of the variable y_2 in Eq. (4.10) is 0. The spontaneous magnetization is in the direction of the y_1 axis, but this will not cause any loss of generality. We therefore obtain (denoting for convenience y_1 with x)

$$Z(\beta, N) = \frac{N\beta}{2\pi} \exp\left(-\frac{N\beta}{2}\right) \left(\frac{2\pi}{\beta}\right)^{N/2}$$
$$\times \int d\theta_1 \ldots d\theta_N dx \exp\left[-\frac{N\beta x^2}{2} + \beta x \sum_{i=1}^{N} \cos\theta_i + \beta K \sum_{i=1}^{N} \cos(\theta_{i+1} - \theta_i)\right].$$
$$(5.31)$$

The integral over the θ values is performed by applying the transfer integral method. For our one-dimensional geometry, this is given by

$$\int d\theta_1 \ldots d\theta_N \exp\left[\beta x \sum_{i=1}^{N} \cos\theta_i + \beta K \sum_{i=1}^{N} \cos(\theta_{i+1} - \theta_i)\right] = \sum_j \lambda_j^N(\beta x, \beta K), \quad (5.32)$$

where $\lambda_j(\gamma, \sigma)$ is the jth eigenvalue of the symmetric integral operator

$$(T\psi)(\theta) = \int d\alpha \exp\left[\frac{1}{2}\gamma(\cos\theta + \cos\alpha) + \sigma\cos(\theta - \alpha)\right]\psi(\alpha). \qquad (5.33)$$

In the thermodynamic limit, only the largest eigenvalue λ_{max} will contribute to the partition function. This eigenvalue can be computed numerically by a suitable discretization of the integral operator (5.33). The rescaled free energy is therefore

$$\phi(\beta) = \beta f(\beta) = \inf_x \tilde{\phi}(\beta, x), \qquad (5.34)$$

with

$$\tilde{\phi}(\beta, x) = -\frac{1}{2}\ln\frac{2\pi}{\beta} + \frac{\beta}{2}(1 + x^2) - \ln\lambda_{max}(\beta x, \beta K), \qquad (5.35)$$

where we have not explicitly written the K dependence of $\tilde{\phi}(\beta, x)$ and $\phi(\beta)$.

Accordingly to the min–max procedure, the microcanonical entropy $s(\varepsilon)$ is given by

$$s(\varepsilon) = \sup_x \tilde{s}(\varepsilon, x), \qquad (5.36)$$

with

$$\tilde{s}(\varepsilon, x) = \inf_\beta\left[\beta\varepsilon - \tilde{\phi}(\beta, x)\right] \qquad (5.37)$$

$$= \inf_\beta\left[\beta\varepsilon + \frac{1}{2}\ln\frac{2\pi}{\beta} - \frac{\beta}{2}(1 + x^2) + \ln\lambda_{max}(\beta x, \beta K)\right]. \qquad (5.38)$$

Again the dependence of $\tilde{s}(\varepsilon, x)$ and $s(\varepsilon)$ on K has not been explicitly written. Expressions (5.35) and (5.38) are used to study canonical and microcanonical thermodynamic phase diagrams, respectively, and then to check whether there is ensemble inequivalence. As for the Ising case, in order to obtain the critical lines on the (T, K) plane, we must expand $\tilde{\phi}(\beta, x)$ and $\tilde{s}(\varepsilon, x)$ in powers of x. The power series expansion of $\lambda_{max}(\beta x, \beta K)$ can be obtained explicitly as a function of modified Bessel functions. We will avoid here the details of the calculation leading to the final result.

Taking into account that our system is invariant under the symmetry $\theta \to -\theta$, the expansion of $\tilde{\phi}(\beta, x)$ will have only even powers. Let us write explicitly the K dependence of the expansion coefficients

$$\tilde{\phi}(\beta, x) = \phi_0(\beta, K) + \phi_1(\beta, K)x^2 + \phi_2(\beta, K)x^4 + O(x^6). \qquad (5.39)$$

The canonical second-order transition line in the (T, K) plane is determined by

$$\phi_1(\beta, K) = 0, \quad \text{with } \phi_2(\beta, K) > 0. \qquad (5.40)$$

Inserting this expansion in Eq. (5.38) and minimizing with respect to β, we obtain

$$\varepsilon = \phi_0'(\beta, K) + \phi_1'(\beta, K) x^2 + \phi_2'(\beta, K) x^4 + O(x^6), \qquad (5.41)$$

where the prime denotes derivation with respect to β. This equation gives an expansion of the form

$$\beta(\varepsilon, K, x) = \beta_0(\varepsilon, K) + \beta_1(\varepsilon, K) x^2 + \beta_2(\varepsilon, K) x^4 + O(x^6). \qquad (5.42)$$

Using this expression, we obtain the following expansion for the energy

$$\tilde{s}(\varepsilon, x) = s_0(\varepsilon, K) + s_1(\varepsilon, K) x^2 + s_2(\varepsilon, K) x^4 + O(x^6), \qquad (5.43)$$

where

$$s_0(\varepsilon, K) = \beta_0(\varepsilon, K)\varepsilon - \phi_0(\beta_0(\varepsilon, K)) \qquad (5.44)$$
$$s_1(\varepsilon, K) = \beta_1(\varepsilon, K)\varepsilon - \phi_0'(\beta_0(\varepsilon, K))\beta_1(\varepsilon, K) - \phi_1(\beta_0(\varepsilon, K)) \qquad (5.45)$$
$$s_2(\varepsilon, K) = \beta_2(\varepsilon, K)\varepsilon - \frac{1}{2}\phi_0''(\beta_0(\varepsilon, K))\beta_1^2(\varepsilon, K) - \phi_0'(\beta_0(\varepsilon, K))\beta_2(\varepsilon, K)$$
$$- \phi_1'(\beta_0(\varepsilon, K))\beta_1(\varepsilon, K) - \phi_2(\beta_0(\varepsilon, K)). \qquad (5.46)$$

The microcanonical second-order transition line is determined by

$$s_1(\varepsilon, K) = 0, \quad \text{with} \quad s_2(\beta, K) < 0. \qquad (5.47)$$

Now we can use the following equalities, which are obtained by inserting (5.42) back into (5.41)

$$\phi_0'(\beta_0(\varepsilon, K)) = \varepsilon \qquad (5.48)$$
$$\phi_1'(\beta_0(\varepsilon, K)) = -\phi_0''(\beta_0(\varepsilon, K))\,\beta_1(\varepsilon, K). \qquad (5.49)$$

Equation (5.48) implies that $\beta_0(\varepsilon, K)$ is the temperature on the second-order transition line. We obtain

$$\tilde{s}(\varepsilon, x) = \beta_0(\varepsilon, K)\,\varepsilon - \phi_0(\beta_0(\varepsilon, K)) - \phi_1(\beta_0(\varepsilon, K))\,x^2$$
$$- \left[-\frac{1}{2}\phi_0''(\beta_0(\varepsilon, K))\,\beta_1^2(\varepsilon, K) + \phi_2(\beta_0(\varepsilon, K)) \right] x^4 + O(x^6). \qquad (5.50)$$

Therefore, the microcanonical second-order transition line is determined by

$$\phi_1(\beta_0(\varepsilon, K)) = 0, \quad \text{with} \quad \frac{1}{2}\phi_0''(\beta_0(\varepsilon, K))\,\beta_1^2(\varepsilon, K) - \phi_2(\beta_0(\varepsilon, K)) < 0. \qquad (5.51)$$

The concavity of $\phi(\beta)$ implies that, again on the critical line, $\phi_0''(\beta_0(\varepsilon, K)) < 0$. This property is valid in models where parity with respect to x implies that $\tilde{\phi}_{\beta x} = 0$ for $x = 0$. Comparing Eqs. (5.40) and (5.51), and taking into account the last observation, we see that, as we expected from the general discussion about ensemble equivalence at the end of Section 2.2, the points belonging to the canonical second-order transition line also belong to the microcanonical one. The microcanonical line, however, goes beyond and includes more points, as we have already seen in several models, since by continuity the inequality in Eq. (5.51) is satisfied also for some points where $\phi_2(\beta_0(\varepsilon, K)) < 0$.

The study of the canonical and microcanonical critical lines involves the analytic determination of $\lambda_{\max}$, which, in the limit $x \to 0$, can be performed using perturbation theory for Hermitian operators as shown by Lori (2008). For both ensembles, we get

$$\frac{I_1(\beta K)}{I_0(\beta K)} = \frac{2 - \beta}{2 + \beta}, \tag{5.52}$$

where I_0 and I_1 are modified Bessel functions. The canonical tricritical point turns out to be

$$K_{\text{CTP}} \simeq -1.705 \quad \text{and} \quad T_{\text{CTP}} \simeq 0.267, \tag{5.53}$$

while microcanonical tricritical point is at

$$K_{\text{MTP}} \simeq -0.182 \quad \text{and} \quad T_{\text{MTP}} \simeq 0.234. \tag{5.54}$$

In Fig. 5.12, we plot the relevant part of the phase diagram on the (K, T) plane. As commented earlier, it is shown how the microcanonical critical line (ending at the microcanonical tricritical point, MTP) extends beyond the canonical critical line, which ends at the CTP.

5.3.3   Ergodicity breaking

We here prove that the XY model (5.30) with short- and long-range interactions shows ergodicity breaking, similarly to the Ising long- plus short-range model (5.1). Let us first study separately the bounds of the short-range term

$$\frac{1}{N} \sum_{i=1}^{N} \cos(\theta_{i+1} - \theta_i). \tag{5.55}$$

The study of maxima and minima of this term can be reduced to those configurations where the spins are aligned parallel or antiparallel to a given direction. There is no loss of generality, due to rotational symmetry, to choose this direction as the x axis. Therefore, only the x component of m is non-zero and in the following we will restrict to $m = m_x \geq 0$. It turns out that, in the thermodynamic limit, both extrema of Eq. (5.55) for a

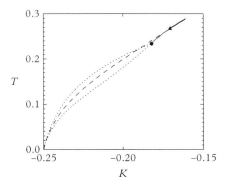

Figure 5.12 *Canonical and microcanonical (K, T) phase diagram of the one-dimensional XY model with mean-field and nearest neighbour interactions. The canonical critical line (bold solid line) ends at the tricritical point CTP indicated by a triangle; then the transition becomes first order (dashed line). The microcanonical second-order transition line coincides with the canonical one at large K (bold solid line). It continues at lower K (light solid line) down to the tricritical point MTP, indicated by a filled circle; then the transition becomes first order, with a branching of the transition line (dotted lines), giving the two extremes of the temperature jump. Reprinted from Campa* et al. *(2009), © 2009, with permission from Elsevier.*

given value of m are attained with a fraction $(1+m)/2$ of rotators parallel to the x axis and a fraction $(1-m)/2$ antiparallel to the x axis. The maximum is 1 and is achieved when the parallel rotators and the antiparallel rotators are grouped in two separated blocks. The minimum $2m-1$ is instead attained when the antiparallel rotators are all isolated, which is possible since we are considering $m \geq 0$.

Now we must to distinguish the two cases $K > 0$ and $K < 0$. In the case of positive K, the minimum of this contribution to the energy density is $-K$ (thus actually independent of m), while for negative K, it is $-K(2m-1)$. We obtain the minimum of the total potential energy per particle just by adding $(1 - m^2)/2$. This is actually the minimum of the total energy per particle ε, since the kinetic energy is positive definite. We finally obtain that the minimum of the energy per particle is

$$\varepsilon_{\min}^{K>0}(m) = \frac{1 - m^2}{2} - K \tag{5.56}$$

for $K > 0$, and

$$\varepsilon_{\min}^{K<0}(m) = \frac{1}{2}(1 - m^2) - K(2m - 1) \tag{5.57}$$

for $K < 0$. Plotting the two functions $\varepsilon_{\min}^{K>0}$ and $\varepsilon_{\min}^{K<0}$, we find the accessible m values for a given energy ε. The functions are plotted in Fig. 5.13 for two representative values: $K = 0.2$ and $K = -0.2$. For positive K, we see that below the energy $-K + 1/2$ the

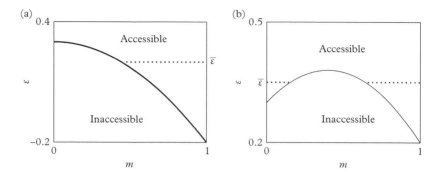

Figure 5.13 *Accessible regions in the (ε, m) plane for the one-dimensional XY model (5.30) with both mean-field and nearest neighbour interactions. Solid curves represent the function $\varepsilon_{\min}^{K>0}(m)$ given by Eq. (5.56) (in a) and $\varepsilon_{\min}^{K<0}(m)$ given by Eq. (5.57) (in b). The accessible and inaccessible regions in parameter space are indicated. (a) $K = 0.2$: no breaking of ergodicity is present since at fixed energy $\bar{\varepsilon}$ the accessible interval of m is connected (dotted line); (b) $K = -0.2$: there is breaking of ergodicity for values of ε between 0.30 and 0.38. For instance, for $\bar{\varepsilon} = 0.35$ there are two disconnected intervals of accessible magnetizations (dotted lines). Reprinted from Campa* et al. *(2009), © 2009, with permission from Elsevier.*

system cannot have all magnetizations, and values of m near 0 are inaccessible. We note, however, that the region of accessible values of m, for any energy, is connected; therefore there is no ergodicity breaking. Nevertheless, we remark that the accessible region in the (ε, m) plane is not convex, as it can happen only for long-range interactions. For negative K, we note, instead, that for a given energy interval ($K + 1/2 < \varepsilon < 2K^2 + K + 1/2$ for $K > -1/4$ and $-K < \varepsilon < 2K^2 + K + 1/2$ for $K < -1/4$), the attainable values of m are separated in two disconnected regions. Therefore for $K < 0$ we have ergodicity breaking.

5.4 α-HMF Model

Let us consider now another a generalization of the HMF model (4.1) which was originally proposed in Anteneodo and Tsallis (1997), Tamarit and Anteneodo (2000). The interaction between two rotators has the same form as in the original model, but now the coupling constant is a weakly decaying function of the distance between the two lattice sites where the rotators respectively sit. Note that we consider the lattice in general dimension d. In Campa *et al.* (2003), the generalized case of n-vector spin models was analysed, including the HMF model (for $n = 2$) and the Ising spins (for $n = 1$). Only periodic boundary conditions have been considered, as we will do here.

The Hamiltonian of the α-HMF model is

$$H_N = \sum_{i=1}^{N} \frac{p_i^2}{2} + \frac{1}{2\tilde{N}} \sum_{i,j=1}^{N} \frac{1 - \cos(\theta_i - \theta_j)}{r_{ij}^{\alpha}}, \tag{5.58}$$

where r_{ij} is the minimal distance between lattice sites i and j (minimal distance convention under periodic boundary conditions). We consider here cases for which the exponent α, determining the decay with distance of the coupling constant between rotators, is smaller than the spatial dimension d, so that the system is long range. The $\tilde{N}$ prefactor is introduced in order to have an extensive energy as explained in Section 2.1.1. We show later that, in any spatial dimension d and for each $\alpha < d$, the thermodynamics of this model is the same as that of the HMF model; in particular, there is a second-order phase transition at the inverse temperature $\beta = 2$.

Let us introduce the matrix of the coupling constants $\mathcal{J}_{ij} = 1/r_{ij}^{\alpha}$. Since the diagonal elements of this matrix are not defined, we assign to them the common arbitrary value b. The prefactor is

$$\tilde{N} = \sum_{j=1}^{N} \mathcal{J}_{ij}, \tag{5.59}$$

which does not depend on i, because we adopt the minimal distance convention for r_{ij}. It is clear that in the thermodynamic limit $N \to \infty$, the prefactor $\tilde{N}$ behaves as $N^{1-\alpha/d}$. The arbitrary parameter b, which appears only in $\tilde{N}$, becomes negligible in the thermodynamic limit. Its introduction is, however, useful for carrying out the computation, as it will become clear.

The most relevant steps of the solution in the canonical ensemble (we refer the reader to Campa *et al.* (2003) for details) can be obtained starting from the partition function

$$Z(\beta, N) = \left(\frac{2\pi}{\beta}\right)^{N/2} \int d\theta_1 \dots d\theta_N \, \exp\left(-\frac{\beta}{2\tilde{N}} \sum_{i,j=1}^{N} \mathcal{J}_{ij}\left[1 - \cos(\theta_i - \theta_j)\right]\right) \tag{5.60}$$

$$= \left(\frac{2\pi}{\beta}\right)^{N/2} \exp\left(-\frac{N\beta}{2}\right) \int d\theta_1 \dots d\theta_N$$

$$\times \exp\left(\sum_{i,j=1}^{N} \left[\cos\theta_i R_{ij} \cos\theta_j + \sin\theta_i R_{ij} \sin\theta_j\right]\right), \tag{5.61}$$

where we have introduced the matrix $R_{ij} = \beta \mathcal{J}_{ij}/(2\tilde{N})$ and used the property $\sum_{ij} \mathcal{J}_{ij} = N\tilde{N}$. In order to use the Hubbard–Stratonovich transformation, we diagonalize the expression in the second exponential. Denoting by R the symmetric matrix with elements R_{ij}, and by V the unitary matrix that reduces R to its diagonal form $D = VRV^T$, we get

$$\sum_{i,j=1}^{N} \cos\theta_i R_{ij} \cos\theta_j = \sum_{i=1}^{N} R_i a_i^2, \tag{5.62}$$

where the quantities R_i are the real eigenvalues of the matrix R, and

$$a_i = \sum_{j=1}^{N} V_{ij} \cos\theta_j. \tag{5.63}$$

An analogous expression holds for the sum with the sines in Eq. (5.61). In order to express $\exp(R_i a_i^2)$ using the Hubbard–Stratonovich transformation, the eigenvalues R_i must be positive. By choosing a sufficiently large value of b, but in any case of order 1, all eigenvalues R_i are positive. We can therefore write

$$\exp\left(\sum_{i=1}^{N} R_i a_i^2\right) = \frac{1}{\sqrt{(4\pi)^N \det R}} \int dy_1 \ldots dy_N \exp\left[\sum_{i=1}^{N}\left(-\frac{y_i^2}{4R_i} + a_i y_i\right)\right] \quad (5.64)$$

$$= \frac{1}{\sqrt{(4\pi)^N \det R}} \int dz_1 \ldots dz_N$$

$$\times \exp\left[-\frac{1}{4}\sum_{i,j} z_i \left(R^{-1}\right)_{ij} z_j + \sum_{i=1}^{N} z_i \cos\theta_i\right]. \quad (5.65)$$

In this N-dimensional Hubbard–Stratonovich transformation, we have used that $\det R = \prod_{i=1}^{N} R_i$, we have introduced the unitary change of integration variables defined by $y_i = \sum_{j=1}^{N} V_{ij} z_j$ and we have indicated with R^{-1} the inverse matrix of R. Using the analogous expression for the term with the sines, we arrive, after performing the integration over the angles θ, to the following expression for the partition function:

$$Z(\beta, N) = \left(\frac{2\pi}{\beta}\right)^{N/2} \frac{\exp(-N\beta/2)}{(4\pi)^N \det R} \int dz_1 \ldots dz_N dz_1' \ldots dz_N'$$

$$\times \exp\left[\frac{1}{4}\sum_{i,j} z_i \left(R^{-1}\right)_{ij} z_j - \frac{1}{4}\sum_{i,j} z_i' \left(R^{-1}\right)_{ij} z_j' + \sum_{i=1}^{N} \ln I_0\left(\sqrt{z_i^2 + z_i'^2}\right)\right]. \quad (5.66)$$

This integral can be computed using the saddle point method, since most eigenvalues R_i vanish in the thermodynamic limit (Campa *et al.*, 2003). Introducing $\rho_i = (z_i^2 + z_i'^2)^{1/2}$, the stationary points are solutions of the consistency equations

$$\frac{1}{2}\sum_{j=1}^{N} \left(R^{-1}\right)_{ij} z_j = \frac{I_1(\rho_i)}{I_0(\rho_i)} \frac{z_i}{\rho_i} \quad (5.67)$$

$$\frac{1}{2}\sum_{j=1}^{N} \left(R^{-1}\right)_{ij} z_j' = \frac{I_1(\rho_i)}{I_0(\rho_i)} \frac{z_i'}{\rho_i}, \quad (5.68)$$

which, after inversion with respect to z_i, z_i', become

$$z_i = 2\sum_{j=1}^{N} R_{ij} \frac{I_1(\rho_j)}{I_0(\rho_j)} \frac{z_j}{\rho_j} \quad (5.69)$$

$$z_i' = 2 \sum_{j=1}^{N} R_{ij} \frac{I_1\left(\rho_j\right)}{I_0\left(\rho_j\right)} \frac{z_j'}{\rho_j}. \tag{5.70}$$

Let us first look for homogeneous solutions, i.e., for solutions in z_i, z_i' that do not depend on i. The previous system reduces to the following pair of equations for $z \equiv z_i$ and $z' \equiv z_i'$,

$$z = \beta \frac{I_1\left(\rho\right)}{I_0\left(\rho\right)} \frac{z}{\rho} \quad \text{and} \quad z' = \beta \frac{I_1\left(\rho\right)}{I_0\left(\rho\right)} \frac{z'}{\rho}, \tag{5.71}$$

where we have used that $2 \sum_{j=1}^{N} R_{ij} = \beta$ for each i. By expressing $z = \rho \cos \gamma$ and $z' = \rho \sin \gamma$, we see that these two equations determine only the modulus $\rho \equiv (z^2 + z'^2)^{1/2}$, leaving the angle γ undetermined. They thus reduce to a single equation for the modulus

$$\rho = \beta \frac{I_1\left(\rho\right)}{I_0\left(\rho\right)}. \tag{5.72}$$

This equation, after the change of variable $\rho = \beta m$, coincides with Eq. (4.13) of the HMF model.

In order for this homogeneous stationary point to be the relevant one for the evaluation of the integral in Eq. (5.66), it is necessary that the exponential has an absolute maximum at this stationary point. We can easily show that the homogeneous stationary point is a local maximum (Campa *et al.*, 2003). Here, we will directly prove that it is also a global maximum with respect to any other possible inhomogeneous stationary point.

Let us then go back to Eqs. (5.69) and (5.70) and suppose that they have inhomogeneous solution, considering as a first case a solution with different values for the moduli $\rho_i \equiv (z_i^2 + z_i'^2)^{1/2}$. If k is the site with the largest modulus ρ_k, let us choose the x axis of the XY rotators such that $z_k = \rho_k$ and $z_k' = 0$. Then we can write

$$z_k = \rho_k = 2 \sum_{j=1}^{N} R_{ij} \frac{I_1\left(\rho_j\right)}{I_0\left(\rho_j\right)} \frac{z_j}{\rho_j} \tag{5.73}$$

$$\leq 2 \sum_{j=1}^{N} R_{ij} \frac{I_1\left(\rho_j\right)}{I_0\left(\rho_j\right)} \tag{5.74}$$

$$< 2 \sum_{j=1}^{N} R_{ij} \frac{I_1\left(\rho_k\right)}{I_0\left(\rho_k\right)} = \beta \frac{I_1\left(\rho_k\right)}{I_0\left(\rho_k\right)}, \tag{5.75}$$

where use has been made of the monotonicity of the function I_1/I_0. We emphasize that the last inequality is strict since, by hypothesis, not all moduli ρ_i are equal. From the properties of the function I_1/I_0, it follows from the last inequality that also the largest modulus ρ_k of this inhomogeneous stationary point is smaller than that of the homogeneous solution. We get the same conclusion for a possible stationary point with all equal moduli but with different directions; in that case inequality (5.75) becomes an equality, but inequality (5.75) is strict. The proof is completed by showing that the exponent in Eq. (5.66), if computed at a stationary point, has a positive derivative with respect to the moduli ρ_i of the stationary point. In fact, it can be easily shown that at a stationary point the exponent in Eq. (5.66), let us call it A, can be written as a function of the moduli ρ_i of the stationary point in the following way

$$A = \sum_{i=1}^{N} \left[-\frac{1}{2} \rho_i \frac{I_1(\rho_i)}{I_0(\rho_i)} + \ln I_0(\rho_i) \right]. \tag{5.76}$$

Its derivative with respect to ρ_i is

$$\frac{\partial A}{\partial \rho_i} = \frac{1}{2} \left[\frac{I_1(\rho_i)}{I_0(\rho_i)} - \rho_i \frac{\partial}{\partial \rho_i} \frac{I_1(\rho_i)}{I_0(\rho_i)} \right]. \tag{5.77}$$

The concavity of the function I_1/I_0, for positive values of the argument, assures that the right-hand side is positive. This concludes the proof.

We conclude this section by noting that for each dimension d and for each decaying exponent $\alpha < d$, the free energy of the α-HMF model with periodic boundary conditions is the same as that of the HMF model (see Mori (2013) for more details).

5.5 Dipolar Interactions in a Ferromagnet

In this last section, we considered dipolar interactions. They are marginal as far as the long-range character of the interaction is concerned. Indeed, in this case, the exponent governing the decay of the interaction potential is $\alpha = d = 3$. Actually, systems with this type of interaction will be treated in greater details in Chapter 15, while here, we will consider particular cases, where simplifying assumptions can be made. In Chapter 15, it will be shown that despite the long-range character (although marginal) of the interaction, its peculiar directionality properties cause a system of dipoles to be extensive. This is valid for both electric and magnetic dipoles, since the structure of the interaction is the same in both cases. However, the treatment will be made having in mind the more common magnetic systems.

Therefore, we here consider magnetic systems, i.e. systems of classical spins residing on the sites of a lattice, that are subject to magnetic dipolar interactions. In particular, we will focus on elongated samples, where it will be possible to introduce some simplifications on the spin ordering. In the second part of the section, we will argue about the possibility of observing directly, in these samples, the manifestation of long-range interactions.

Let us begin with the interaction energy between two magnetic dipoles $\mathbf{m}_1$ and $\mathbf{m}_2$

$$\frac{\mu_0}{4\pi} \frac{\mathbf{m}_1 \cdot \mathbf{m}_2 |\mathbf{r} - \mathbf{r}'|^2 - 3 \left[(\mathbf{r} - \mathbf{r}') \cdot \mathbf{m}_1 \right] \left[(\mathbf{r} - \mathbf{r}') \cdot \mathbf{m}_2 \right]}{|\mathbf{r} - \mathbf{r}'|^5}, \tag{5.78}$$

which will be our starting point also in the detailed treatment of Chapter 15. As mentioned earlier, we consider the case in which the magnetic moments arise from spins. The relation between the magnetic moment and the spin is given by

$$\mathbf{m} = 2\mu_B \mathbf{S}, \tag{5.79}$$

where μ_B is the Bohr magneton, and $\mathbf{S}$ the spin in units of $\hbar$. Then, the Hamiltonian of a systems of spins is

$$H = \left(\frac{\mu_0}{4\pi}\right) 2\mu_B^2 \sum_{i \neq j} \frac{1}{r_{ij}^3} \left(\mathbf{S}_i \cdot \mathbf{S}_j - 3\frac{(\mathbf{S}_i \cdot \mathbf{r}_{ij})(\mathbf{S}_j \cdot \mathbf{r}_{ij})}{r_{ij}^2}\right), \tag{5.80}$$

where r_{ij} is the modulus of the vector $\mathbf{r}_{ij}$ joining the sites of the spins $\mathbf{S}_i$ and $\mathbf{S}_j$. Now we shift for convenience to atomic units, in which $\mu_B = 1/2$. Then, the last Hamiltonian is written as

$$H = \frac{1}{2} \sum_{i \neq j} \frac{1}{r_{ij}^3} \left(\mathbf{S}_i \cdot \mathbf{S}_j - 3\frac{(\mathbf{S}_i \cdot \mathbf{r}_{ij})(\mathbf{S}_j \cdot \mathbf{r}_{ij})}{r_{ij}^2}\right). \tag{5.81}$$

Note that we have also redefined the Hamiltonian, $H \rightarrow H/\alpha^2$, where α is the fine structure constant.

It is often the case that computations based on this Hamiltonian are performed with a procedure in which the sum is restricted to the nearest neighbours, while the contribution of the other terms are evaluated with the help of macroscopic magnetostatics, i.e. with a continuous approximation (see, for example, Landau and Lifshitz, 1984, or Akhiezer *et al.*, 1968). Details on this procedure will be given in Chapter 15; here it suffices to anticipate that the terms approximated by continuous magnetostatics give rise to a contribution proportional to the components of the total magnetization of the sample, with shape-dependent coefficients. In the remaining of this section, we will adopt this standard approximation, together with other simplified assumptions that are possible for spins in elongated samples.

5.5.1 Simplified Hamiltonian in elongated ferromagnets

We consider elongated samples in the direction of the z axis and we assume, in the analytical calculations only, that the transversal components of the spins are zero on average; therefore we will keep only the longitudinal components S_i^z in the Hamiltonian (5.81), which then reduces to

$$H = \frac{1}{2} \sum_{i \neq j} \left(\frac{S_i^z S_j^z}{r_{ij}^3} - 3\frac{S_i^z S_j^z z_{ij}^2}{r_{ij}^5}\right). \tag{5.82}$$

As said earlier, the contribution to Eq. (5.82) not coming from nearest neighbours is evaluated by a continuum approximation. For the nearest neighbours, we use the exact expression, and for this it is crucial to define the geometry of the lattice.

5.5.1.1   Simple cubic lattice

We will consider the extreme case of a sample made of a single lattice cell in the xy plane. The sample will therefore have a needle-like structure elongated in the z direction. Let us start first with a simple cubic lattice with four spins in a basis xy layer, transversal to the z axis of the sample: see graphs (a) and (d) in Fig. 5.14. We will make a further approximation, for this and the lattices considered later. Since the coupling between neighbouring spins in different layers is ferromagnetic, and since a ferromagnetic coupling arises also from the continuous approximation, we will consider the former included in the latter, and thus we will restrict the nearest neighbour coupling only between spins belonging to the same layer, which instead is antiferromagnetic. Thus our system of spins could be described by the Hamiltonian

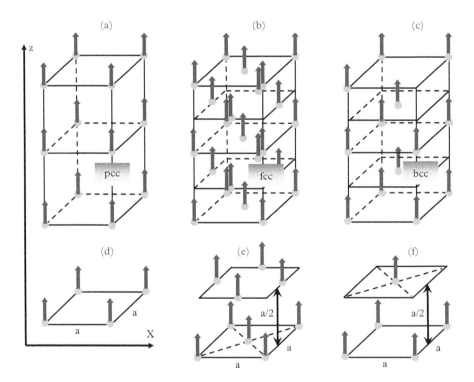

Figure 5.14 *Schematic drawings of parallelepiped samples with different geometries. (a) A simple cubic lattice, while (b) and (c) present, respectively, a face-centred cubic (fcc) and body-centred cubic (bcc) lattices. (d–f) The corresponding basic cell of the different lattices.*

$$\mathcal{H} = -\frac{K}{2} \sum_{i,I} \left(S_{i,I}^z S_{i+1,I}^z - 1 \right) - \frac{\mathcal{J}}{2n} \left(\sum_{i,I} S_{i,I}^z \right)^2, \tag{5.83}$$

where n is the total number of spins in the lattice while i is the index of spins inside the specific layer I. Making the assumption that, due to ferromagnetic coupling between nearest neighbours in different layers, $S_{i,I}^z$ is roughly independent of I, we see that this Hamiltonian recovers model (5.1), with $K/\mathcal{J}$ playing the role of K in that model.

The value of K depends on the lattice structure. On the other hand, the value of the mean field constant $\mathcal{J}$ depends on the volume per spin v_0 and on the sample geometry; its derivation for general geometries will be given in Chapter 15. Here, we only mention that it can be cast in the form $\mathcal{J} = 4\pi(1 - 3N_z)/(3v_0)$, where N_z is the so-called demagnetizing factor

$$N_z = \frac{1 - \ell^2}{2\ell^3} \left(\ln \frac{1 + \ell}{1 - \ell} - 2\ell \right), \qquad \text{in which} \qquad \ell = \sqrt{1 - \frac{a^2}{L^2}}, \tag{5.84}$$

as shown by Akhiezer *et al.* (1968) for a prolate ellipsoid of revolution; here a is the edge length of the cubic lattice cell, while L is the length of the sample in the z direction. We can easily check that N_z is equal to 1/3 if the sample length L coincides with the basis dimension a, while it tends to 0 for a very elongated needle. It means that the mean field coefficient $\mathcal{J}$ is controlled by changing the sample length along the z direction.

To draw a careful comparison with numerical experiments, we must perform simulations based on the complete Hamiltonian (5.81) (and therefore without neglecting the transversal components of the spins) together with the associated torque equation, given in general by

$$\frac{d\mathbf{S_i}}{dt} = \mathbf{S_i} \times \mathbf{H}_i, \tag{5.85}$$

where $\mathbf{H}_i = -\partial H/\partial \mathbf{S}_i$ is the effective magnetic field acting on the spin $\mathbf{S}_i$. However, a numerical simulation with a lattice would be unable to reproduce exactly a needle, and we should consider the averaged value of the demagnetization factor of the simulated parallelepiped shaped sample via the integral (Landau and Lifshitz, 1984)

$$\bar{N}_z = -\frac{1}{4\pi V} \int_V d\vec{r} \int_V d\vec{r}_1 \frac{\partial^2}{\partial z^2} \left(\frac{1}{|\vec{r} - \vec{r}_1|} \right), \tag{5.86}$$

where V is the volume of the sample. The dependence on the length of the samples of the demagnetization factor in both cases (5.84) and (5.86) is presented in Fig. 5.15.

As it directly follows from Section 5.1, the magnetized states appear if $K/\mathcal{J} > -1/2$. In the case of a simple cubic lattice with the simplest basis of four spins, we have $K = -2/a^3$. Thus in the present case of a simple basis of four spins and a large aspect ratio $L/a \gg 1$, this condition is satisfied since we get $K/\mathcal{J} = -3/(2\pi) > -1/2$. However, the above

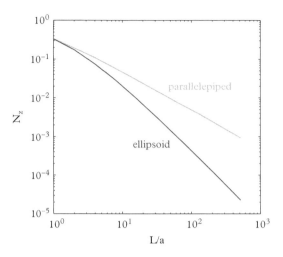

Figure 5.15 *Demagnetizing factor N_z as a function of the aspect ratio L/a. The lower curve corresponds to a needle approximated by an ellipsoid via Eq. (5.84), while the upper curve is computed for the paralleliped shape using Eq. (5.86).*

condition is not met if we consider a larger basis. Indeed, for wide lattices, each spin has four neighbours (instead of two in the four-spin basis system) and thus the effective coupling constant reads $K = -4/a^3$: consequently, even for an infinitely large aspect ratio $L/a \gg 1$, the value $K/\mathcal{J}$ is around (-1) and no magnetized states could be realized in a simple cubic lattice. Numerical simulations for the case of a 3×3-type basis do confirm this prediction (Miloshevich *et al.*, 2013).

5.5.1.2   Face-centred cubic lattice

The situation is slightly more complicated if we consider a face-centred cubic lattice: the simplest realization of this lattice is when the two neighbouring layers have 4 and 5 spins, as shown in Figs. 5.14b and 5.14e. In this case, the spins in the lattice corners are coupled only with one nearest neighbour within the transversal layer, while the spin in the centre is coupled with 4 neighbouring spins. In this situation, we must average the coupling constants to get the effective coupling $K = -32\sqrt{2}/(9a^3)$. In order to compute $\mathcal{J}$, we should estimate the volume per spin. In large face-centred cubic samples $v_0 = a^3/4$. At variance, in case of 4- and 5-spin basis, v_0 is slightly different, and we get $v_0 = (9/10)(a^3/4) = 5a^3/18$. From these considerations, it follows that for large sample lengths, we have $\mathcal{J} = 24\pi/(5a^3)$, which leads to $K/\mathcal{J} = -20\sqrt{2}/(27\pi) \simeq -0.3$, a value for which magnetized states, magnetization flips and temperature jumps do appear.

For wider samples in face-centred cubic lattices, however, the magnetization condition $K/\mathcal{J} > -1/2$ is not satisfied. Indeed, in such cases, each spin interacts with 4 neighbours inside the transversal layers and thus the effective coupling constant is $K = -8\sqrt{2}/a^3$, while $\mathcal{J} < 16\pi/(3a^3)$. We finally get a value $K/\mathcal{J} < -3/(\sqrt{2}\pi) \simeq -0.7$, below the threshold.

5.5.1.3 Body-centred cubic lattice

The body-centred cubic lattice presented in Figs. 5.14c and 5.14f shows that we have 4 and 1 spins in the neighbouring layer. In the layer with 4 spins, the coupling constant is $1/a^3$ and each spin interacts with two neighbours, while in the layer with a single spin there is no interaction. Therefore, the averaged effective coupling constant is $K = -(4/5)2/a^3$ while the averaged volume per spin is $v_0 = 4a^3/5$. Consequently, the effective mean field coefficient for large aspect ratios $L/a \gg 1$ is $\mathcal{J} = 5\pi/(3a^3)$. From this result, we automatically deduce that $K/\mathcal{J} = -24/(25\pi) \simeq -0.3$, a value above the threshold which allows, as for a face-centred small basis sample, the possibility of magnetized states and spin flips.

Besides that, magnetized body-centred cubic lattices could be achieved even in the case of wide samples but with a large aspect ratio. Indeed, in this case, each spin has 4 neighbours within the transversal layer, leading to $K = -4/a^3$; the volume per spin being $v_0 = a^3/2$, the mean field coefficient could therefore reach the value $\mathcal{J} = 8\pi/(3a^3)$ when the longitudinal dimension is much larger than the transversal one. We get finally $K/\mathcal{J} = -3/(2\pi) > -1/2$. This means that magnetized states can in principle be found in body-centred cubic lattices even in the thermodynamic limit along both transversal and longitudinal directions, in contrast to what we have in simple cubic and face-centred cubic cases. This is confirmed by numerical simulations (Miloshevich *et al.*, 2013).

In conclusion, using a standard treatment of a dipolar Hamiltonian, we have shown how to map it into a Hamiltonian with long- and short-range interactions. Remarkably, this transformation makes it possible to get very interesting predictions on the possibility for magnetized needles to exist. We get a value above the threshold for simple cubic lattices and face-centred cubic lattices, when the needle is thin enough. In contrast, body-centred cubic lattices could be magnetized even in the thermodynamic limit when considering wide samples. These results have been confirmed by numerical simulations (Miloshevich *et al.*, 2013) and experimental verifications are at hand.

5.5.2 Dynamical effects in layered ferromagnets

As already noted, dipolar interactions are marginal as far as the long-range character of the interaction is concerned. Indeed, in this case, the exponent governing the decay of the interaction potential is $\alpha = d = 3$. Despite this, we will show here that magnetic dipolar interactions offer a relatively simple way of observing a direct manifestation of long-range interactions in samples of suitable shape.

Generally, a Heisenberg exchange interaction between electronic spins is much higher (usually a few order of magnitudes) than magnetic dipolar energy, and therefore the only role played by dipolar interaction is the introduction of anisotropies of the spontaneous magnetization. On the other hand, magnetic ordering in nuclear spins requires experiments to be performed at nano-Kelvin temperatures (Oja and Lounasmaa, 1997). Here, we describe an example in which the dipolar interaction between electronic spins acquires, through the careful preparation of the sample, an effective strength in a suitable range of temperatures, such that its long-range character gives rise to an interesting dynamical effect related to a phase transition (Campa *et al.*, 2007b).

Compounds of the type $(C_\nu H_{2\nu+1}NH_3)_2 CuCl_4$ have been considered in Sato and Sievers (2004, 2005) and Wrubel *et al.* (2005), the focus being on the observation of intrinsic localized modes (discrete breathers). These compounds, organized in a face-centred orthorhombic crystal, are layered spin structures, in which the weak magnetic interlayer interaction is antiferromagnetic for $\nu > 1$ and ferromagnetic for $\nu = 1$. The relevant variables are the $s = 1/2$ spins of the Cu^{2+} ions, which are placed in two-dimensional layers. The hard axis of the magnetic interaction is orthogonal to the layers, which therefore are the easy planes. However, also the in-plane interaction is anisotropic: an easy and a 'second easy' axis can be determined. The cited works focused on the anti-ferromagnetic case $\nu = 2$. Here, we are interested in the ferromagnetic case $\nu = 1$, i.e. on the compound $(C_1 H_3 NH_3)_2 CuCl_4$, called bis(Methylammonium) tetrachloro-copper (Chapuis *et al.*, 1983).

When writing the Hamiltonian of the system, we adopt a coordinate frame in which the hard axis is the x axis, while the z axis and the y axis are the easy and 'second easy' axis belonging to the two-dimensional layers, respectively. The Hamiltonian reads therefore

$$
H = -W \sum_{I,<i,j>} \left(s^z_{Ii} s^z_{Ij} + \eta s^y_{Ii} s^y_{Ij} + \xi s^x_{Ii} s^x_{Ij} \right) - w \sum_{I,<i,j>} \mathbf{s}_{Ii} \cdot \mathbf{s}_{I+1,j}
$$
$$
+ \sum_{Ii \neq Jj} \frac{2\tilde{\mu}_B^2}{r^3} \left(\mathbf{s}_{Ii} \cdot \mathbf{s}_{Jj} - 3 \frac{(\mathbf{s}_{Ii} \cdot \mathbf{r})(\mathbf{s}_{Jj} \cdot \mathbf{r})}{r^2} \right), \tag{5.87}
$$

where the first, second and third sum represent the intralayer exchange interaction, the interlayer exchange interaction and the dipolar interaction, respectively. The quantity $\tilde{\mu}_B$ is defined by $\tilde{\mu}_B = \sqrt{(\mu_0/4\pi)}\mu_B$, with μ_B the Bohr magneton. The crystal layers are parallel to the yz plane. The capital indices number the layers, while the lowercase indices denote the spins. In the first sum $< i,j >$ refers to nearest neighbours within the same layer and in the second sum to nearest neighbours in adjacent layers; the last sum is extended over all pairs of spins of the system. The intralayer exchange constant W is much larger than the interlayer one w, for example $W \simeq 10^4 w$ (Dupas *et al.*, 1977). The parameters ξ and η, both smaller than 1, are related to the out-of-plane and in-plane anisotropies. In the dipolar interaction, r is the modulus of $\mathbf{r}$, the vector between the sites of the spins $\mathbf{s}_{Ii}$ and $\mathbf{s}_{Jj}$.

The large W/w ratio determines the existence, for a given shape of the sample, of a temperature range in which all the spins of a single layer are ferromagnetically ordered, such that the spin vector $\mathbf{s}_{Ii}$ can be considered as independent of the index i, $\mathbf{s}_I \equiv \mathbf{s}_{Ii}$, while full three-dimensional ordering has not yet been reached. This situation arises if the temperature is below the single-layer ordering temperature, of the order of W, and if in addition the number n of spins in each single layer is such that $nw < W$. Under such conditions the thermodynamic and dynamical properties of the system will be determined by the interlayer exchange constant w and by the long-range dipolar inter-action. Exploiting the fact that all spins in a layer are ordered, the Hamiltonian can be cast in a one-dimensional form. To this purpose, we must use a procedure employed in the treatment of dipolar forces, in which the short-range contribution (here includ-ing the interaction between nearest neighbours) and the long-range contribution are treated separately (Landau and Lifshitz, 1984). The latter contribution gives rise to shape-dependent terms. Here, we consider a rod-shaped sample, with short sides along the out-of-plane x axis and the in-plane y axis, and long sides along the in-plane z axis. With this shape, the long-range dipolar interaction produces an interaction term in the Hamiltonian which is typical of an ellipsoidal shape (Akhiezer *et al.*, 1968; Landau and Lifshitz, 1984). The magnetization in the sample varies only along the x axis. If we de-note by N the number of layers of the sample, the Hamiltonian is then transformed to (Akhiezer *et al.*, 1968; English *et al.*, 2003)

$$H_N = n\left[B_x \sum_{\tilde{J}=1}^{N} \left(s_{\tilde{J}}^x\right)^2 + B_y \sum_{\tilde{J}=1}^{N} \left(s_{\tilde{J}}^y\right)^2 - 2\omega_{\mathrm{ex}} \sum_{\tilde{J}=1}^{N-1} \left(s_{\tilde{J}}^y s_{\tilde{J}+1}^y + s_{\tilde{J}}^z s_{\tilde{J}+1}^z\right) \right.$$

$$\left. - \frac{\omega_M}{N} \left(\sum_{\tilde{J}=1}^{N} s_{\tilde{J}}^z\right)^2 + \frac{\omega_M}{2N} \left(\sum_{\tilde{J}=1}^{N} s_{\tilde{J}}^y\right)^2 \right]. \tag{5.88}$$

The first two sums come from the first sum in Eq. (5.87), and describe the intralayer exchange interaction. Here $B_x = 4W(1-\xi)$ and $B_y = 4W(1-\eta)$. Since for this compound the out-of-plane anisotropy is much larger than the in-plane one, we have that $B_x \gg B_y$; a constant additive term involving W has been neglected. The third sum in Eq. (5.88) is a combination of the interlayer exchange interaction (second sum in (5.87)) and the nearest neighbour interlayer dipolar interaction, while the intralayer nearest neighbour dipolar interaction produces a constant term that has been neglected. In this sum, ω_{ex} is given by

$$\omega_{\mathrm{ex}} = 2w - \frac{2\tilde{\mu}_B^2}{r_b^3}\left(2 - \frac{3r_a^2}{2r_b^2}\right), \tag{5.89}$$

where r_a and r_b are the distances between nearest neighbour spins within the same layer and in adjacent layers, respectively. The last two mean-field terms in Eq. (5.88) come from the long-range part of the dipolar interaction. This contribution is proportional to a combination of the square of the components of the magnetization density vector, with the coefficients of the combination depending on sample shape. In this case, where the magnetization varies along the x axis, we must consider the average magnetization density vector (Akhiezer *et al.*, 1968). In these mean-field terms, ω_M is given by $\omega_M = (4\pi/3)(2\tilde{\mu}_B^2/v_0)$, where v_0 is the volume of the unit cell of the lattice. In the effective Hamiltonian (5.88), we have neglected all terms including the x component of the spins which emerge from dipolar or exchange interlayer forces, since B_x is much larger than B_y, ω_M and ω_{ex}.

Hamiltonian (5.88) governs the dynamics of the vector spin $\mathbf{s}_{\tilde{j}}$ through the torque equation

$$\hbar \frac{d\mathbf{s}_{\tilde{j}}}{dt} = \mathbf{s}_{\tilde{j}} \times \mathbf{H}_{\tilde{j}}, \qquad (5.90)$$

where $\mathbf{H}_{\tilde{j}} = -\partial H_N / \partial \mathbf{s}_{\tilde{j}}$ is an effective magnetic field acting on the spin $\mathbf{s}_{\tilde{j}}$. The numerical results that we are going to present shortly are based on the integration of the last equation. To make a connection with the analytical results, we introduce an approximation in the Hamiltonian. We define $\mathbf{S}_{\tilde{j}} \equiv 2\mathbf{s}_{\tilde{j}}$, $S_{\tilde{j}}^y = \sqrt{1 - (S_{\tilde{j}}^x)^2} \sin \theta_{\tilde{j}}$ and $S_{\tilde{j}}^z = \sqrt{1 - (S_{\tilde{j}}^x)^2} \cos \theta_{\tilde{j}}$, which yields the equations of motion in terms of the angular variables $\theta_{\tilde{j}}$ and $S_{\tilde{j}}^x$. Noting that B_x is much larger than all the other parameters, which implies that $\sqrt{1 - (S_{\tilde{j}}^x)^2} \approx 1$, the equations of motion are simplified as

$$\hbar \frac{d\theta_{\tilde{j}}}{dt} = -nB_x S_{\tilde{j}}^x. \qquad (5.91)$$

This dynamics is consistent with the following effective Hamiltonian, which includes only the angles $\theta_{\tilde{j}}$ and their time derivatives

$$H_N' = \frac{2H_N}{n\omega_M} = \frac{1}{2} \sum_{\tilde{j}=1}^{N} \left[\frac{d\theta_{\tilde{j}}}{dt} \right]^2 - \frac{\omega_{ex}}{\omega_M} \sum_{\tilde{j}=1}^{N-1} \cos(\theta_{\tilde{j}+1} - \theta_{\tilde{j}}) + \frac{B_y}{2\omega_M} \sum_{\tilde{j}=1}^{N} \sin^2 \theta_{\tilde{j}}$$
$$- \frac{1}{2N} \left(\sum_{\tilde{j}=1}^{N} \cos \theta_{\tilde{j}} \right)^2 + \frac{1}{4N} \left(\sum_{\tilde{j}=1}^{N} \sin \theta_{\tilde{j}} \right)^2, \qquad (5.92)$$

where we have introduced a dimensionless time via the transformation

$$t \rightarrow tn\sqrt{B_x \omega_M}/\hbar. \qquad (5.93)$$

We note the similarity of this Hamiltonian with the Hamiltonian (5.30) of the chain of rotators coupled by a nearest neighbour interaction and a mean-field one studied in Section 5.3. With respect to Hamiltonian (5.30), the one given in Eq. (5.92) is not invariant under global rotations, because of the presence of the term proportional to B_y and of the difference between the coefficients of the two mean-field terms. However, if $B_y > 0$, which is the case experimentally, these differences do not change qualitatively (Lori, 2008) the phase diagram obtained in Section 5.3.

Let us now discuss the dynamical effect that could in principle be observed in experiments with this type of samples. We have shown that the XY model with nearest neighbour and mean-field interactions has, in particular, a second-order ferromagnetic phase transition for all positive values of the nearest neighbour coupling. We have also shown that, below the critical energy, there is another energy threshold, below which ergodicity is broken, with an inaccessible range of values of the magnetization around 0 (in contrast to the case $K > 0$ of Section 5.3, the inaccessibility of a magnetization range around 0 must be considered a breaking of ergodicity, since now the lack of invariance under global rotation can forbid the passage from $m_x > 0$ to $m_x < 0$). For the particular values of the parameters in Hamiltonian (5.92), we find (Campa *et al.*, 2007b) that the

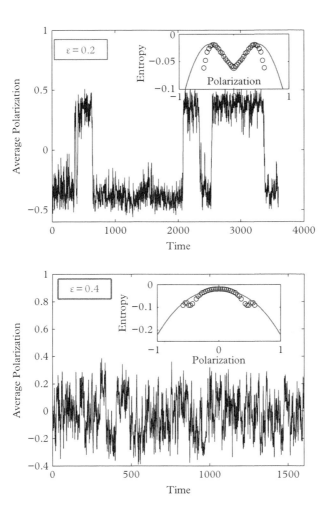

Figure 5.16 *Numerical simulations of the one-dimensional spin chain model, Eqs. (5.88) and (5.90), with two different energies: $\varepsilon = 0.2$ (upper) and $\varepsilon = 0.4$ (lower), i.e. below and above the second-order phase transition energy $\varepsilon_c \simeq 0.376$ predicted for Hamiltonian (5.92). The number of layers in the simulations is $N = 100$. The insets show the corresponding entropy curves (solid lines) obtained for Hamiltonian (5.92) and the data obtained from the probability distribution function of the magnetization (open circles) for Hamiltonian (5.88). Reprinted from Campa et al. (2009), © 2009, with permission from Elsevier.*

critical energy is $\varepsilon_c \simeq 0.376$, and the threshold below which there is breaking of ergo-dicity is $\varepsilon = -0.3$. Besides that, because of the lack of global rotational invariance, the spontaneous magnetization has only two possible directions, i.e., $\theta_{\bar{\jmath}} = 0$ and $\theta_{\bar{\jmath}} = \pi$. If we consider the dynamics of such a system, at an energy between the threshold for ergo-dicity breaking and the critical energy, and with a finite number of degrees of freedom, we should therefore observe that the modulus of the magnetization will fluctuate around that of the spontaneous magnetization, and from time to time will flip between the two directions of the magnetization. We expect that the flips will be more and more rare as the system size increases and approaches the ergodicity breaking threshold (Celardo *et al.*, 2006). The sudden flips are due to the fact that in contrast to short-range systems, the formation of domains with different directions of the magnetization is not possible.

Numerical experiments have been performed using Hamiltonian (5.88) and the torque equation (5.90) (Campa *et al.*, 2007b). We here present results concerning the two energies $\varepsilon = 0.2$ and $\varepsilon = 0.4$ (in the adimensional values computed after the transformation to Hamiltonian (5.92)). The first value is below the critical energy but still far enough from the ergodicity breaking threshold to observe magnetization flips on a reasonable time scale. The second value is above the critical energy, and the system should fluctuate around zero magnetization. In Figs. 5.16, we show the results of the numerical simulations that confirm the expectations.

From the numerical data, it is also possible to obtain the probability distribution function for the magnetization, shown in the insets of Figs. 5.16 by the open circles. The logarithm of the probability distributions obtained numerically is compared with that computed analytically at the corresponding energy values on the basis of Hamiltonian (5.92). The comparison is then made between the numerical distribution functions obtained by direct simulations of Hamiltonian (5.88) and the expression $\tilde{s}(\varepsilon, x)$ given by Eq. (5.38), obtained for Hamiltonian (5.92). The agreement is good although not perfect. The effects which lead to disagreement are twofold. First, we should have obtained the expression of $\tilde{s}(\varepsilon, x)$ for Hamiltonian (5.88), which is in principle possible. Second, we have taken x as playing the role of the spontaneous magnetization m. However, we have already emphasized several times that the parameter x coming from the Hubbard–Stratonovich transformation coincides with the spontaneous magnetization only at equilibrium. The reasonably good agreement that we obtain justifies a posteriori this identification, which is valid only close to equilibrium.

In summary, our main purpose here was to see that a system which can be in principle implemented in the laboratory can be used to observe the peculiarities predicted for long-range interacting systems. We have emphasized the importance of sample shape. For instance: for spherical or approximately spherical samples, spin flips will not be observed, since the mean-field terms in the Hamiltonian will not introduce any anisotropy.

6

Quantum Long-Range Systems

6.1 Introduction

In Chapter 1, we saw that the two fundamental properties that ensure the existence of a well-behaved thermodynamic limit and the equivalence of the ensembles are the *stability* of the interaction potential and its *temperedness*. For convenience, we report here the two inequalities expressing these two properties, restricting ourselves, as everywhere throughout this book, to the case of two-body interactions.

Temperedness in d dimensions holds when there exist $R > 0$, $C > 0$ and $\alpha > d$ such that the potential $V(\mathbf{r})$ satisfies

$$V(\mathbf{r}) \le C|\mathbf{r}|^{-\alpha}, \tag{6.1}$$

for $|\mathbf{r}| > R$. We have seen that a short-range potential automatically satisfies temperedness, since in that case we have the stronger inequality in which the left-hand side of Eq. (6.1) is replaced by the absolute value of $V(\mathbf{r})$.

On the other hand, in a classical system, the potential is stable when there exists $B > 0$ such that, for all N and for all configurations, the inequality

$$U(\mathbf{r}_1, \dots, \mathbf{r}_N) = \sum_{1 \le i < j \le N} \sum V(\mathbf{r}_i - \mathbf{r}_j) \ge -NB \tag{6.2}$$

is satisfied. Attractive potentials that allow the collapse of the system in a finite volume for $N \to \infty$ violate stability since, in that case, the absolute value of the negative potential energy of the system will not grow slower than N^2.

In quantum systems, the stability conditions must be expressed differently from Eq. (6.2). If we denote with $\hat{H}_N$ the Hamiltonian operator of the N particles system, and with $\Psi_N(\mathbf{r}_1, \dots, \mathbf{r}_N)$ an acceptable wavefunction, stability is expressed by

$$\inf_{\Psi} \langle \Psi_N | \hat{H}_N | \Psi_N \rangle \ge -NA, \tag{6.3}$$

with $A > 0$ a given constant, and where the usual bra-ket notation of quantum mechanics has been used. The left-hand side of Eq. (6.3) is nothing but the energy E_0 of

Physics of Long-Range Interacting Systems. First Edition. A. Campa *et al.*
© A. Campa, T. Dauxois, D. Fanelli, and S. Ruffo 2014. Published in 2014 by Oxford University Press.

the ground state of the system. It is easy to see that classical stability implies quantum stability. In fact, since the kinetic energy operator is positive definite, we have the relation

$$E_0 \geq \min_{\{\mathbf{r}_i\}} U(\mathbf{r}_1, \ldots, \mathbf{r}_N). \tag{6.4}$$

In contrast, quantum stability does not imply classical stability. It is then possible that unstable systems in classical mechanics become stable when they are treated quantum mechanically.

We can argue that with quantum mechanics the major modification in the statistical behaviour of a system will have to do with the collapsed configurations, because of the uncertainty principle. For example, two particles attracting each other with a potential that at small distances r diverges as $-1/r^\alpha$ will have a ground state for the bound system with a finite negative energy if $\alpha < 2$ in three dimensions (Landau and Lifshitz, 1991). Since the two important cases of the gravitational interaction and of the Coulomb interaction between particles of opposite sign are of this form with $\alpha = 1$, we infer that quantum mechanics could induce a strong modification for these systems. However, physical considerations suggest to us that in this respect we do not have to worry for the case of gravitational systems. In fact, before quantum mechanical effects, other interactions will enter into play; therefore, in a model of particles interacting with a $-1/r$ potential, we implicitly assume that we are representing only configurations where other interactions are not acting. Therefore, in the following, we concentrate on the case of charged particles.

The interaction among N charged particles in the non-relativistic regime is given by the sum of the two-body Coulomb potential between each pair of particles

$$U(\mathbf{r}_1, \ldots, \mathbf{r}_N) = \sum\sum_{1 \leq i < j \leq N} \frac{q_i q_j}{|\mathbf{r}_i - \mathbf{r}_j|}, \tag{6.5}$$

where q_i is the charge of the ith particle, including its sign. It is known that the exact expression of the Coulomb potential requires, in front of the $q_i q_j/r$ term, a coefficient that depends on the system of units adopted. Later we will use the SI system, for which the coefficient is $1/(4\pi\varepsilon_0)$, where ε_0 is vacuum permittivity. However, in this chapter we are interested in the study of Coulomb systems from the point of view of stability and temperedness. For this the units and the corresponding coefficient obviously do not play any role, and for simplicity of notation we adopt the Gaussian units, for which the coefficient is 1.

Clearly, in a realistic situation, we are concerned with systems in which there are a few different kinds of particles and, correspondingly, there are only a few different values for the charge. The inclusion of relativistic and radiative effects requires additional terms, which are proportional to increasing powers of the ratios v_i/c between the velocities of the particles and the speed of light (Landau and Lifshitz, 1980). We will limit our discussion to the non-relativistic case, since it is clear that the problem of the statistical mechanics description of a system of charged particles, in particular the problem of the stability and of the existence of the thermodynamics limit, must be solved first of all with reference to this setting.

We see that the Coulomb potential poses problems at both short and long distances. Unless we consider a system with all charges of the same sign, the singularity of the potential at the origin violates stability. On the other hand, temperedness is always violated. As a matter of fact, classically the violation of stability, before posing a problem on the statistical behaviour of a many-body system, would be a problem for the dynamics of two charges of opposite sign, since the loss of energy by the emission of radiation would lead the two particles to collapse. Obviously, Hamiltonian (6.5), which does not contain radiative effects, when considered for the case of a positive and a negative charge, dynamically prevents the collapse unless the angular momentum is zero. Therefore, the statement must be intended as the existence of configurations with unbounded negative energy. When quantum effects are taken into account, we know that the situation is radically different since, as we noted earlier, there is a ground state with a finite negative energy, e.g. the system composed of a proton and an electron leading to the stability of the hydrogen atom. We can hope that the passage from classical to quantum mechanics has a similar beneficial effect for the statistical mechanics of many-body Coulomb systems. We will see that this is indeed the case, although we should emphasize that the road from the mechanical stability of the two-body system to the stability of the many-body system as expressed by Eq. (6.3) is by no means obvious.

In this chapter, we will review the basic results concerning the equilibrium properties of Coulomb systems in $d = 3$ dimensions, as obtained several decades ago. Most of the results will be cited without proofs, which are often quite lengthy. In general, we will provide only a few details of the derivation. Exceptions are made for particularly simple cases. The reader interested in the full derivations is referred to the cited bibliography.

6.2   Classical Coulomb Systems

An attractive long-range potential is by definition tempered since $V < 0$ at least at large distances. Even if it avoids the collapse of the system, e.g. by the presence of a hard core, it is nevertheless unstable, since the non-integrability does not allow the inequality (6.2) to be satisfied. For example, the gravitational potential, even if modified at short distances with a hard core, is not stable: it is not difficult to see that the total potential energy would scale as $-N^{5/3}$. In fact, from Eq. (2.3) we obtain that the absolute value of the energy per particle would scale, when $d = 3$ and $\alpha = 1$, as L^2, where L is the linear size of the system. On the other hand, L scales as $N^{1/3}$; then we conclude, since the gravitational potential is attractive, that the energy per particle would scale as $-N^{2/3}$, and the total energy would scale as $-N^{5/3}$. The conclusion is not so trivial for Coulomb systems. As for the short distance singularity, the long-range character of the Coulomb potential could violate stability only if there were charges of different signs. However, it is argued that screening effects prevent the divergence of the potential energy per particle with the size of the system (Brydges and Martin, 1999). In fact, we show now that a Coulomb system with a regularized short-range behaviour is stable already in the classical case, and therefore *a fortiori* in the quantum case, despite the long-rangedness.

Let us consider a modified Coulomb potential of the form

$$V_a(\mathbf{r}) = \frac{1 - e^{-ar}}{r}, \tag{6.6}$$

in which $r \equiv |\mathbf{r}|$ and a a given positive constant; at long distance this potential behaves like the ordinary Coulomb potential. We analyse the situation in which there are several different types of particles, with both charge signs present. The stability of this system can be easily proven by expressing the total potential energy via the Fourier transform of the potential (6.6) as shown by Fisher and Ruelle (1966). We begin by expressing the potential energy of the system, given by

$$U_a(\{\mathbf{r}_i\}) = \sum \sum_{1 \le i < j \le N} q_i q_j V_a(\mathbf{r}_i - \mathbf{r}_j), \tag{6.7}$$

in the form

$$U_a(\{\mathbf{r}_i\}) = \frac{1}{2} \sum_{i=1}^{N} \sum_{j=1}^{N} q_i q_j V_a(\mathbf{r}_i - \mathbf{r}_j) - \frac{1}{2} \sum_{i=1}^{N} q_i^2 V_a(0) \tag{6.8}$$

$$= \frac{1}{2} \sum_{i=1}^{N} \sum_{j=1}^{N} q_i q_j V_a(\mathbf{r}_i - \mathbf{r}_j) - \frac{1}{2} a \sum_{i=1}^{N} q_i^2. \tag{6.9}$$

Since the absolute value of the last term on the right-hand side is bounded by

$$a \sum_{i=1}^{N} q_i^2 \le a N q_0^2, \tag{6.10}$$

where q_0 is the largest charge, in absolute value, in the system, stability follows if we show that the double sum on the right-hand side is non-negative definite. Using the Fourier transform of V_a

$$\tilde{V}_a(\mathbf{k}) = \frac{1}{(2\pi)^3} \int d\mathbf{r} \, e^{-i\mathbf{k}\cdot\mathbf{r}} \, V_a(\mathbf{r}) = \frac{1}{2\pi^2} \frac{a^2}{k^2(k^2 + a^2)}, \tag{6.11}$$

with $k \equiv |\mathbf{k}|$, the double sum in Eq. (6.9) can be expressed as

$$\frac{1}{2} \sum_{i=1}^{N} \sum_{j=1}^{N} q_i q_j V_a(\mathbf{r}_i - \mathbf{r}_j) = \frac{1}{2} \sum_{i=1}^{N} \sum_{j=1}^{N} q_i q_j \int d\mathbf{k} \, \tilde{V}_a(\mathbf{k}) e^{i\mathbf{k}\cdot(\mathbf{r}_i - \mathbf{r}_j)} \tag{6.12}$$

$$= \frac{1}{2} \int d\mathbf{k} \left(\sum_{i=1}^{N} q_i e^{-i\mathbf{k}\cdot\mathbf{r}_i} \right)^* \tilde{V}_a(\mathbf{k}) \left(\sum_{j=1}^{N} q_j e^{-i\mathbf{k}\cdot\mathbf{r}_j} \right) \tag{6.13}$$

$$= \frac{1}{2} \int d\mathbf{k} \left| \sum_{i=1}^{N} q_i e^{-i\mathbf{k}\cdot\mathbf{r}_i} \right|^2 \tilde{V}_a(\mathbf{k}) \ge 0, \tag{6.14}$$

where $(\cdot)^*$ stands for the complex conjugate. The conclusion (6.14) follows from the positivity of $\tilde{V}_a(\mathbf{k})$ given in Eq. (6.11).

We have therefore come to the conclusion that instability could be due only to a collapsed configuration. It is easy to see that the singularity at short distances actually leads to that. Let us consider a system with an equal number N of two different types of particles, one positively and one negatively charged. If we pair the positive and negative charges, it is not difficult to imagine a configuration in which the potential energy is on the same order of N times the potential energy of a pair. But the energy of a pair is not lower bounded, and therefore the system is not stable.

Thus, stability of Coulomb systems could hold only in quantum mechanics. In the next section, we will show when quantum Coulomb systems are stable.

6.3 The Problem of Stability of Quantum Coulomb Systems

The Hamiltonian operator for an N-particles Coulomb system is

$$\hat{H}_N = -\sum_{i=1}^{N} \frac{\hbar^2}{2m_i} \nabla_i^2 + \sum\sum_{1 \leq i < j \leq N} \frac{q_i q_j}{|\mathbf{r}_i - \mathbf{r}_j|} = \hat{T}_N + \hat{U}_N. \tag{6.15}$$

We begin by giving a short description of the results that prove that a system in which there is no exclusion principle, i.e. in which none of the particle types is a fermion, is not stable.

6.3.1 Systems without exclusion principle

Let us consider a system with two types of particles of opposite sign, and in particular a system of N positively and N negatively charged particles, with the negative charge equal, in absolute value, to the positive one, so that the system is globally neutral. A lower and an upper bound for the ground state energy of such a system can be easily found. For the upper bound, we can consider the configuration in which each positively charged particle is paired with a negatively charged particle, each pair being in the ground state; moreover, the configuration is one in which the pairs are sufficiently apart that their mutual interaction can be neglected (the interaction potential between neutral pairs in the ground state being short-ranged). Denoting by $-\varepsilon$ the ground state energy of each pair $(\varepsilon > 0)$, the energy of such configuration is thus equal to $-N\varepsilon$. Therefore, the ground state energy $E_0(2N)$ of the $2N$-particle system satisfies

$$E_0(2N) \leq -N\varepsilon. \tag{6.16}$$

For the lower bound, we can use a scaling property of the Hamiltonian (6.15). From its structure, it is evident that, if we define (Fisher and Ruelle, 1966)

$$\Psi_N^\lambda(\mathbf{r}_1,\dots,\mathbf{r}_N) = \lambda^{\frac{3N}{2}}\,\Psi_N(\lambda\mathbf{r}_1,\dots,\lambda\mathbf{r}_N) \tag{6.17}$$

and

$$\hat{H}_N^\lambda = \frac{1}{\lambda^2}\hat{T}_N + \frac{1}{\lambda}\hat{U}_N, \tag{6.18}$$

then we have

$$\langle\Psi_N^\lambda|\hat{H}_N^\lambda|\Psi_N^\lambda\rangle = \langle\Psi_N|\hat{H}_N|\Psi_N\rangle. \tag{6.19}$$

Therefore $\hat{H}_N^\lambda$ has the same spectrum as $\hat{H}_N$, in particular the same ground state energy. The Hamiltonian (6.15), specialized to the case of a system of N particles of charge q and mass m_a with coordinates $\mathbf{r}_i^a$, and N particles of charge $-q$ and mass m_b with coordinates $\mathbf{r}_i^b$, can be written as

$$
\begin{aligned}
\hat{H} = &-\sum_{i=1}^N \frac{\hbar^2}{2m_a}\nabla_i - \sum_{i=j}^N \frac{\hbar^2}{2m_b}\nabla_j - q^2\sum_{i=1}^N\sum_{j=1}^N \frac{1}{|\mathbf{r}_i^a - \mathbf{r}_j^b|} \\
&+ q^2\sum_{1\le i<j\le N}\frac{1}{|\mathbf{r}_i^a - \mathbf{r}_j^a|} + q^2\sum_{1\le i<j\le N}\frac{1}{|\mathbf{r}_i^b - \mathbf{r}_j^b|}
\end{aligned} \tag{6.20}
$$

$$
\ge -\sum_{i=1}^N \frac{\hbar^2}{2m_a}\nabla_i - \sum_{i=j}^N \frac{\hbar^2}{2m_b}\nabla_j - q^2\sum_{i=1}^N\sum_{j=1}^N \frac{1}{|\mathbf{r}_i^a - \mathbf{r}_j^b|}. \tag{6.21}
$$

The inequality sign must be understood in the operatorial meaning, and it is clearly due to the positive definiteness of the interaction energy between the charges of the same sign. Use of the property (6.19) can be made by writing the right-hand side of (6.21) as

$$
-\sum_{i=1}^N \frac{\hbar^2}{2m_a}\nabla_i - \sum_{i=j}^N \frac{\hbar^2}{2m_b}\nabla_j - q^2\sum_{i=1}^N\sum_{j=1}^N \frac{1}{|\mathbf{r}_i^a - \mathbf{r}_j^b|}
$$

$$
= N\sum_{i=1}^N\sum_{j=1}^N \left\{ \frac{1}{N^2}\left(-\frac{\hbar^2}{2m_a}\nabla_i - \frac{\hbar^2}{2m_b}\nabla_j \right) - \frac{1}{N}\frac{q^2}{|\mathbf{r}_i^a - \mathbf{r}_j^b|} \right\}. \tag{6.22}
$$

Each term in the curly brackets is the rescaled Hamiltonian of a pair of oppositely charged particles, i.e. in the notation just introduced each such term is $\hat{H}_1^N$. It has the same ground state energy as $\hat{H}_1$, namely ε. We therefore have

$$E_0(2N) \ge -N^3\varepsilon. \tag{6.23}$$

The last expression shows that the energy of a quantum system, made of two types of particles of opposite charge and globally neutral, is lower bounded, in contrast to the classical case. It is not difficult to infer that also with an unbalance between positive and negative charges the energy is lower bounded. In fact, adding a particle of either sign to a system with N_+ positive charges and N_- negative charges can only decrease the ground state energy. This can be deduced thinking that the added particle can be placed very far from all the others, and this does not increase the total energy. Thus, if, for example, $N_+ - N_- = \Delta N > 0$, we can add ΔN negative charges, obtaining a globally neutral system with $2N_+$ particles, with a ground state energy lower than the original system and lower bounded by $-N_+^3 \varepsilon$. The extension to the case in which there are more charge values is also simple.

Taken together, Eqs. (6.16) and (6.23) lead to

$$- N^3 \varepsilon \le E_0(2N) \le -N\varepsilon, \tag{6.24}$$

which might even be compatible with a stability of the system. However, improved upper and lower bounds have been obtained, i.e. an upper bound with a scaling in N with a larger exponent and a lower bound with a smaller exponent. In particular, the improved upper bound for the ground state energy, obtained by Dyson (1967), is

$$E_0(2N) \le -CN^{\frac{7}{5}}. \tag{6.25}$$

This expression proves that, without exclusion principle, a Coulomb system is unstable also in quantum mechanics. The proof of this result is lengthy and will not be given here. It is nevertheless instructive to see how the power 7/5 can be inferred, using an heuristic argument also provided in Dyson (1967).

Let us consider again a globally neutral system, with N positive and N negative charges, which are in a volume characterized by a linear length scale L. The localization in such a volume will give to the system a kinetic energy contribution of the order of $N\hbar^2/(mL^2)$ if, for simplicity purposes, we consider equal masses for positive and negative charges. If each charge is, on the average, screened by a cloud of charges that together have a total unit charge of the opposite sign, and if this cloud extends on a volume with linear length scale Λ, the total Coulomb energy of the system will be of the order of $-Nq^2/\Lambda$. However, the forming of charge clouds gives rise to an additional kinetic energy contribution, since it requires a component, in the wavefunction of the particles, which varies on a scale of Λ. If the clouds are formed by n particles, the relative variation of each particle wavefunction will be of the order $1/n$. Then each particle of the cloud produces a kinetic energy contribution of the order of $\hbar^2/(mn^2\Lambda^2)$. Multiplying by the number n of particles in each cloud, and by the number N of clouds, we have a total kinetic energy contribution, due to screening, equal to $N\hbar^2/(mn\Lambda^2)$; since n is of the order of $N\Lambda^3/L^3$ if we assume that the density is approximately constant, this kinetic energy can also be written as $\hbar^2 L^3/(m\Lambda^5)$. We have therefore an estimate of the total energy of the system (6.20) given by

$$N\frac{\hbar^2}{mL^2} - N\frac{q^2}{\Lambda} + \frac{\hbar^2 L^3}{m\Lambda^5}. \tag{6.26}$$

We can now look for the extremum of this expression with respect to Λ and L, which on physical grounds we expect to be a minimum; furthermore, from the previous evaluations, we know that this minimum will be negative; i.e. the system is bound. We do not need the exact expression giving Λ and L at the minimum, since it is sufficient to have their scaling properties with N. Taking the derivative of Eq. (6.26) with respect to Λ, while N and L are kept fixed, we get $N \sim L^3/\Lambda^4$. Similarly, from the derivative of Eq. (6.26) with respect to L, while N and Λ are fixed, we get $N \sim L^5/\Lambda^5$. Consequently, we find that Λ scales as $N^{-2/5}$, while L scales as $N^{-1/5}$. We therefore find that the three terms in Eq. (6.26) have the same scaling with N and finally that the total energy (6.26) scales as $N^{7/5}$, in agreement with the bound (6.25).

The proof of the improved lower bound (Lenard and Dyson, 1967) is also lengthy, and we only give here the result, again encouraging the interested reader to consult the cited reference. The bound is

$$E_0(2N) \geq -AN^{\frac{5}{3}}, \tag{6.27}$$

where A denotes a new constant. We find, in addition, that both the kinetic and the potential energy have again the same scaling.

We have seen that, in the passage from a classical to a quantum treatment, the ground state energy is no more lower unbounded, as shown by Eq. (6.27), and this is clearly due to the uncertainty principle: the configurations in which the particles tend to stay close together have a sufficiently large positive kinetic energy, which produces a higher energy with respect to the classical case. However, this is not sufficient for having a stable system, as proven by the upper bound (6.25). We may hope that the exclusion principle, which prevents that any single particle state is occupied by more than one particle, could help in this respect; in particular, we should expect that both total kinetic and potential energies scale with N. This is what has been actually proven, as explained next.

6.3.2   Systems with exclusion principle

Lenard and Dyson (1967, 1968) have given the proof of the stability of Coulomb systems, described by the Hamiltonian (6.15), in the two following situations:

(i) The system is composed of a finite number s of fermion species, with charges that can be of either sign;

(ii) The system is composed of a finite number s of fermion species, all of which with charges of a given sign, e.g. negative, and of charges of the opposite sign, e.g. positive, with arbitrary statistics.

The proof is based on the spatial antisymmetry of the wavefunction with respect to the coordinates of the same species of fermion, expression of the exclusion principle. Therefore the distinct spin state of each fermion type is considered, in the proof, as a distinct fermion species. Thus, we have at least $s = 2$.

We recognize immediately in case (ii) the situation that occurs in ordinary matter, with negative electrons (fermions) and positively charged nuclei that can be either fermions or bosons. Again we will not enter into the lengthy calculations necessary for the proofs, but we will only give very few details of the derivation.

Let us deal with the first case, with s distinct species of fermions. Obviously, the theorem is not trivial only if there are charges of both signs. It is not difficult to prove (Lenard and Dyson, 1967) that, independently of the statistics obeyed by the particles, the following bound holds for the expected value of the potential energy $\hat{U}_N$ of the Hamiltonian (6.15),

$$U_N = \langle \Psi_N \,|\, \hat{U}_N \,|\, \Psi_N \rangle > -Nq^2 K, \tag{6.28}$$

where q is the largest absolute value of the charges in the systems, and K is the expectation value

$$K = \frac{1}{N} \langle \Psi_N \,|\, \sum_{i=1}^{N} \frac{1}{R_i} \,|\, \Psi_N \rangle, \tag{6.29}$$

in which

$$R_i(\mathbf{r}_1, \ldots, \mathbf{r}_N) = \min_{1 \le j \le N, j \ne i} |\mathbf{r}_j - \mathbf{r}_i|. \tag{6.30}$$

In practice, the bound (6.28) is obtained by substituting the total potential energy with an imaginary interaction in which each particle interacts attractively with its nearest neighbour, with charges of maximum value. Incidentally, we note that the instability of systems without the exclusion principle implies that in that case, in the ground state, K scales with N at least as $N^{2/3}$ (see Eq. (6.27)). Instead, in our systems with only fermions, the key inequality, which we cite without the proof given in Lenard and Dyson (1967), concerns a relation between the average kinetic energy and the expectation value K. If m is the largest particle mass in the system, we can write

$$\frac{\hbar^2}{2m} w \equiv -\frac{1}{N} \frac{\hbar^2}{2m} \langle \Psi_N \,|\, \sum_{i=1}^{N} \nabla_i^2 \,|\, \Psi_N \rangle \le \frac{1}{N} \langle \Psi_N \,|\, \hat{T}_N \,|\, \Psi_N \rangle, \tag{6.31}$$

where m is the largest particle mass in the system and the expectation value K. The relation just mentioned gives a lower bound for the kinetic energy of the system expressed as a function of K (Lenard and Dyson, 1967)

$$K < C s^{\frac{1}{3}} \sqrt{w}, \tag{6.32}$$

with a new constant C. Putting together Eqs. (6.29), (6.31) and (6.32), we obtain the following w-dependent lower bound for the energy:

$$\langle \Psi_N \,|\, \hat{H}_N \,|\, \Psi_N \rangle > N \left(\frac{\hbar^2}{2m} w - Cq^2 s^{\frac{1}{3}} \sqrt{w} \right). \tag{6.33}$$

Minimizing with respect to w, we have the lower bound for the ground state energy

$$E_0(2N) > -NC^2 s^{\frac{2}{3}} \frac{mq^4}{2\hbar^2}, \tag{6.34}$$

which implies stability.

As we said earlier, this proof of stability has been obtained also in the case that can represent ordinary matter (Lenard and Dyson, 1968), namely negative fermions and positive charges of indefinite statistics.

6.4   The Thermodynamic Limit of Coulomb Systems

We know that stability alone does not imply the existence of a well-behaved thermo-dynamic limit, which is assured if both stability and temperedness hold. We remind that by the existence of the thermodynamic limit we mean that the ratio between the loga-rithm of the partition function of the statistical ensembles and the number of particles tends to a finite limit when the number of particles goes to infinity. A Coulomb system composed of particles with charges all of the same sign provides a trivial example in which stability is satisfied while temperedness is not, and in which the thermodynamic limit does not exist.

Temperedness for Coulomb systems is never satisfied but, despite this, it can be shown that the thermodynamic limit, under some conditions, exists as proven by Le-bowitz and Lieb (1969, 1972), assuming the situation in which stability holds. As in the previous sections of this chapter, we will only give here the main results.

The use of temperedness in the proof of the existence of the thermodynamic limit in the canonical ensemble is essentially based on the following fact: if we divide the volume V available to the N particles into subdomains $V_1, \ldots, V_n$, containing $N_1, \ldots, N_n$ particles respectively, we then have for any inverse temperature β

$$Z_N(\beta, V) \geq \prod_{i=1}^{n} Z_{N_i}(\beta, V_i). \tag{6.35}$$

We can interpret the passage from the volumes V_i to the volume V that includes all of them, as a particular step of the repeating passages needed to perform the thermo-dynamic limit. If the interaction between particles belonging to different volumes V_i can be neglected, Eq. (6.35) translates in a not increasing sequence for the free energy per particle $f(\beta, V)$. Since stability imposes a lower bound, we find that $f(\beta, V)$ has a limit for $N \to \infty$. The temperedness is the property that allows us to neglect the interaction between particles in different subdomains V_i.

Since the Coulomb interaction is not tempered, Eq. (6.35) is not useful. However, we can employ another inequality that uses the same partition of the volume V, i.e.

$$Z_N(\beta, V) \geq e^{-\beta U_c} \prod_{i=1}^{n} Z_{N_i}(\beta, V_i), \tag{6.36}$$

where U_c is the average value of the interaction energy between particles that are in different volumes V_i. As remarked by Lebowitz and Lieb, the key of the proof is in the fact that for spherical subdomain U_c vanishes, if each subdomain V_i is globally neutral. Then, we substantially apply the same argument on the existence of a limit for $f(\beta, V)$. Note that there are some technical points involved in the proof, related to the construction of a sequence of spherical domains.

The conclusion is that globally neutral Coulomb systems have a proper thermodynamic limit for the canonical partition function, despite the non-temperedness of the interaction. It can also be shown that the different ensembles are equivalent.

The situation is somewhat different for systems with a charge unbalance. The non-existence of the limit for systems with charges of only a sign is trivial, since there is no way to screen the interaction between faraway particles. Therefore, there must be a transition, as a function of the global charge of the system, from a situation where the thermodynamic limit exists, as when the global charge is zero, to a situation where it does not exist, as when there are charges of only one sign. Lebowitz and Lieb have shown that the ratio $Q/V^{2/3}$ acts as control parameter: here, Q is the global charge of the system. If the limit of this ratio is 0, the free energy per particle exists and is the same as in the globally neutral system with $Q = 0$; if the limit is infinite, the thermodynamic limit does not exist, and the free energy per particle diverges; if the limit is finite, the free energy per particles tends to a shape-dependent limit, which is given by the free energy of the globally neutral system plus the electrostatic energy of a surface layer of charge Q. This can be easily understood by simple considerations of electrostatics. If a charge Q is uniformly distributed on a spherical volume, we easily compute the electrostatic energy W to be $2Q^2/(3V^{1/3})$, while if the charge is distributed uniformly on the surface of the sphere, the energy W is $Q^2/(2V^{1/3})$. In the first case, increasing the volume V implies that the charge Q increases proportionally to V, and therefore the electrostatic interaction per unit volume W/V diverges as $V^{2/3}$. In the second case, Q increases proportionally to $V^{2/3}$, and therefore W/V is a constant. We thus expect that if Q increases as $V^{2/3}$ the thermodynamic limit exists, although in a shape-dependent manner. We also understand that if Q increases slower, the thermodynamic limit of the free energy per particle will be the same as that of a globally neutral system.

To conclude our brief overview of the statistical behaviour of quantum long-range systems, in which we have concentrated on the Coulomb interaction, let us stress the following important facts, emerged in this chapter, concerning the role of the quantum effects.

- Quantum mechanics is essential in the stability property of Coulomb systems with respect to the short-range singularity of the potential. The uncertainty principle prevents the collapse of two particles in the same point. In a many-body system, this principle provides a lower bound to the energy, in contrast to the classical case, but alone it is not sufficient to have stability, since this bound is proportional to $N^{7/5}$. However, when also the exclusion principle plays a role, stability is recovered for systems in which all the particles with charges of a given sign are fermions.

- In contrast, we have seen that already in the classical case the screening effect assures that the long-range nature of the Coulomb interaction does not affect stability.

- Finally, we have showed that, despite the absence of temperedness, the thermodynamic limit exists for systems that are globally neutral, or even in systems where the total charge does not diverge faster than $V^{2/3}$. This result is valid both in the classical and in the quantum case.

Part II

Dynamical Properties

7

BBGKY Hierarchy, Kinetic Theories and the Boltzmann Equation

Experience shows that a many-body system, initially prepared in an arbitrary state, evolves, generally quite rapidly, until an equilibrium state is reached. Which equilibrium state is chosen by the system is determined by the external thermodynamic parameters imposed on it. For example, a given amount of gas pumped inside a container of volume V in contact with the environment at temperature T will quickly reach a uniform density and will exert a pressure p on the walls of the container: p is determined from the equilibrium equation of state. In Chapter 1, we discussed the difficulties in treating a system through the study of its equations of motion, and how we are led to the concept of statistical ensembles. We know also that the constant values of the observables in a system at equilibrium allow their computation with an average over proper distribution functions in the Γ space of the system.

In this approach, there is no trace of the transient states of the system before it reaches equilibrium and of the time scale over which this dynamical process takes place. As a matter of fact, the time variable is totally neglected. Therefore, to study the out-of-equilibrium dynamics of a many-body system, in particular its approach to equilibrium, we need new concepts and tools. It is important to emphasize that the study of the out-of-equilibrium dynamics is even more important for long-range than for short-range systems since, as will be described in the following chapters, long-range interactions may lead to transient states with macroscopic lifetimes, possibly diverging in the thermodynamic limit.

Since we cannot use the equations of motion as we would do for a system with few degrees of freedom, we must resort again to statistical methods. This is done with the help of kinetic theories, where one derives an equation for the one-particle distribution function $f_1(\mathbf{q}, \mathbf{p}, t)$, which is proportional to the probability density to find a particle around the position $(\mathbf{q}, \mathbf{p})$ in the phase space at time t. As for the first part of the book, in this introductory chapter we introduce the basic concepts focusing on short-range systems, which are simpler. From the next chapter, we will devote our attention to long-range systems and to the kinetic equations on the basis of their study.

Considering not only the statistical properties of the equilibrium state, but also the dynamical processes, makes matters more complicated in general. We note that dynamics

Physics of Long-Range Interacting Systems. First Edition. A. Campa *et al.*
© A. Campa, T. Dauxois, D. Fanelli, and S. Ruffo 2014. Published in 2014 by Oxford University Press.

must be taken into account also in the equilibrium states when one is interested in time correlation functions of observables at equilibrium.

Two approaches can be used to systematically derive kinetic equations; they have in common the starting point of the full N-body dynamics. However, the route to arrive to an equation for f_1 is different. Here, we will make use of the so-called BBGKY hierarchy proposed in parallel by Bogoliubov (1946), Born and Green (1949), Kirkwood (1946) and Yvon (1935). Then we will concentrate on the most celebrated kinetic equation, namely the Boltzmann equation, which is suitable for dilute short-range systems. It was the first kinetic equation for a Hamiltonian N-body system (Boltzmann, 1872). The other approach uses the so-called Klimontovich equation. It will be employed in the next chapter to derive the kinetic equations for long-range systems. Both approaches lead to the same equations, for both short- and long-range systems, and it is a matter of taste which one to prefer. We found it useful to describe both of them in this book.

7.1 The BBGKY Hierarchy

Let us start with the Liouville equation. We already derived it in Chapter 1, see Eq. (1.8), but for convenience we shortly rederive it again here, with a somewhat different notation that is more suitable for the forthcoming discussion.

We consider a system with N identical particles of mass m, whose coordinates and momenta are $(\mathbf{q}_i, \mathbf{p}_i)$, $i = 1, \ldots, N$, and with Hamiltonian

$$H = \sum_{i=1}^{N} \frac{p_i^2}{2m} + \sum\sum_{1 \le i < j \le N} V\left(|\mathbf{q}_i - \mathbf{q}_j|\right) = \sum_{i=1}^{N} \frac{p_i^2}{2m} + \sum\sum_{1 \le i < j \le N} V_{ij}, \tag{7.1}$$

where $p_i = |\mathbf{p}_i|$, and where the short-hand notation introduced in the rightmost side will prove useful in the following, where we will also make use of the notation $x_i \equiv (\mathbf{q}_i, \mathbf{p}_i)$. The N-body density distribution function is denoted now with $f_N(x_1, \ldots, x_N, t)$, where we have explicitly put in evidence with a subscript the number of particles on which the function depends; $f_N(x_1, \ldots, x_N, t)dx_1 \ldots dx_N$ is proportional to the number of systems that at time t are in the infinitesimal volume $dx_1 \ldots dx_N$ around the position $x_1, \ldots, x_N$ in the phase space. The time evolution of the density distribution function $f_N(x_1, \ldots, x_N, t)$ is expressed by the continuity equation

$$\frac{\partial}{\partial t} f_N + \sum_{i=1}^{N} \frac{\partial}{\partial \mathbf{q}_i} \left(\dot{\mathbf{q}}_i f_N\right) + \sum_{i=1}^{N} \frac{\partial}{\partial \mathbf{p}_i} \left(\dot{\mathbf{p}}_i f_N\right) = 0. \tag{7.2}$$

Using the Hamilton equations of motion

$$\dot{\mathbf{q}}_i = \frac{\mathbf{p}_i}{m} \quad \text{and} \quad \dot{\mathbf{p}}_i = -\sum_{j \ne i}^{N} \left(\nabla_i V_{ij}\right), \tag{7.3}$$

where we have adopted the usual notation $\mathbf{\nabla}_i \equiv \partial/\partial \mathbf{q}_i$, we obtain the Liouville equation

$$\frac{\partial f_N}{\partial t} + \sum_{i=1}^{N} \frac{\mathbf{p}_i}{m} \cdot \mathbf{\nabla}_i f_N - \sum_{i=1}^{N} \sum_{j \neq i}^{N} \left(\mathbf{\nabla}_i V_{ij} \right) \cdot \frac{\partial f_N}{\partial \mathbf{p}_i} = 0. \tag{7.4}$$

This equation shows that the evolution of f_N is equivalent to that of an incompressible fluid. In fact, the equation can be rewritten as

$$\frac{\mathrm{d} f_N}{\mathrm{d} t} = 0, \tag{7.5}$$

meaning that when $x_1, \ldots, x_N$ are taken to depend on time through the Hamilton equations of motion, the total time derivative of f_N is 0. In other words, f_N is constant when moving in Γ space according to the equations of motion. Such a total time derivative is sometimes called convective derivative.

The Liouville equation preserves the normalization, and from now on we suppose that f_N is normalized to 1. Besides, since the Hamiltonian (7.1) is invariant with respect to particle permutations, the Liouville dynamics preserves the transformation properties of f_N with respect to these permutations. As it is natural for identical particles, we therefore consider only functions f_N, which are completely invariant at all times with respect to permutations, i.e. symmetric with respect to the exchange of any two particles. The exact solution of the Liouville equation (7.4) is equivalent to the exact solution of the equations of motion (7.3); we already know that this is an unreachable goal.

Starting from f_N, it is possible to define distribution functions that depend on a reduced number of particles. Precisely, the s-particle reduced distribution functions are defined by the following partial integrations of f_N,

$$f_s(x_1, \ldots, x_s, t) = \frac{N!}{(N-s)!} \int \mathrm{d}x_{s+1} \ldots \mathrm{d}x_N \, f_N(x_1, \ldots, x_N, t), \tag{7.6}$$

for $s = 1, \ldots, N-1$. The complete symmetry of f_N with respect to the exchange of any two particles makes it immaterial on which set of $N-s$ particles we integrate to obtain f_s. Each f_s is normalized to $N!/(N-s)!$, i.e.

$$\int \mathrm{d}x_1 \ldots \mathrm{d}x_s f_s(x_1, \ldots, x_s, t) = \frac{N!}{(N-s)!} \tag{7.7}$$

and, for $s > 1$, the function f_s is symmetric with respect to the exchange of any two of its s particles. The time evolution of f_s is obtained by performing the corresponding partial integration in the Liouville equation (7.4). The computation is straightforward, and it exploits the symmetry properties of the Hamiltonian and of the N-body density distribution function f_N. We also suppose that f_N vanishes at the boundary of the domain of definition, which implies that

$$\int dx_i \, \frac{\mathbf{p}_i}{m} \cdot \boldsymbol{\nabla}_i f_N = 0 \qquad \text{and} \qquad \int dx_i \, (\boldsymbol{\nabla}_i V_{ij}) \cdot \frac{\partial f_N}{\partial \mathbf{p}_i} = 0. \tag{7.8}$$

We then have

$$\frac{\partial f_s}{\partial t} = -\sum_{i=1}^{s} \frac{\mathbf{p}_i}{m} \cdot \boldsymbol{\nabla}_i f_s + \sum_{i=1}^{s} \sum_{j \neq i}^{s} (\boldsymbol{\nabla}_i V_{ij}) \cdot \frac{\partial f_s}{\partial \mathbf{p}_i}$$

$$+ \frac{N!}{(N-s)!} \sum_{i=1}^{s} \sum_{j=s+1}^{N} \int dx_{s+1} \ldots dx_N \, (\boldsymbol{\nabla}_i V_{ij}) \cdot \frac{\partial f_N}{\partial \mathbf{p}_i}. \tag{7.9}$$

Because of the permutational symmetry of f_N, for each i all terms with different values of j, in the integral of the right-hand side, are equal. There are $(N - s)$ such terms for each i, and we therefore get

$$\frac{\partial f_s}{\partial t} = -\sum_{i=1}^{s} \frac{\mathbf{p}_i}{m} \cdot \boldsymbol{\nabla}_i f_s + \sum_{i=1}^{s} \sum_{j \neq i}^{s} (\boldsymbol{\nabla}_i V_{ij}) \cdot \frac{\partial f_s}{\partial \mathbf{p}_i} + \sum_{i=1}^{s} \int dx_{s+1} \, (\boldsymbol{\nabla}_i V_{i,s+1}) \cdot \frac{\partial f_{s+1}}{\partial \mathbf{p}_i} \tag{7.10}$$

for $s = 1, \ldots, N - 1$. The second term on the right-hand side is not present for $s = 1$. Together with the Liouville equation (7.4) these are N coupled integro-differential equations, in which the time evolution of each f_s is coupled to f_{s+1} (except the Liouville equation itself, which is a closed integral equation for f_N). They constitute the *BBGKY hierarchy*. We cannot obtain a closed system by cutting the hierarchy, keeping only the terms with $s = 1, 2, \ldots, s_0 < N$ for any value of s_0; the only closed system is that given by the $N - 1$ equations (7.10) plus the Liouville equation (7.4).

However, we can hope to have a tractable dynamical problem only by cutting the hierarchy at a very small value of s, at the expenses of introducing some degree of approximation. Although the knowledge of only the first few f_s, even if exact, is not as complete as the knowledge of f_N, it is sufficient for the computation of the expectation values of the usual observables. In fact, let us suppose to have an observable which is a sum of identical one-particle functions

$$O_1(x_1, \ldots, x_N) = \sum_{i=1}^{N} a(x_i), \tag{7.11}$$

or an observable which is a sum of identical two-particle functions

$$O_2(x_1, \ldots, x_N) = \sum_{i=1}^{N} \sum_{j \neq i}^{N} b(x_i, x_j). \tag{7.12}$$

Then, the expectation value $\langle O_1 \rangle$ is given by

$$\langle O_1 \rangle(t) = \int dx_1 \ldots dx_N \, O(x_1, \ldots, x_N) f_N(x_1, \ldots, x_N, t) \tag{7.13}$$

$$= \int dx_1 \ldots dx_N \sum_{i=1}^{N} a(x_i) \, f_N(x_1, \ldots, x_N, t) \tag{7.14}$$

$$= \sum_{i=1}^{N} \int dx_i \, a(x_i) \int dx_1 \dots dx_{i-1} \, dx_{i+1} \dots dx_N \, f_N(x_1, \dots, x_N, t) \quad (7.15)$$

$$= \sum_{i=1}^{N} \int dx_i \, a(x_i) \, \frac{(N-1)!}{N!} f_1(x_i, t) \quad (7.16)$$

$$= \int dx_1 \, a(x_1) f_1(x_1, t), \quad (7.17)$$

while the expectation value $\langle O_2 \rangle$ can be similarly simplified as

$$\langle O_2 \rangle (t) = \int dx_1 \, dx_2 \, b(x_1, x_2) f_2(x_1, x_2, t), \quad (7.18)$$

The purpose of kinetic theories is to obtain a closed equation for $f_1(x_1, t)$, and to describe in this way the statistical dynamics of a many-body system. This necessarily implies some kind of approximations, which will depend on the type of system under consideration. The kinetic equations that we will treat in this book can be derived by truncating the hierarchy at $s = 2$; in this way, one obtains a closed system of two equations. Then, introducing further approximations, one expresses $f_2(x_1, x_2, t)$ in terms of $f_1(x_1, t)$, arriving in this way to a closed equation for the latter, the so-called kinetic equation. The explicit form of these approximations depends on the features of the system under study; however, in all cases, the kinetic equation is nonlinear, while the Liouville equation and the equations of the BBGKY hierarchy are linear: this is the price to pay to close the system.

For convenience, we rewrite here the first two equations of the hierarchy, expliciting the dependence of f_1 and f_2 on the phase-space variables

$$\frac{\partial}{\partial t} f_1(x_1, t) = -\frac{\mathbf{p}_1}{m} \cdot \mathbf{\nabla}_1 f_1(x_1, t) + \int dx_2 \, (\mathbf{\nabla}_1 V_{12}) \cdot \frac{\partial}{\partial \mathbf{p}_1} f_2(x_1, x_2, t) \quad (7.19)$$

$$\frac{\partial}{\partial t} f_2(x_1, x_2, t) = -\frac{\mathbf{p}_1}{m} \cdot \mathbf{\nabla}_1 f_2(x_1, x_2, t) - \frac{\mathbf{p}_2}{m} \cdot \mathbf{\nabla}_2 f_2(x_1, x_2, t)$$

$$+ (\mathbf{\nabla}_1 V_{12}) \cdot \left(\frac{\partial}{\partial \mathbf{p}_1} - \frac{\partial}{\partial \mathbf{p}_2} \right) f_2(x_1, x_2, t)$$

$$+ \sum_{i=1}^{2} \int dx_3 \, (\mathbf{\nabla}_i V_{i3}) \cdot \frac{\partial}{\partial \mathbf{p}_i} f_3(x_1, x_2, x_3, t). \quad (7.20)$$

When f_N assumes a stationary state, obviously also the reduced distribution functions f_s will be stationary. In Chapter 1, we studied the equilibrium statistical ensembles, in which f_N has particular forms, depending on $(x_1, \dots, x_N)$ only through the Hamiltonian H. It is not difficult to obtain the equilibrium form of the reduced one-particle distribution function $f_1(x_1)$ in those cases. For both the microcanonical and the canonical distributions function, the form assumed is

$$f_1(\mathbf{q}_1, \mathbf{p}_1) = n \left(\frac{\beta}{2\pi m} \right)^{\frac{3}{2}} e^{-\beta \frac{p_1^2}{2m}}, \tag{7.21}$$

where $n = N/V$ is the number density of the system, and β is its inverse temperature. The derivation of Eq. (7.21) is trivial in the case of the canonical ensemble, and just a little bit more difficult for the microcanonical ensemble. They are both reported in Appendix C.

It is useful to write Eqs. (7.19) and (7.20) also in another equivalent form, which is obtained by introducing the correlation functions. These functions are based on the following physical consideration. In the thermodynamic limit, we expect that for large separations the reduced distribution functions for $s \geq 2$ are well approximated by a product of one-particle distribution functions, i.e.

$$f_s(x_1, \dots, x_s, t) \approx \prod_{i=1}^{s} f_1(x_i, t) \tag{7.22}$$

when $|\mathbf{q}_i - \mathbf{q}_j|$ is sufficiently large for all couples (i, j) among the s particles. The physical meaning of this factorization is clear. When the particles are sufficiently far apart they are uncorrelated, and the probability to find them in those positions is equal to the product of the separate probability to find each of them in its position. This will be no more true when the particles are close, and by continuity we expect a passage from the uncorrelated form, Eq. (7.22), to the correlated form. The correlation functions $g_s(x_1, \dots, x_s, t)$ express this passage, being related to the deviation of the complete distribution function from the uncorrelated form. We report here their definition for $s = 2$

$$f_2(x_1, x_2, t) = f_1(x_1, t)f_1(x_2, t) + g_2(x_1, x_2, t) \tag{7.23}$$

and for $s = 3$

$$f_3(x_1, x_2, x_3, t) = f_1(x_1, t)f_1(x_2, t)f_1(x_3, t) + f_1(x_1, t)g_2(x_2, x_3, t) \tag{7.24}$$
$$+ f_1(x_2, t)g_2(x_1, x_3, t) + f_1(x_3, t)g_2(x_1, x_2, t) + g_3(x_1, x_2, x_3, t).$$

While for $s = 2$ we separate from $f_2(x_1, x_2, t)$ the uncorrelated form, for $s = 3$ we separate not only the totally uncorrelated form, but also the partially uncorrelated forms, where one of the particles is uncorrelated to the other two. It is easy to see that, in the thermodynamic limit $N \to \infty$, the leading term of the normalization of the reduced distribution functions $f_s(x_1, \dots, x_s)$ is provided by the completely uncorrelated term. For example, integrating both sides of Eq. (7.23) we obtain, using Eq. (7.7),

$$N(N-1) = N^2 + \int dx_1 dx_2 \, g_2(x_1, x_2, t), \tag{7.25}$$

showing that, while the normalization of f_2 goes like N^2, that of g_2 goes like N. Analogous evaluations can be done for $s > 2$.

Substituting (7.23) and (7.24) into Eqs. (7.19) and (7.20), we arrive at the two equations

$$\frac{\partial}{\partial t} f_1(x_1, t) = -\frac{\mathbf{p}_1}{m} \cdot \mathbf{\nabla}_1 f_1(x_1, t)$$

$$+ \int dx_2 \, (\mathbf{\nabla}_1 V_{12}) \cdot \frac{\partial}{\partial \mathbf{p}_1} \, (f_1(x_1, t) f_1(x_2, t) + g_2(x_1, x_2, t)) \qquad (7.26)$$

$$\frac{\partial}{\partial t} g_2(x_1, x_2, t) = -\frac{\mathbf{p}_1}{m} \cdot \mathbf{\nabla}_1 g_2(x_1, x_2, t) - \frac{\mathbf{p}_2}{m} \cdot \mathbf{\nabla}_2 g_2(x_1, x_2, t)$$

$$+ (\mathbf{\nabla}_1 V_{12}) \cdot \left(\frac{\partial}{\partial \mathbf{p}_1} - \frac{\partial}{\partial \mathbf{p}_2} \right) (f_1(x_1, t) f_1(x_2, t) + g_2(x_1, x_2, t))$$

$$+ \int dx_3 \, (\mathbf{\nabla}_1 V_{13}) \cdot \frac{\partial}{\partial \mathbf{p}_1} \, (f_1(x_1, t) g_2(x_2, x_3, t) + f_1(x_3, t) g_2(x_1, x_2, t))$$

$$+ \int dx_3 \, (\mathbf{\nabla}_2 V_{23}) \cdot \frac{\partial}{\partial \mathbf{p}_2} \, (f_1(x_2, t) g_2(x_1, x_3, t) + f_1(x_3, t) g_2(x_1, x_2, t))$$

$$+ \int dx_3 \left[(\mathbf{\nabla}_1 V_{13}) \cdot \frac{\partial}{\partial \mathbf{p}_1} + (\mathbf{\nabla}_2 V_{23}) \cdot \frac{\partial}{\partial \mathbf{p}_2} \right] g_3(x_1, x_2, x_3, t). \qquad (7.27)$$

The kinetic equations that are suitable for the different situations are all obtained by neglecting in Eq. (7.27) the three-particle correlation function $g_3(x_1, x_2, x_3, t)$. It is sufficient to have a closed system of two integro-differential equations. Depending on the type of system under consideration and on the approximation adopted, further terms on the right-hand sides of Eqs. (7.26) and (7.27) of this system can be neglected. For example, eliminating the two-particle correlation functions, we have then an equation for the one-particle distribution function $f_1(x_1, t)$.

In this chapter, we will concentrate on the celebrated Boltzmann equation, which describes the dynamics of diluted systems with short-range interactions. However, we will first show a general feature of systems with short-range interactions, which can be derived by analysing the terms of Eq. (7.26).

We rewrite Eq. (7.26) in the form

$$\frac{\partial}{\partial t} f_1(x_1, t) + \frac{\mathbf{p}_1}{m} \cdot \mathbf{\nabla}_1 f_1(x_1, t) - (\mathbf{\nabla}_1 V[f_1, t]) \cdot \frac{\partial}{\partial \mathbf{p}_1} f_1(x_1, t)$$

$$= \int dx_2 \, (\mathbf{\nabla}_1 V_{12}) \cdot \frac{\partial}{\partial \mathbf{p}_1} g_2(x_1, x_2, t), \qquad (7.28)$$

where $V[f_1, t]$ is the averaged two-body interaction potential

$$V[f_1, t] = \int dx_2 \, V_{12} f_1(x_2, t). \qquad (7.29)$$

This equation shows that the effects of the interactions on the evolution of the one-particle distribution function can be divided into two terms: the last one on the left-hand-side, corresponding to a mean-field potential, and the term on the right-hand side, which depends on the two-particle correlation. A crucial point is that the relative weight of the two terms is different between systems with short-range or long-range interactions, as can be seen by the following reasoning.

Both the integral defining the mean-field potential (7.29) and the integral with the correlation function effectively extends, about x_1, to a distance corresponding to the range of the interaction potential V. For a short-range interacting system, in most situations and even out-of-equilibrium, the characteristic length of variation of the one-particle distribution f_1 is much larger than the range of the potential, and therefore the gradient of Eq. (7.29) will be very small. On the other hand, the correlation function and its gradient will have a range similar to that of the interaction potential, so that the right-hand side of Eq. (7.28) will not be negligible. For a long-range interacting system, the range of the potential extends throughout the whole system. In this case, unless the system is homogeneous, the mean-field term will be larger than the term with the correlation function which has a shorter range than f_1.

7.2   The Boltzmann Equation and the Rapid Approach to Equilibrium due to Collisions

The Boltzmann equation is suitable for dilute short-range interacting systems, where the average interparticle distance is large compared to the effective range R_0 of the two-body potential; two particles effectively interact only when they come close within a distance R_0. Exploiting the dilution of the system, the Boltzmann equation can be derived by writing an expression for the last term on the right-hand side of Eq. (7.19) on the basis of the following hypotheses: only binary collisions take place; they are local in space; molecular chaos holds, i.e. before colliding two molecules are completely uncorrelated. The derivation can be made more rigorous starting from the BBGKY hierarchy, as we will do now, although we will not use arguments that claim full mathematical rigour.

We can use the average density $n = N/V$ formally as a small expansion parameter, keeping in mind that the physical expression of dilution is given by $nR_0^3 \ll 1$. The normalization properties (7.7) implies that the reduced distribution functions f_s are proportional to n^s. The BBGKY hierarchy can thus be closed, keeping terms up to n^2, and we can either employ Eqs. (7.19) and (7.20), neglecting in the latter the term with f_3, or, equivalently, employ Eqs. (7.26) and (7.27), neglecting in the latter all three-particle terms, i.e., the terms with the integral in dx_3. In this way, we finally end up with the Boltzmann equation. The discussion of Boltzmann equation can be found in Huang (1987) and Balian (1992).

7.2.1   The derivation of the Boltzmann equation

In this case, it is convenient to work with Eqs. (7.19) and (7.20), since we can start by simply neglecting the last term in Eq. (7.20). Then, the latter becomes a closed equation for $f_2(x_1, x_2, t)$. Let us see the physical meaning of this truncation, comparing the order of magnitude of the various terms on the right-hand sides of Eqs. (7.19) and (7.20), which determine the time variation of $f_1(x_1, t)$ and $f_2(x_1, x_2, t)$.

We begin by considering the distribution function themselves and the integrals on the right-hand sides of both equations. The spatial integration is effectively extended up to the effective range of the potential R_0 from $\mathbf{q}_1$ in the first equation, and from $\mathbf{q}_1$ or $\mathbf{q}_2$ in the second one. Then, since f_2 is of the order of $nf_1 = (N/V)f_1$, the spatial integration produces a term of the order of $nR_0^3 f_1$, namely much smaller than f_1. Analogously, the spatial integration in the second equation produces a term of the order of $nR_0^3 f_2$, much smaller than f_2.

Secondly, we estimate that the gradient of the potential times the derivative with respect to the momentum gives a factor of the order of $1/t_c$, where t_c is the effective duration of a collision. Then the integrals on the right-hand sides of the two equations give, respectively, terms of the order of $nR_0^3 f_1 / t_c$ and $nR_0^3 f_2 / t_c$.

Finally, we evaluate the order of the terms outside the collision integrals. The only such term in the equation for f_1 is of the order of f_1/t_1, where $t_1 \gg t_c$ is the time taken by a particle to traverse a distance over which f_1 varies significantly. In the equation for f_2, there is an analogous term of the order of f_2/t_1 and a collisional term of the order of f_2/t_c.

Summarizing:

- The terms on the right-hand side of Eq. (7.19) are of the order of f_1/t_1 and $nR_0^3 f_1 / t_c$ (the collisional integral). The ratio of both is $nR_0^3 t_1 / t_c$, whose value depends on the values of the (small) density and of the range of the interaction.

- On the other hand, the order of the terms on the right-hand side of Eq. (7.20) are f_2/t_1, f_2/t_c and, for the term corresponding to the collisional integral, we get $nR_0^3 f_2 / t_c \ll f_2 / t_c$. We, therefore, see that the evolution of f_2 due to collisions is mainly determined by the two-body collisions between the two particles, while collisions with the other particles, represented by the integral term, contribute much less.

- In the spirit of keeping only the dominating terms in the evolution of f_1, we then neglect the integral collision in the equation for f_2 and we obtain a close system of two equations, which for convenience we write as

$$\frac{\partial}{\partial t} f_1(x_1, t) = -\frac{\mathbf{p}_1}{m} \cdot \mathbf{\nabla}_1 f_1(x_1, t) + \int dx_2 \, (\mathbf{\nabla}_1 V_{12}) \cdot \frac{\partial}{\partial \mathbf{p}_1} f_2(x_1, x_2, t) \qquad (7.30)$$

$$\frac{\partial}{\partial t} f_2(x_1, x_2, t) = -\frac{\mathbf{p}_1}{m} \cdot \mathbf{\nabla}_1 f_2(x_1, x_2, t) - \frac{\mathbf{p}_2}{m} \cdot \mathbf{\nabla}_2 f_2(x_1, x_2, t)$$

$$+ (\mathbf{\nabla}_1 V_{12}) \cdot \left(\frac{\partial}{\partial \mathbf{p}_1} - \frac{\partial}{\partial \mathbf{p}_2} \right) f_2(x_1, x_2, t). \qquad (7.31)$$

We emphasize that the procedure mentioned in the last point means assuming that the dynamics of the system is completely determined by binary collisions.

The next step is to put, in Eq. (7.31), the time derivative equal to 0. This is equivalent to say that f_2 goes to equilibrium much faster than f_1. Then, the latter equation reduces to

$$(\nabla_1 V_{12}) \cdot \left(\frac{\partial}{\partial \mathbf{p}_1} - \frac{\partial}{\partial \mathbf{p}_2} \right) f_2(x_1, x_2, t) = \frac{\mathbf{p}_1}{m} \cdot \nabla_1 f_2(x_1, x_2, t) + \frac{\mathbf{p}_2}{m} \cdot \nabla_2 f_2(x_1, x_2, t). \quad (7.32)$$

We use this equation to substitute the integrand in the collision integral on the right-hand side of Eq. (7.30), to obtain

$$\frac{\partial}{\partial t} f_1(x_1, t) = -\frac{\mathbf{p}_1}{m} \cdot \nabla_1 f_1(x_1, t) + \int dx_2 \left(\frac{\mathbf{p}_1}{m} \cdot \nabla_1 + \frac{\mathbf{p}_2}{m} \cdot \nabla_2 \right) f_2(x_1, x_2, t), \quad (7.33)$$

where we have exploited the fact that

$$\int dx_2 \ (\nabla_1 V_{12}) \cdot \frac{\partial}{\partial \mathbf{p}_2} f_2(x_1, x_2, t) = 0, \quad (7.34)$$

which makes it possible to substitute the term $\frac{\partial}{\partial \mathbf{p}_1} f_2$ with the term $\left(\frac{\partial}{\partial \mathbf{p}_1} - \frac{\partial}{\partial \mathbf{p}_2} \right) f_2$ inside the collision integral in Eq. (7.30). Introducing the centre-of-mass coordinates, which replace $\mathbf{q}_1$ and $\mathbf{q}_2$ with

$$Q = \frac{1}{2} (\mathbf{q}_1 + \mathbf{q}_2) \quad (7.35)$$

$$\mathbf{q} = \mathbf{q}_1 - \mathbf{q}_2 \quad (7.36)$$

we derive the equality

$$\frac{1}{m} \mathbf{p}_1 \cdot \nabla_1 f_2(x_1, x_2, t) + \frac{1}{m} \mathbf{p}_2 \cdot \nabla_2 f_2(x_1, x_2, t)$$

$$= \frac{1}{m} (\mathbf{p}_1 - \mathbf{p}_2) \cdot \frac{\partial}{\partial \mathbf{q}} f_2(x_1, x_2, t) + \frac{1}{2m} (\mathbf{p}_1 + \mathbf{p}_2) \cdot \frac{\partial}{\partial Q} f_2(x_1, x_2, t). \quad (7.37)$$

At this point, we neglect the gradient with respect to the centre-of-mass coordinate Q. Then, we expect that the variation of f_2 with $\mathbf{q}$, when $|\mathbf{q}|$ is of the order of R_0 (this is the range of $\mathbf{q}$ that contributes in the collision integral of Eq. (7.19)), is much larger than the variation with Q. We finally get

$$\frac{\partial}{\partial t} f_1(x_1, t) = -\frac{\mathbf{p}_1}{m} \cdot \nabla_1 f_1(x_1, t) + \int dx_2 \frac{(\mathbf{p}_1 - \mathbf{p}_2)}{m} \cdot \frac{\partial}{\partial \mathbf{q}} f_2(x_1, x_2, t). \quad (7.38)$$

Let us consider only the spatial integral in the collisional term of the last equation

$$\int d\mathbf{q}_2 \frac{(\mathbf{p}_1 - \mathbf{p}_2)}{m} \cdot \frac{\partial}{\partial \mathbf{q}} f_2(x_1, x_2, t) = \int d\mathbf{q} \frac{(\mathbf{p}_1 - \mathbf{p}_2)}{m} \cdot \frac{\partial}{\partial \mathbf{q}} f_2(\mathbf{q}, \mathbf{p}_1, \mathbf{p}_2, t), \quad (7.39)$$

where on the right-hand side we have made a change of coordinates, and where, according to the approximation to neglect the variation of f_2 with respect to the centre-of-mass coordinate $\mathbf{Q}$, we have indicated only its dependence on the relative coordinate $\mathbf{q}$. Denoting with $q_\parallel$ and $\mathbf{q}_\perp$ the components of $\mathbf{q}$ that are respectively parallel and perpendicular to $\mathbf{p}_1 - \mathbf{p}_2$, the right-hand side of Eq. (7.39) becomes

$$\int dq_\parallel d\mathbf{q}_\perp \frac{|\mathbf{p}_1 - \mathbf{p}_2|}{m} \frac{\partial}{\partial q_\parallel} f_2(q_\parallel, \mathbf{q}_\perp, \mathbf{p}_1, \mathbf{p}_2, t). \tag{7.40}$$

The integral over $q_\parallel$ is equal to the variation of f_2 due to the interaction of the two particles. More precisely, for given $\mathbf{p}_1$, $\mathbf{p}_2$ and $\mathbf{q}_\perp$, it is equal to the distribution function f_2 with those momenta and that transversal coordinate after a two-body collision minus the distribution function f_2 with the same momenta and the same transversal coordinate before a two-body collision; both cases occur when $|q_\parallel|$ (and thus $|\mathbf{q}|$) is much larger than R_0, i.e. for $q_\parallel$ large and negative before the collision while large and positive after the collision. The last expression can thus be written as

$$\int d\mathbf{q}_\perp \frac{|\mathbf{p}_1 - \mathbf{p}_2|}{m} \left[f_2(q_\parallel = +\infty, \mathbf{q}_\perp, \mathbf{p}_1, \mathbf{p}_2, t) - f_2(q_\parallel = -\infty, \mathbf{q}_\perp, \mathbf{p}_1, \mathbf{p}_2, t) \right], \tag{7.41}$$

where we have formally indicated with $\pm\infty$ the large values of $q_\parallel$.

It is at this point that we use the famous molecular chaos hypothesis introduced by Boltzmann. As we remarked earlier, for large interparticle separation, we expect that $f_2(x_1, x_2, t) \sim f_1(x_1, t) f_1(x_2, t)$. The assumption of molecular chaos consists of extending this factorization also at small distances, in particular at distances just above the range of interaction R_0; physically, this means that the two particles entering a collision are uncorrelated. As a consequence, we can make the substitution

$$f_2(q_\parallel = -\infty, \mathbf{q}_\perp, \mathbf{p}_1, \mathbf{p}_2, t) \to f_1(q_{1\parallel}, \mathbf{q}_{1\perp}, \mathbf{p}_1, t) f_1(q_{2\parallel} = +\infty, \mathbf{q}_{2\perp}, \mathbf{p}_2, t), \tag{7.42}$$

where, on the right-hand-side, we have used the initial coordinates $\mathbf{q}_1$ and $\mathbf{q}_2$ instead of the centre-of-mass coordinates. In principle, we could not make the same substitution for the other f_2 term inside brackets in Eq. (7.41), since that term refers to f_2 soon after a two-body collision and, in that case, we expect the two particles to be correlated. However, we can use the fact that Eq. (7.31) is the Liouville equation of an isolated two-particle system, for which the vanishing of the convective derivative (see Eq. (7.5)) holds. Then, if we denote by $\mathbf{p}_1'$ and $\mathbf{p}_2'$ the momenta that are transformed to $\mathbf{p}_1$ and $\mathbf{p}_2$ by a binary collision, we can write

$$f_2(q_\parallel = +\infty, \mathbf{q}_\perp, \mathbf{p}_1, \mathbf{p}_2, t) = f_2(q_\parallel = -\infty, \mathbf{q}_\perp, \mathbf{p}_1', \mathbf{p}_2', t)$$
$$\to f_1(q_{1\parallel}, \mathbf{q}_{1\perp}, \mathbf{p}_1', t) f_1(q_{2\parallel} = +\infty, \mathbf{q}_{2\perp}, \mathbf{p}_2', t), \tag{7.43}$$

where in the second step we have used the fact that before the collision the particles are uncorrelated. After these manipulations, we have transformed Eq. (7.38) in

$$\frac{\partial}{\partial t} f_1(x_1, t) = -\frac{\mathbf{p}_1}{m} \cdot \nabla_1 f_1(x_1, t) + \int d\mathbf{p}_2 d\mathbf{q}_\perp \frac{|\mathbf{p}_1 - \mathbf{p}_2|}{m} \mathcal{A}[f_1], \tag{7.44}$$

with $\mathcal{A}[f_1]$ defined by

$$\mathcal{A}[f_1] \equiv f_1(q_{1\|}, \mathbf{q}_{1\perp}, \mathbf{p}'_1, t) f_1(q_{2\|} = +\infty, \mathbf{q}_{2\perp}, \mathbf{p}'_2, t)$$
$$- f_1(q_{1\|}, \mathbf{q}_{1\perp}, \mathbf{p}_1, t) f_1(q_{2\|} = +\infty, \mathbf{q}_{2\perp}, \mathbf{p}_2, t). \tag{7.45}$$

We can further transform Eq. (7.44), proceeding in four steps. In the first step we exploit the invariance of the dynamics by time reversal and by parity inversion. We note that $\mathbf{p}'_1$ and $\mathbf{p}'_2$ are the momenta that are transformed by a binary collision, for a given transversal coordinate $\mathbf{q}_\perp$, to $\mathbf{p}_1$ and $\mathbf{p}_2$. Then time reversal invariance implies that the momenta $-\mathbf{p}_1$ and $-\mathbf{p}_2$ are transformed by a binary collision, for the same $\mathbf{q}_\perp$, to the momenta $-\mathbf{p}'_1$ and $-\mathbf{p}'_2$. From this, invariance by parity inversion implies that the momenta $\mathbf{p}_1$ and $\mathbf{p}_2$ for that $\mathbf{q}_\perp$ are transformed by a binary collision to $\mathbf{p}'_1$ and $\mathbf{p}'_2$. Therefore we can interpret in this way the momenta $\mathbf{p}'_1$ and $\mathbf{p}'_2$ that appear in $\mathcal{A}[f_1]$ in Eq. (7.45) and in Eq. (7.44), i.e. the final momenta of two particles entering the collision, for the given $\mathbf{q}_\perp$, with momenta $\mathbf{p}_1$ and $\mathbf{p}_2$. In the second step we make a change of coordinate in the plane of $\mathbf{q}_\perp$, using polar coordinates (b, ϕ), such that $d\mathbf{q}_\perp = b db d\phi$. In the third step we note that the final momenta $\mathbf{p}'_1$ and $\mathbf{p}'_2$ depend on $\mathbf{p}_1, \mathbf{p}_2$ and $b = |\mathbf{q}_\perp|$. In the fourth and final step, noting that the interaction range R_0 is much smaller than the distance over which f_1 varies significantly, we can evaluate the functions f_1 in Eq. (7.45) all at the same point $\mathbf{q}_1$. Thus, Eq. (7.44) can be written as

$$\frac{\partial}{\partial t} f_1(x_1, t) = -\frac{\mathbf{p}_1}{m} \cdot \nabla_1 f_1(x_1, t) + \frac{1}{m} \int d\mathbf{p}_2 db d\phi \, b \, |\mathbf{p}_1 - \mathbf{p}_2| \tag{7.46}$$
$$\times \left[f_1(\mathbf{q}_1, \mathbf{p}'_1(\mathbf{p}_1, \mathbf{p}_2, b), t) f_1(\mathbf{q}_1, \mathbf{p}'_2(\mathbf{p}_1, \mathbf{p}_2, b), t) - f_1(\mathbf{q}_1, \mathbf{p}_1, t) f_1(\mathbf{q}_1, \mathbf{p}_2, t) \right].$$

We have arrived at the celebrated Boltzmann equation. It is often written in an equivalent form, in which the dependence of the final momenta $\mathbf{p}'_1$ and $\mathbf{p}'_2$ on the initial momenta $\mathbf{p}_1, \mathbf{p}_2$ and on b, called the impact parameter, is expressed through the differential cross-section $d\sigma/d\Omega$: this function is determined by the two-body potential $V(q)$, and the solid angle Ω denotes the direction of the vector $\mathbf{p}'_1 - \mathbf{p}'_2$ with respect to that of the vector $\mathbf{p}_1 - \mathbf{p}_2$. Since

$$\frac{d\sigma}{d\Omega} d\Omega = b db d\phi, \tag{7.47}$$

the Boltzmann equation becomes

$$\frac{\partial}{\partial t} f_1(x_1, t) = -\frac{\mathbf{p}_1}{m} \cdot \nabla_1 f_1(x_1, t) + \frac{1}{m} \int d\mathbf{p}_2 d\Omega \, |\mathbf{p}_1 - \mathbf{p}_2| \frac{d\sigma}{d\Omega} \tag{7.48}$$
$$\times \left[f_1(\mathbf{q}_1, \mathbf{p}'_1, t) f_1(\mathbf{q}_1, \mathbf{p}'_2, t) - f_1(\mathbf{q}_1, \mathbf{p}_1, t) f_1(\mathbf{q}_1, \mathbf{p}_2, t) \right],$$

where it is understood that $\mathbf{p}_1'$ and $\mathbf{p}_2'$ depend on $\mathbf{p}_1 - \mathbf{p}_2$ and Ω as determined by the two-body potential $V(\mathbf{q})$, and that the differential cross-section $d\sigma/d\Omega$ depends on both Ω and $|\mathbf{p}_1 - \mathbf{p}_2|$. In Appendix D, we give some details about the differential cross-section $d\sigma/d\Omega$, in particular explaining the relation in Eq. (7.47).

The Boltzmann equation is an integro-differential equation, like the other kinetic equations that we will encounter later in this book.

7.2.2   The H-theorem

This important theorem shows that the evolution determined by the Boltzmann equation (7.48) drives the system towards equilibrium. This is done by defining a functional of the distribution function $f_1(\mathbf{q}_1, \mathbf{p}_1, t)$ and showing that its time derivative is never positive. Thus, the distribution function will approach the form that makes this time derivative vanish. The H functional is defined by

$$H(t) = \int d\mathbf{q}_1 \, d\mathbf{p}_1 \, f_1(\mathbf{q}_1, \mathbf{p}_1, t) \ln [f_1(\mathbf{q}_1, \mathbf{p}_1, t)] \,. \tag{7.49}$$

Using Eq. (7.48), its time derivative is

$$\frac{dH(t)}{dt} = \int d\mathbf{q}_1 \, d\mathbf{p}_1 \, \{1 + \ln [f_1(\mathbf{q}_1, \mathbf{p}_1, t)]\} \frac{\partial}{\partial t} f_1(\mathbf{q}_1, \mathbf{p}_1, t) \tag{7.50}$$

$$= - \int d\mathbf{q}_1 \, d\mathbf{p}_1 \, \{1 + \ln [f_1(\mathbf{q}_1, \mathbf{p}_1, t)]\} \frac{\mathbf{p}_1}{m} \cdot \nabla_1 f_1(\mathbf{q}_1, \mathbf{p}_1, t)$$

$$+ \frac{1}{m} \int d\mathbf{q}_1 \, d\mathbf{p}_1 \, d\mathbf{p}_2 \, d\Omega \, \frac{d\sigma}{d\Omega} |\mathbf{p}_1 - \mathbf{p}_2| \, \{1 + \ln [f_1(\mathbf{q}_1, \mathbf{p}_1, t)]\}$$

$$\times \left[f_1(\mathbf{q}_1, \mathbf{p}_1', t) f_1(\mathbf{q}_1, \mathbf{p}_2', t) - f_1(\mathbf{q}_1, \mathbf{p}_1, t) f_1(\mathbf{q}_1, \mathbf{p}_2, t) \right] \,. \tag{7.51}$$

The first integral on the right-hand side vanishes, since the integrand can be transformed to $(\mathbf{p}_1/m) \cdot \nabla_1 \{f_1(\mathbf{q}_1, \mathbf{p}_1, t) \ln [f_1(\mathbf{q}_1, \mathbf{p}_1, t)]\}$. The second integral can be first transformed using the equality

$$\int d\mathbf{p}_1 \, d\mathbf{p}_2 \, d\Omega \, \frac{d\sigma}{d\Omega} |\mathbf{p}_1 - \mathbf{p}_2| \, \{1 + \ln [f_1(\mathbf{q}_1, \mathbf{p}_1, t)]\}$$

$$\times \left[f_1(\mathbf{q}_1, \mathbf{p}_1', t) f_1(\mathbf{q}_1, \mathbf{p}_2', t) - f_1(\mathbf{q}_1, \mathbf{p}_1, t) f_1(\mathbf{q}_1, \mathbf{p}_2, t) \right]$$

$$= \int d\mathbf{p}_1 \, d\mathbf{p}_2 \, d\Omega \, \frac{d\sigma}{d\Omega} |\mathbf{p}_2 - \mathbf{p}_1| \, \{1 + \ln [f_1(\mathbf{q}_1, \mathbf{p}_2, t)]\}$$

$$\times \left[f_1(\mathbf{q}_1, \mathbf{p}_1', t) f_1(\mathbf{q}_1, \mathbf{p}_2', t) - f_1(\mathbf{q}_1, \mathbf{p}_1, t) f_1(\mathbf{q}_1, \mathbf{p}_2, t) \right] \,, \tag{7.52}$$

obtained by simply interchanging the dummy variables $\mathbf{p}_1$ and $\mathbf{p}_2$. The right-hand side can therefore be written as

$$\frac{1}{2} \int d\mathbf{p}_1 d\mathbf{p}_2 d\Omega \, \frac{d\sigma}{d\Omega} |\mathbf{p}_1 - \mathbf{p}_2| \, \{2 + \ln [f_1(\mathbf{q}_1, \mathbf{p}_1, t) f_1(\mathbf{q}_1, \mathbf{p}_2, t)]\}$$
$$\times \left[f_1(\mathbf{q}_1, \mathbf{p}_1', t) f_1(\mathbf{q}_1, \mathbf{p}_2', t) - f_1(\mathbf{q}_1, \mathbf{p}_1, t) f_1(\mathbf{q}_1, \mathbf{p}_2, t) \right]. \tag{7.53}$$

Interchanging $(\mathbf{p}_1, \mathbf{p}_2)$ and $(\mathbf{p}_1', \mathbf{p}_2')$, we obtain the inverse collision, which has the same differential cross-section. Therefore the last expression is equal to

$$\frac{1}{2} \int d\mathbf{p}_1' d\mathbf{p}_2' d\Omega \, \frac{d\sigma}{d\Omega} |\mathbf{p}_1' - \mathbf{p}_2'| \, \{2 + \ln [f_1(\mathbf{q}_1, \mathbf{p}_1', t) f_1(\mathbf{q}_1, \mathbf{p}_2', t)]\}$$
$$\times \left[f_1(\mathbf{q}_1, \mathbf{p}_1, t) f_1(\mathbf{q}_1, \mathbf{p}_2, t) - f_1(\mathbf{q}_1, \mathbf{p}_1', t) f_1(\mathbf{q}_1, \mathbf{p}_2', t) \right]. \tag{7.54}$$

Using that $d\mathbf{p}_1' d\mathbf{p}_2' = d\mathbf{p}_1 d\mathbf{p}_2$, we have that each one of the last two expressions is equal to

$$\frac{1}{4} \int d\mathbf{p}_1 d\mathbf{p}_2 d\Omega \, \frac{d\sigma}{d\Omega} |\mathbf{p}_1 - \mathbf{p}_2|$$
$$\times \left\{ \ln [f_1(\mathbf{q}_1, \mathbf{p}_1, t) f_1(\mathbf{q}_1, \mathbf{p}_2, t)] - \ln [f_1(\mathbf{q}_1, \mathbf{p}_1', t) f_1(\mathbf{q}_1, \mathbf{p}_2', t)] \right\}$$
$$\times \left[f_1(\mathbf{q}_1, \mathbf{p}_1', t) f_1(\mathbf{q}_1, \mathbf{p}_2', t) - f_1(\mathbf{q}_1, \mathbf{p}_1, t) f_1(\mathbf{q}_1, \mathbf{p}_2, t) \right]. \tag{7.55}$$

Substituting in Eq. (7.51), we finally have

$$\frac{dH(t)}{dt} = \frac{1}{4m} \int d\mathbf{q}_1 d\mathbf{p}_1 d\mathbf{p}_2 d\Omega \, \frac{d\sigma}{d\Omega} |\mathbf{p}_1 - \mathbf{p}_2|$$
$$\times \left\{ \ln [f_1(\mathbf{q}_1, \mathbf{p}_1, t) f_1(\mathbf{q}_1, \mathbf{p}_2, t)] - \ln [f_1(\mathbf{q}_1, \mathbf{p}_1', t) f_1(\mathbf{q}_1, \mathbf{p}_2', t)] \right\}$$
$$\times \left[f_1(\mathbf{q}_1, \mathbf{p}_1', t) f_1(\mathbf{q}_1, \mathbf{p}_2', t) - f_1(\mathbf{q}_1, \mathbf{p}_1, t) f_1(\mathbf{q}_1, \mathbf{p}_2, t) \right]. \tag{7.56}$$

The right-hand side is never positive, since the logarithm is a monotonic increasing function, and therefore $(x - y) [\ln(y) - \ln(x)] < 0$ whenever $x \neq y$.

Then, $f_1(\mathbf{q}_1, \mathbf{p}_1, t)$ will rapidly converge, in a time of the order of magnitude of the collision time t_c, to a function that makes $dH(t)/dt = 0$. This can happen only for a function where, for each $\mathbf{q}_1$, we have the equality

$$f_1(\mathbf{q}_1, \mathbf{p}_1', t) f_1(\mathbf{q}_1, \mathbf{p}_2', t) = f_1(\mathbf{q}_1, \mathbf{p}_1, t) f_1(\mathbf{q}_1, \mathbf{p}_2, t) \tag{7.57}$$

for each $(\mathbf{p}_1, \mathbf{p}_2)$ and $(\mathbf{p}_1', \mathbf{p}_2')$, where the former and the latter are the initial and the final momenta, respectively, of any possible binary collision. This means that $\ln [f_1(\mathbf{q}_1, \mathbf{p}_1, t)]$ is a function of the total momentum and the total energy, the two conserved quantities in binary collisions, so that we have

$$\ln [f_1(\mathbf{q}_1, \mathbf{p}_1, t)] = -c_1(\mathbf{q}_1, t) \frac{p_1^2}{2m} + \mathbf{c}_2(\mathbf{q}_1, t) \cdot \mathbf{p}_1 + c_3(\mathbf{q}_1, t), \tag{7.58}$$

where $c_1(\mathbf{q}_1, t)$ and $c_3(\mathbf{q}_1, t)$ are two arbitrary scalar functions of $\mathbf{q}_1$ and t, while $\mathbf{c}_2(\mathbf{q}_1, t)$ is an arbitrary vector function. Posing

$$\beta \equiv c_1, \qquad \mathbf{p}_0 \equiv \frac{m}{c_1}\mathbf{c}_2, \quad \text{and} \quad n \equiv \left(\frac{2\pi m}{c_1}\right)^{\frac{3}{2}} e^{-\frac{mc_2^2}{2c_1} + c_3}, \tag{7.59}$$

we obtain

$$f_1(\mathbf{q}_1, \mathbf{p}_1, t) = n(\mathbf{q}_1, t) \left(\frac{\beta(\mathbf{q}_1, t)}{2\pi m}\right)^{\frac{3}{2}} \exp\left[-\beta(\mathbf{q}_1, t)\frac{(p_1 - p_0(\mathbf{q}_1, t))^2}{2m}\right], \tag{7.60}$$

which has the form of the equilibrium distribution function (7.21), but with the density n and the inverse temperature β arbitrary functions of the position and time; furthermore, for each $\mathbf{q}_1$, the Gaussian velocity distribution function is centred on an arbitrary average velocity. The Boltzmann collision integral vanishes for functions of the form (7.60), and only for this form. Obviously, a function like this one is not a stationary distribution function of the Boltzmann equation, since the streaming term, i.e. the first term on the right-hand side of Eq. (7.48), is not 0. However, we can expect that this term will gradually transform n, β and $\mathbf{p}_0$ in constants, the latter equal to 0 for a system globally at rest. This statement must be taken with care. In fact, if at a particular time f_1 has the form expressed in Eq. (7.60), this form will not be maintained at the immediately following times, under the action of the streaming term; however, the action of the collision integral (that acts on a much shorter time scale) will soon restore that form. Consequently, f_1 will reach the form (7.21), which is therefore the only stationary solution of the Boltzmann equation.

We note that if at time $t = 0$ the function f_1 is uniform in space, i.e. it depends only on the momentum $\mathbf{p}_1$, this uniformity is conserved by the Boltzmann equation. For this reason, the H-theorem is very often presented, in textbooks, using uniform distribution functions. In that case, the theorem directly shows that f_1 will reach the uniform equilibrium distribution function (7.21), remaining uniform for all times.

In conclusion, the evolution of f_1 determined by the Boltzmann equation will make $H(t)$ to decrease monotonically until it reaches its minimum possible value, compatible with the average density and average kinetic energy of the system. It is then natural to assume that $-H$ represents the entropy of the system, since this function will reach the maximum value compatible with the macroscopic constraints.

7.2.3 H-theorem and irreversibility

The Boltzmann equation and the H-theorem introduce us to a feature of profound conceptual importance embodied in kinetic equations. This is a property that belongs also to the other kinetic equations (except the simplest of all, the Vlasov equation, which in fact is not considered a genuine kinetic equation), which we will study in the following chapters and which are suited for long-range interacting systems.

The equations of motion (7.3) are invariant under time reversal, and this is reflected in the same invariance property of the Liouville equation (7.4) and of the equations of the BBGKY hierarchy (7.10). In contrast, the Boltzmann equation, as can be checked in Eq. (7.48), is not time reversal invariant, and this lack of invariance is due to the collisional integral. Coherently with this, we have found a functional, precisely the H functional, which monotonically decreases during the dynamical evolution of the distribution function f_1 as determined by the Boltzmann equation. On the one hand, this is a very satisfying property, since one of the main purposes of statistical mechanics is, starting from the microscopic laws, to describe and explain the behaviour of macroscopic systems which is manifestly irreversible, according to experience; moreover, an irreversible behaviour can be a natural consequence of evolution equations that are not time reversal invariant. On the other hand, this raises a difficult conceptual problem. In fact, while it can be argued that obtaining kinetic equations not time reversal invariant has been a consequence of the approximations introduced in their derivation, we should be aware of where and how this crucial step has been taken, and we should make sure that it is a good representation of physical reality.

For the Boltzmann equation irreversibility has been inserted by using the molecular chaos hypothesis, in which the two particles undergoing a binary collision are considered to be completely uncorrelated up to just before the collision. Obviously soon after the collision, both particles are strongly correlated, and this correlation will diminish while time elapses after the collision, because of further collisions of each of the two partners with other particles. We then understand that, assuming only collisions between uncorrelated particles, we are making the physical assumption that two particles that have collided can collide again only when the correlation built from the previous collision has been erased. Intuitively, this is a very reasonable assumption in dilute fluids with short-range interactions, the systems for which the Boltzmann equation is suited, and we expect that this is physically realized in a real system. On the other hand, it is also clear that this assumption violates time reversal invariance: reversing the velocities at a given instant will make particles to begin a collision when they had finished it before time reversal, and therefore to begin a collision while they are strongly correlated; but this is not described by the Boltzmann equation. Despite this, the overall picture is made coherent by the fact that the motion of f_1 obtained by reversing the time in a solution of the Boltzmann equation is extremely improbable, so that it can be considered impossible from any reasonable physical point of view; therefore we do not have to worry if this reversed motion is not a possible solution of the Boltzmann equation. Summarizing, we can expect the following. Given an initial f_1, we cannot compute with full exactness its real time evolution; however, the evolution evaluated with the Boltzmann equation will always be very close to the real (and unknown) one. In particular, in the real dynamics, $\mathrm{d}H/\mathrm{d}t$ will not be a negative definite function, but the real $H(t)$ will always stay very close to that obtained with the Boltzmann equation.

In this chapter, we have described the physical explanation of the coexistence of microscopic reversibility and macroscopic irreversibility, and the way in which this coexistence is coherently inserted inside a statistical mechanics picture, which starts from the microscopic laws and describes the dynamics of the macroscopic behaviour. Although some technical details can differ between different kinetic equations (and the quality of the approximation obtained by the kinetic description may vary), the heart of the argument and its validity remain the same.

8

Kinetic Theory of Long-Range Systems: Klimontovich, Vlasov and Lenard–Balescu Equations

We have seen in the previous chapter that a kinetic equation is a closed equation for the one-particle distribution function; the exact evolution of this function is determined by an equation that involves also the two-particle distribution function (two equivalent expressions are given in Eq. (7.19) and in Eq. (7.28)). With the help of suitable assumptions and approximations, it is possible to transform the term depending on the two-particle function into another term, the collisional term, which depends on the one-particle function, obtaining a closed equation. We have shown that Boltzmann's theory assumes that the particles interact only through binary collisions whose duration is very short with respect to the average time between collisions, and it describes diluted gases with short-range interactions. These assumptions are not suitable for systems that interact through long-range forces, and therefore the derivation of a kinetic equation for these systems must use different approximations. As a consequence, the collisional term for long-range systems must be different from the Boltzmann collisional term.

To take into account collisions determined by long-range Coulomb and gravitational forces, Landau (1936) and Chandrasekhar (1942) modified Boltzmann's collisional term. Mean-field collective effects were first considered by Vlasov (1945) and led to the Vlasov equation, later revisited by Landau (1946) himself. A treatment of collision terms in the context of the Vlasov–Landau approach was developed by Lenard (1960) and Balescu (1960) (see Feix and Bertrand (2005) for a review).

In this chapter, we will give a brief introduction to the kinetic theory of long-range interactions. As anticipated, in doing so we will not follow the approach based on the Liouville equation and the BBGKY hierarchy (Willis and Picard, 1974; Balescu, 1975; Huang, 1987; Kandrup, 1991), which has been employed in Chapter 7 with reference to short-range systems. Here we will use the approach that starts from the Klimontovich equation (Klimontovich *et al.*, 1967; Lifshitz and Pitaevskii, 1981; Nicholson, 1983); as we will see, this equation is an equivalent formulation of the equations of motion of the N-body system. We will then derive the Vlasov and the Lenard–Balescu equations. The former is a widely used equation for long-range systems, above all for plasma, in which

Physics of Long-Range Interacting Systems. First Edition. A. Campa *et al.*

the effect of the interaction is taken into account only through a mean-field potential. As a matter of fact, it should not be considered strictly a kinetic equation, since the discreteness of the system is completely wiped out. Despite this approximation, the Vlasov equation is often a very good approximation of the dynamics, since, as we briefly mentioned in the previous chapter, in long-range systems the mean-field term can be the dominant one. On the other hand, the Lenard–Balescu equation includes the collisions due to the discrete nature of the system, and it represents an improved approximation with respect to the Vlasov equation. The collisional term of the Lenard–Balescu equation will be shown to be of the order $1/N$ with respect to the mean-field term of the Vlasov equation, where N is the number of particles of the system.

For pedagogical reasons and for simplicity, we will limit ourselves to one-dimensional systems with periodic spatial coordinates. Within this choice, we can fix the periodicity to 2π without losing generality. With obvious changes, the computations presented in this chapter can be extended to 3D systems with coordinates in $\mathbb{R}^3$.

8.1   Derivation of the Klimontovich Equation

The one-dimensional Hamiltonian that we will analyse takes the general form

$$H_N = \sum_{j=1}^{N} \frac{P_j^2}{2} + U(\{\Theta_j\}), \tag{8.1}$$

where the canonical coordinates Θ_j and P_j refer to particle j. As anticipated the system is periodic in space, the periodicity being set to 2π. Then we assume, as usual, that U is a sum of two-body potentials

$$U(\Theta_1, \dots, \Theta_N) = \sum_{i<j}^{N} V(\Theta_i - \Theta_j), \tag{8.2}$$

where $V(\Theta)$ is an even function, i.e. $V(-\Theta) = V(\Theta)$. We do not use here Kac's scaling of the potential, which would amount to inserting a prefactor equal to $1/N$ in the previous formula. The reason is that we wish to keep the analysis as general as possible, ideally embracing both short- and long-range interactions. We will introduce for convenience the Kac's scaling in the derivation of the Lenard–Balescu equation, later in the chapter. In the final section, we will comment on the physical meaning of Kac's prescription.

The state of the above system, made of N mutually interacting particles, can be described by the *discrete* time-dependent density function

$$f_d(\theta, p, t) = \frac{1}{N} \sum_{j=1}^{N} \delta\left(\theta - \Theta_j(t)\right) \delta\left(p - P_j(t)\right), \tag{8.3}$$

where δ denotes the Dirac function and (θ, p) are the Eulerian coordinates in the phase space. The dynamics of the Lagrangian coordinates (Θ_i, P_i) of the N particles is ruled by the $2N$ equations of motions

$$\dot{\Theta}_j = P_j, \tag{8.4}$$

$$\dot{P}_j = -\frac{\partial U}{\partial \Theta_j}. \tag{8.5}$$

By differentiating with respect to time the density function (8.3) and by making use of Eqs. (8.4) and (8.5), we find

$$\frac{\partial f_d(\theta, p, t)}{\partial t} = -\frac{1}{N} \sum_j P_j \frac{\partial}{\partial \theta} \delta\left(\theta - \Theta_j(t)\right) \delta\left(p - P_j(t)\right)$$

$$+\frac{1}{N} \sum_j \frac{\partial U}{\partial \Theta_j} \frac{\partial}{\partial p} \delta\left(\theta - \Theta_j(t)\right) \delta\left(p - P_j(t)\right). \tag{8.6}$$

Recalling that $a\delta(a-b) = b\delta(a-b)$, an obvious property of the Dirac function, we can rewrite equation (8.6) as

$$\frac{\partial f_d(\theta, p, t)}{\partial t} = -\frac{1}{N} \sum_j p \frac{\partial}{\partial \theta} \delta\left(\theta - \Theta_j(t)\right) \delta\left(p - P_j(t)\right)$$

$$+\frac{1}{N} \sum_j \frac{\partial v}{\partial \theta} \frac{\partial}{\partial p} \delta\left(\theta - \Theta_j(t)\right) \delta\left(p - P_j(t)\right), \tag{8.7}$$

where

$$v(\theta, t) = N \int d\theta' dp' \, V(\theta - \theta') f_d(\theta', p', t). \tag{8.8}$$

Hence, by recalling the formal expression of the density function (8.3) we get the Klimontovich equation

$$\frac{\partial f_d}{\partial t} + p\frac{\partial f_d}{\partial \theta} - \frac{\partial v}{\partial \theta} \frac{\partial f_d}{\partial p} = 0. \tag{8.9}$$

It is important to emphasize that the above derivation is *exact* even for a finite number of particles N. The Klimontovich equation contains information on the orbit of every single particle, since f_d depends on the $2N$ Lagrangian coordinates of each particle, namely (Θ_i, P_i). Therefore, the Klimontovich equation is mathematically equivalent to the full system of equations of motion (8.4) and (8.5) of the N-body system. Obviously, this is exceedingly complicated for our purposes. Nevertheless, it constitutes a useful starting point for eventually deriving simpler, though approximate, equations that describe the average properties of the system. The forthcoming sections are entirely devoted to discussing this issue.

Before proceeding, we now show how with a similar procedure it is possible to obtain the Liouville equation, which we derived in the previous chapter. The Liouville equation governs the time evolution of the probability density in the full $2N$ dimensional phase space $(\theta_1, p_1, \ldots, \theta_N, p_N)$. To obtain the Liouville equation, we define the distribution function

$$F_d(\theta_1, p_1, \theta_2, p_2, \ldots, \theta_N, p_N, t) = \prod_{j=1}^{N} \delta\left(\theta_i - \Theta_j(t)\right) \delta\left(p_j - P_j(t)\right), \qquad (8.10)$$

where $(\Theta_i(t), P_i(t))$ are again the Lagrangian coordinates. Deriving (8.10) with respect to time, and making use of the equations of motion, we obtain

$$\frac{\partial F_d}{\partial t} + \sum_{i=1}^{N} p_i \frac{\partial F_d}{\partial \theta_i} - \sum_{i=1}^{N} \frac{\partial U(\theta_1, \ldots, \theta_N)}{\partial \theta_i} \frac{\partial F_d}{\partial p_i} = 0. \qquad (8.11)$$

This is the Liouville equation for F_d. It is the same equation as Eq. (7.4) (in that equation the total potential U as a sum of two-body potentials was explicitly given), although here it has been derived for a particular singular distribution function, which describes a single point in the $2N$ dimensional space, with Eulerian coordinates $(\theta_1, p_1, \ldots, \theta_N, p_N)$. The Liouville equation for a generic distribution function is obtained by an averaging procedure which enables one to introduce a smooth density $\rho(\theta_1, p_1, \theta_2, p_2, \ldots, \theta_N, p_N, t)$, which can be shown to obey the same equation (8.11). Such a smoothing procedure is similar to that that leads to the Vlasov equation, and that we will describe in the next section. However, while the Liouville equation for ρ is exact (as the Klimontovich equation for f_d), the Vlasov equation is approximate. In general, finite N corrections are to be accounted for, as we will substantiate in the following.

8.2   Vlasov Equation: Collisionless Approximation of the Klimontovich Equation

To determine $f_d(\theta, p, t)$, which amounts to asking whether a particle is to be found at a given location (θ, p) in Eulerian phase space, we need to solve the equations of motion (8.4) and (8.5) with specific initial conditions $(\{\Theta_i(0), P_i(0)\})$. In general, this is a difficult task and typically not doable for nonlinear systems. Alternatively, we can define an averaged one-particle density function. The average is performed over an infinite set of realizations, prepared according to a specific prescription. We could, for instance, consider a large number of initial conditions, close to the same macroscopic state. Let us denote by $f_{in}(\{\Theta_i(0), P_i(0)\})$ the density of such initial macroscopic state. The average one-particle density function f_0 is obtained as

$$f_0(\theta, p, t) \equiv \langle f_d(\theta, p, t) \rangle = \int \prod_{i} d\Theta_i(0) dP_i(0) f_{in}(\{\Theta_i(0), P_i(0)\}) f_d(\theta, p, t), \qquad (8.12)$$

where the dependence of f_d on $(\{\Theta_i(0), P_i(0)\})$ comes from the solution of the equations of motion that enter the definition of f_d (see Eq. (8.3)). Note that the discrete and the smooth one-particle distribution functions, Eq. (8.3) and Eq. (8.12), respectively, are normalized to 1, in contrast to the one-particle reduced distribution function used in the previous chapter, which was normalized to N. The two choices are equally possible, and of course they have no relevance on the physical arguments described here.

The equation for the time evolution of the smoothed distribution f_0 is found by averaging again over f_{in}. Before carrying out the explicit calculation, we introduce the fluctuations δf around the smooth distribution

$$f_d(\theta, p, t) = f_0(\theta, p, t) + \frac{1}{\sqrt{N}} \delta f(\theta, p, t). \tag{8.13}$$

Clearly, f_0 is not sensitive to the detailed microscopic property of the initial state and depends on it only through f_{in}. In contrast, δf, like f_d, depends on all the Lagrangian variables of the initial state. Obviously, by averaging δf over f_{in} returns exactly 0. The introduction of the prefactor $1/\sqrt{N}$ accounts for the typical size of relative fluctuations. Indeed, $f_d - f_0$ is the difference between a singular distribution, containing Dirac deltas, and a smooth function; therefore, the statement that this difference is of the order $1/\sqrt{N}$ must be interpreted physically. Its meaning is the following: if we integrate $f_d - f_0$ in θ and p in a volume which is small compared to the total available volume, but large enough to contain many particles, then the value of this integral, both at equilibrium and out-of-equilibrium, is of the order $1/\sqrt{N}$.

Inserting Eq. (8.13) into Eq. (8.8) leads to

$$v(\theta, t) = \langle v \rangle (\theta, t) + \frac{1}{\sqrt{N}} \delta v(\theta, t), \tag{8.14}$$

where the first term comes from the average over f_{in}, and coincides with

$$\langle v \rangle (\theta, t) = N \int d\theta' dp' V(\theta - \theta') f_0(\theta', p', t), \tag{8.15}$$

while the second term defines δv, which depends on all the details of the initial state. If we had used Kac's scaling, the N prefactor in formula (8.15) (as, on the other hand, in formula (8.8)) would be absent, making the average potential intensive. This latter scaling is commonly used in the literature (Spohn, 1991).

Inserting both expressions (8.13) and (8.14) into the Klimontovich equation (8.9), we get

$$\frac{\partial f_0}{\partial t} + p \frac{\partial f_0}{\partial \theta} - \frac{\partial \langle v \rangle}{\partial \theta} \frac{\partial f_0}{\partial p} = -\frac{1}{\sqrt{N}} \left(\frac{\partial \delta f}{\partial t} + p \frac{\partial \delta f}{\partial \theta} - \frac{\partial \delta v}{\partial \theta} \frac{\partial f_0}{\partial p} - \frac{\partial \langle v \rangle}{\partial \theta} \frac{\partial \delta f}{\partial p} \right)$$

$$+ \frac{1}{N} \frac{\partial \delta v}{\partial \theta} \frac{\partial \delta f}{\partial p}. \tag{8.16}$$

Performing an average over f_{in} of the above equation yields

$$\frac{\partial f_0}{\partial t} + p\frac{\partial f_0}{\partial \theta} - \frac{\partial \langle v \rangle}{\partial \theta}\frac{\partial f_0}{\partial p} = \frac{1}{N}\left\langle \frac{\partial \delta v}{\partial \theta}\frac{\partial \delta f}{\partial p} \right\rangle. \tag{8.17}$$

It is important to stress that Eq. (8.17) is still exact. Up to now we have not involved any assumption on either the long- or short-range properties of the potential V. In standard kinetic theory, Eq. (8.17) would correspond to the first equation of the BBGKY hierarchy expressed in the form of Eq. (7.28). For short-range interactions, the right-hand side of Eq. (8.17) or of Eq. (7.28), after a suitable approximation, would result, as we have studied, in the collisional term of the Boltzmann equation, while the third term of the left-hand side would be negligible at equilibrium and in most situations out-of-equilibrium. In contrast, for long-range interactions, it turns out that the right-hand side, as anticipated, is of order $1/N$ (Landau, 1936; Lenard, 1960; Balescu, 1960). Indeed, the scaling with $1/\sqrt{N}$ in Eqs. (8.13) and (8.14) is appropriate for long-range interactions. Therefore, for long-range interactions, the leading contribution is the third term on the left-hand side and, in the limit $N \to \infty$, we eventually end up with the *Vlasov equation*

$$\frac{\partial f_0}{\partial t} + p\frac{\partial f_0}{\partial \theta} - \frac{\partial \langle v \rangle}{\partial \theta}\frac{\partial f_0}{\partial p} = 0. \tag{8.18}$$

Let us consider the case in which f_0 does not depend on θ. Due to the periodicity of the potential $V(\theta)$, we immediately see from Eq. (8.15) that $\langle v \rangle = $ const. Therefore, the Vlasov equation reduces to $\partial f_0/\partial t = 0$: i.e. f_0 is a stationary solution of the Vlasov equation.

Going back to Eq. (8.17), to separate the effects of granularity (finite value of N) from those due to *collective* dynamics, grouped on its left-hand side, the right-hand side is usually referred to as the *collisional* term. For this reason, Eq. (8.18) is also called the *collisionless Boltzmann equation*. It is, however, important to underline that there are no true collisions for long-range systems: granular effects, discreteness effects or finite N corrections would be more appropriate names within this framework; for this reason, the Vlasov equation is not strictly a kinetic equation.

By subtracting Eq. (8.17) from Eq. (8.16), we get

$$\frac{\partial \delta f}{\partial t} + p\frac{\partial \delta f}{\partial \theta} - \frac{\partial \delta v}{\partial \theta}\frac{\partial f_0}{\partial p} - \frac{\partial \langle v \rangle}{\partial \theta}\frac{\partial \delta f}{\partial p} = \frac{1}{\sqrt{N}}\left[\frac{\partial \delta v}{\partial \theta}\frac{\partial \delta f}{\partial p} - \left\langle \frac{\partial \delta v}{\partial \theta}\frac{\partial \delta f}{\partial p} \right\rangle\right]. \tag{8.19}$$

For times much shorter than $\sqrt{N}$ (or equivalently for $N \to \infty$), it is safe to drop the right-hand side of Eq. (8.19), which contains quadratic terms in the fluctuations. With this simplification we obtain an equation which is linear in δf. Considering the case in which f_0 does not depend on θ, i.e. when $\langle v \rangle = $ const and f_0 is a stationary solution of the Vlasov equation, the fluctuating part of f_d, i.e. δf, obeys (Vlasov, 1945; Landau, 1946; Braun and Hepp, 1977; Spohn, 1991; Messer and Spohn, 1982) the *linearized Vlasov equation*

$$\frac{\partial \delta f}{\partial t} + p\frac{\partial \delta f}{\partial \theta} - \frac{\partial \delta v}{\partial \theta}\frac{\partial f_0}{\partial p} = 0. \tag{8.20}$$

This equation, which can be alternatively derived via a direct linearization of the Vlasov equation (8.18), will be useful in the remainder of the chapter. This linear equation can be studied using the Fourier transform, from which we obtain a dispersion relation depending on the function f_0, or with the spatiotemporal Fourier–Laplace transform. The solutions of the dispersion relation obtained with the Fourier transform analysis determine whether f_0 is a stable or an unstable stationary solution (Nicholson, 1983). The study of the linearized Vlasov equation will be described in some details in Chapter 13, dedicated to hot plasma. In fact, as a final remark, it is worth recalling that the Vlasov equation (8.18) is widely applied in gravitational systems (Binney and Tremaine, 1987) and in plasma physics (Nicholson, 1983).

8.3   The Lenard–Balescu Equation

We have so far concentrated on the collisionless dynamics as described by the Vlasov equation. When using this equation, we implicitly assume that the particles of the system move under the influence of the average potential generated by all the other particles. This means that the acceleration of all the particles is given by a single function, i.e. by $-\partial \langle v \rangle/\partial \theta$. This assumption is not valid for arbitrarily long times. 'Collisions' contribute to the particles' acceleration, and this latter does not solely depend on a single mean-field function: 'collisions' perturb particles away from the trajectories they would have taken if the distribution of particles in the system was perfectly smooth. For long-range systems, however, for times shorter than N (thus for times that can be very long), the collisionless approximation holds, as we argued earlier. For larger times, the Vlasov equation is no longer valid and the effect of 'collisions' must be properly accounted for. As a consequence, particles will deviate from the orbits determined by the Vlasov equation on a characteristic time that is called *relaxation time*.

Let us then go back to the exact Eq. (8.17) and focus on its right-hand side. At the next level of approximation, i.e. the level $1/N$, we can rewrite the right-hand side of such equation by making explicit use of the solutions for δv and δf, as determined under the collisionless dynamics, given by the linearized Vlasov equation (8.20). If we restrict to the case in which f_0 does not depend on θ and is a stable stationary solution of the Vlasov equation, our kinetic equation takes the form

$$\frac{\partial f_0}{\partial t} = \frac{1}{N}\left\langle \frac{\partial \delta v}{\partial \theta}\frac{\partial \delta f}{\partial p}\right\rangle. \tag{8.21}$$

We emphasize that, assuming the right-hand side to be given by the solution of the linearized equation (8.20), we have introduced an approximation, making the equation no more strictly exact. With this implicit assumption, Eq. (8.21) becomes the Lenard–Balescu equation, and it is valid up to a time when the right-hand side makes f_0 leave its stability basin as determined by the Vlasov equation. As already anticipated, we can expect this time to be of order N. To solve the Lenard–Balescu equation, it is necessary to estimate the correlation function on the right-hand side. This task can be accomplished by using the spatiotemporal Fourier–Laplace transform. Employing this technique, we can determine the correlation expressing the collisional term, as a function of its value at the initial time. We are, therefore, brought to solve an initial value problem.

The spatiotemporal Fourier–Laplace transform of the fluctuation of the density δf is defined by

$$\widetilde{\delta f}(k, p, \omega) = \int_0^{2\pi} \frac{d\theta}{2\pi} \int_0^{+\infty} dt\, e^{-i(k\theta - \omega t)}\, \delta f(\theta, p, t), \tag{8.22}$$

associated with a similar expression for the fluctuation of the potential δv. The Fourier–Laplace transform is defined by the last equation only for Im(ω) sufficiently large, while for the remaining part of the complex ω plane it is defined by its analytic continuation. Since the coordinates take values in $[0, 2\pi]$, the wavenumber k takes only discrete values, precisely the integers. The inverse Fourier–Laplace transform is given by

$$\delta f(\theta, p, t) = \sum_{k=-\infty}^{+\infty} \int_C \frac{d\omega}{2\pi}\, e^{i(k\theta - \omega t)}\, \widetilde{\delta f}(k, p, \omega), \tag{8.23}$$

where the Laplace contour C in the complex ω plane must pass above all poles of the integrand.

If we multiply Eq. (8.20) by $e^{-i(k\theta - \omega t)}$ and integrate over θ from 0 to 2π and over t from 0 to ∞, we obtain

$$-\widehat{\delta f}(k, p, 0) - i\omega\, \widetilde{\delta f}(k, p, \omega) + ikp\, \widetilde{\delta f}(k, p, \omega) - ik\, \widetilde{\delta v}(k, \omega) f_0'(p) = 0, \tag{8.24}$$

where the first term is the spatial Fourier transform of the initial value

$$\widehat{\delta f}(k, p, 0) = \int_0^{2\pi} \frac{d\theta}{2\pi}\, e^{-ik\theta}\, \delta f(\theta, p, 0), \tag{8.25}$$

and it arises from the integration by parts in the computation of the Fourier–Laplace transform of $\partial \delta f / \partial t$. Solving the above equation for $\widetilde{\delta f}(k, p, \omega)$, we obtain

$$\widetilde{\delta f}(k, p, \omega) = \frac{k f_0'(p)}{pk - \omega}\, \widetilde{\delta v}(k, \omega) + \frac{\widehat{\delta f}(k, p, 0)}{i(pk - \omega)}, \tag{8.26}$$

where on the right-hand side the first 'collective' term depends on the perturbation of the potential, while the second term depends on the initial condition. As anticipated, for the derivation of the Lenard–Balescu equation, we introduce Kac's scaling; therefore, we assume to have $V \to V/N$ for the two-body potential in Eq. (8.2). As a consequence, the fluctuation of the potential $\delta v(\theta, t)$ is given by

$$\delta v(\theta, t) = \int_0^{2\pi} d\theta' \int_{-\infty}^{+\infty} dp \, V(\theta - \theta') \, \delta f(\theta', p, t), \tag{8.27}$$

From this equation we obtain the Fourier–Laplace transform of the fluctuation of the potential as

$$\widetilde{\delta v}(k, \omega) = 2\pi \, \widehat{V}(k) \int_{-\infty}^{+\infty} dp \, \widetilde{\delta f}(k, p, \omega), \tag{8.28}$$

where $\widehat{V}(k)$ is the spatial Fourier transform of the two-body potential, i.e.

$$\widehat{V}(k) = \int_0^{2\pi} \frac{d\theta}{2\pi} \, e^{-ik\theta} \, V(\theta). \tag{8.29}$$

As mentioned earlier, the potential $V(\theta)$ is even in θ, i.e. $V(-\theta) = V(\theta)$; then, $\widehat{V}(k)$ is real. Inserting in Eq. (8.26) the expression (8.28) of the Fourier–Laplace transform of the fluctuations of the potential $\widetilde{\delta v}(k, \omega)$, and integrating over the p variable, gives

$$\int_{-\infty}^{+\infty} dp \, \widetilde{\delta f}(k, p, \omega) \left[1 - 2\pi k \widehat{V}(k) \int_{-\infty}^{+\infty} dp' \, \frac{f_0'(p')}{(p'k - \omega)} \right] = \int_{-\infty}^{+\infty} dp \, \frac{\widehat{\delta f}(k, p, 0)}{i(pk - \omega)}, \tag{8.30}$$

where the expression in square brackets on the left-hand side is called the plasma response dielectric function

$$\widetilde{D}(\omega, k) = 1 - 2\pi k \widehat{V}(k) \int_{-\infty}^{+\infty} dp \, \frac{f_0'(p)}{(pk - \omega)}. \tag{8.31}$$

We note that for real values of ω the integrand in the last expression has a pole, and therefore the integral is not well defined. It is then interpreted with the help of the Plemelj formula, which for $\gamma > 0$ is given by

$$\lim_{\gamma \to 0} \frac{1}{x - x_0 \pm i\gamma} = \mathcal{P} \frac{1}{x - x_0} \mp i\pi \delta(x - x_0), \tag{8.32}$$

where $\mathcal{P}$ denotes the principal value, and where the equality of the two sides of the expression is intended to hold when they appear in an integration over x. The meaning of the procedure is the following. The singularity that is at $x = x_0$ when $\gamma = 0$ is avoided when performing the integration over the real x axis, deforming the integration contour in such a way to circulate around the singularity along an infinitesimal semicircle above (below) the singularity when the sign on the left-hand side is plus (minus). Equivalently, the singularity in x_0 is displaced at $x_0 \mp i0$. Applying this procedure to the dielectric function Eq. (8.31), we must decide which sign to use. Since we have seen that the Fourier–Laplace transform is defined for $\text{Im}(\omega) > 0$, the sign to use is the lower one of Eq. (8.32), corresponding to $\omega \to \omega + i0$ when ω is real. In conclusion, the dielectric function is given by Eq. (8.31) for $\text{Im}(\omega) > 0$ and by the analytic continuation for $\text{Im}(\omega) \leq 0$. In particular, for real ω, we have

$$\widetilde{D}(\omega, k) = 1 - 2\pi k \widehat{V}(k) \mathcal{P} \int_{-\infty}^{+\infty} dp \, \frac{f_0'(p)}{pk - \omega} - 2i\pi^2 \widehat{V}(k) f_0' \left(\frac{\omega}{k}\right), \qquad \omega \text{ real.} \qquad (8.33)$$

Going back to Eq. (8.28), it can be cast, using Eq. (8.30) and Eq. (8.31), in the form

$$\widetilde{\delta v}(k, \omega) = \frac{2\pi \widehat{V}(k)}{\widetilde{D}(\omega, k)} \int_{-\infty}^{+\infty} dp \, \frac{\widehat{\delta f}(k, p, 0)}{i(pk - \omega)}. \qquad (8.34)$$

Note that the Laplace contour $\mathcal{C}$ for the inversion formula must pass above all zeroes of the dielectric function $\widetilde{D}(k, \omega)$. We can consider that these zeros will all be located in the half-plane $\text{Im}(\omega) < 0$, since otherwise the problem of the $1/N$ perturbations to a stable stationary solution of the Vlasov equation would not make sense (Lifshitz and Pitaevskii, 1981).

We can then use the above expressions to compute the collisional term which appears on the right-hand side of Eq. (8.21). Forgetting temporarily the factor $1/N$ and the derivative with respect to p, we have

$$\left\langle \frac{\partial \delta v}{\partial \theta} \delta f \right\rangle = \left\langle \sum_{k=-\infty}^{+\infty} \int_{\mathcal{C}} \frac{d\omega}{2\pi} \, ik \, e^{i(k\theta - \omega t)} \, \widetilde{\delta v}(k, \omega) \right.$$

$$\left. \times \sum_{k'=-\infty}^{+\infty} \int_{\mathcal{C}'} \frac{d\omega'}{2\pi} \, e^{i(k'\theta - \omega' t)} \, \widetilde{\delta f}(k', p, \omega') \right\rangle \qquad (8.35)$$

$$= \frac{1}{(2\pi)^2} \sum_{k=-\infty}^{+\infty} \sum_{k'=-\infty}^{+\infty} \int_{\mathcal{C}} d\omega \int_{\mathcal{C}'} d\omega' \, ik \, e^{i[(k+k')\theta - (\omega + \omega')t]}$$

$$\times \left\langle \widetilde{\delta v}(k, \omega) \, \widetilde{\delta f}(k', p, \omega') \right\rangle, \qquad (8.36)$$

which relies on evaluating the correlation $\langle \widetilde{\delta v}(k, \omega) \, \widetilde{\delta f}(k', p, \omega') \rangle$. The factor k in the last equation shows that the $k = 0$ component of $\widetilde{\delta v}(k, \omega)$ does not contribute. This is somehow expected, since the constant component of δv cannot contribute to the force. Using Eq. (8.26), we get

$$\left\langle \widetilde{\delta v}(k, \omega) \, \widetilde{\delta f}(k', p, \omega') \right\rangle = \frac{k' f_0'(p)}{pk' - \omega'} \left\langle \widetilde{\delta v}(k, \omega) \, \widetilde{\delta v}(k', \omega') \right\rangle$$

$$+ \frac{\left\langle \widetilde{\delta v}(k, \omega) \, \widehat{\delta f}(k', p, 0) \right\rangle}{i(pk' - \omega')}. \qquad (8.37)$$

The first term on the right-hand side corresponds to the self-correlation of the potential, while the second one to the correlation between the fluctuations of the potential and of the distribution at time $t = 0$. In the following, we consider separately the two terms of the last equation.

From Eq. (8.34), taking the statistical average, we have

$$\langle \widetilde{\delta v}(k,\omega)\widetilde{\delta v}(k',\omega')\rangle = \frac{(2\pi)^2\, \widehat{V}(k)\widehat{V}(k')}{\widetilde{D}(\omega,k)\widetilde{D}(\omega',k')}$$
$$\times \int_{-\infty}^{+\infty} dp \int_{-\infty}^{+\infty} dp' \, \frac{\langle \widehat{\delta f}(k,p,0)\widehat{\delta f}(k',p',0)\rangle}{i(pk-\omega)i(p'k'-\omega')} \quad (8.38)$$
$$= \frac{2\pi\, \delta_{k,-k'}\, \widehat{V}(k)\widehat{V}(-k)}{\widetilde{D}(\omega,k)\widetilde{D}(\omega',-k)}$$
$$\times \int_{-\infty}^{+\infty} dp \int_{-\infty}^{+\infty} dp' \, \frac{\left[f_0(p)\delta(p-p') + \mu(k,p,p')\right]}{(pk-\omega)(p'k+\omega')}, \quad (8.39)$$

where we have replaced the autocorrelation of the fluctuation of the distribution at $t = 0$ (see Appendix E) with the expression

$$\langle \widehat{\delta f}(k,p,0)\widehat{\delta f}(k',p',0)\rangle = \frac{\delta_{k,-k'}}{2\pi}\left[f_0(p)\delta(p-p') + \mu(k,p,p')\right]. \quad (8.40)$$

Here, the first term quantifies the single-particle contribution to the correlation, while in the second term the function $\mu(k,p,p')$ weights the contribution of different particles. This last function is smooth, but otherwise arbitrary, since it is related to initial conditions of our initial value problem. However, it can be shown (Lifshitz and Pitaevskii, 1981) that, going back in the time domain with the inverse Fourier–Laplace transform, the contribution of this function to the correlation decays in time. It is therefore legitimate to just consider the first term in Eq. (8.40), yielding

$$\langle \widetilde{\delta v}(k,\omega)\widetilde{\delta v}(k',\omega')\rangle = \frac{2\pi\, \delta_{k,-k'}\, \widehat{V}(k)\widehat{V}(-k)}{\widetilde{D}(\omega,k)\widetilde{D}(\omega',-k)} \int_{-\infty}^{+\infty} dp \, \frac{f_0(p)}{(pk-\omega)(pk+\omega')}. \quad (8.41)$$

Considering again only the contributions that, after integration in ω and ω', do not decay in time, it can be shown (Lifshitz and Pitaevskii, 1981), through the use of the Plemelj formula (8.32), that $[(pk-\omega)(pk+\omega')]^{-1}$ can be substituted by $(2\pi)^2\delta(\omega+\omega')\delta(\omega-pk)$. In addition, using the properties $\widetilde{D}(\omega,k) = \widetilde{D}^*(-\omega,-k)$ and $\widehat{V}(k) = \widehat{V}(-k)$, we finally end up with the result

$$\langle \delta v(k,\omega)\delta v(k',\omega')\rangle = (2\pi)^3\, \delta_{k,-k'}\, |\widehat{V}(k)|^2\, \frac{\delta(\omega+\omega')}{|\widetilde{D}(\omega,k)|^2} \int_{-\infty}^{+\infty} dp\, f_0(p)\delta(\omega-pk). \quad (8.42)$$

We now consider the second term of Eq. (8.37). Using again Eq. (8.34), we have

$$\frac{\langle \widetilde{\delta v}(k,\omega)\widehat{\delta f}(k',p,0)\rangle}{i(pk'-\omega')} = \frac{2\pi\, \widehat{V}(k)}{\widetilde{D}(\omega,k)} \int_{-\infty}^{+\infty} dp' \, \frac{\langle \widehat{\delta f}(k,p',0)\widehat{\delta f}(k',p,0)\rangle}{i(p'k-\omega)i(pk'-\omega')}. \quad (8.43)$$

As for the analysis of the first term of Eq. (8.37), we substitute the initial time correlation in the last integral with the first term in Eq. (8.40), and afterwards we replace $[(pk - \omega)(pk + \omega')]^{-1}$ with $(2\pi)^2 \delta(\omega + \omega')\delta(\omega - pk)$. We eventually obtain

$$\frac{\langle \widetilde{\delta v}(k, \omega, k) \, \widehat{\delta f}(k', p, 0) \rangle}{i(pk' - \omega')} = \frac{(2\pi)^2 \, \delta_{k,-k'} \, \widehat{V}(k)}{\widetilde{D}(\omega, k)} f_0(p) \delta(\omega + \omega')\delta(\omega - pk). \tag{8.44}$$

From Eq. (8.42), we get the contribution to (8.36) of the first term of Eq. (8.37). Exploiting the presence of the factors $\delta_{k,-k'}$ and $\delta(\omega + \omega')$, this contribution is

$$2i\pi \sum_{k=-\infty}^{+\infty} \int_C d\omega \, \frac{k^2 f_0'(p)}{pk - \omega} \frac{|\widehat{V}(k)|^2}{|\widetilde{D}(\omega, k)|^2} \int_{-\infty}^{+\infty} dp' \, f_0(p')\delta(\omega - p'k)$$

$$= 2i\pi \sum_{k=-\infty}^{+\infty} \int_C d\omega \, \frac{|k| f_0'(p)}{pk - \omega} f_0\left(\frac{\omega}{k}\right) \frac{|\widehat{V}(k)|^2}{|\widetilde{D}(\omega, k)|^2} \tag{8.45}$$

$$= 2i\pi \sum_{k=-\infty}^{+\infty} \int_{-\infty}^{+\infty} d\omega \, |k| f_0'(p) f_0\left(\frac{\omega}{k}\right) \frac{|\widehat{V}(k)|^2}{|\widetilde{D}(\omega, k)|^2}\left[\mathcal{P}\frac{1}{pk - \omega} - i\pi \delta(\omega - pk)\right] \tag{8.46}$$

$$= 4\pi^2 \sum_{k=1}^{+\infty} \int_{-\infty}^{+\infty} d\omega \, k f_0'(p) f_0\left(\frac{\omega}{k}\right) \frac{|\widehat{V}(k)|^2}{|\widetilde{D}(\omega, k)|^2}\delta(\omega - pk). \tag{8.47}$$

Passing from the second to the third line we use the Plemelj formula, which allows us to integrate on the real ω axis. Note that in this case the rule of singularity encircling is the opposite; i.e. it is $\omega \to \omega - i0$, since the ω of the $(pk - \omega)$ term in the denominator in the first line comes from the integration in ω', which gives $\omega = -\omega'$ (Lifshitz and Pitaevskii, 1981). In the third line, a change of the variable of integration from ω to $-\omega$ in the terms with negative k, and the use of the property $\widetilde{D}(\omega, k) = \widetilde{D}^*(-\omega, -k)$, makes it possible to obtain the expression in the fourth line. The contribution of the second term of (8.37) to Eq. (8.36) is obtained from (8.44). By exploiting again the factors $\delta_{k,-k'}$ and $\delta(\omega + \omega')$, this contribution is

$$i \sum_{k=-\infty}^{+\infty} \int_C d\omega \, \frac{k\widehat{V}(k)}{\widetilde{D}(\omega, k)} f_0(p) \, \delta(\omega - pk)$$

$$= i \sum_{k=-\infty}^{+\infty} \int_C d\omega \, \frac{k\widehat{V}(k)}{|\widetilde{D}(\omega, k)|^2} \widetilde{D}^*(\omega, k) f_0(p)\delta(\omega - pk) \tag{8.48}$$

$$= -4\pi^2 \sum_{k=1}^{+\infty} \int_{-\infty}^{+\infty} d\omega \, \frac{k|\widehat{V}(k)|^2}{|\widetilde{D}(\omega, k)|^2} f_0(p) f_0'\left(\frac{\omega}{k}\right) \delta(\omega - pk), \tag{8.49}$$

where in the passage from the second to the third line we use the expression (8.33) and again the property $\widetilde{D}(\omega, k) = \widetilde{D}^*(-\omega, -k)$.

We now have all elements for the determination of the right-hand side of the Lenard–Balescu equation (8.21). Inserting again the $1/N$ factor and the derivative with respect to p, we obtain the following result which is valid only in one dimension:

$$\frac{\partial f_0}{\partial t} = \frac{4\pi^2}{N} \sum_{k=1}^{+\infty} \frac{\partial}{\partial p} \int_{-\infty}^{+\infty} d\omega \, \frac{k \left| \widehat{V}(k) \right|^2}{\left| \widetilde{D}(\omega, k) \right|^2} \delta\left(\omega - pk\right) \left[f_0\left(\frac{\omega}{k}\right) f_0'(p) - f_0(p) f_0'\left(\frac{\omega}{k}\right) \right] = 0.$$
(8.50)

We thus see that, in one dimension, the Lenard–Balescu operator vanishes: the diffusion term (first term on the right-hand side) is exactly balanced by the friction term (second term on the right-hand side). Consequently the collisional evolution is due to terms of higher order in $1/N$, and the Vlasov equation is valid for a time longer than previously expected. This remark was made long ago in plasma physics by Kadomtsev and Pogutse (1970).

The previous computation can be performed for a general (long-range) two-body potential in d dimensions. The result is the general Lenard–Balescu equation (Balescu, 1963; Chavanis, 2006d)

$$\frac{\partial f(\mathbf{v}, t)}{\partial t} = \frac{\pi (2\pi)^d}{m^2} \frac{\partial}{\partial \mathbf{v}} \cdot \int d\mathbf{v}_1 d\mathbf{k} \, \mathbf{k} \frac{\left| \hat{u}(|\mathbf{k}|) \right|^2}{\left| \widetilde{D}(\mathbf{k} \cdot \mathbf{v}_1, \mathbf{k}) \right|^2} \delta\left(\mathbf{k} \cdot (\mathbf{v} - \mathbf{v}_1)\right)$$
$$\times \left[\mathbf{k} \cdot \left(f(\mathbf{v}_1, t) \frac{\partial f(\mathbf{v}, t)}{\partial \mathbf{v}} - f(\mathbf{v}, t) \frac{\partial f(\mathbf{v}_1, t)}{\partial \mathbf{v}_1} \right) \right],$$
(8.51)

where the boldface variables are d-dimensional vectors, and where $\hat{u}(|\mathbf{k}|)$ is the real d-dimensional Fourier transform of the two-body potential; m is the mass of the particles and f is normalized to 1. The dielectric function $\widetilde{D}(\omega, \mathbf{k})$ in this general case is given by

$$\widetilde{D}(\omega, \mathbf{k}) = 1 + \frac{(2\pi)^d}{m} \hat{u}(|\mathbf{k}|) \int d\mathbf{v} \, \frac{\mathbf{k} \cdot \frac{\partial f(\mathbf{v})}{\partial \mathbf{v}}}{\omega - \mathbf{k} \cdot \mathbf{v}}.$$
(8.52)

The $1/N$ 'smallness' of the right-hand side of Eq. (8.51) is incorporated in the two-body potential. From this general expression we see that, while in $d = 1$ the collisional term vanishes, it is instead present for $d > 1$. Thus, e.g., in a two-dimensional Coulombian plasma or in three-dimensional Newtonian interactions and in plasma physics, the Lenard-Balescu collisional term gives a contribution already at order $1/N$. Nicholson (1983) derived its expression for general potentials for homogeneous cases. Chavanis (2006b, 2006c) extended the analysis to the inhomogeneous case (2008b, 2008c), without, however, taking into account collective effects.

The approach that neglects the collective effects in Eq. (8.51) corresponds to the Landau approximation, often used in plasma physics; it amounts to taking $\widetilde{D}(\omega, \mathbf{k}) = 1$, so that the structure of the Landau equation does not depend on the potential.

We note that in a one-dimensional framework, the Lenard–Balescu and the Landau equations coincide: since they give a vanishing contribution, they both reduce to the Vlasov equation. The collisional evolution is due to terms of higher order in $1/N$. Hence, the collisional relaxation time scales as N^δ with $\delta > 1$. As a consequence, the system can remain frozen in a stationary solution of the Vlasov equation for a very long time, larger than N. Only non-trivial three-body correlations can induce a further evolution of the system, as also remarked in Chavanis (2001). A detailed discussion of the derivation of Vlasov, Landau and Lenard–Balescu equations and of the approximations involved can be found in Chavanis (2008a). Finally, we would like to emphasize that the theoretical tools described in this chapter have also led to the development of kinetic theories for point vortices in two-dimensional hydrodynamics (Dubin and O'Neil, 1999; Chavanis, 2002b) and for non-neutral plasmas confined by a magnetic field (Dubin and Zin, 2001; Dubin, 2003).

From the above analysis we argue that when N is very large, the dynamics of long-range systems is approximated very well by the Vlasov equation. In theoretical works, the relation between the N-particle dynamics and the solution of the Vlasov equation has been studied, for the case of mean-field systems, by Neunzert (1975, 1978), Braun and Hepp (1977) and Spohn (1991). From these works we derive that, for any chosen value of ε and system size N, there exists a time t_0 up to which the dynamics of the original Hamiltonian and that of its Vlasov description coincide within an error bounded by ε, while the time t_0 increases at least as $\ln N$.

8.4   The Boltzmann Entropy and the Mean-Field Approximation

In this section, we derive an expression of the Boltzmann entropy that is often used in the study of long-range systems. In Chapter 1 we gave the definition of the Boltzmann entropy at equilibrium, see Eqs. (1.13) and (1.14). In that case, the N-body distribution function $f_N(x_1, \ldots, x_N)$ (or rather its unnormalized form used in the computation of the microcanonical partition function) has the particular δ function form, as shown in the integrand of Eq. (1.13). This definition can be generalized to arbitrary distributions. It is expressed by

$$S = -\int dx_1 \ldots dx_N \, f_N(x_1, \ldots, x_N) \ln f_N(x_1, \ldots, x_N). \tag{8.53}$$

We can check that this expression reduces to the previous one in the case of the microcanonical equilibrium distribution function. It is easier, for this purpose, to use, rather than the singular expression with a δ function, the other one in which the distribution function is constant on the portion of Γ space between two hypersurfaces corresponding to two very close energies, and 0 elsewhere. As we noted in Chapter 1, this alternative definition leads to exactly the same results in the thermodynamic limit. Therefore, let us suppose that the unnormalized distribution function, used in the computation of the microcanonical partition function, is equal to 1 in a portion of Γ with hypervolume A. Then, from Eq. (1.14) we get $S = \ln A$ (we neglect here the factor $1/N!$, which gives rise to an unimportant additive constant in the entropy). On the other hand, in this case we have $f_N(x_1, \ldots, x_N) = 1/A$ in the mentioned portion of Γ, and 0 elsewhere. Thus, also Eq. (8.53) gives $S = \ln A$.

In long-range systems, the mean-field approximation is generally a very good one. In this approximation, correlations between particles are neglected, expressing $f_N(x_1, \ldots, x_N)$ as

$$f_N(x_1, \ldots, x_N) = \prod_{j=1}^{N} f(x_j),$$ (8.54)

where $f(x)$ is the normalized one-particle distribution function. Substituting this form in Eq.(8.53) we get

$$S = -N \int \mathrm{d}x f(x) \ln f(x).$$ (8.55)

In the following chapters, we will make use several times of this expression, as a functional to maximize, given some constraints, to obtain the equilibrium one-particle distribution function. In this calculation, the factor N appearing in the last equation is not relevant and it will be neglected, so that

$$S = - \int \mathrm{d}x f(x) \ln f(x).$$ (8.56)

Analogously, this expression can be used independently of the normalization chosen for $f(x)$.

8.5 The Kac's Prescription in Long-Range Systems

As promised already in Chapter 2, we comment here on the physical consequences of Kac's prescription. We consider here for simplicity only Kac's prescription for purely mean-field system, i.e. the $1/N$ scaling of the interaction.

We begin with the dynamics. It is clear that the relative weights of the mean-field term of the Vlasov equation and of the collisional terms of the kinetic equations do not depend on Kac's scaling. On the other hand, the absolute weights of the terms do depend on it. However, it is not difficult to see that this induces only, starting from identical initial conditions, a difference in the time scale of the dynamics. To understand this point, let us consider very general equations of motion determined by a potential with Kac's scaling. The equations for the coordinates $(q_1, \ldots, q_N)$ will be of the form

$$\frac{\mathrm{d}^2 q_i}{\mathrm{d}t^2} = \frac{1}{N} F_i(q_1, \ldots, q_N).$$ (8.57)

By making the change of time units such that $t \rightarrow \sqrt{N}t$, we get rid of the factor $1/N$ in the above equations. Now let us suppose that, with Kac's scaling, the time scale of the dynamics of the system for large N does not depend on N, at least in leading order, so that, e.g., the system, starting from a given initial state, reaches a stationary state or the equilibrium state in a time that does not depend on N for N sufficiently large. We then conclude that without Kac's scaling the system will go through the same states during its evolution, but the dynamics will be $\sqrt{N}$ times faster. We should not infer from this that the evolution will then be very fast, as it happens in short-range systems. On the contrary, even without Kac's scaling the time scale to reach a stationary or equilibrium state can be macroscopic.

We now consider the equilibrium states. We have seen in the first part of the book that Kac's prescription, although it cannot avoid the lack of additivity, it does restore the extensivity. As a consequence of this fact, the values of the intensive thermodynamic quantities (the temperature, the energy per particle, the entropy per particle, the free energy per particle, etc.) determining the behaviour of the system do not depend on N. In particular, their values at the phase transitions points do not depend on N (independently from the equivalence or inequivalence of the ensembles). Making the same change of time units mentioned earlier, it is then not difficult to understand that, considering the phase transition points, the value of, e.g., the energy per particle and of the temperature in those points will scale like N. In fact, making the above scaling of the time in the Hamiltonian H of the system, we see that we get rid of the $1/N$ scaling in the interaction potential, at the expenses of the change of units $H \to NH$.

In conclusion, Kac's prescription is indeed a convenient mathematical tool to deal with long-range systems. However, the conclusions derived for a system in which this scaling has been introduced can be easily transformed quantitatively to determine the behaviour of the system without Kac's scaling.

9

Out-of-Equilibrium Dynamics and Slow Relaxation

This chapter is devoted to discussing the peculiar out-of-equilibrium behaviour of systems subject to long–range interactions. We shall start by presenting N-body numerical simulations of the Hamiltonian Mean-Field (HMF) model to demonstrate the existence of the so-called *quasi-stationary* states (QSS). As we will prove, the system gets in fact trapped into intermediate regimes, whose lifetime diverges with the system size. Aiming at providing a quantitative interpretation of such phenomenon, we will turn to considering a kinetic theory approach, which formally applies to the continuum limit. In doing so, we will argue that QSSs represent stable equilibria of the underlying continuum picture, different from the Boltzmann–Gibbs equilibrium state, an observation which opens up the perspective to a formal statistical mechanics approach.

9.1 Numerical Evidence of Quasi-stationary States

Consider the HMF model introduced in Section 4.1, which is defined by the Hamiltonian

$$H_N = \sum_{i=1}^{N} \frac{p_i^2}{2} + \frac{\varepsilon_{\mathcal{J}}}{2N} \sum_{i,j} \left[1 - \cos(\theta_i - \theta_j) \right], \tag{9.1}$$

where $\theta_i \in [0, 2\pi[$ is the angle of the ith unit mass particle on a circle and p_i the corresponding conjugated momentum.

In the previous chapter, it was proven that for any one-dimensional long-range system, including the specific HMF case under study, Vlasov stable homogeneous distribution functions do not evolve on time scales of order smaller than or equal to N.

This result is an illuminating explanation of the numerical strong disagreement which was reported in Antoni and Ruffo (1995) and Latora *et al.* (2001) between constant energy molecular dynamics simulations and canonical statistical mechanics calculations, for the ferromagnetic case with $\varepsilon_{\mathcal{J}} = 1$ (in the following we will restrict to this case).

Physics of Long-Range Interacting Systems. First Edition. A. Campa *et al.*
© A. Campa, T. Dauxois, D. Fanelli, and S. Ruffo 2014. Published in 2014 by Oxford University Press.

Indeed, for energies slightly below the second-order phase transition specific energy $\varepsilon = 3/4$ (see Fig. 9.1), numerical simulations in the microcanonical ensemble reveals that the system is trapped for a long time, whose duration increases with the number N of particles, in a state far from that predicted by equilibrium statistical mechanics. Since the latter was initially derived using the canonical ensemble, these results had been initially interpreted as the fingerprint of inequivalence between microcanonical and canonical ensembles. However, as previously discussed in Chapter 4, both ensembles lead to the same results for the HMF model, where only a second-order phase transition occurs. More careful numerical experiments by Yamaguchi *et al.* (2004) have later shown the tendency of the simulation data points, i.e. of the out-of-equilibrium states in which the system is trapped, to lie on the continuation to lower energies of the supercritical branch with zero magnetization of the caloric curve. These states have been called quasi-stationary states (QSS) and systematic simulations have determined a $N^{1.7}$ scaling law for the duration of the QSS, at the end of which the system eventually evolves towards the Boltzmann–Gibbs equilibrium state.

In Fig. 9.2, we display the time evolution of the magnetization, $m(t)$, with increasing particle number, showing the increase of the duration of the QSS: the power-law increase is emphasized by the choice of the logarithmic scale in the abscissa. Since this scaling law has been found when the system is initially prepared in a state that is a stable stationary solution of the Vlasov equation (in particular, for initial states homogeneous in θ), this numerical evidence is in agreement with the result derived earlier, which Vlasov stable homogeneous distribution functions do not evolve on time scales of order smaller than

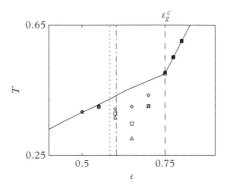

Figure 9.1 *Caloric curve of the HMF Hamiltonian (9.1) with $\varepsilon_{\mathfrak{I}} = 1$. The solid line is the equilibrium result in both the canonical and the microcanonical ensemble. The second-order phase transition is revealed by the kink at $\varepsilon_g^c = 3/4$. The three values of the energy indicated by the vertical lines are the Vlasov stability thresholds for the homogeneous Gaussian (dashed), a distribution (9.31) with power-law tails with exponent $v = 8$ (dash-dotted) and water-bag (dotted) initial momentum distribution (9.19). The Gaussian stability threshold coincides with the phase transition energy. The points are the results of constant energy (microcanonical) simulations for the Gaussian (losanges), the power law (squares) and the water bag (triangles). Simulations were performed with $N = 5000$.*

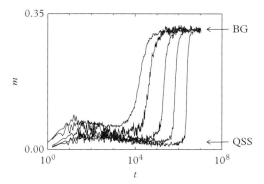

Figure 9.2 *Time evolution of the modulus of the magnetization m(t) for different particle numbers: $N = 10^3$, 2×10^3, 5×10^3, 10^4 and 2×10^4 from left to right ($\varepsilon = 0.69$). In all cases, an average over several samples has been taken. Two values of the magnetization, indicated by horizontal arrows, can be identified in this figure: the upper one (labelled BG) corresponds to the expected equilibrium result for the magnetization, while the lower one, labelled QSS, represents the value of the magnetization in the quasi-stationary state. Reprinted from Campa et al. (2009), © 2009, with permission from Elsevier.*

or equal to N. Obviously, even if the initial state is Vlasov stable, finite N effects will eventually drive the system away from it and towards the Boltzmann–Gibbs equilibrium state.

If the initial condition is Vlasov unstable, a rapid evolution will take place. Simulations have been performed also in this case (Latora *et al.*, 2001), showing that, after an initial transient, the systems remains trapped in other types of QSS, with different scaling laws in N for their duration. The existence of an infinite number of Vlasov stable distributions is actually the key point to explain the out-of-equilibrium QSS observed in the HMF dynamics. Let us show that the system evolves through other stable stationary states. In order to check the stationarity and the stability of an initial distribution $f_0(\theta, p)$, it is possible to study the temporal evolution of the magnetization, which is constant if the system is stable and stationary. Other possible macrovariables are the moments of the distribution function. It can be easily shown that any distribution of the form $f(\theta, p, 0) = f_0(\theta, p) = F(p^2/2 - m_x \cos\theta - m_y \sin\theta) \equiv f_0(e)$, where F stands for a generic function, is a stationary solution of the Vlasov equation, provided m_x and m_y are determined self-consistently from Eq. (8.15).

Obviously, this does not yet mean that the above solution is also stable. This latter observation has suggested studying numerically the stability of f_0 by checking the stationarity of the first few moments $\mu_n = \langle e^n \rangle_N$. Since the stationarity of the moments is a necessary condition for stability, the vanishing, for a long time lapse, of the time derivatives $\dot{\mu}_n = d\mu_n/dt$, for $n = 1, 2$ and 3, has been used as a numerical suggestion that the system is in a QSS and that the distribution $f_0(\theta, p)$ is a stable stationary solution of the Vlasov equation. In contrast, large derivatives clearly indicate a non-stationary state.

Figure 9.3 presents the temporal evolution of these quantities, together with the temporal evolution of the modulus m of the magnetization, for power-law and Gaussian initial distributions in the case of an energy ε in the interval $[\varepsilon_{pl}^c, \varepsilon_g^c]$. In case (a), the stationarity holds throughout the computed time since we note that the three quantities $\dot{\mu}_n$ have vanishingly small fluctuations around 0. In contrast, in case (b), the system is first in an unstable stationary state ($\tau < 0.0005$), before becoming non-stationary ($0.0005 < \tau < 0.003$) and finally reaches stable stationary states ($\tau > 0.003$). Consequently the system evolves among different Vlasov stationary states. In the stable case, Fig. 9.3a, the magnetization m stays around 0 before taking off around $\tau = 20$ to reach the equilibrium value m^*. In the unstable case, Fig. 9.3b shows that after experiencing unstable stationary and non-stationary states, the system presents a slow quasi-stationary

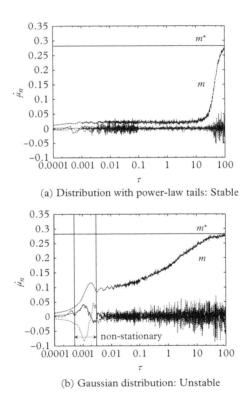

(a) Distribution with power-law tails: Stable

(b) Gaussian distribution: Unstable

Figure 9.3 *Time evolution of the magnetization m. The time is rescaled as $\tau = t/N$. The quantities $\dot{\mu}_n$ ($n = 1, 2, 3$), which detect the stationarity, are also plotted (multiplied by a factor 100 for graphical purposes). The equilibrium value of the magnetization is m^*. Panel (a) corresponds to a stable homogeneous initial distributions (9.31) with power-law tails, while panel (b) shows an unstable initial condition with gaussian tails. In both panels, energy is $\varepsilon = 0.7$. All numerical curves are obtained by averaging over 20 initial conditions with $N = 10^4$. Reprinted from Campa et al. (2009), © 2009, with permission from Elsevier.*

evolution across an infinite number of stationary and stable Vlasov states. In Yamaguchi *et al.* (2004), a careful numerical study has shown that this slow characteristic timescale associated with this final relaxation towards the Boltzmann–Gibbs equilibrium is proportional to $N^{1.7}$. However, this law might be dependent on the energy ε and on the initial distributions. The previous chapter made it possible, however, to claim that this relaxation timescale is larger than N at least. This is, thus, a very slow process in comparison to the relaxation from an initially unstable state. Recently, this slow evolution of the HMF system through different stable stationary states of the Vlasov equation has been more systematically studied, with the aim of determining how the Vlasov stable solutions characterizing the system during the out-of-equilibrium dynamics can be parametrized (Campa *et al.*, 2007a, 2008a). It has been found that starting from a homogeneous distribution given by a q-Gaussian with compact support

$$f_{\mathrm{T}}(p) \sim [1 - \alpha(1-q)p^2]^{\frac{1}{1-q}}, \tag{9.2}$$

with α positive and $q < 1$, at an energy slightly below the second-order phase transition energy, the evolution of the system during the QSS is well approximated by distribution functions of the same type, but with varying q, until the system heads towards Boltzmann–Gibbs equilibrium. Note that the q-Gaussian (9.2) decays with a power-law tail for $q > 1$, while we recover the Gaussian for $q = 1$.

Let us stress that the above scenario is consistent with what happens generically for systems with long-range interactions (Lynden–Bell and Wood, 1968; Chavanis, 2002b; Barré *et al.*, 2006). In a first stage, called *violent relaxation*, the system goes from a generic initial condition, which is not necessarily Vlasov stable, towards a Vlasov stable state. This is a fast process happening usually on a fast timescale, *independent* of the number of particles. In a second stage, named *collisional relaxation*, finite N effects come into play and the Vlasov description is no more valid for the discrete systems. The timescale of this second process is strongly dependent on N. One generally considers that it is a power law N^{δ}. A typical example is the Chandrasekhar relaxation timescale for stellar systems, which is proportional to $N/\ln N$. This scenario of the typical evolution of long-range systems is summarized in Fig. 9.4.

It is important to remark that, recently, Caglioti and Rousset (2008) rigorously proved that for a wide class of potentials, particle systems starting close to a Vlasov stable distribution remain close to it for times that scale at least like $N^{1/8}$: this result is consistent with the power law conjectured for collisional relaxation. Unfortunately, apart from a recent progress by Hauray and Jabin (2007), very few rigorous results exist in the case of singular potentials, which would be of paramount importance for Coulomb and gravitational interactions.

Stronger divergences with system size in long-range systems are observed in connection with metastable states (Antoni *et al.*, 2004; Chavanis, 2005; Mukamel *et al.*, 2005) where the relaxation time increases exponentially with N.

In summary, quasi-stationary states observed in the N-particle dynamics of the HMF model are nothing but Vlasov stable stationary states, which evolve because of *collisional*, finite N, effects. There is an *infinity* of Vlasov stable homogeneous (zero

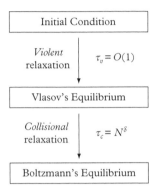

Figure 9.4 *Schematic description of the typical dynamical evolution of systems with long-range interactions. τ_v and τ_c are the violent relaxation and the collisional relaxation timescales, respectively.*

magnetization) states corresponding to different initial velocity distributions $f_0(t = 0, p)$, whose stability domain in energy are different. Homogeneous distributions, q-Gaussians in momentum, are Vlasov stable stationary states in a certain energy region where QSS are observed in the HMF model. However, they are not special in any respect, among an *infinity* of others. In the HMF model at finite N, all of them converge sooner or later to the Boltzmann–Gibbs equilibrium. However, the relaxation time is shown numerically to diverge with a power law N^δ, with $\delta \simeq 1.7$ for the homogeneous water-bag state (Yamaguchi *et al.*, 2004; Moyano and Anteneodo, 2006). The value of this exponent and its relation with kinetic equations is still debated.

We do not treat the important and debated issue of the effect of a heat bath or of noise on the QSS lifetime. An interesting study of this can be found in, e.g., Gupta and Mukamel (2010), who have shown that the presence of energy conserving stochastic processes in the HMF model induces a faster relaxation towards Boltzmann-Gibbs equilibrium.

Let us stress the important point that follows from the analysis described in this section. For finite N, as large as it might be, the long-range system will first settle to a Vlasov stable stationary state, from which it will eventually be driven towards the Boltzmann–Gibbs equilibrium state by finite size effects. On the other hand, if the thermodynamic limit $N \to \infty$ is taken while the system is in the Vlasov stable state, it will never reach the Boltzmann–Gibbs equilibrium state. This means that the infinite time limit $t \to \infty$ and the thermodynamic limit $N \to \infty$ do not commute.

9.2 Fokker–Planck Equation for the Stochastic Process of a Single Particle

Let us now consider the relaxation properties of a test particle, initially with a momentum p_1 and an angle θ_1, immersed in a homogeneous background of N particles; the latter is consequently a thermal bath, or a reservoir, for the test particle. The description of the motion for a so-called test particle in a system with identical particles is a classical problem in kinetic theory. Initially the test particle is in a given microscopic state, while the other particles are distributed according to a Vlasov stable distribution f_0 and the test particle is assumed not to affect the reservoir. The interaction with the fluid induces a complicated stochastic process.

For simplicity and for pedagogical reason we consider here the case of the HMF model. Since in the following we need to refer to the response dielectric function and to the autocorrelation of the fluctuation of the potential, we show here the expressions that these two quantities have for the HMF model, where the particular form of the potential induces a semplification. In fact, the spatial Fourier transform of the two-body potential, Eq. (8.29), becomes $\widehat{V}(k) = \left(2\delta_{k,0} - \delta_{k,1} - \delta_{k,-1}\right)/2$, so that formula (8.42) becomes

$$
\begin{aligned}
&\langle \delta v(k,\omega)\delta v(k',\omega')\rangle \\
&= 2\pi^3 \delta_{k,-k'} \left(4\delta_{k,0} + \delta_{k,1} + \delta_{k,-1}\right) \frac{\delta(\omega+\omega')}{\left|\widetilde{D}(\omega,k)\right|^2} \int_{-\infty}^{+\infty} dp\, f_0(p)\delta(\omega - pk),
\end{aligned}
\tag{9.3}
$$

while the response dielectric function simplifies to

$$
\widetilde{D}(\omega,k) = 1 + \pi k \left(\delta_{k,1} + \delta_{k,-1}\right) \int_{-\infty}^{+\infty} dp\, \frac{f_0'(p)}{(pk-\omega)}.
\tag{9.4}
$$

We see that for the HMF model the dielectric function $\widetilde{D}(\omega, k)$ and the autocorrelation of the Fourier–Laplace transform of the fluctuation of the potential $\delta v(k,\omega)$ have components only for $k = \pm 1$ (actually $\delta v(k,\omega)$ has also a $k = 0$ component, which, however, does not have any effect on the dynamics). In the following, referring to these two quantities, we implicitly assume to treat the HMF model and the above expressions.

The test particle distribution is initially not in equilibrium, and not necessarily close to it. However, it is natural to expect that the distribution of the test particle will eventually correspond to the distribution of the bath generated by all the other particles. How it evolves from the initial Dirac distribution $f_1(\theta, p, 0) = \delta(\theta - \theta_1)\delta(p - p_1)$ towards the equilibrium distribution is thus of high interest. We will show that the distribution is a solution of a Fokker–Planck equation that can be derived analytically. This equation describes the dynamical coupling with the fluctuations of the density of particles, which induce fluctuations in the potential: this is the origin of the underlying stochastic process.

We analyse therefore the relaxation properties of a test particle, indexed by 1, surrounded by a background system of $(N-1)$ particles with a homogeneous in angle distribution $f_0(p)$. The averaged potential $\langle v\rangle$ still vanishes for a homogeneous distribution, so that the particle only feels the fluctuations of the potential that, according to Eq. (8.14), is given by $\delta v(\theta)/\sqrt{N}$. The potential felt by the test particle at the position $\theta_1(t)$ is therefore $\delta v(\theta)/\sqrt{N}$ computed at $\theta = \theta_1(t)$. We thus expect that the instantaneous force on the test particle will be of order $1/\sqrt{N}$. The equations of motion of the test particle are therefore

$$\frac{d\theta_1}{dt} = p_1 \quad \text{and} \quad \frac{dp_1}{dt} = -\frac{1}{\sqrt{N}} \left.\frac{\partial \delta v(\theta, t)}{\partial \theta}\right|_{\theta=\theta_1(t)}, \tag{9.5}$$

the integration of which leads to

$$\theta(t) = \theta(0) + p(0)\,t - \frac{1}{\sqrt{N}} \int_0^t du_1 \int_0^{u_1} du_2 \frac{\partial \delta v}{\partial \theta}(\theta(u_2), u_2) \tag{9.6}$$

$$p(t) = p(0) - \frac{1}{\sqrt{N}} \int_0^t du \frac{\partial \delta v}{\partial \theta}(\theta(u), u), \tag{9.7}$$

where, again for simplicity, we have indicated directly inside the dependence of δv the test particle variable and we have omitted the index 1. The key point of this approach is that we do not limit the study to the usual ballistic approximation, in order to have an expansion exact at order $1/N$. Therefore it is of paramount importance here to treat accurately the essential *collective effects*.

By introducing iteratively the expression for the variable θ on the right-hand side of Eq. (9.7) and by expanding the derivatives of the potential, one gets the result at order $1/N$ of the momentum dynamics

$$p(t) = p(0) - \frac{1}{\sqrt{N}} \int_0^t du \frac{\partial \delta v}{\partial \theta}(\theta(0) + p(0)u, u)$$

$$+ \frac{1}{N} \int_0^t du \frac{\partial^2 \delta v}{\partial \theta^2}(\theta(0) + p(0)u, u) \int_0^u du_1 \int_0^{u_1} du_2 \frac{\partial \delta v}{\partial \theta}(\theta(0) + p(0)u_2, u_2). \tag{9.8}$$

As the changes in the momentum are small, of order $1/\sqrt{N}$, the description of the momentum dynamics is well represented by a stochastic process governed by a Fokker–Planck equation (Van Kampen, 1992). If we denote by $f_1(p, t)$ the distribution function at time t of the test particle momentum, then the general form of this equation is

$$\frac{\partial f_1(p, t)}{\partial t} = -\frac{\partial}{\partial p}\left[A(p, t)f_1(p, t)\right] + \frac{1}{2}\frac{\partial^2}{\partial p^2}\left[B(p, t)f_1(p, t)\right], \tag{9.9}$$

with

$$A(p, t) = \lim_{\tau \to 0} \frac{1}{\tau}\langle(p(t + \tau) - p(t))\rangle_{p(t)=p} \tag{9.10}$$

$$B(p, t) = \lim_{\tau \to 0} \frac{1}{\tau}\langle(p(t + \tau) - p(t))^2\rangle_{p(t)=p}, \tag{9.11}$$

where the expectation values are conditioned by $p(t) = p$. This equation is therefore characterized by the time behaviour of the first two moments, called Fokker–Planck coefficients. We approximate these two coefficients by an expression which is valid in the range of time t defined by $1 \ll t \ll N$. In this time range, using a generalization of formula (9.3) that takes into account that the initial coordinate of the test particle is given, it is possible, as shown in Appendix F, to obtain

$$A(p,t) \sim \frac{1}{N} \left(\frac{\mathrm{d}D}{\mathrm{d}p}(p) + \frac{1}{f_0} \frac{\partial f_0}{\partial p} D(p) \right) \tag{9.12}$$

$$B(p,t) \sim \frac{2}{N} D(p), \tag{9.13}$$

where the diffusion coefficient is given by

$$D(p) = 2 \operatorname{Re} \int_0^{+\infty} \mathrm{d}t \, e^{ipt} \left\langle \widehat{\delta v}(1,t) \widehat{\delta v}(-1,0) \right\rangle = \pi^2 \frac{f_0(p)}{\left| \tilde{D}(p,1) \right|^2}. \tag{9.14}$$

Substituting (9.12) and (9.13) in the general form of the Fokker–Planck equation (9.9), we end up with

$$\frac{\partial f_1(p,t)}{\partial t} = \frac{1}{N} \frac{\partial}{\partial p} \left[D(p) \left(\frac{\partial f_1(p,t)}{\partial p} - \frac{1}{f_0} \frac{\partial f_0}{\partial p} f_1(p,t) \right) \right]. \tag{9.15}$$

We thus recover what has been established in plasma physics (Ichimaru, 1973): the evolution of the velocity distribution $f_1(p,t)$ of the test particle is governed by a Fokker–Planck equation that takes a form similar to the Lenard–Balescu equation (8.50), provided that we replace the distribution $f_0(p,t)$ of the bath by $f_1(p,t)$. The integro-differential equation is thus transformed in the Fokker–Planck differential equation. Similar results in higher dimension were obtained later by Chavanis (2006c,d).

For one-dimensional systems, it was shown in Section 8.3 that the Lenard–Balescu collision term cancels out so that the distribution function does not evolve on a time scale of order N. Since, on the other hand, the Fokker–Planck equation (9.15) shows that the relaxation time of a test particle towards the distribution of the bath is of order N, this implies that we can assume that the distribution of the particles f_0 is stationary when one studies the relaxation of a test particle. This is true for any distribution function $f_0(p)$ that is a stable stationary solution of the Vlasov equation. This is not true in higher dimensions, except for the Maxwell distribution.

Equation (9.15) emphasizes that the momentum distribution of particle 1 evolves on timescales of order N. We will thus introduce the timescale $\tau = t/N$, so that the Fokker–Planck equation can be rewritten as

$$\frac{\partial f_1}{\partial \tau} = \frac{\partial}{\partial p} \left[D(p) \left(\frac{\partial f_1}{\partial p} - \frac{1}{f_0} \frac{\partial f_0}{\partial p} f_1 \right) \right], \tag{9.16}$$

valid for times τ at least of order 1. We see that the time derivative of f_1 vanishes if the distribution function f_1 of the test particle is equal to the quasi-stationary distribution f_0 of the surrounding bath. We expect that f_1, governed by this Fokker–Planck equation, will converge to f_0 in a time τ of order 1: this means that the distribution function f_1 of the test particle converges towards the quasi-stationary distribution f_0 of the surrounding bath. Thus, it does not converge towards the equilibrium Gaussian distribution, in complete agreement with the result that f_0 is stationary for timescales of order N. This result is not valid in higher dimensions (Chavanis, 2006d). It is important to stress here that collective effects are taken into account through the presence of the dielectric response function $\tilde{D}(p, 1)$ in the denominator of the diffusion coefficient given in Eq. (9.14).

Analysing the stochastic process of equilibrium fluctuations in the particular case of homogeneous Gaussian distribution, Bouchet (2004) derived the diffusion coefficient of a test particle in an equilibrium bath. This result is recovered when considering a homogeneous Gaussian distribution in expression (9.14) since one gets

$$D(p) = \pi^2 \, \frac{\frac{1}{2\pi} \sqrt{\frac{\beta}{2\pi}} \, e^{-\beta p^2/2}}{\left[1 - \frac{\beta}{2} + \frac{1}{2} \beta^{3/2} \, p \, e^{-\beta p^2/2} \int_0^{p\sqrt{\beta}} dt \, e^{t^2}/2 \right]^2 + \frac{1}{8} \pi \beta^3 p^2 \, e^{-\beta p^2}} \,, \tag{9.17}$$

plotted in Fig. 9.5. It is important to stress that such an expression leads to a diffusion coefficient with Gaussian-like tails

$$D(p) \sim \sqrt{\frac{\pi \beta}{8}} \, \exp\left(-\frac{\beta p^2}{2}\right). \tag{9.18}$$

The above general derivation (Bouchet and Dauxois, 2005) makes it possible, however, to study *any* arbitrary distribution. Let us carry on here the calculation of the diffusion coefficient for the Vlasov-stable water-bag distribution

$$f_{\text{wb}}(p) = \frac{1}{4\pi p_0} \left[\Theta(p + p_0) - \Theta(p - p_0) \right]. \tag{9.19}$$

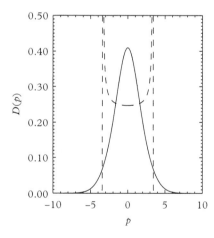

Figure 9.5 *Diffusion coefficient $D(p)$ in the case $\varepsilon = 2$ for a Boltzmann thermal bath (solid line) and a water-bag distribution (9.19) (dashed line). Reprinted from Campa et al. (2009), © 2009, with permission from Elsevier.*

It has an interesting behaviour, since the dielectric response $\tilde{D}(\omega, 1)$ has zeroes on the real axis, in contrast to any even distributions *strictly* decreasing for positive values of the frequency ω. We get

$$D(p) = \pi^2 \, \frac{\dfrac{1}{4\pi p_0}\left[\Theta(p + p_0) - \Theta(p - p_0)\right]}{\left[1 - \dfrac{1}{2}\dfrac{1}{p_0^2 - p^2}\right]^2 + \left[\dfrac{\pi}{4p_0}\left(\delta(p + p_0) - \delta(p - p_0)\right)\right]^2}, \tag{9.20}$$

which is also plotted in Fig. 9.5. We can obtain similar results for q-exponential distributions (9.2) (see Chavanis *et al.* (2005), for details) or power-law tails distributions as reported by Yamaguchi *et al.* (2007) .

9.3 Long-Range Temporal Correlations and Diffusion

Since the Fokker–Planck equation for the single particle distribution function $f_1(p)$ has a variable diffusion coefficient, the relaxation towards the Boltzmann distribution can be slowed down, especially if the diffusion coefficient decreases rapidly with momentum. It is thus important to study this Fokker–Planck equation for different distribution functions of the bath as performed earlier for 2D vortices by Chavanis (2001). One consequence is that velocity correlation functions can decrease algebraically with time instead of exponentially, a behaviour which might lead to anomalous diffusion as we will show later.

By introducing the appropriate change of variable $x = x(p)$, defined by

$$dx/dp = 1/\sqrt{D(p)}, \tag{9.21}$$

and the associated distribution function $\widehat{f_1}$, defined by $\widehat{f_1}(\tau, x)dx = f_1(\tau, p)dp$, we can map (Marksteiner *et al.*, 1996; Farago, 2000; Bouchet and Dauxois, 2005; Chavanis and Lemou, 2007; Micciché, 2008) the Fokker–Planck equation (9.16) to the constant diffusion coefficient Fokker–Planck equation

$$\frac{\partial \widehat{f_1}}{\partial \tau} = \frac{\partial}{\partial x}\left(\frac{\partial \widehat{f_1}}{\partial x} + \frac{\partial \psi}{\partial x}\widehat{f_1}\right), \tag{9.22}$$

where the potential $\psi(x)$ is given by

$$\psi(x) = -\ln\left(\sqrt{D(p(x))}f_0(p(x))\right). \tag{9.23}$$

Using the property $\tilde{D}(p, 1) \overset{|p| \to \infty}{\sim} 1$, which implies, by Eq. (9.14), that

$$D(p) \overset{|p| \to \infty}{\sim} \pi^2 f_0(p), \tag{9.24}$$

we get

$$\psi(x) \overset{x \to \pm\infty}{\sim} -\frac{3}{2} \ln f_0(p(x)). \tag{9.25}$$

From this, we derive that for many classes of distribution functions f_0, the potential $\psi(x)$ is asymptotically equivalent to a logarithm. In fact, we have

$$\psi(x) \overset{x \to \pm\infty}{\sim} \alpha \ln |x|, \tag{9.26}$$

with $\alpha = 3$ if $f_0(p)$ decreases to zero more rapidly than algebraically for large p, and $\alpha < 3$ if $f_0(p)$ decreases to zero algebraically; more precisely, $\alpha = 3\nu/(2 + \nu)$ if $f_0(p)$ decays at large p as $p^{-\nu}$.

For weakly confining potentials $\psi(x)$, i.e. when $\nu < 1$ and thus $\alpha < 1$, Eq. (9.22) has a non-normalizable ground state. The example of the heat equation, which corresponds to $\psi(x) = 0$, describes a diffusive process leading to an asymptotic self-similar evolution. In such a case, the spectrum of the Fokker–Planck equation is purely continuous. By contrast, a strongly confining potential $\psi(x)$, for instance the Ornstein–Uhlenbeck process with a quadratic potential, would lead to exponentially decreasing distributions and autocorrelation functions, linked to the existence in the spectrum of a gap above the ground state. The logarithmic potential (9.26) is a limiting case between the two behaviours. The normalizable ground state is unique and coincides with the bottom of the continuous spectrum. The absence of a gap forbids *a priori* any exponential relaxation.

To illustrate this result, we evaluate the asymptotic behaviour explicitly in two cases.

- Let us first consider distribution functions $f_0(p)$ with fast (more than algebraically) decreasing tails, so that

$$f_0(p) \overset{|p| \to \infty}{\sim} C \exp(-\beta p^\delta), \tag{9.27}$$

which includes not only the Gaussian ($\delta = 2$) and exponential tails ($\delta = 1$), but also stretched-exponential ones with any arbitrary positive exponent δ. From the change of variable $dx/dp = 1/\sqrt{D(p)}$, asymptotic analysis leads to

$$p(x) \overset{|x| \to \infty}{\sim} (2 \ln |x|/\beta)^{1/\delta} \tag{9.28}$$

and to

$$\psi(x) \overset{x \to \pm\infty}{\sim} 3 \ln |x|. \tag{9.29}$$

These estimates are sufficient (see Bouchet and Dauxois, 2005, for details) to evaluate the long-time behaviour of the momentum autocorrelation function

$$\langle p(\tau)p(0) \rangle \overset{\tau \to +\infty}{\propto} \frac{(\ln \tau)^{2/\delta}}{\tau}, \tag{9.30}$$

which proves the existence of a long-range temporal momentum autocorrelation for all values of δ, and therefore also in the case of Boltzmann equilibrium, $\delta = 2$.

• Let us now consider a distribution function $f_0(p)$ with algebraic tails

$$f_0(p) \overset{|p| \to \infty}{\sim} C|p|^{-\nu}. \tag{9.31}$$

In this case, we have $p(x) \overset{|x| \to \infty}{\sim} C'x^{2/(2+\nu)}$ and the asymptotic behaviour (9.26) with, as we said, $\alpha = 3\nu/(2+\nu)$. We consider only cases where $\nu > 3$, to ensure that the second moment of the distribution f_0 (i.e. the average kinetic energy) does exist. For $\nu > 3$ we have that $9/5 < \alpha < 3$. The result for the momentum autocorrelation function is

$$\langle p(\tau)p(0) \rangle \overset{\tau \to +\infty}{\propto} \tau^{(3-\nu)/(2+\nu)}, \tag{9.32}$$

which characterizes an algebraic asymptotic behaviour.

From the momenta autocorrelation, we usually derive the angle diffusion

$$\sigma_\theta^2(\tau) = \langle (\theta(\tau) - \theta(0))^2 \rangle = 2D_\theta\, \tau, \tag{9.33}$$

where D_θ is defined via the Kubo formula

$$D_\theta = \int_0^{+\infty} \mathrm{d}\tau\, \langle p(\tau)p(0) \rangle. \tag{9.34}$$

However, since the exponent $(3-\nu)/(2+\nu) = -1 + 5/(2+\nu)$ is larger than -1, the asymptotic result (9.32) shows that the integral (9.34) diverges. The asymptotic result (9.30) leads also to a divergent integral, although less singular. It is thus natural to expect anomalous diffusion for the angles. This extremely small anomaly (logarithmic) for distribution functions with Gaussian or stretched exponential tails induced difficulties to detect numerically anomalous diffusion (Latora *et al.*, 1999; Yamaguchi *et al.*, 2003).

Note that, generalizing the theory of Potapenko *et al.* (1997), Chavanis and Lemou (2005) studied how the structure and the progression of the distribution function tails, also called fronts, depends on the behaviour of the diffusion coefficient for large velocities. They showed that the progression of the front is extremely slow (logarithmic) in that case so that the convergence towards the equilibrium state is peculiar.

Using the time rescaling $\tau = t/N$, which introduces a factor $1/N^2$, the angle diffusion can be rewritten as

$$\frac{\sigma_\theta^2(\tau)}{N^2} = \int_0^\tau d\tau_1 \int_0^\tau d\tau_2 \, \langle p(\tau_1)p(\tau_2) \rangle \tag{9.35}$$

$$= 2 \int_0^\tau ds \int_0^{\tau-s} d\tau_2 \, \langle p(s+\tau_2)p(\tau_2) \rangle, \tag{9.36}$$

in which the new variable $s = \tau_1 - \tau_2$ has been introduced to take advantage of the division of the square domain into two isoscale triangles corresponding to $s > 0$ and $s < 0$. In the quasi-stationary states, the integrand $\langle p(s+\tau_2)p(\tau_2) \rangle$ does not depend on τ_2 (the QSS evolves on a timescale much larger than N) and hence diffusion can be simplified (Yamaguchi *et al.*, 2003) as

$$\frac{\sigma_\theta^2(\tau)}{N^2} = 2 \int_0^\tau ds \, (\tau - s) \, \langle p(s)p(0) \rangle. \tag{9.37}$$

A distribution with power-law tails (9.31) will therefore correspond to

$$\langle (\theta(\tau) - \theta(0))^2 \rangle \overset{\tau \to +\infty}{\propto} \tau^{1 + \frac{5}{2+\nu}}. \tag{9.38}$$

A comparison of this predicted (Bouchet and Dauxois, 2005) anomalous diffusion for angles with direct numerical computation of the HMF dynamics is a tough task because of the scaling with N of the time dependence of the autocorrelation function. However, it has been confirmed by numerical simulations by Yamaguchi *et al.* (2007). For the stable case ($\varepsilon = 0.7$) with power-law tails ($\nu = 8$), the theory predicts that the correlation function decays algebraically with the exponent $-1/2$, according to Eq. (9.32). Figure 9.6a shows that the theoretical prediction agrees well with numerical computations. The expression (9.38) can be rewritten in that case as $\sigma_\theta^2(\tau) \sim \tau^{3/2}$: Figure 9.6b, in which the four curves for the four different values of N almost collapse, attests also the validity of this prediction. This is a clear example of anomalous diffusion.

For the stretched exponential distribution function (9.27), we end up with

$$\langle (\theta(\tau) - \theta(0))^2 \rangle \overset{\tau \to +\infty}{\propto} \tau (\ln \tau)^{2/\delta + 1}. \tag{9.39}$$

As for power-law tails, the diffusion is again anomalous, although with a logarithmically small anomaly. Consequently, the anomalous diffusion for angles also occurs for the Gaussian distribution, which corresponds to the special case $\delta = 2$. This weak anomalous diffusion, i.e. normal diffusion with logarithmic corrections, has also been confirmed by Yamaguchi *et al.* (2007). Since Gaussian distributions correspond to equilibrium distributions in the microcanonical and the canonical ensembles, it is important to realize that anomalous diffusion might thus be encountered for both equilibrium and out-of-equilibrium initial conditions.

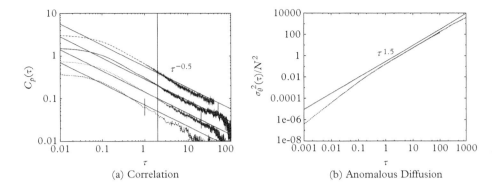

(a) Correlation   (b) Anomalous Diffusion

Figure 9.6 *Check of the theoretical prediction for stable initial distributions with power-law tails, in the case $\varepsilon = 0.7$. Points are numerically obtained by averaging $20, 20, 10$ and 5 realizations for $N = 10^3, 10^4, 2 \times 10^4$ and 5×10^4, respectively. In (a), four curves represent the correlation functions of momenta, while the straight lines with the slope $-1/2$ represent the theoretical prediction. The curves and the lines are multiplied from the original vertical values by $2, 4$ and 8 for $N = 10^4, 2 \times 10^4$ and 5×10^4 for graphical purposes. Similarly, (b) presents the diffusion of angles, while the straight line with the slope $3/2$ is theoretically predicted. The four curves for the four different values of N are reported and almost collapse. Reprinted from Campa* et al. *(2009), © 2009, with permission from Elsevier.*

9.4 Lynden–Bell's Entropy

9.4.1 The principle

In the preceding section, we discussed the existence of stationary stable and unstable one-particle distributions for the Vlasov equation. We have also argued that quasi-stationary states, as revealed by direct simulations, can be possibly explained in terms of 'attractive' Vlasov equilibria. Motivated by this working assumption, one wishes to gain analytical insight into the QSS by resorting to the Vlasov scenario. The statistical approach proposed by Lynden–Bell (1967) enables us to explicitly calculate equilibrium solutions of the Vlasov equation which are compatible with the imposed dynamical constraints.

Indeed, after an initial rapid evolution, the single-particle distribution f becomes progressively more filamented and stirred at smaller and smaller scales, without attaining any final stable equilibrium. However, averaging over larger windows in phase space which encompass the above filamentations, we obtain a coarse-grained function $\bar{f}$ that is smooth and likely to converge towards an equilibrium state. To illustrate this concept more precisely let us restrict to a two-dimensional phase space.

Imagine to discretize the one-particle distribution function into a set of k quantized levels $\eta_i, i = 1, \ldots, k$. The area corresponding to the ith level,

$$\gamma(\eta_i) = \int d\theta \, dp \, \delta(f(\theta, p, t) - \eta_i), \tag{9.40}$$

is a constant of the motion through the Vlasov dynamics. It can be defined as the 'mass' of the level η_i. In the limit of a continuous distribution of levels, an infinity of conservation laws is hence recovered. More generally, it can be proven, using an integration by parts in Eq. (8.18), that any functional of the form $\int d\theta\, dp\, C(f)$ is conserved through the Vlasov dynamics: these quantities are the so-called Casimirs. The Boltzmann entropy (see Eq. (8.56))

$$S = -\int d\theta\, dp\, f \ln f, \tag{9.41}$$

is a particular Casimir. This specifically implies that, in terms of the *fine-grained* one-particle distribution, Gibbs entropy cannot increase. For this reason, and to clarify the convergence to equilibrium, we can look at the dynamics from the viewpoint of the *coarse-grained* distribution

$$\bar{f}(\theta, p) = \sum_{i=1}^{k} \rho(\theta, p, \eta_i)\, \eta_i, \tag{9.42}$$

where $\rho(\theta, p, \eta_i)\, d\theta\, dp$ is the probability of finding level η_i in the phase-space macrocell $D_{\text{macro}} = [\theta, \theta + d\theta] \times [p, p + dp]$.

Lynden–Bell's proposal consists in evaluating $\rho(\theta, p, \eta_i)$ via a maximum entropy principle (Robert, 1990, 1991; Robert and Sommeria, 1991; Michel and Robert, 1994; Ellis, 1999). Formally, $\rho(\theta, p, \eta_i)$ can be calculated for any number of levels k but, for simplicity purposes, we shall here restrict to the case where only two levels $\eta_1 = 0$ and $\eta_2 = f_0$ are allowed. At the fine-grained level this corresponds to assuming at $t = 0$ a distribution of the *water-bag* type, as often employed in astrophysics and plasma physics. The initial water bag gets distorted by the Vlasov dynamics: the area occupied by the Vlasov 'fluid' with level f_0 is, in fact, stretched and folded due to the hyperbolicity originated by the nonlinear dynamics of the Vlasov equation (see e.g. Fig. 2 in Lynden–Bell (1967)). However, as it follows from Liouville's theorem, such area is conserved all along the system evolution. Let us divide each macrocell D_{macro} into ν microcells of volume ω. Consider then a microscopic configuration such that the i-th macrocell is occupied by n_i microcells with level f_0 and $\nu - n_i$ with level 0. The total number of occupied microcells is labelled $\mathcal{N}$, and the total 'mass' reads $M = \mathcal{N}\omega f_0$. This latter is also equal to the normalization of the fine-grained distribution $M = \int d\theta\, dp\, f(\theta, p)$. The $\mathcal{N}$ occupied microcells are first placed into macrocells. There are $\mathcal{N}! / \prod_i n_i!$ ways to perform this task. Within the ith cell, we can distribute the first of the n_i occupied microcells in ν ways, the second in $\nu - 1$ and so on. The number of ways of assigning the n_i occupied microcells is thus $\nu!/(\nu - n_i)!$. Then, the total number of microstates W compatible with the macrostate where n_i microcells are occupied in macrocell i results from the product of these two quantities

$$W(\{n_i\}) = \frac{\mathcal{N}!}{\prod_i n_i!} \times \prod_i \frac{\nu!}{(\nu - n_i)!}. \tag{9.43}$$

The first factor is calculated exactly as for a Boltzmann gas (Huang, 1987), because the occupied microcells are *distinguishable*, while the second factor reminds a Fermi–Dirac statistics. In fact it descends from an *exclusion principle*, which is the consequence of fluid incompressibility: every microcell cannot be occupied more than once by a fluid element of level f_0. Apart from this latter constraint, fluid microcells distribute freely among the different macrocells: this, in turn, corresponds to assuming the system inherently ergodic. This is a crucial assumption, which does not rigorously apply to the true Vlasov scenario, and it is sometime referred to as to the *efficient mixing* hypothesis. Indeed, it has been found that dynamical effects can hinder mixing (Antoniazzi *et al.*, 2007a,b; Bachelard *et al.*, 2008a; Levin *et al.*, 2008a,b). Using Stirling's approximation and expressing n_i in terms of the average probability to find level f_0 in cell i, $\rho_i(f_0) = n_i/\nu$, we obtain, apart from an additive constant,

$$\ln W = -\nu \sum_i \left[\rho_i \ln \rho_i + (1 - \rho_i) \ln(1 - \rho_i) \right], \tag{9.44}$$

which can also be rewritten in terms of the coarse-grained distribution function, $\rho_i = \bar{f}_i/f_0$. Taking the continuum limit

$$\sum_i \rightarrow \int \frac{\mathrm{d}\theta\,\mathrm{d}p}{\omega\,\nu}, \tag{9.45}$$

we finally obtain the entropy functional

$$s_{\mathrm{LB}}[\bar{f}] = -\int \frac{1}{\omega}\,\mathrm{d}\theta\,\mathrm{d}p \left[\frac{\bar{f}}{f_0} \ln \frac{\bar{f}}{f_0} + \left(1 - \frac{\bar{f}}{f_0}\right) \ln \left(1 - \frac{\bar{f}}{f_0}\right) \right]. \tag{9.46}$$

Following the standard procedure, inspired by large deviations theory (Touchette, 2008), we can maximize s_{LB} subject to the relevant constraints: the conservation of energy E, mass M and other global invariants like momentum (or angular momentum for higher dimensions).

9.4.2   Application to the HMF model

Let us turn to considering the case of the HMF for which the Vlasov equation takes the form

$$\frac{\partial f}{\partial t} + p \frac{\partial f}{\partial \theta} - \frac{\mathrm{d}V}{\mathrm{d}\theta} \frac{\partial f}{\partial p} = 0, \tag{9.47}$$

where $f(\theta, p, t)$ is the *fine-grained* one-particle distribution function and

$$V(\theta)[f] = 1 - m_x[f] \cos(\theta) - m_y[f] \sin(\theta), \tag{9.48}$$

$$m_x[f] = \int_{-\pi}^{+\pi} \int_{-\infty}^{+\infty} f(\theta, p, t) \, \cos\theta \, d\theta \, dp, \tag{9.49}$$

$$m_y[f] = \int_{-\pi}^{+\pi} \int_{-\infty}^{+\infty} f(\theta, p, t) \, \sin\theta \, d\theta \, dp. \tag{9.50}$$

The globally conserved quantities are the energy

$$h[f] = \iint \frac{p^2}{2} f(\theta, p, t) \, d\theta \, dp - \frac{m_x^2 + m_y^2 - 1}{2}, \tag{9.51}$$

and the momentum

$$P[f] = \iint pf(\theta, p, t) \, d\theta \, dp. \tag{9.52}$$

As already mentioned, the initial distribution should be of the *water-bag* kind for (9.46) to apply. In particular, we will restrict the analysis to a rectangular water bag in the (θ, p) plane. The distribution f takes only two distinct values, namely $f_0 = 1/(4\Delta\theta\Delta p)$, if the angles (velocities) lie within an interval centred around 0 and of half-width $(\Delta\theta \; \Delta p)$, and 0 otherwise. Note that thus the 'mass' M is normalized to 1 and the momentum $P[f]$ is 0. A one-to-one relation exists between the parameters $\Delta\theta$ and Δp and the initial values of magnetization and energy since we have

$$m_0 = \frac{\sin(\Delta\theta)}{\Delta\theta} \qquad \text{and} \qquad \varepsilon = \frac{(\Delta p)^2}{6} + \frac{1 - (m_0)^2}{2}. \tag{9.53}$$

While $h[f] = \varepsilon$ and $P[f] = 0$ are constants of the motion, the magnetization $m = \sqrt{m_x^2 + m_y^2}$ evolves with time and eventually attains a value that we can predict via the theory. The Lynden–Bell maximum entropy principle yields the constrained variational problem

$$s_{LB}(\varepsilon) = \max_{\bar{f}} \left(s(\bar{f}) \Big| h(\bar{f}) = \varepsilon; P(\bar{f}) = 0; \int d\theta \, dp \, \bar{f} = 1 \right). \tag{9.54}$$

The problem is solved by introducing three Lagrange multipliers $\beta/(\omega f_0)$, $\lambda/(\omega f_0)$ and $\mu/(\omega f_0)$ for energy, momentum and mass normalization, respectively. Using

$$\delta s_{LB} - \beta\delta h - \lambda\delta P - \mu\delta \left(\iint d\theta \, dp \, \bar{f} \right) = 0, \tag{9.55}$$

we recover the following analytical form of the distribution

$$\bar{f}(\theta, p) = \frac{f_0}{1 + \exp\left[\beta(p^2/2 - m_x[\bar{f}]\cos\theta - m_y[\bar{f}]\sin\theta) + \lambda p + \mu\right]}. \tag{9.56}$$

Note that this distribution differs from the Boltzmann–Gibbs one because of the 'fermionic' denominator. Inserting expression (9.56) into the energy, momentum and normalization constraints and making use of the definition of the magnetization, it immediately yields the condition $\lambda = 0$. Moreover, defining $x = e^{-\mu}$ and $\mathbf{z} = (\cos\theta, \sin\theta)$ allows us to obtain the following implicit equations in the unknowns β, x, m_x and m_y

$$f_0 \frac{x}{\sqrt{\beta}} \int d\theta \, e^{\beta\mathbf{m}\cdot\mathbf{z}} \, F_0\left(xe^{\beta\mathbf{m}\cdot\mathbf{z}}\right) = 1 \tag{9.57}$$

$$f_0 \frac{x}{2\beta^{3/2}} \int d\theta \, e^{\beta\mathbf{m}\cdot\mathbf{z}} F_2\left(xe^{\beta\mathbf{m}\cdot\mathbf{z}}\right) = \varepsilon + \frac{m^2 - 1}{2} \tag{9.58}$$

$$f_0 \frac{x}{\sqrt{\beta}} \int d\theta \, \cos\theta \, e^{\beta\mathbf{m}\cdot\mathbf{z}} \, F_0\left(xe^{\beta\mathbf{m}\cdot\mathbf{z}}\right) = m_x \tag{9.59}$$

$$f_0 \frac{x}{\sqrt{\beta}} \int d\theta \, \sin\theta \, e^{\beta\mathbf{m}\cdot\mathbf{z}} \, F_0\left(xe^{\beta\mathbf{m}\cdot\mathbf{z}}\right) = m_y \tag{9.60}$$

with $F_0(y) = \int \exp(-v^2/2)/(1 + y\exp(-v^2/2))dv$ and $F_2(y) = \int v^2 \exp(-v^2/2)/(1 + y\exp(-v^2/2))dv$. Once the initial condition and therefore its energy, momentum and m_0 are being fixed (thus fixing also the value of f_0), the above system of equations can be solved numerically, returning the values of the Lagrange multipliers β and μ, and also m_x and m_y, where Lynden–Bell's entropy is extremal. Finally, the distribution $\bar{f}$ can be calculated via formula (9.56), *with no adjustable parameter*.

Multiple stationary solutions are in principle possible. To identify the global maximum, which in turn corresponds to the Vlasov equilibrium state, we must punctually evaluate the entropy in correspondence of the selected stationary points. Depending on the predicted value of $m = \sqrt{m_x^2 + m_y^2}$, we can ideally distinguish between two distinct regimes: the homogeneous case corresponds to $m \simeq 0$ (non-magnetized), while the non-homogeneous (magnetized) setting is found for $m \neq 0$ distributions. Indeed the predictions of the Lynden–Bell equilibrium (9.56) are derived scanning the parameter plane (m_0, ε); we recall that these two parameters univocally identify the initially selected water-bag distribution. The underlying scenario as first recognized in Antoniazzi *et al.* (2007b) is depicted in Fig. 9.7. When fixing the initial magnetization and decreasing the energy density, the system undergoes a phase transition, from homogeneous to magnetized QSS. The plane can be then formally divided into two zones respectively associated with an ordered non-homogeneous phase, $m_{QSS} \neq 0$ (lower part of Fig. 9.7), and a disordered homogeneous state, $m_{QSS} = 0$ (upper part). These regions are separated by a transition line, collection of all the critical points (m_0^c, ε^c), which can be segmented into two distinct parts. The dashed line corresponds to a second-order phase transition, meaning that the magnetization is continuously modulated, from 0 to positive values, when passing the curve from top to bottom. Conversely, the full line refers to a first-order phase transition: here, the magnetization experiences a sudden jump when crossing the critical value (m_0^c, ε^c). First-order and second-order lines merge together in a tricritical point approximately located at $(m_0, \varepsilon) = (0.2, 0.61)$. Antoniazzi *et al.* also

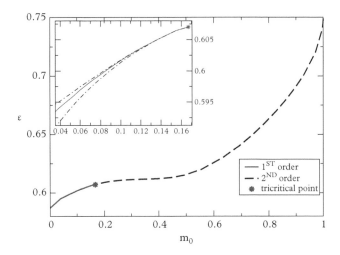

Figure 9.7 *Theoretical phase diagram in the control parameter plane (m_0, ε) for a rectangular water-bag initial profile. The dashed line $\varepsilon_c(m_0)$ stands for the second-order phase transition, while its continuation as a full line refers to the first-order phase transition. The full dot is the tricritical point. Inset: zoom of the first-order transition region. Dash-dotted lines represent the limits of the metastability region. In region (II), delimited from above by the upper dash-dotted line and from below by the full line, the homogeneous phase is fully stable and the inhomogeneous phase is metastable. In region (III), located below the full line and above the lower dash-dotted line, the homogeneous phase is metastable and the inhomogeneous phase is fully stable.*

noted that around the first-order transition line a metastability region exists: different solutions of the variational problem coexist as reported in the inset of Fig. 9.7.

Such a rich out-of-equilibrium scenario results from a straightforward application of Lynden-Bell's theory and proves extremely accurate versus direct simulations, based on the discrete formulation. The existence of phase transition is, in particular, numerically confirmed. The values of the critical parameters (ε^c, m_0^c) and the order of the transition are correctly predicted. Lynden–Bell's theory constitutes indeed a powerful analytical tool which enables us to characterize the global properties of QSSs and unravel their unexpected richness.

As a special case let us consider the energy value $\varepsilon = 0.69$. The maximum entropy state has zero magnetization for $m_0 < m_{crit} = 0.897$, a value at which Lynden–Bell's theory predicts a second-order phase transition, as reported in Antoniazzi *et al.* (2007a,b). Interpreting $\bar{f}$ as the distribution in the quasi-stationary state (QSS), we obtain in this case

$$\bar{f} = f_{\mathrm{QSS}}(p) = \frac{f_0}{1 + \exp\left[\beta p^2/2 + \mu\right]}, \tag{9.61}$$

with β and μ to be determined from the knowledge of m_0. Velocity profiles predicted by (9.61) are displayed in Fig. 9.8 for different values of the initial magnetization.

Although not a single free parameter is used, one finds an excellent qualitative agreement. The presence of two symmetric bumps in the velocity distributions is not predicted by Lynden–Bell's theory and is a consequence of a collective phenomenon, which leads to the formation of two *clusters* in the (θ, p) plane. This is shown by the direct simulation of the Vlasov equation (9.47) presented in Fig. 9.9. The bumps represent an intrinsic peculiarity of QSS and have been characterized dynamically in Bachelard *et al.* (2008a).

The Lynden–Bell approach turns out to be a good way of attacking the problem of quasi-stationary states. It gives predictions for both averages and distributions functions, which compare quite well with numerical simulations. When the approach fails to

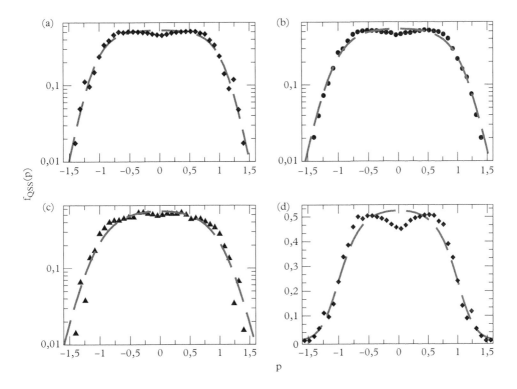

Figure 9.8 *Velocity distribution functions in the quasi-stationary state for the HMF model with $\varepsilon = 0.69$ and different values of m_0. Symbols refer to numerical simulations, while dashed solid lines stand for the theoretical profile (9.61). (a–c) present the three cases $m_0 = 0.3$, $m_0 = 0.5$ and $m_0 = 0.7$ in lin–log scale, while (d) shows the case $m_0 = 0.3$ in lin–lin scale. The numerical curves are computed from one single realization with $N = 10^7$ at time $t = 100$. Reprinted from Campa et al. (2009), © 2009, with permission from Elsevier.*

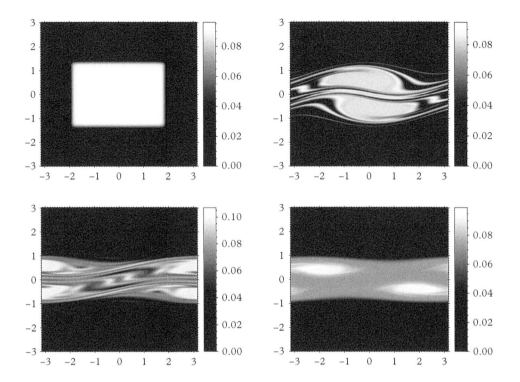

Figure 9.9 *Simulation of the Vlasov equation (9.47) that starts from a* water-bag *initial condition with* $\varepsilon = 0.69$ *and* $m_0 = 0.5$. *The final snapshot (lower right panel) is the quasi-stationary state. In the plots, the horizontal and vertical axes refer to* θ *and* p, *respectively. Reprinted from Campa* et al. *(2009),* © *2009, with permission from Elsevier.*

describe detailed features, there are viable ways of modifying it by taking into account dynamical properties.

9.5   Lynden–Bell's Entropy: Beyond the Single Water-Bag Case Study

The Lynden–Bell maximum entropy approach was applied to describe the generalized case of a multi-level initial distribution in Asslani *et al.* (2012). The theoretical predictions have then been tested for the two water-bags case study. We shall here revisit the results of Asslani *et al.* (2012) and specifically start with a discussion of the generalized notation.

As for the case discussed earlier, we divide the phase space into a very large number of micro-cells, each of volume $\tilde{\omega}$. The micro-cells define a hyperfine support that provides an adequate representation for the fine-grained function f, provided the mass of the phase element that occupies each cell is given. Consider n levels of phase density f_j, $j = 1, \ldots, n$. Then the phase element mass is $\tilde{\omega} f_j$ or 0. Lynden–Bell suggested to group these micro-cells into coarse-grained macro-cells, still very small, but sufficiently large to contain many micro-cells. Let us call ν the number of micro-cells inside the macro-cell, the latter having therefore a volume $\nu \tilde{\omega}$. Define n_{ij} the number of elements with phase density f_j that populate cell i, located in (θ_i, p_i). Clearly $\sum_i n_{ij} = N_j$, where N_j stands for number of micro-cells occupied by level f_j.

The Lynden–Bell entropy can be rigorously derived via the following steps. First, we quantify the number of ways of assigning the micro-cells to all $\sum_j n_{ij}$ phase elements confined in the micro-cell i. A simple combinatorial argument yields

$$\frac{\nu!}{(\nu - \sum_j n_{ij})!}. \tag{9.62}$$

Then, we need to calculate the total number of microstates W compatible with the single macrostate defined by the numbers n_{ij}. W is the product of (9.62) with the total number of ways of splitting the pool of available N_j elements into groups of n_{ij}, which leads to:

$$W = \Pi_j \frac{N_j!}{\Pi_i (n_{ij})!} \times \Pi_i \frac{\nu!}{(\nu - \sum_j n_{ij})!}. \tag{9.63}$$

Finally, the entropy $s_{LB} = \ln W$ can be cast in the form

$$s_{LB} = -\sum_j \sum_i \frac{n_{ij}}{\nu} \ln \frac{n_{ij}}{\nu} - \sum_i \left(1 - \sum_j \frac{n_{ij}}{\nu}\right) \ln \left(1 - \sum_j \frac{n_{ij}}{\nu}\right), \tag{9.64}$$

where we have rescaled S by ν and neglected some unimportant constant contributions. The term $\sum_i \left(1 - \sum_j n_{ij}/\nu\right) \ln \left(1 - \sum_j n_{ij}/\nu\right)$ reflects the exclusion principle being imposed in the combinatorial analysis. Indeed, two elements of phase cannot overlap, each micro-cell being solely occupied by one of the available density levels, including 0.

We now introduce the probability density ρ_{ij} of finding the level of phase density f_j in cell i as

$$\rho_{ij} = \frac{n_{ij}}{\nu}. \tag{9.65}$$

Note that $\sum_i \sum_j \rho_{ij} = 1$, as it should be. By inserting Eq. (9.65) into the entropy expression (9.64), we get

$$s_{LB} = -\sum_j \sum_i \rho_{ij} \ln \rho_{ij} - \sum_i \left(1 - \sum_j \rho_{ij}\right) \ln \left(1 - \sum_j \rho_{ij}\right). \tag{9.66}$$

Following Lynden–Bell (1967), we define the coarse-grained distribution function $\bar{f}$ in (θ_i, p_i) as

$$\bar{f}(\theta_i, p_i) = \sum_j \frac{n_{ij}}{v} f_j = \sum_j \rho_{ij} f_j. \tag{9.67}$$

The density ρ_{ij} and the coarse-grained distribution $\bar{f}(\theta_i, p_i)$ are the two main quantities upon which the description relies. However, these are not independent quantities. Let us write the density as $\rho_{ij} = \alpha_j h_i$, assuming that h_i depends on the ith cell, while the other contribution, α_j, on the jth level. By inserting this ansatz into the definition (9.67) for $\bar{f}(\theta_i, p_i)$ we get

$$\bar{f}(\theta_i, p_i) = \sum_j \alpha_j h_i f_j = h_i \left(\sum_j \alpha_j f_j \right), \tag{9.68}$$

for all i, from which we can write

$$h_i = \bar{f}(\theta_i, p_i) \tag{9.69}$$

together with the normalization condition

$$\sum_j f_j \alpha_j = 1. \tag{9.70}$$

Hence, summarizing, we can rewrite (9.65) as

$$\rho_{ij} = \bar{f}(\theta_i, p_i) \alpha_j, \tag{9.71}$$

which admits a simple interpretation. The probability of finding an element of phase density f_j in cell i is given by the probability of finding any element in such a cell, $\bar{f}(\theta_i, p_i)$, times the probability that the selected element is actually of type f_j. Reasoning along this lines, α_j can be seen as the relative fraction of phase-space volume that hosts the elements of phase density f_j.

Finally, one can obtain a compact expression for ρ_{ij} that explicitly evidences all allowed levels

$$\rho_{ij} = \sum_L \bar{f}(\theta_i, p_i) \alpha_L \delta_{Lj}. \tag{9.72}$$

By taking the continuum limit both in the spatial variable $(\theta_i, p_i) \to (\theta, p)$ and in the level distribution $f_j \to \eta$, we obtain the generalized density function $\rho(\theta, p, \eta)$. Operating under this conditions, (9.67) can be rewritten as

$$\bar{f}(\theta, p) = \int_{\text{levels}} \rho(\theta, p, \eta) \eta \, d\eta, \tag{9.73}$$

while (9.72) takes the form

$$\rho(\theta, p, \eta) = \int_{\text{levels}} dx \bar{f}(\theta, p) \alpha(x) \delta(x - \eta), \tag{9.74}$$

where $\alpha(x)$ is the volume of the set of points (θ, p) such that $f(\theta, p) = x$.

In the following, we will be concerned with the intermediate situation where the levels are discrete in number. In this case, by using the spatially continuous version of Eq. (9.71) in the entropy (9.66), we get

$$s_{\text{LB}}(\bar{f}) = -\int d\tau' \left[\sum_j \alpha_j \bar{f}(\theta, p) \ln \alpha_j \bar{f}(\theta, p) \right. $$
$$\left. + \left(1 - \sum_j \alpha_j \bar{f}(\theta, p) \right) \ln \left(1 - \sum_j \alpha_j \bar{f}(\theta, p) \right) \right], \tag{9.75}$$

where $d\tau' = d\theta dp/(\bar{\omega}\nu)$ and the $\sum_j$ cumulates the contribution of all levels that refer to cell i.

The equilibrium coarse-grained distribution function $\bar{f}$ maximizes the entropy functional $s_{LB}(\bar{f})$, while imposing the constraints of the dynamics (energy, momentum and normalization), as well as the phase-space volumes α_j associated with each of the allowed levels.

Following the notation introduced above, the arbitrary integer n quantifies the total number of distinct levels to be allowed for, when considering the generalized initial distribution function f_{init}. Arguably, by accounting for a large enough collection of independent and discrete levels, we can approximately mimic any smooth profile. A pictorial representation of the family of initial conditions to which we shall refer to in the following, when discussing the specific case study $n = 3$, is depicted in Fig. 9.10.

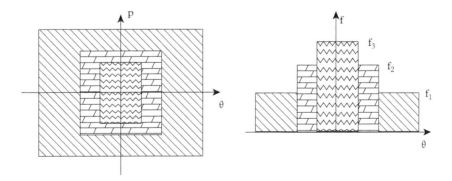

Figure 9.10 *Pictorial representation of a three-level (n = 3) water-bag initial condition. Reprinted with permission from Asslani et al. (2012), © 2012 by the American Physical Society.*

Mathematically, the initial distribution function f_{init} can be written as:

$$f_{init}(\theta, p) = \begin{cases} f_j & \text{if } \theta \in \Theta_j \text{ and } p \in P_j, \\ f_0 = 0 & \text{elsewhere.} \end{cases} \tag{9.76}$$

$\Gamma_j = [\Theta_j, P_j], j = 1, \ldots, n$ identifies the domain in phase space associated with level f_j and has area α_j.

We have already seen that the normalization condition (9.70) links together the $2n$ constants, f_j and α_j, which are to be assigned to fully specify the initial condition. In other words, only $2n-1$ scalars are needed to completely parametrize the initial condition. Importantly, the single water-bag limit is readily recovered once the phase-space support of the levels indexed with j other than $j = 1$ shrinks and eventually fades out. This condition implies requiring $\alpha_j \to 0$ for $j > 1$. Moreover, by making use of the normalization condition (9.70), we get $\alpha_1 = 1/f_1$. Lynden–Bell's entropy becomes therefore

$$s_{LB}(\bar{f}) = -\int d\tau' \left[\frac{\bar{f}}{f_1} \ln \frac{\bar{f}}{f_1} + \left(1 - \frac{\bar{f}}{f_1}\right) \ln \left(1 - \frac{\bar{f}}{f_1}\right) \right], \tag{9.77}$$

which coincides with the fermionic-like functional (9.46) that is known to apply to the single water-bag case, provided f_1 is replaced with the symbol f_0 and disregarding unimportant normalization factors.

The QSS distribution function $\bar{f}_{eq}(\theta, p)$ for the HMF model, relative to the generalized n-levels water-bag initial condition, is found by maximizing Lynden–Bell's entropy, under the constrains of the dynamics. This, in turn, implies solving a variational problem. The solution is relative to the microcanonical ensemble, since the Vlasov equation implies that we work with fixed total energy.

Let us start by recalling the generic n-level entropy which was shown to take the functional form

$$s_{LB}[\bar{f}] = -\int d\theta\, dp \left[\sum_{j=1}^{n} \bar{f}\alpha_j \ln(\bar{f}\alpha_j) + \left(1 - \sum_{j=1}^{n} \bar{f}\alpha_j\right) \ln\left(1 - \sum_{j=1}^{n} \bar{f}\alpha_j\right) \right]. \tag{9.78}$$

The conserved quantities are, respectively, the energy

$$h[\bar{f}] = \int \frac{p^2}{2} \bar{f}(\theta, p)\, d\theta\, dp - \frac{m[\bar{f}]^2 - 1}{2} \equiv \varepsilon_n, \tag{9.79}$$

and the total momentum

$$P[\bar{f}] = \int \bar{f}(\theta, p) p\, d\theta\, dp \equiv P_n. \tag{9.80}$$

The scalar quantity ε_n relates to the geometric characteristics of the bounded domains that define our initial condition. As we will be dealing with patches Γ_j symmetric with respect to the origin, we can immediately realize that $P_n = 0$.

The n volumes of phase space, each deputed to hosting one of the considered levels, are also invariant of the dynamics. We must therefore account for the conservation of n additional quantities, the volumes $\Omega_j[\bar{f}]$ for $j = 1, \ldots, n$, defined as

$$\Omega_j[\bar{f}] = \int \bar{f}(\theta, p) \alpha_j \, d\theta \, dp. \tag{9.81}$$

Moreover, using the normalization condition for the coarse-grained distribution function $\bar{f}(\theta, p)$, we get $\Omega_j[\bar{f}] = \alpha_j$. Equivalently, by imposing the above constraints on the hypervolumes, we also guarantee the normalization of the distribution function, which physically amounts to imposing the conservation of the mass.

Summing up, the variational problem that needs to be solved to eventually recover the stationary distribution $\bar{f}_{eq}(\theta, p)$ reads

$$\max_{\bar{f}} \{s_{\mathrm{LB}}[\bar{f}] \mid h[\bar{f}] = \varepsilon_n; P[\bar{f}] = P_n; \Omega_j[\bar{f}] = \alpha_j\}, \tag{9.82}$$

where the entropy functional $s_{\mathrm{LB}}[\bar{f}]$ is given by Eq. (9.78). This immediately translates into

$$\delta s_{\mathrm{LB}} - \beta \delta h - \lambda \delta P - \sum_{j=1}^{n} \left(\mu_j / \alpha_j \right) \delta \Omega_j = 0, \tag{9.83}$$

where β, λ and $\left(\mu_j / \alpha_j \right)$ stands for the Lagrange multipliers associated, respectively, with energy, momentum and volumes (or equivalently mass) conservations.

A straightforward calculation yields the expression

$$\bar{f}_{eq}(\theta, p) = \frac{1}{B + A \exp \left[\beta' \left(p^2/2 - \mathbf{m}[\bar{f}_{eq}] \cdot \mathbf{z} \right) + \lambda' p + \mu' \right]}, \tag{9.84}$$

where

$$B = \sum_{j=1}^{n} \alpha_j, \qquad A = \left(\prod_{j=1}^{n} \alpha_j^{\alpha_j} \right)^{\frac{1}{B}} \tag{9.85}$$

and

$$\beta' = \frac{\beta}{B}, \qquad \lambda' = \frac{\lambda}{B}, \qquad \mu' = \frac{\sum_{j=1}^{n} \mu_j}{B}, \qquad \mathbf{z} = [\cos \theta, \sin \theta]. \tag{9.86}$$

Recalling that $\alpha_1 = 1/f_1$, the above solution is clearly consistent with Eq. (9.56) obtained for the single water-bag case, upon proper rescaling of the Lagrange multipliers as explained just above Eq. (9.55).

Note that the equilibrium distribution $\bar{f}_{eq}$ depends on $\mathbf{m}$, which is, in turn, a function of $\bar{f}_{eq}$ itself. The two components of the magnetization, respectively m_x and m_y, are therefore unknowns of the problem, implicitly dependent on $\bar{f}_{eq}$. This latter is parametrized in terms of the Lagrange multipliers. Their values need to be self-consistently singled out. As a first simplification, we observe that the specific symmetry of the selected initial condition, $P_n = 0$, implies that $\lambda = 0$. Hence, just the two residual Lagrange multipliers are to be computed: the Lynden–Bell inverse temperature β and the cumulative chemical potential μ'. Being only interested in μ' to solve for $\bar{f}_{eq}$ and not on the complete collection of μ_j, we can hereafter focus just on the conservation of the global mass, i.e. the normalization. There are finally four unknowns, m_x, m_y, β, μ', which enter the following system of implicit equations for the constraints

$$\varepsilon = \frac{\tilde{A}}{2\beta'^{3/2}} \int d\theta \, e^{\beta' \mathbf{m} \cdot \mathbf{z}} F_2(y) - \frac{m^2 - 1}{2} \tag{9.87}$$

$$1 = \frac{\tilde{A}}{\sqrt{\beta'}} \int d\theta \, e^{\beta' \mathbf{m} \cdot \mathbf{z}} F_0(y) \tag{9.88}$$

$$m_x = \frac{\tilde{A}}{\sqrt{\beta'}} \int d\theta \, e^{\beta' \mathbf{m} \cdot \mathbf{z}} F_0(y) \cos\theta \tag{9.89}$$

$$m_y = \frac{\tilde{A}}{\sqrt{\beta'}} \int d\theta \, e^{\beta' \mathbf{m} \cdot \mathbf{z}} F_0(y) \sin\theta, \tag{9.90}$$

expressed as a function of the Fermi integrals

$$F_h(y) = \int dp \frac{p^h e^{-p^2/2}}{1 + y e^{-p^2/2}}, \tag{9.91}$$

with $y = \tilde{A} B \exp(\beta' \mathbf{m} \cdot \mathbf{z})$ and $\tilde{A} = A^{-1} \exp(-\mu')$. The system of equations (9.87), (9.88), (9.89) and (9.90) can be solved numerically to obtain a value for the involved Lagrange multipliers, as well as for the magnetization components, by varying the parameters that encode for the initial condition. It has been numerically checked that in the limit of a single water bag, $\alpha_{j>1} \to 0$, the expected solution is indeed recovered.

Consider now the specific choice $n = 2$. For the case of a single water bag, out-of-equilibrium transitions have been found (Antoniazzi *et al.*, 2007b), which separates between homogeneous and magnetized phases. A natural question is thus to understand what is going to happen if one additional level is introduced in the initial condition. The level f_1 is associated with a rectangular domain Γ_1 of respective widths $\Delta\theta_1$ and Δp_1. The level f_2 refers instead to an adjacent domain Γ_2, whose external perimeter is delimited by a rectangle of dimensions $\Delta\theta_2$ and Δp_2. The corresponding surface is thus $\Delta\theta_2 \Delta p_2 - \Delta\theta_1 \Delta p_1$.

To explore the parameter space, we monitor the dependence of m on f_1, which therefore acts as a control parameter. To this end, we proceed by fixing the quantity $\Delta f \equiv f_2 - f_1$, the difference in height of the considered levels. Furthermore, we specify

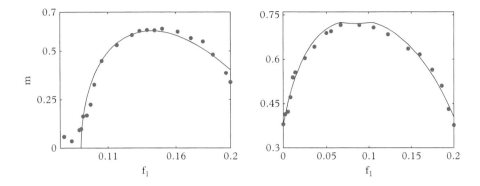

Figure 9.11 *The analytical predictions (solid line) for the QSS magnetization as a function of f_1 in the two-level water-bag case, are compared to the numerical simulations (filled circles) performed for $N = 10^4$. The comparison is drawn for two distinct values of the difference in height of the considered levels Δf: (Left) $\Delta f = 0.1$ and (right) $\Delta f = 0.2$. $\alpha_1 = 5$ and α_2 follows the normalization condition (9.92). Numerical values of m are computed as a time average over a finite time window where the QSS holds. The data are further mediated over four independent realizations. Expected uncertainties are about the size of the circle.*

the quantity α_1, while α_2 is calculated so to match the normalization constraint, which for $n = 2$ reads

$$\alpha_1 f_1 + \alpha_2 f_2 = 1. \tag{9.92}$$

Out-of-equilibrium phase transitions are predicted to occur and observed via direct simulations of the inspected system in its QSS phase. The comparison between theory and simulations is reported in Fig. 9.11. Filled symbols refer to the simulations, while the solid lines stand for the theory, for two distinct choice of Δf. The agreement is certainly satisfying and points to the validity of the Lynden–Bell interpretative framework, beyond the case of the single water bag.

9.6   The Core-Halo Solution

The Lynden–Bell approach allows us to gain insight into the behaviour of the system, and returns predictions which are justified from first principles. However, and despite the overall agreement with the results of numerical simulations, deviations are punctually observed when the ergodic hypothesis, which underlies the maximization procedure, does not apply. Dynamical effects are therefore to be properly accounted for, so as to improve the agreement between the theory and the numerics. Working along these lines, Pakter and Levin (2011) have proposed the so-called core–halo solution to characterize the QSS. We will shortly comment about this alternative approach with reference to the relevant HMF case study (see Levin et al., 2014, for other examples).

The mechanism of core–halo formation is similar to the process of evaporative cooling. Macroscopic density waves are formed and propagate in the embedding medium. Resonant particles can gain a large amount of energy at the detriment of the collective motion, thus escaping from the inner core and forming a diffuse halo. On the other hand, the loss of energy damps the macroscopic oscillations, so that the remaining particles become condensed into the low energy states, and so yielding a dense core. Because of the incompressibility of the Vlasov dynamics, the core cannot completely freeze, by collapsing into the minimum of the potential. At variance, the inner core approaches the maximum allowed phase-space density, as determined by the selected initial condition.

In the case of the HMF, the oscillations of the magnetization m can drive a selected bunch of rotors towards the high energy states. As a consequence, the rotors abandon the bulk and populate the surrounding halo. At the same time, the oscillations of the magnetization get damped on a characteristic timescale, which corresponds to the time needed for the halo to be established. The particles trapped in the inner core pack up to the limiting density $f_0 = 1/(\Delta\theta\Delta p)$, for the case of a rectangular water-bag initial condition. In the final (quasi-) stationary states, the distribution is that of a fully degenerate Fermi gas, and the core energy extends up to the Fermi energy ε_F. This latter is an unknown quantity that needs to be determined self-consistently. To this end, Pakter and Levin (2011) put forward the following ansatz for the core–halo single-particle distribution function $f(\theta, p)$

$$f(\theta, p) = f_0 \left(\Theta(\varepsilon_F - \varepsilon) + \chi \Theta(\varepsilon_h - \varepsilon)\Theta(\varepsilon - \varepsilon_F) \right), \tag{9.93}$$

where $\varepsilon = p^2/2 + 1 - m\cos\theta$ is the single-particle energy, χ is the ratio between the halo and the core phase-space densities and m is the value of the magnetization in the QSS state. ε_h is the maximum energy of the halo particles and it is determined from the short-time dynamics of the spins driven by the oscillations of the magnetization. To estimate ε_h one needs to obtain an equation for the evolution of magnetization. To this end, following Pakter and Levin (2011), we take the second derivative of m_x to get

$$\ddot{m}_x = m_x\langle\sin^2\theta\rangle - \langle p^2\cos\theta\rangle. \tag{9.94}$$

To truncate the hierarchy of equations necessary to fully characterize the dynamical evolution of, respectively, $\langle\sin^2\theta\rangle$ and $\langle p^2\cos\theta\rangle$, we can assume at sufficiently small times

$$\langle\sin^2\theta\rangle \simeq \frac{1}{2} \tag{9.95}$$

$$\langle p^2\cos\theta\rangle \simeq \langle p^2\rangle\langle\cos\theta\rangle = \left(2u - 1 + m_x^2\right)m_x, \tag{9.96}$$

where use has been made of the conservation of the energy (we denote here by u the given energy per particle), together with the obvious condition $m = m_x$, which follows from rotational invariance. After a simple manipulation, we obtain the equation

$$\ddot{m}_x = -m_x \left(2u + m_x^2 - \frac{3}{2} \right), \tag{9.97}$$

which can be integrated numerically to provide the temporal evolution of $m_x(t)$ for short times, when the halo is being established. For a given initial distribution, we can determine the maximum energy attained by a group of non-interacting test particles, which are injected with an initial condition that matches the water-bag profile imposed at $t = 0$. Their dynamical evolution is governed by $\ddot{\theta}_i = -m_x(t) \sin \theta_i$ with $m_x(t)$ determined by Eq. (9.97) with $m_x(0) = m_0$ and $\dot{m}_x(0) = 0$. Pakter and Levin solved this equation for two periods of oscillation of m_x: ε_h corresponds, hence, to the maximum energy obtained by any of the test particles, as recorded upon numerical integration of the above simplified system. Once ε_h has been determined via the test particles' dynamics, the other parameters in the core–halo representation of the solutions can be deduced by imposing

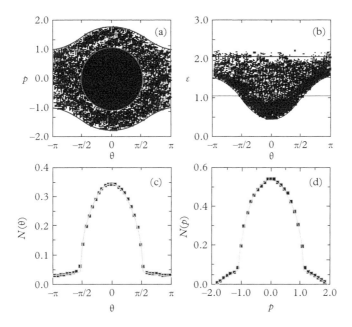

Figure 9.12 *Snapshots of (a) the phase space and (b) rotors energies, as a function of the angle θ. The figure is taken from Pakter and Levin (2011), and the snapshot refers to $t = 10000$, using $N = 20000$ particles. The solid curves correspond to the calculated Fermi energy ε_F and to the maximum halo energy ε_h as estimated numerically, via the procedure described in the text. The Fermi energy curve encloses the high density region. The maximum halo energy delimits the extent of the particle distribution in the phase space. (c) and (d) show the angle and the momentum distributions, respectively. Solid curves are the theoretical predictions and the symbols are the results of molecular dynamics simulations averaged over 20 runs. The initial distribution has $m_0 = 0.8$ and $u = 0.55$. Reprinted with permission from Pakter and Levin (2012), © 2012 by the American Physical Society.*

the condition of conservation of the norm and energy, together with the self-consistent definition of the magnetization. In formulae

$$\int d\theta \, dp \, f(\theta, p) = 1 \tag{9.98}$$

$$\int d\theta \, dp \, f(\theta, p) \varepsilon(\theta, p, M) = u$$

$$\int d\theta \, dp \, f(\theta, p) \cos \theta = m.$$

This is a close set of algebraic equations which can be analytically solved to calculate the unknown parameters ε_F, χ and m. Let us recall, however, that ε_h needs to be preliminarily determined via the semi-analytical procedure illustrated earlier. In Fig. 9.12 a snapshot of the phase space and that of the distribution of rotors energy are shown. From visual inspection, we can appreciate the core–halo structure attained by the system in its QSS phase, which motivates the above treatment. The energy ε_h delimits the region of space made accessible to the system, while the Fermi energy restricts the extent of the core–halo. In Figs. 9.12c and 9.12d, the theoretical and numerical position and momentum distributions are compared, returning a very good agreement. This method also makes it possible to further characterize the out-of-equilibrium phase transitions, first predicted by the Lynden–Bell maximum entropy approach.

Part III

Applications

10

Gravitational Systems

The statistical mechanics of self-gravitating systems has crucial implications in distinct astrophysical contexts ranging from the large-scale structure formation to the dynamical evolution of globular clusters. Clearly, gravity governs the dynamics of these problems, while associated distances, boundary conditions and spatial symmetries turn out to be sensibly different, and specific to a given field of applications. On the other hand, and on a different scale, statistical mechanics also contribute to elucidating the process of non-linear gravitational instability of dark matter which is supposedly responsible for galaxies formation and evolution. In Fig. 10.1, a beautiful image of the spiral galaxy NGC 1232 is reported.

We begin this chapter by discussing the statistical and dynamical behaviour of particles interacting each other with Newtonian gravity, the so-called self-gravitating systems. It is often stated that the thermodynamic limit of such systems is not well defined (Ruelle, 1969), since the partition function in the canonical ensemble and the density of states in the microcanonical ensemble diverge due to the singularity of the Newtonian potential at the origin. Furthermore, when the thermodynamic limit is taken, divergences of the specific thermodynamic potentials with the number of particles appear: they are a consequence of the non-additivity due to the long-range nature of the Newtonian gravitational potential. The addition of a regularization of the potential (i.e. a finite lower bound) or a repulsion at short distances and the consideration of a finite volume V can make, for finite N, the partition function and the density of states finite. The repulsion can be physically justified, since particles have a natural core. However, due to the long-range nature of the interaction, the potential is *thermodynamically unstable* (Ruelle, 1969) i.e. the potential energy cannot be bounded from below by $-CN$, with $C > 0$. In fact, as explained in Section 6.2, even with the inclusion of a hard core at short distance, the value of the energy of a gravitational system would scale as $-N^{5/3}$. This scaling would hold also with a short-range repulsion, while with a simple short-range regularization we would have a scaling like $-N^2$ (since the regularization avoids the divergence of the energy at short distance but it does not forbid collapsed configurations). Therefore in all these cases the scaling with N of the energy implies that it is impossible to define intensive thermodynamic potentials and to avoid the divergence of the grand-partition function. A possible way out from this catch is to introduce the *mean-field limit* (Spohn, 1991) rescaling the interaction by $1/N$. In this limit, all thermodynamic

Physics of Long-Range Interacting Systems. First Edition. A. Campa *et al.*
© A. Campa, T. Dauxois, D. Fanelli, and S. Ruffo 2014. Published in 2014 by Oxford University Press.

Figure 10.1 *A spectacular image of the large spiral galaxy NGC 1232. The image is based on three exposures in ultra-violet, blue and red light, respectively. The central core contains older stars of reddish colour, while the spiral arms are populated by young, blue stars and many star-forming regions. Courtesy of ESO.*

potentials (free energy, entropy, grand-canonical potential) are finite. There is no clear physical justification of this limit; however, in its favour, we can quote that, after scaling back the coupling constant, the comparison with numerical simulations is often satisfactory.

Although the $N \to \infty$ limit can be performed with the mean-field scaling, the statistical ensemble might give inequivalent results (Thirring, 1970). Indeed, it was in the context of self-gravitating systems that the possibility of realizing negative specific heat in the microcanonical ensemble was discussed by Emden (1907), Eddington (1926), Chandrasekhar (1942), Antonov (1962) and Lynden–Bell and Wood (1968). The connection of this phenomenon with *ensemble inequivalence* was made by Hertel and Thirring (1971): as discussed in Chapter 2, specific heat can be negative only in the microcanonical ensemble.

Section 10.1 will make an overview of the statistical mechanics treatment of self-gravitating systems; Section 10.2 will describe simplified one-dimensional models. After Section 10.3, which will bridge the gap between the relevant general relativity setting and the customarily adopted classical picture, we shall devote Sections 10.4 to 10.8 to presenting a general overview of the cosmological setting. As we shall comment, this provides the background for understanding the importance of classical Newtonian self-gravitating systems in the context of structure formation, where the medium extends to infinity. Structure formation in the Universe is a rich and complex problem that touches many sides of physics, from theories of the origin of primordial fluctuations, to detailed calculations of galaxy dynamics in the present epoch, including radiation pressure from stars and absorption in clouds, star birth and star mass loss. It is of great topical interest because of the recent and still improving observational data on the inhomogeneities in the cosmic microwave background. Structure formation is also a challenging problem of classical physics, i.e. how a small perturbation of a spatially almost uniform Universe develops, first linearly and then non-linearly, to the pronounced structures we see today. Finally, in Section 10.9 we will describe some results of numerical simulations in 3D.

10.1 Equilibrium Statistical Mechanics of Self-Gravitating Systems

The gravitational problem in three dimensions is particularly difficult to tackle. In addition to the non-additivity stemming from the long-range character of the interaction, we must also face the divergence of the potential which needs to be properly regularized at small distances. Consider N particles in a domain of volume V and assume each particle to have the same mass m. Although we have shown in Chapter 2 that the construction of the canonical ensemble is not well defined when the potential is long range, let us consider the canonical partition function, which can be cast in the form

$$Z_N = \frac{1}{\lambda^{3N} N!} \int_V \prod_{i=1}^{N} d\mathbf{r_i} \, \exp\left[-\beta U(\mathbf{r}_1, \dots, \mathbf{r}_N)\right], \tag{10.1}$$

where

$$U(\mathbf{r}_1, \dots, \mathbf{r}_N) = -Gm^2 \sum_{i<j}^{N} V(|\mathbf{r}_1 - \mathbf{r}_N|), \tag{10.2}$$

with

$$V(r) = \frac{1}{r}. \tag{10.3}$$

Here β stands for the inverse temperature, G labels the gravitational constant, and $\lambda = \hbar(2\pi\beta/m)^{1/2}$ the De Broglie thermal wavelength. In Eq. (10.1), we have used the factor $1/h^{3N}$, often employed also in a classical setting to make the partition function dimensionless.

Clearly, Z_N diverges because of the singularity at the origin. In order to regularize the potential, within the realm of classical physics and without involving quantum mechanical concepts, we can introduce an *ad hoc* cut-off (Binney and Tremaine, 1987): it can be hypothesized that either the potential has a hard core, which refers to the particles' size, or the potential can be smoothed near the origin. In both cases, there is a minimum of $U(r)$. As a consequence, the inequality $U \geq -Gm^2 C \equiv -C'$ makes it possible to easily determine a finite upper bound of the configurational partition function

$$Z_N \leq \frac{V^N}{\lambda^{3N} N!} \exp[\beta C' N(N-1)/2]. \tag{10.4}$$

Although we have succeeded in bounding the partition function at finite N, it is clear from Eq. (10.4) that the free energy per particle will be bounded by a quantity which diverges with N. This divergence can be eliminated if we perform the *mean-field limit*, i.e. if we rescale the gravitational potential by the overall factor $1/N$. As we discussed at the end of Chapter 8, this is equivalent to rescaling by the factor N the temperature in the canonical ensemble. We have also seen that, from the point of view of the dynamical evolution of the system, the rescaling of the temperature is also equivalent to rescaling the time by $1/\sqrt{N}$. In the following, our choice will be to rescale the potential by $1/N$ as usually done in the mean-field limit (Braun and Hepp, 1977; Messer and Spohn, 1982; Kiessling, 1989).

As we have seen in Section 8.4, the entropy defined in terms of the N-body distributions function reduces to the Boltzmann entropy defined in terms of the one-particle distributions function $f(\mathbf{r}, \mathbf{p}, t)$ if correlations can be neglected, which is possible in the mean-field limit provided some conditions are imposed on the potential. In the following, we will analyse the statistical ensembles using the Boltzmann entropy

$$S = -\int \mathrm{d}\mathbf{r}\mathrm{d}\mathbf{p}\, f \ln f. \tag{10.5}$$

Let us study the microcanonical and canonical problems.

In the microcanonical ensemble, the reduction of the potential energy as driven by the presence of pairs of close particles should be balanced by an analogous increase of the kinetic energy, being the total energy conserved. This process leads to an increase of the accessible phase volume in the direction of momentum, and hence to an entropy increase. Since this process extends to the limit of zero distance among the particles, it induces a divergence of the density of states, and consequently of the Boltzmann entropy. This mechanism was first elucidated by Antonov (1962) when he set down to calculating the local maxima of the Boltzmann entropy, for a system of N particles, confined in a spherical box of external radius r_e so to prevent evaporation. The total mass of the system is assumed to be $M = Nm$ and the associated energy E. The entropy S in formula (10.5) is subject to the constraints

$$N = \int \mathrm{d}\mathbf{r}\mathrm{d}\mathbf{p}\, f(\mathbf{r}, \mathbf{p}, t) \tag{10.6}$$

$$E = \int \mathrm{d}\mathbf{r}\mathrm{d}\mathbf{p}\, f(\mathbf{r}, \mathbf{p}, t) \frac{p^2}{2m} - \frac{1}{2} \int \mathrm{d}\mathbf{r} \int \mathrm{d}\mathbf{r}'\, Gm^2 \frac{n(\mathbf{r})n(\mathbf{r}')}{|\mathbf{r} - \mathbf{r}'|}, \tag{10.7}$$

where

$$n(\mathbf{r}) = \int \mathrm{d}\mathbf{p}\, f(\mathbf{r}, \mathbf{p}) \tag{10.8}$$

is the number density. We are here dealing with the simpler case where the Universe expansion is being neglected, which implies working in the original spatial coordinates r. The above integrals are taken over the 6-dimensional phase space, since f is the one-particle distribution function. The problem is to maximize the entropy S keeping the energy E and the number N constant. Consequently, introducing the associated Lagrange multipliers that we denote, respectively, by β and α, we obtain the most probable density by cancelling the first variation, which leads to

$$\delta S - \beta \delta E - \alpha \delta N = 0. \tag{10.9}$$

After some algebra, we can write the energy variation δE as

$$\delta E = \int \mathrm{d}\mathbf{r} \mathrm{d}\mathbf{p} \, \delta f \left(\frac{p^2}{2m} + m\phi(\mathbf{r}, t) \right), \tag{10.10}$$

where the mean gravitational field $\phi(\mathbf{r}, t)$ follows from the Poisson equation

$$\Delta\phi(\mathbf{r}, t) = 4\pi \, Gn(\mathbf{r}), \tag{10.11}$$

and reads

$$\phi(\mathbf{r}, t) = -G \int \mathrm{d}\mathbf{r}' \, \frac{n(\mathbf{r}')}{|\mathbf{r} - \mathbf{r}'|}. \tag{10.12}$$

Taking advantage from the above expression for δE and substituting into (10.9), we eventually get

$$-\int \delta f \left[\log f + 1 + \beta \left(\frac{p^2}{2m} + m\phi \right) + \alpha \right] = 0, \tag{10.13}$$

which returns

$$\log f + 1 + \beta \left(\frac{p^2}{2m} + m\phi \right) + \alpha = 0, \tag{10.14}$$

and finally

$$f = A \exp(-\beta\varepsilon), \tag{10.15}$$

where $\varepsilon = (p^2/(2m) + m\phi)$ and $A = \exp(-\alpha - 1)$. As expected, Eq. (10.15) is the Maxwell–Boltzmann distribution, expressed as a function of the self-consistent potential ϕ. This latter is itself defined via Eq. (10.12), which involves again the unknown distribution function f. The procedure to break this implicit dependence is to integrate Eq. (10.15) so to get, through the use of the Poisson equation (10.11), an equation for ϕ, solve it and then substitute in (10.15) to obtain the explicit solution for f. It was noted by Antonov (see e.g. discussion in Lynden–Bell and Wood (1968)) that only spherically symmetric solutions can correspond to local maxima. When developing the calculation, we can therefore limit the discussion to this latter class, which in turn implies dealing with the well-known equation for the isothermal gas sphere

$$\frac{1}{r^2} \frac{\mathrm{d}}{\mathrm{d}r} \left(r^2 \frac{\mathrm{d}\phi}{\mathrm{d}r} \right) = 4\pi \, GB \exp(-\beta m\phi), \tag{10.16}$$

where $B = A(2\pi/\beta)^{3/2}$. Following Lynden–Bell and Wood (1968), we can rescale the variables as

$$v_1 = m\beta(\phi - \phi(0)) \tag{10.17}$$
$$r_1 = (4\pi G n_0 \beta)^{1/2} r, \tag{10.18}$$

where $n_0 = n(0) = B\exp[-\beta m\phi(0)]$ is the number density at the origin. By taking advantage of (10.17) and (10.18), Eq. (10.16) reduces to the Lamé–Emden equation

$$\frac{d^2 v_1}{dr_1^2} + \frac{2}{r_1}\frac{dv_1}{dr_1} - \exp(v_1) = 0. \tag{10.19}$$

The solution of this equation can be computed numerically and this makes it possible in turn to estimate all relevant thermodynamical quantities, including the total energy E and total mass M within the sphere of radius r_e, as well as the total entropy S and the surface pressure p. These results were all known before Antonov asked himself about stability of the obtained solutions. He then settled down to elucidate whether such solutions correspond to global or, conversely, local entropy maxima. As a consequence of this analysis, he could prove that 'equilibrium' states can exist only in association with local entropy maxima (Antonov, 1962), since there is not an absolute maximum. These local maxima are found if the condition $\Gamma = -Er_e/GM^2 \leq 0.335$ is met, i.e. if the energy E is sufficiently large (for a given r_e) or, conversely, if the radius of the confining sphere is sufficiently small (for a given value of E). This conclusion was also elaborated by Lynden–Bell and Wood (1968). The stability of the extremal solution of the Antonov variational problem has been investigated by several authors (Lynden–Bell and Wood, 1968; Katz, 1978; Padmanabhan, 1990; Chavanis, 2006a) who found that, above a critical density contrast $n(0)/n(r_e) = n_{core}/n_{halo}$, all extrema are unstable, namely they are not local maxima. Lynden-Bell and Wood termed this phenomenon *gravothermal catastrophe*, which is also known as Antonov instability: for specific values of the control parameters no equilibrium is allowed. The system takes a core–halo structure, the core becoming progressively more dense and presumably evolving towards a black hole singularity.

As a side comment, we mention that the above analysis does not apply for $N = 2$: a subtle derivation (Padmanabhan, 1990; Chabanol *et al.*, 2000) reveals that the entropy integral diverges only when $N \geq 3$. It is indeed remarkable that this also corresponds to the transition from an integrable ($N = 2$) to a non-integrable ($N = 3$) gravitational system.

Kiessling (1989) studied the thermodynamical stability of an N-body point mass system confined in a spherical box, within the canonical picture, so as to gain insight into the Antonov problem. He assumed a dedicated softening at small scales so to regularize the Newtonian potential and showed that in the limit where the softening vanishes (pure Newtonian interaction) the canonical equilibrium measure is a superposition of Dirac measure at any temperatures, meaning that the system is in a collapsed point mass state. Based on the analysis of the discrete model, he showed that the one-particle density function is proportional to the Dirac distribution in the continuum limit and therefore the equilibrium state is the collapsed one in the canonical ensemble. This result strictly applies to a self-gravitating system in contact with a heat bath and does not contradict the microcanonical mean field Antonov analysis, as no global entropy maximum is found to occur.

Miller and collaborators (Klinko and Miller, 2000; Youngkins and Miller, 2000) proposed a modified version of the mean-field spherical shell models by incorporating the influence of rotation and studied in particular the case where each particle has specific angular momentum of the same magnitude ℓ. They rigorously proved the existence of an upper bound of the energy and demonstrated that a phase transition occurs in both the microcanonical and canonical ensembles, when ℓ falls below a critical value. A detailed discussion on phase transitions in self-gravitating systems, in both the canonical and microcanonical ensemble, is not the aim of this chapter, and can be found in Padmanabhan (1990) and Chavanis (2006a).

Self-gravitating systems were historically the first physical systems for which ensemble inequivalence was brought into evidence via the phenomenon of negative specific heat as we discussed in Section 2.2. The possibility of finding a negative specific heat in gravitational systems was emphasized by Emden (1907) and Eddington (1926). An early remark on the possibility of having conditions that can in principle yield a negative specific heat can be found in the seminal review paper on statistical mechanics by Maxwell (1876). For a long time this was considered a paradox, until the controversial point was clarified in Thirring (1970) and Hertel and Thirring (1971): the specific heat can be negative only in the microcanonical ensemble.

For self-gravitating systems at constant energy (i.e. in the microcanonical ensemble) a simple physical argument which justifies the presence of a negative specific heat has been given by Lynden–Bell (1999). It is based on the virial theorem, which, for the gravitational potential, states that

$$2\langle K \rangle + \langle U \rangle = 0, \tag{10.20}$$

where K and U are the kinetic and potential energy, respectively. Recalling that the total energy E is constant, we get that

$$E = \langle K \rangle + \langle U \rangle = -\langle K \rangle, \tag{10.21}$$

where in the second identity we have used the virial theorem (10.20). Since the kinetic energy K defines the temperature, we get

$$C_V = \frac{\partial E}{\partial T} \propto \frac{\partial E}{\partial K} < 0. \tag{10.22}$$

Losing its energy, the system becomes hotter.

Because of these peculiarities, ultimately stemming from the long-range and singularity properties of the force, it is hence difficult to directly apply the conventional tools of statistical mechanics in the canonical ensemble to the study of gravitational systems.

Moreover, although for gases it is natural to confine them in a box, this is not completely justified for gravitational systems (apart for globular clusters as discussed below). In addition, a sound model of structure formation should encapsulate the role of Universe expansion and so reconsider the problem in a wider context (Padmanabhan, 2009). In the following, we shall discuss possible approaches to the problem of structure formation and evolution.

10.2   Self-Gravitating Systems in Lower Dimensions

In the attempt to progress towards a comprehensive understanding of the physics of gravitational clustering, it is natural to look for simplified toy models. These latter have the merit of capturing the essence of the problem while allowing for straightforward numerical investigations. Besides their applied interest for the problem at hand, simple self-gravitating models constitute ideal testbed for statistical mechanics studies, notably with reference to the field of long-range interacting systems. An obvious simplification results when mapping the original 3D model into the analogous 1D setting. In one dimension, in fact, particles experience pair forces independent of their separation, this property yielding an efficient and accurate, up to round-off errors, numerical scheme. The original formulation, termed the 'sheet model', applies to finite masses distribution, but variants have been proposed to develop the analogy with the 3D infinite space problem (Hohl and Feix, 1967; Severne and Luwel, 1986; Rouet *et al.*, 1990; Reidl and Miller, 1991; Rouet *et al.*, 1991; Yano and Gouda, 1998; Tsuchiya and Gouda, 2000; Aurell *et al.*, 2001; Fanelli and Aurell, 2002). In the following, we shall review the main results, respectively, focusing on the static and expanding Universe.

Consider a point particle picture and focus on the special case of a stratified perturbation: velocity has one component only and velocity and density vary with respect to this direction.

If the effect of the overall expansion is neglected, the Vlasov–Poisson dynamics is equivalent, at a discrete level, to the Hamiltonian

$$H = \sum_{i=1}^{N} \frac{p_i^2}{2m} + 2\pi G m^2 \sum_{i>j}^{N} |x_i - x_j|, \tag{10.23}$$

which describes the gravitational interaction of N *infinite parallel sheets* (here on referred to as particles) carrying the same mass m, as presented in Fig. 10.2; x_i is the particle position and p_i is the momentum conjugate to x_i. The particular form of the interaction potential is the one satisfying the Poisson equation in one dimension, while the three-dimensional case leads to the conventional $1/r$ potential. When particles reach simultaneously the same spatial position, they cross each other or, equivalently, collisions are elastic.

The beauty of the Hamiltonian (10.23) is that it makes it possible to implement an exact, up to round-off errors, event-driven numerical scheme. In fact, in between two

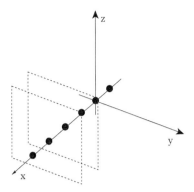

Figure 10.2 *Visualization of a stratified perturbation. The infinite and parallel sheets (or equivalently particles represented by filled black circles) move in the x direction, carrying mass m. When they reach simultaneously the same position they cross each other, experiencing a finite change in the accelerations.*

successive collisions, the acceleration of each particle is constant, and proportional to the net mass difference on its right and left. When a crossing occurs, the particles involved experience a sudden change in their accelerations. Hence, a particle trajectory is re-covered by connecting continuously arcs of parabola, in correspondence of the locations of crossing.

The event-driven scheme is based on the capability of computing the next crossing time between two adjacent particles. The algorithm starts by ordering the particles with ascending positions. Then, the collision times between consecutive pairs is computed, and the results stored in an array of size $N-1$. The minimum is selected and the particles are evolved until this time is reached. At this moment, the two colliding particles share the same spatial position and their velocities are exchanged. It is clear that in such a scheme the most time-consuming operation is the search of the minimal collision time. The original implementation due to Eldridge and Feix (1962) required an $\mathcal{O}(N)$ operations count for each collision. An optimized scheme, based on the concept of a *heap* (a rooted tree to store the data) which takes $\mathcal{O}(\log N)$ operations per collision, was proposed in Noullez *et al.* (2001) and allows a significant enhancement of statistics over previous investigations.

Since the algorithm was invented, the one-dimensional Hamiltonian (10.23) has been widely studied, with special regards to possible cosmological implication (Konishi and Kaneko, 1992). In particular, the model served as a playground to investigate

   i. The validity of the Vlasov description for large N values;

  ii. The adequacy of the Lynden–Bell theoretical picture; and

 iii. The subsequent relaxation as driven by the discreteness.

A milestone result in the analysis of the relaxation issue is the exact statistical mechanics solution obtained by Rybicki (1971). As previously recalled, the full three-dimensional problem does not admit a state of thermodynamic equilibrium. At variance, the case of a one-dimensional self-gravitating system approaches an asymptotic distribution. The method set down by Rybicki builds on a previous analysis by Salzburg (1965) for a 1D self-gravitating equal mass system confined in a box of assigned length L. This latter work was itself an extension of Lenard (1962) and Prager (1961) studies on one-dimensional plasma models. The calculations by Salzburg were not directly applicable to the system of interest for stellar dynamics application, as this latter requires working in a free space setting, without entrapping the system within external walls. Rybicki derivation moves from the canonical calculation which is then extended to the (more relevant) microcanonical setting. The possibility of performing the canonical configurational integrations rests on the possibility of recasting the system into an equivalent setting where the particles' coordinates are ordered in ascending positions, namely $x_1 \leq x_2 \leq \cdots \leq x_N$. Then the potential takes the simple form

$$V = \lambda \sum_{i>j} (x_i - x_j) \tag{10.24}$$

without absolute values and where $\lambda = 2\pi \, Gm^2$. After some manipulations, we obtain

$$V = \lambda \sum_{\ell=1}^{N-1} \ell(N - \ell)(x_{\ell+1} - x_\ell). \tag{10.25}$$

A simple interpretation of this formula, as originally proposed by Rybicki, goes as follows. Consider the work done in reducing an interval $x_{\ell+1} - x_\ell$ to 0, while keeping rigid connections between the members of each group of particles to the left and to the right, so that all other intervals stay constant. Since the force is independent of the distance, the sought quantity corresponds to calculating the work done in moving two sheets of mass ℓm and $(N - \ell)m$ into coincidence over a distance equal to $x_{\ell+1} - x_\ell$, namely $-\lambda\ell(N - \ell)(x_{\ell+1} - x_\ell)$. Each interval may in turn be reduced to 0 and eventually all particles will be sitting on the same point so that $V = 0$. The interest of Eq. (10.25) is that it expresses the potential as the sum of independent contributions, so enabling ones to perform the configurational integrations. This is the crucial trick. In the following, we shall highlight the main steps of this latter calculation.

The canonical one-particle distribution function $f_c(p, x)$ is defined as

$$f_c(p, x) = \frac{1}{zN!} \int \mathrm{d}\mathbf{p} \int \mathrm{d}\mathbf{x} \, \delta(\bar{x})\delta(\bar{p}) \exp(-\beta H) \frac{1}{N} \sum_n \delta(p - p_n)\delta(x - x_n), \tag{10.26}$$

where

$$z = \frac{1}{N!} \int \mathrm{d}\mathbf{p} \int \mathrm{d}\mathbf{x} \, \delta(\bar{x})\delta(\bar{p}) \exp(-\beta H). \tag{10.27}$$

Note that in the above formulae we have omitted the Planck's constant, which, as mentioned in Chapter 1, is often used in the definition of the ensembles. In fact, it cancels out in all relevant expressions derived below. $\bar{x}$ and $\bar{p}$ represent, respectively, the coordinate and momentum centre of mass. These quantities are invariant of the dynamics and are set to 0 in the current setting. This is, in turn, equivalent to selecting a frame of reference in which the centre of mass is at rest in the origin.

By invoking the separability of the Hamiltonian (10.23), we can factorize space and momentum contribution as $f_c(p, x) = n_c(x)\theta_c(p)$. Moreover, the kinetic contribution is quadratic in p, thus allowing for a straightforward estimate of the corresponding $\theta_c(p)$ expression. Following Rybicki (1971), we eventually obtain

$$\theta_c(p) = \left[\frac{\beta N}{2\pi m(N-1)}\right]^{1/2} \exp\left[-\frac{\beta N p^2}{2m(N-1)}\right], \tag{10.28}$$

where the unusual $N/(N-1)$ factor stems from the fact that N particles must share the thermal energy of $N-1$ degrees of freedom, since one degree of freedom is lost by virtue of the centre-of-mass constraint. The derivation of the spatial contribution is more lengthy, and the reader can refer to Rybicki (1971) for a detailed account on the whole procedure. Here we limit ourselves to recalling the remarkably compact form of the Fourier transform $\bar{n}_c(k)$ of $n_c(x)$, namely

$$\bar{n}_c(k) = \prod_{\ell=1}^{N-1} \frac{\ell^2}{\ell^2 + k^2/(N\beta\lambda)^2}. \tag{10.29}$$

The Fourier transform in Eq. (10.29) can be inverted in closed form, as outlined in Rybicki (1971). $\bar{n}_c(k)$ as a function of the complex variable k displays, in fact, $(2N-2)$ simple poles, equally spaced by $N\beta\lambda$ along the imaginary axis, ranging from $-iN(N-1)\beta\lambda$ to $+iN(N-1)\beta\lambda$, with the exception of $k = 0$. The inverse transform reads

$$n_c(x) = \frac{1}{2\pi} \int_{-\infty}^{+\infty} dx\,\bar{n}_c(k)\exp(-ikx), \tag{10.30}$$

where the contour of integration is deformed upward (resp. downward) for $x < 0$ (resp. $x > 0$) so as to enclose the poles on the positive (resp. negative) portion of the imaginary axis. Upon estimating the associated residues, and following the subsequent algebraic manipulation, the density in real space takes the final form

$$n_c(x) = N\beta\lambda \sum_{\ell=1}^{N-1} A_\ell^N \exp(-N\beta\lambda\ell\,|x|), \tag{10.31}$$

where

$$A_\ell^N = \frac{(-1)^{\ell-1}[(N-1)!]^2\ell}{(N-1-\ell)!(N-1+\ell)!}. \tag{10.32}$$

Combining Eqs. (10.28) and (10.31) eventually yields the distribution

$$f_c(p, x) = \left[\frac{(\beta N)^{3/2} \lambda}{\sqrt{2\pi m(N-1)}} \right] \sum_{\ell=1}^{N-1} A_\ell^N \exp\left(-\frac{\beta N p^2}{2m(N-1)} - N\beta\lambda\ell|x| \right). \tag{10.33}$$

This solution applies to a system in contact with a thermal bath, forcing a constant temperature $T = 1/\beta$. We can readily obtain the microcanonical solution from the canonical one, the former being supposedly related to more realistic settings. Performing the Laplace transform, we are hence led to the following expression for the microcanonical one-particle distribution function

$$f_{MC}(p, x) = \frac{N\lambda}{E} \left(\frac{N}{2\pi m(N-1)E} \right)^{1/2} \frac{\Gamma(3N/2 - 3/2)}{\Gamma(3N/2 - 3)}$$
$$\times \sum_{\ell=1}^{N-1} A_\ell^N \left(1 - \frac{N p^2}{2m(N-1)E} - \frac{N\lambda\ell|x|}{E} \right)_+^{(3N/2)-4}. \tag{10.34}$$

Here E denotes the total energy, and the notation $(\cdot)_+$ is defined by

$$(u)_+ = u, \qquad u \geq 0 \tag{10.35}$$
$$= 0, \qquad u \leq 0. \tag{10.36}$$

Particularly interesting in this respect is the calculation of the density and momentum distributions that are readily recovered by integrating (10.34), respectively, over p and x. The explicit expressions read, respectively,

$$n_{MC}(x) = \frac{N\lambda}{2E} (3N - 5) \sum_{l=1}^{N-1} A_l^N \left(1 - \frac{N\lambda l|x|}{E} \right)_+^{(3N/2)-7/2} \tag{10.37}$$

and

$$\theta_{MC}(p) = \frac{\Gamma(3N/2 - 3/2)}{\Gamma(3N/2 - 2)} \left(\frac{N}{2\pi m(N-1)E} \right)^{1/2} \left(1 - \frac{N p^2}{2m(N-1)E} \right)_+^{(3N/2)-3}. \tag{10.38}$$

Performing the large N limit in these expressions while assuming the total mass $M = N\sigma$ conserved, we can in turn, obtain closed formulae which formally apply to the continuum limit. Note that the continuum limit is here performed *after* the infinite time limit. As previously mentioned, the two limits do not commute and a substantially different behaviour can in principle be expected when taking them in the opposite ordering. To complete the review of Rybicki's theory, it is useful to introduce the rescaled variables

$$\eta = \frac{p}{mV} \tag{10.39}$$

$$\xi = \frac{x}{L} \tag{10.40}$$

$$K = kL, \tag{10.41}$$

where the characteristic velocity V and length L read

$$L = \frac{2E}{3\pi GM^2} \tag{10.42}$$

$$V^2 = \frac{4E}{3M}. \tag{10.43}$$

Following Rybicki, we can define the scaled distribution $f^*(\eta, \xi) = mVLf(mV\eta, L\xi)$, which in the limit $N \to \infty$ leads to

$$f^*_{MC} = \frac{1}{2\sqrt{\pi}} \exp(-\eta^2)\mathrm{sech}^2\xi. \tag{10.44}$$

Numerical studies on one-dimensional self–gravitating system have been performed on several occasions during the past decades, because of their interest as potential description of the motion of stars along a direction normal to the disk of highly flattened galaxies (Camm, 1950). Moreover, one-dimensional systems became popular as a testbed for the adequacy of violent relaxation theories, as we shall be mentioning in the following. Particularly fascinating are the investigations aimed at resolving the fine details of the thermalization. This task became indeed feasible thanks to the aforementioned Rybicki's prediction, which in turn sets the correct reference equilibrium solution. In one of the earliest numerical studies, it was conjectured by Hohl and Feix (1967) that it should take a time of the order of $N^2 t_c$ to equilibrate. Here, t_c is the characteristic time for the system to complete an oscillation, while N refers to the system size. Later on, Luwel *et al.* (1986) elaborated on the relevance of the initial condition, suggesting that for a specific class of initial conditions that do not excite core–halo structures in phase space, relaxation takes place rapidly, within Nt_c. Proving this scenario has turned out to be difficult and controversial as discussed in Reidl and Miller (1992): standard statistical tests applied to relatively short regimes of the evolution of a one-dimensional self-gravitating system might signal an achieved convergence towards equilibrium, while the system is instead just temporarily trapped in an intermediate phase. As a corollary of these speculations, it was shown by Miller and collaborators that the system experiences long lived correlations that persist for times longer than $2N^2 t_c$. These contradictory evidences were later on reconciled by Tsuchiya *et al.* (1994, 1996), who pointed to the existence of two types of relaxations, involving, respectively, the macroscopic and microscopic level of description. The relaxation proceeds at smaller scales while the macroscopic observables seem to have attained their final profile, and changes of this quasi-equilibria, as driven by microscopic effects, take place on a longer time scale. This is, of course, reminiscent of the distinction between violent and collisional relaxation described in Chapter 9. The same authors further discussed the coupling between microscopic and macroscopic descriptions by making reference to the concept of itinerant stage; this is again very similar to the quasi-stationary states described earlier in this book. Accordingly, the one-body distribution stays in a quasi-equilibrium for some time, and then changes to other quasi-equilibria, compatible with the constraints of the dynamics. As we will comment in the following, and despite an incomplete mixing in phase space, the macroscopic picture is successfully captured within the Lynden–Bell interpretative scenario at least for a selected class of initial conditions. This is a recent achievement by Yamaguchi (2008), who gained novel insight into a long-standing problem on the validity of the violent relaxation picture, which dates back to the early sixties.

In Yamaguchi (2008) it is shown that the Lynden–Bell statistics is capable of reproducing the energy distribution as seen in direct numerical experiments for initial states near the virial equilibrium. Dynamically accelerated high-energy sheets drive the segmentation into core and halo structures, a regime that is only approximately described within the Lynden–Bell scenario. Indeed, the Lynden–Bell statistics returns a profile which solely matches the core distribution. To take one leap forward, Yamaguchi proposed a modified Lynden–Bell theory by supplying two external parameters: the core energy U_{core} and R, which measures the ratio between particles inside the halo and the total number of particles, for a fixed value of the threshold energy. This latter is a parameter which allows us to separate the particles belonging to the core from those that populate the surrounding halo. The parameters U_{core} and R are not deduced on the basis of a self-consistent calculation, but assigned by inspection of direct N-body simulations, for any fixed threshold in energy. Indeed, the two parameters are linearly dependent and so the actual unknowns to be externally supplied reduce to one. The predictions inspired by the Lynden–Bell approach and the modified strategy proposed by Yamaguchi (2008) are compared in Figs. 10.3 and 10.4. In these figures the virial ratio r is defined as twice the ratio between the initial kinetic and potential energy of the system, respectively.

The one-dimensional-sheet model admits a sound and straightforward physical interpretation, as far as finite mass distributions are concerned (particles in a box). It is, in turn, more cumbersome to justify its extension to the infinite space problem. In fact, the extrapolation of the finite version of the model (which can be simulated in silico) to its infinite version must be handled with care and, as such, it has often been overlooked in the relevant literature. In particular, the definition of the force driving the particles dynamics is subtle and boils down to the so-called 'Jeans swindle' concept. This can be seen as the subtraction of a compensating negative mass background in the estimate of the potential, which Kiessling (1989) justifies as a prescription for the calculation of the force in the infinite volume limit. In the context of static 1D toy models, a few studies (Aurell *et al.*, 2001; Valageas, 2006a) have accommodated for such an effect, by naively combining into the force the contributions from the sheets composing the (finite) system under scrutiny to the additional effect stemming from the uniform negative background. Aurell *et al.* (2001) discuss with some detail the problem of taking the infinite system limit and consequently assume in their analysis a finite number of particles distributed and perturbed off a perfect lattice, modelling a finite localized perturbation embedded in an otherwise uniform Universe. This procedure, correct from first principle, is shown to yield a set of governing equations for the individual displacements which has been recently proven valid also for the infinite lattice with perturbations which do not break the translational invariance (Gabrielli *et al.*, 2009). In this latter paper, the problem of

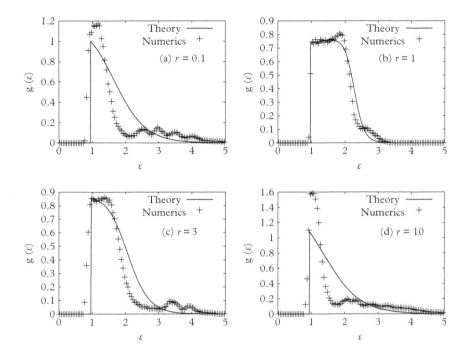

Figure 10.3 *Energy distributions for virial ratio r equal to 0.1 (a), 1 (b), 3 (c) and 10 (d). In each panel the solid line represents the Lynden–Bell distribution and points refer to the distribution collected numerically and averaged over* 100 *realizations. Here* N = 1000. *Reprinted with permission from Yamaguchi (2008), © 2008 by the American Physical Society.*

defining a consistent dynamical picture of an infinite one-dimensional Universe is approached by employing the definition of a smooth screening in the potential definition, which is sent to 0 at the end of the calculation. Further details can be found in the quoted references, as well as a comprehensive description of the algorithmic aspects.

Before discussing the expanding setting for which similar conceptual problems do arise, we shall also briefly mention the studies on simple one-dimensional self-gravitating spherical shell models. The latter have been often invoked to provide a first-order representation of globular clusters, which are localized assemblies of stars, spherically distributed in space and interacting through gravity (Chandrasekhar, 1943; Meylan and Heggie, 1997). Globular clusters are hence gravitationally bound objects: they orbit around a parent galaxy, so experiencing tidal forces which cause stars to escape from the system. A core–halo structure is found in real clusters: the dense inner core is surrounded by a diluted sea, the density profile displaying a sharp outer cut-off, as suggested by King (1962, 1965, 1966). Indeed, the velocity of the particles in the bulk can increase without limit, due to the singularity of the gravitational potential, and the system undergoes the well-known Antonov gravitational catastrophe.

In a simple toy-model version (Hénon, 1964), the shells move radially and the transversal component of the dynamics is neglected. The evaporation can eventually be

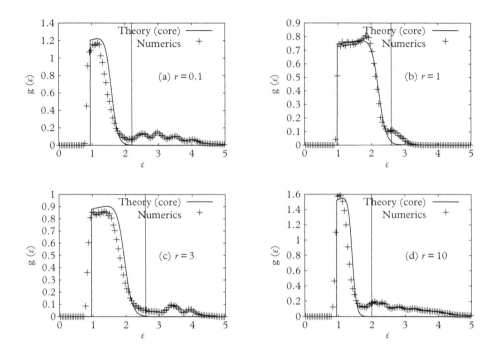

Figure 10.4 *Energy distribution for virial ratio r equal to 0.1 (a), 1 (b), 3 (c) and 10 (d). In each panel the solid line represents the modified Yamaguchi's distribution and points refer to the distribution collected numerically and averaged over 100 realizations. Here N = 1000. The vertical lines stand for the threshold in energy that has been set in the analysis. Reprinted with permission from Yamaguchi (2008), © 2008 by the American Physical Society.*

taken into account by implementing absorbing boundary conditions. Conversely, we can focus on reflecting conditions, being in particular concerned with the late time relaxation as driven by mutual encounters. More complex scenarios that allow for a rotation of individual shells, so as to resolve the transversal degrees of freedom, are being proposed (Klinko and Miller, 2000). In all these cases, event–driven numerical schemes can be implemented, which solve exactly the dynamics between successive crossings. Explicit statistical mechanics calculations, which can be directly compared to the numerical investigations, are possible.

10.3 From General Relativity to the Newtonian Approximation

In General Relativity, events in the space–time manifold are labelled by the four coordinates $x^\mu \equiv (t, \mathbf{r})$, with $\mu = 0, 1, 2, 3$. Obviously this choice implies that the velocity of light c is set to unity. The infinitesimal interval is given by

$$d\tau^2 = -g_{\mu\nu}dx^\mu dx^\nu, \tag{10.45}$$

where $g_{\mu\nu}$ is the covariant symmetric metric tensor. The contravariant metric tensor is defined by the identity $g_{\mu\nu}g^{\nu\lambda} = \delta_\mu^\lambda$, where δ_μ^λ is the unit tensor. If space–time is flat, $g_{\mu\nu}$ reduces to the diagonal Minkowski form $\eta_{\mu\nu} = \text{diag}(-1, +1, +1, +1)$. Consider a particle moving under the influence of purely gravitational forces, otherwise freely. According to the *Principle of Equivalence*, there exist specific coordinates ξ^α, in which the equations of motion take the simple form

$$\frac{d^2\xi^\alpha}{d\tau^2} = 0, \tag{10.46}$$

where $d\tau^2 = -\eta_{\alpha\beta}d\xi^\alpha d\xi^\beta$. Now, suppose that we are using generic coordinates x^μ: the previous coordinates ξ^α are functions of the x^μ. By applying the chain rule, we can rewrite Eq. (10.46) as

$$0 = \frac{d}{d\tau}\left(\frac{\partial\xi^\alpha}{\partial x^\mu}\frac{dx^\mu}{d\tau}\right) = \frac{\partial\xi^\alpha}{\partial x^\mu}\frac{d^2x^\mu}{d\tau^2} + \frac{\partial^2\xi^\alpha}{\partial x^\mu\partial x^\nu}\frac{dx^\mu}{d\tau}\frac{dx^\nu}{d\tau}. \tag{10.47}$$

Multiplying Eq. (10.47) by $\partial x^\lambda/\partial\xi^\alpha$ and rearranging the terms, we end up with the equations of motion

$$\frac{d^2x^\lambda}{d\tau^2} + \Gamma^\lambda_{\mu\nu}\frac{dx^\mu}{d\tau}\frac{dx^\nu}{d\tau} = 0, \tag{10.48}$$

where $\Gamma^\lambda_{\mu\nu}$ is the *affine connection* defined by

$$\Gamma^\lambda_{\mu\nu} = \frac{\partial x^\lambda}{\partial\xi^\alpha}\frac{\partial^2\xi^\alpha}{\partial x^\mu\partial x^\nu}. \tag{10.49}$$

So far, the discussion has shown that the affine connection is related to the gravitational force, while the proper time interval $d\tau$ between two events with a given infinitesimal coordinate separation is controlled by the metric tensor. A standard calculation, not detailed here (see Weinberg, 1972), establishes the following relation between the field $\Gamma^\lambda_{\mu\nu}$ and the derivatives of $g_{\mu\nu}$:

$$\Gamma^\lambda_{\mu\nu} = \frac{1}{2}g^{\sigma\lambda}\left(\frac{\partial g_{\nu\sigma}}{\partial x^\mu} + \frac{\partial g_{\mu\sigma}}{\partial x^\nu} - \frac{\partial g_{\nu\mu}}{\partial x^\sigma}\right). \tag{10.50}$$

This means that $g_{\mu\nu}$ can be interpreted as the gravitational potential.

To make a bridge with Newton theory we consider a particle moving slowly in a weak stationary gravitational field. If the particle is sufficiently slow, we can assume that

$$\frac{d\mathbf{r}}{d\tau} \ll \frac{dt}{d\tau}, \tag{10.51}$$

and write Eq. (10.48) as

$$\frac{d^2 x^\lambda}{d\tau^2} + \Gamma^\lambda_{00} \left(\frac{dt}{d\tau} \right)^2 = 0. \tag{10.52}$$

If we make now the assumption that the gravitational field is stationary, all time derivatives of $g_{\mu\nu}$ vanish, and therefore Eq. (10.50) reads

$$\Gamma^\lambda_{00} = -\frac{1}{2} g^{\lambda\sigma} \frac{\partial g_{00}}{\partial x^\sigma}. \tag{10.53}$$

Moreover, due to the weakness of the gravitational field, we may adopt a nearly Minkowskian coordinate system in which

$$g_{\alpha\beta} = \eta_{\alpha\beta} + h_{\alpha\beta}, \qquad |h_{\alpha\beta}| \ll 1. \tag{10.54}$$

Hence, considering the first order in $h_{\alpha\beta}$, we get

$$\Gamma^\alpha_{00} = -\frac{1}{2} \eta^{\alpha\beta} \frac{\partial h_{00}}{\partial x^\beta}, \tag{10.55}$$

and coming back to the equation of motion (10.52), we have

$$\begin{cases} \dfrac{d^2 \mathbf{r}}{d\tau^2} = \dfrac{1}{2} \left(\dfrac{dt}{d\tau} \right)^2 \boldsymbol{\nabla} h_{00} \\[4mm] \dfrac{d^2 t}{d\tau^2} = 0. \end{cases} \tag{10.56}$$

By combining the previous equations, we find

$$\frac{d^2 \mathbf{r}}{dt^2} = \frac{1}{2} \boldsymbol{\nabla} h_{00}, \tag{10.57}$$

while the corresponding Newtonian equation reads

$$\frac{d^2 \mathbf{r}}{dt^2} = -\boldsymbol{\nabla} \phi, \tag{10.58}$$

where ϕ is the standard gravitational potential. Therefore, we are led to conclude that

$$h_{00} = -2\phi + \text{const.} \tag{10.59}$$

Furthermore, the coordinate system must become Minkowskian at large distances, so h_{00} vanishes at infinity. Assuming $\phi = 0$ as $\mathbf{r} \to \infty$, the constant in Eq. (10.59) is 0. Therefore, coming back to the metric we get

$$g_{00} = -(1 + 2\phi).$$ (10.60)

On the surface of the Earth and of the Sun, the gravitational potential ϕ, in these units in which $c = 1$, is of the order 10^{-9} and 10^{-6}, respectively. As a consequence the distortion in $g_{\mu\nu}$ produced by the gravitation is very slight (Weinberg, 1972).

Consider now Einstein's field equations

$$R_{\mu\nu} - \frac{1}{2}g_{\mu\nu}R = -8\pi\, G T_{\mu\nu},$$ (10.61)

where $T_{\mu\nu}$ is the symmetric energy-momentum tensor, G is the gravitational constant and $R_{\mu\nu}$ is the Ricci tensor, which, in terms of the affine connections, reads

$$R_{\mu\nu} = \frac{\partial \Gamma^\lambda_{\mu\lambda}}{\partial x^\nu} - \frac{\partial \Gamma^\lambda_{\mu\nu}}{\partial x^\lambda} + \Gamma^\eta_{\mu\lambda}\Gamma^\lambda_{\nu\eta} - \Gamma^\eta_{\mu\nu}\Gamma^\lambda_{\lambda\eta}.$$ (10.62)

Moreover, R is the scalar curvature defined by

$$R = g^{\mu\nu}R_{\mu\nu} = g_{\mu\nu}R^{\mu\nu}.$$ (10.63)

Contracting the field Eqs. (10.61) with $g^{\mu\nu}$ gives

$$R = 8\pi\, G T^\mu_\mu \equiv 8\pi\, G T.$$ (10.64)

Therefore, the field equations can also be written

$$R_{\mu\nu} = -8\pi\, G \left(T_{\mu\nu} - \frac{1}{2}g_{\mu\nu} T \right).$$ (10.65)

In the weak limit, the Ricci tensor to first order in h becomes

$$R_{\mu\nu} = \frac{\partial \Gamma^\lambda_{\mu\lambda}}{\partial x^\nu} - \frac{\partial \Gamma^\lambda_{\mu\nu}}{\partial x^\lambda}.$$ (10.66)

In particular, from Eq. (10.55) we obtain for $\mu = \nu = 0$

$$R_{00} = \frac{1}{2}\nabla^2 h_{00}.$$ (10.67)

In the same limit (and assuming also $v \ll c$), the energy-momentum tensor, given in general by

$$T_{\mu\nu} = p g_{\mu\nu} + (p + \rho)U_\mu U_\nu,$$ (10.68)

where p is the pressure, ρ is the energy density and U^μ is the local value of $dx^\mu/d\tau$, has only the component $T_{00} = \rho_m$, where ρ_m is the mass density (all other components are negligible). Therefore, in this limit, Eq. (10.65) gives, for $\nu = \mu = 0$,

$$R_{00} = -8\pi G \left(T_{00} + \frac{1}{2} T \right) = -8\pi G \left(T_{00} - \frac{1}{2} T_{00} \right) = -4\pi G T_{00}. \qquad (10.69)$$

We therefore obtain

$$\nabla^2 \phi = -\frac{1}{2} \nabla^2 h_{00} = -R_{00} = 4\pi G T_{00} = 4\pi G \rho_m; \qquad (10.70)$$

the leftmost and rightmost terms give the Poisson equation for ϕ. The solution of the Poisson equation for a set of point masses m_i, $i = 1, \ldots, N$, i.e. when

$$\rho_m = \sum_{i=1}^{N} m_i \delta(\mathbf{r} - \mathbf{r}_i), \qquad (10.71)$$

is

$$\phi(\mathbf{r}) = -\sum_{i=1}^{N} \frac{G m_i}{|\mathbf{r} - \mathbf{r}_i|}. \qquad (10.72)$$

10.4   The Cosmological Problem

At large scales, the Universe can be assumed to be isotropic and homogeneous: this is the *Cosmological Principle* (Weinberg, 1972; Zeldovich and Novikov, 1974; Peebles, 1980). The most compelling evidence of that is the observed near isotropy of the microwave background radiation, a present relic of an early period of the life of the Universe. Furthermore, Hubble's law of the relative motion of galaxies can be successfully interpreted by assuming homogeneity and isotropy of the whole Universe. The Cosmological Principle leads to the Robertson–Walker metric for the space–time manifold. Choosing coordinates t (cosmological time) and (r, θ, ϕ) (coordinates of the space-like slice), the interval $d\tau^2$ between infinitesimally close space–time points is

$$d\tau^2 = dt^2 - a^2(t) \left[dr^2 \frac{1}{1 - kr^2} + r^2 d\theta^2 + r^2 \sin^2 \theta d\phi^2 \right], \qquad (10.73)$$

where the cosmic scale factor $a(t)$ is a so-far undetermined function of time, and k is the curvature parameter. In proper units, k takes the values $-1, 0, 1$, corresponding to negative, zero and positive curvature, respectively (Weinberg, 1972; Zeldovich and Novikov, 1974; Peebles, 1980). Inserting the Robertson-Walker metric tensor in Eq. (10.50), we get the corresponding affine connections which, substituted in Eq. (10.62), give the Ricci tensor. Using the energy-momentum tensor (10.68), we arrive at the following form of the time–time component of the Einstein's field equations (Weinberg, 1972; Zeldovich and Novikov, 1974; Peebles, 1980)

$$3\ddot{a} = -4\pi G(\rho + 3p)a, \qquad (10.74)$$

with the space–space component given by

$$a\ddot{a} + 2\dot{a}^2 + 2k = 4\pi G(\rho - p)a^2, \qquad (10.75)$$

the space–time components being 0. Eliminating $\ddot{a}$ yields the equation

$$\dot{a}^2 + k = \frac{8\pi G}{3}\rho a^2.$$ (10.76)

Equation (10.76), together with the energy conservation and the equation of state, characterizes a so-called Friedmann Universe.

It can be shown by a simple argument of consistency that a general solution of Eq. (10.76) contains a singularity, leading to infinite density and zero cosmic scale factor (Weinberg, 1972; Zeldovich and Novikov, 1974; Peebles, 1980). In fact, at present time t_0, since we observe red shift (see below), $a(t_0) > 0$ and $\dot{a}(t_0) > 0$: hence, using Eq. (10.74), it follows that $a(t)$ versus t must be concave downward and must have taken the value $a = 0$ at some finite time in the past.

The very special moment of time $t = 0$ is the so-called Big Bang and its existence is a necessary consequence of the homogeneity and isotropy assumptions. The present time (t_0) is the time elapsed since this singular condition in which all the energy was concentrated in a single point in space. Following the standard use, we label a quantity with the subscript '0' when it refers to the present epoch. The different stages of Universe evolution can also be chronologically ordered using the red shift, defined as

$$z(t) = \frac{a_0}{a(t)} - 1,$$ (10.77)

so that $z(t_0) = z_0 = 0$ (Weinberg, 1972; Zeldovich and Novikov, 1974; Peebles, 1980). There are three possible scenarios predicted by the theory, depending on the value of the curvature k. If $k = 0$, which corresponds to a flat manifold, the Universe will continue expanding forever. The same qualitative solution holds for $k = -1$, but with a faster growth rate. However, if $k = 1$ the expansion will eventually cease, and the Universe will start to contract back to the singular state (Fig. 10.5). Downward concavity is therefore compatible with both an endless growing Universe and with a Universe which contracts in a finite time.

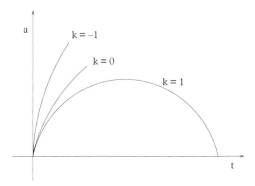

Figure 10.5 *Solution of Einstein's equations for a Robertson–Walker Universe with curvatures $k = +1$, $k = 0$ and $k = -1$.*

The curvature parameter is determined by the overall mass density. In fact, Eq. (10.76) can be rewritten as

$$\left(\frac{\dot{a}}{a_0}\right)^2 - \frac{8\pi}{3}G\rho\left(\frac{a}{a_0}\right)^2 = H_0^2\left(1 - \frac{\rho_0}{\rho_{0cr}}\right) = -\frac{k}{a_0^2}, \tag{10.78}$$

where

$$H_0 = \frac{\dot{a}_0}{a_0}, \tag{10.79}$$

is the Hubble constant at present time, and

$$\rho_{0cr} = \frac{3H_0^2}{8\pi\,G}. \tag{10.80}$$

Therefore $\rho_0 < \rho_{0cr}$ (resp. $\rho_0 > \rho_{0cr}$) corresponds to $k = -1$ (resp. $k = 1$). The solution with $k = 0$ is characterized by $\rho_0 = \rho_{0cr}$. This critical solution is generally known as the Einstein–de Sitter model. The scaling factor grows as a power law of cosmological time, since we have $a/a_0 = (3H_0t/2)^{2/3}$.

Knowledge of the current matter density would allow the determination of the global geometry of the Universe. However, the estimated amount of baryonic matter which corresponds to normal matter (protons, neutrons, ...) does not give correct predictions of the dynamics of cosmic objects (Weinberg, 1972). Fast rotation of hydrogen clouds far outside the luminous disc of spiral galaxies, as well as the high-velocity dispersion of galaxies in clusters, indicate the presence of deep gravitational potential wells. Neither in individual galaxies nor in clusters can the strength of the gravitational field be explained by the luminous matter. These empirical observations motivated the introduction of the concept of *dark matter* in cosmology: according to this assumption most of the mass of the Universe is constituted by a dark component, only interacting by gravity (Weinberg, 1972; Zeldovich and Novikov, 1974; Peebles, 1980). Following this picture, it is believed that the luminous component of the matter can be used as a tracer of the dark component, but itself carries a smaller fraction of the total mass. The list of dark matter candidates is far from exhaustive and the range of particles that could constitute the non-baryonic component of matter is limited only by theorists' imagination.

The details of this general picture are still a fully open problem. Indeed, it is now accepted that a new form of energy, the dark energy, must be included in the description (Caldwell *et al.*, 2001). Mathematically, this is equivalent to consider in the field equations a non-zero cosmological constant, Λ (Weinberg, 1972), originally introduced by Einstein to make General Relativity consistent with the assumption of a static Universe, considered at that time the actual condition.

10.5   Particle Dynamics in Expanding Coordinates

The typical length scale of the Universe expansion is given in terms of the Hubble constant as c/H_0, which, in the present epoch, is about 4000 Mpc. For smaller length scales, we can make the Newtonian approximation, which amounts to neglect space–time curvature and to consider velocities much smaller than c. This is the so-called *background model* (Peebles, 1980), which takes into account just the cosmic scale factor $a(t)$, and where, for each time t, the mass density due to the background $\rho_b(t)$ is assumed to be uniform. In this context, it is possible to point out the motion and the distribution of matter with respect to the mean and isotropic case. In this model, the proper coordinate $\mathbf{r}(t)$ (implicitly defined in Eq. (10.73)) satisfies Newton's equation. It is convenient to introduce the following definition of $\mathbf{x}$

$$\mathbf{r} = a(t)\mathbf{x}, \tag{10.81}$$

which is the position in the comoving frame. Practically, Eq. (10.81) represents a change of variables from the proper locally Minkowski coordinates to the expanding coordinates $\mathbf{x}$. In these latter coordinates, the proper velocity of a particle is

$$\dot{\mathbf{r}} = a(t)\dot{\mathbf{x}} + \mathbf{x}\dot{a}(t). \tag{10.82}$$

The Lagrangian for a system of point particles reads therefore

$$\mathcal{L} = \sum_{i=1}^{N} \left\{ \frac{1}{2} m_i (a\dot{\mathbf{x}}_i + \mathbf{x}_i \dot{a})^2 + \frac{Gm_i}{2} \sum_{j\neq i} \frac{m_j}{a\,|\,\mathbf{x}_i - \mathbf{x}_j\,|} \right\}. \tag{10.83}$$

By applying the canonical transformations

$$\mathcal{L} \to \mathcal{L} - \frac{d\Psi}{dt}, \qquad \Psi = \frac{1}{2} a\dot{a} \sum_{i=1}^{N} m_i x_i^2, \tag{10.84}$$

the Lagrangian (10.83) is transformed into

$$\mathcal{L} = \sum_{i=1}^{N} \left\{ \frac{1}{2} m_i a^2 \dot{\mathbf{x}}_i^2 + \frac{Gm_i}{2} \sum_{i\neq j} \frac{m_j}{a\,|\,\mathbf{x}_i - \mathbf{x}_j\,|} - \frac{1}{2} a\ddot{a} \sum_{i=1}^{N} m_i x_i^2 \right\}. \tag{10.85}$$

The potential of this Lagrangian is

$$\varphi(\mathbf{x}, t) = \phi(\mathbf{x}, t) + \frac{1}{2} a\ddot{a} x^2 \tag{10.86}$$

$$= \phi(\mathbf{x}, t) - \frac{2\pi G\rho_b a^2 x^2}{3}, \tag{10.87}$$

where we have used Eq. (10.74) with $p = 0$ and $\rho = \rho_b$. This potential satisfies the Poisson equation

$$\nabla_x^2 \varphi = 4\pi \, Ga^2 \left(\rho_m(\mathbf{x}, t) - \rho_b(t) \right). \tag{10.88}$$

The source of φ is now $\rho_m - \rho_b(t)$: this is consistent with what we could have expected, since if no inhomogeneities are present, $\rho_m - \rho_b(t)$ vanishes, and each particle remains undisturbed in the comoving system. The Lagrange equations take the form

$$a^2 \ddot{\mathbf{x}}_i + 2a\dot{a}\dot{\mathbf{x}}_i = -\sum_{i \neq j} \frac{Gm_j(\mathbf{x}_i - \mathbf{x}_j)}{a \, |\mathbf{x}_i - \mathbf{x}_j|^3} + \frac{4\pi}{3} G\rho_b a^2 \mathbf{x}. \tag{10.89}$$

As an effect of the coordinate change a friction term has appeared.

We can write these equations also in terms of the proper peculiar velocity $\mathbf{u}_i = a\dot{\mathbf{x}}_i$, which represents the velocity measured by an observer moving with velocity $\dot{a}\mathbf{x}$. In terms of $\mathbf{u}_i$ the left-hand side of Eq. (10.89) rewrites $a\dot{\mathbf{u}}_i + \dot{a}\mathbf{u}_i$.

10.6   The Vlasov Equation for an Expanding Universe

Consider now a distribution of identical, *collisionless* particles (dark matter), moving in the potential φ determined by the smooth space density function. As usual, the distribution function in space and momentum is given by

$$dN = f(\mathbf{x}, \mathbf{p}, t) \, d\mathbf{x}d\mathbf{p}, \tag{10.90}$$

where $\mathbf{p}$ is the momentum conjugate to $\mathbf{x}$, and the mass density is

$$\rho(\mathbf{x}, t) = \frac{m}{a^3} \int d\mathbf{p} f(\mathbf{x}, \mathbf{p}, t) = \rho_b(t) \left[1 + \delta(\mathbf{x}, t) \right]. \tag{10.91}$$

Here, m is common mass of each particle, $\rho_b(t) \propto a^{-3}(t)$ and δ is the dimensionless density contrast. According to the Liouville theorem $f(\mathbf{x}, \mathbf{p}, t)$ is constant along a particle trajectory in phase space. Thus, from the equations of motion, we get

$$\frac{\partial f}{\partial t} + \frac{\mathbf{p}}{ma^2} \cdot \nabla f - m\nabla\varphi \cdot \frac{\partial f}{\partial \mathbf{p}} = 0. \tag{10.92}$$

This is the Vlasov equation, which together with the Poisson equation (10.88) gives a complete description of the dynamics, with $\rho(\mathbf{x}, t)$ specified by Eq. (10.91). Finally, the particle velocities $\mathbf{u}(\mathbf{x}, t)$ are given, in terms of $f(\mathbf{x}, \mathbf{p}, t)$, as

$$\rho(\mathbf{x}, t)\mathbf{u}(\mathbf{x}, t) = \frac{1}{a^4} \int d\mathbf{p} \, \mathbf{p} f(\mathbf{x}, \mathbf{p}, t). \tag{10.93}$$

The Vlasov–Poisson equations provide a complete picture for the process of large-scale structure formation. Finite size effects can in principle result in a departure from the ideal Vlasov setting and in a 'collisional' term on the right-hand side of equation (10.92). This collisional forcing drives the system relaxation to equilibrium. Based on the preceding discussion on the kinetic theory limit of a discrete ensemble of interacting entities, the collisional term formally vanishes as the number of particles is large, as it is certainly the case for large galactic systems. In other words, and recalling the paradigmatic description of quasi-stationary states (QSS) discussed in Chapter 9, galaxies are in a long lasting out-of-equilibrium configuration, being correctly described by the Vlasov scenario. In particular, they are reminiscent of the initial condition, a general property that we appreciated, e.g. with reference to the HMF model.

As a final comment, it should, however, be stressed that the nonlinear regime of the Vlasov dynamics remains to be fully understood: large simulations of nonlinear gravitational instabilities are a numerically difficult problem, and approximations must be made. For these reasons, since a long time, there has been an interest in simplified models of structure formation, including in particular 1D models.

10.7 From Vlasov–Poisson Equations to the Adhesion Model

Consider the Vlasov–Poisson system, namely Eqs. (10.92) and (10.88). It is well known (Vergassola *et al.*, 1994) that they admit special solutions of the form

$$f(\mathbf{x}, \mathbf{p}, t) = \frac{a^3 \rho(\mathbf{x}, t)}{m} \delta^d(\mathbf{p} - ma\mathbf{u}(\mathbf{x}, t)), \tag{10.94}$$

where d is the dimension of space and $\delta^d(.)$ the d-dimensional delta function. We will refer to this class as to single-speed solutions, because to each given $(\mathbf{x}, t)$ corresponds a well-defined velocity $\mathbf{u}$. Assuming (10.94), after some manipulations, it follows from Eqs. (10.91) and (10.93) that

$$\begin{cases} \partial_t \rho + 3\dfrac{\dot{a}}{a}\rho + \dfrac{1}{a}\nabla \cdot (\rho \mathbf{u}) = 0 \\[2ex] \partial_t \mathbf{u} + \dfrac{\dot{a}}{a}\mathbf{u} + \dfrac{1}{a}(\mathbf{u} \cdot \nabla)\mathbf{u} = \mathbf{g} \\[2ex] \nabla \cdot \mathbf{g} = -4\pi\, Ga(\rho - \rho_b), \end{cases} \tag{10.95}$$

where we have introduced $\mathbf{g} = -\nabla\varphi/a$, such that $\nabla \times \mathbf{g} = 0$. It should be stressed that the system (10.95) is valid as long as the distribution function $f(\mathbf{x}, \mathbf{u})$ is of the form (10.94), i.e. when the solution stays single-stream. After a finite time, however, the distribution function becomes multi-stream, because of particles crossing (caustic formation), and the pressureless and dissipationless hydrodynamical system (10.95) should be consistently modified.

Let us focus on the time before the first particle crossing. Then assume the so-called *condition of parallelism*, by requiring that the peculiar velocity is a potential field, which remains parallel to the gravitational peculiar acceleration field (Peebles, 1980; Vergassola *et al.*, 1994; Buchert *et al.*, 1999)

$$\mathbf{g} = F(t)\mathbf{u}, \tag{10.96}$$

where $F(t)$ is a positive, time-dependent, proportionality coefficient. The relation (10.96) is well justified in the linear, as well as in the weakly non-linear regimes and makes it possible to treat analytically the problem. From the linear theory, it follows that (Peebles, 1980; Vergassola *et al.*, 1994; Buchert *et al.*, 1999)

$$F(t) = 4\pi \, G\rho_b b/\dot{b}, \tag{10.97}$$

where b is a new time variable related to the growing mode of the density field in the linear regime. Hence, defining the new velocity field $\mathbf{v} = \mathbf{u}/(a\dot{b})$, the system (10.95) reduces to

$$\partial_b\mathbf{v} + (\mathbf{v} \cdot \nabla)\mathbf{v} = 0 \qquad \mathbf{v} = -\nabla\tilde{\psi}, \tag{10.98}$$

where $\tilde{\psi} = \varphi/(a^2\dot{b}F(t))$, which is the multidimensional Burgers equation (Vergassola *et al.*, 1994). The inviscid Burgers equation describes the free motion of fluid particles subject to zero forcing, and it is equivalent to the Zeldovich approximation (Shandarin and Zeldovich, 1989). Again, it is worth recalling that this picture is correct as long as the solution stays single-stream. After caustic formation, we may think that the resulting change in the gravitational force is modelled by an effective diffusive term: this leads to the adhesion model (Gurbatov *et al.*, 1989). This should represent the effect of the gravitational sticking, not captured by the Zeldovich approximation. Thus, it is customary to introduce a term in the form $\mu\nabla^2\mathbf{v}$, on the right-hand side of the equation (10.98). In order for the diffusion term to have a smoothing effect only in those regions where the particles crossing takes place, the viscosity μ should be small. The adhesion model hence reads

$$\begin{cases} \partial_b\mathbf{v} + (\mathbf{v} \cdot \nabla)\mathbf{v} = \mu\nabla^2\mathbf{v} \\[2mm] \mathbf{v} = -\nabla\tilde{\psi} \\[2mm] \partial_b\rho + \nabla \cdot (\rho\mathbf{v}) = 0. \end{cases} \tag{10.99}$$

The limit when viscosity μ tends to 0 is often performed: as it is well known, this is not equivalent to setting μ to 0 from the outset, but it is instead a regularization of (10.98).

Although numerical experiments suggest qualitative agreement, no theory is to our knowledge currently available, which quantifies the exact relationship between the correct Vlasov-Poisson approach, Eqs. (10.88) and (10.92), and the adhesion model (10.99).

10.8   The One-Dimensional Expanding Universe

Beyond the static picture, if we want to address the problem of large-scale structure formation by focusing on the evolution of a stratified perturbation, the effect of the expansion must be properly included into the model. As outlined earlier, this leads to two extra contributions in the particles' equation of motion, namely a friction and a background term. In this respect, a very interesting model, proposed by Rouet and Feix (Rouet *et al.*, 1990, 1991) and now called the RF model, has the important merit of producing a final tractable expression for computing algebraically the particle crossing times, and contains all the right ingredients of the dynamics. In developing the calculations, the authors assume a Einstein–de Sitter 3D cosmology behaviour for the scaling factor, namely $a(t) \propto t^{2/3}$, while considering a one-dimensional version for the time–time component for the Einstein field equations. This corresponds to imposing a Hubble expansion sourced by a mean density, three times the physical mass density of the system. In practice, it implies overestimating the role of friction, and as a consequence this artificially makes the process of particle sticking more pronounced than what it really is. To overcome this limitation a modified version of the RF model was proposed by Fanelli and Aurell (2002) and termed the quintic (Q) model for reasons that will be transparent in the following. The key ingredients of the derivation are reported below.

As already pointed out in the preceding discussion, the Newtonian equations of motion for N particles interacting via gravity follow from the Lagrangian

$$\mathcal{L} = \sum_i \frac{1}{2} m_i \dot{r}_i^2 - m_i \phi(r_i, t),$$ (10.100)

where $\nabla_r^2 \phi = 4\pi G \rho$ (to be now intended in the one-dimensional case). In the point particle picture, the density profile reads

$$\rho(x_i, t) = \sum_{x_j} m_j \frac{1}{a^3} \delta(x_i - x_j),$$ (10.101)

where x_i is the comoving coordinate of the *i*th particle, in the direction of which the density and velocities vary.

Expressing (10.100) as a function of the comoving coordinate, x_i, and assuming (10.101), the equation of motion of the *i*th particle reads

$$\frac{d^2 x_i}{dt^2} + 2\frac{\dot{a}}{a}\frac{dx_i}{dt} - 4\pi G\rho_b(t)x_i = a^{-3}E_{\text{grav}}(x_i, t),$$ (10.102)

where $\rho_b(t)$ is the mean mass density at time t and

$$E_{\text{grav}}(x_i, t) = -2\pi G \sum_j m_j \operatorname{sign}(x_i - x_j).$$ (10.103)

From the equation of continuity, we have

$$\rho_b(t) = \rho_0\, a(t)^{-3}.$$
(10.104)

By performing a suitable non-linear transformation of the time variable, it is possible to concentrate all the time dependence in the coefficient of the friction term. The choice is

$$\mathrm{d}t = a^{3/2}\mathrm{d}\tau,$$
(10.105)

where τ has dimension of time (Rouet *et al.*, 1990, 1991). Eq. (10.102) is thus transformed into

$$\frac{\mathrm{d}^2 x_i}{\mathrm{d}\tau^2} + \frac{\dot{a}\sqrt{a}}{2}\frac{\mathrm{d}x_i}{\mathrm{d}\tau} - 4\pi\, G\rho_0 x_i = E_{\mathrm{grav}}(x_i,\tau).$$
(10.106)

Note that although the time variable is now τ, $\dot{a}$ denotes the derivative of a with respect to the original time t. As already mentioned, in a flat Einstein–de Sitter Universe, the scale factor $a(t)$ grows with time as a power law ($a(t) \propto t^{2/3}$) and therefore Eq. (10.106) takes the form

$$\frac{\mathrm{d}^2 x_i}{\mathrm{d}\tau^2} + \frac{1}{3t_0}\frac{\mathrm{d}x_i}{\mathrm{d}\tau} - \frac{2}{3t_0^2}x_i = E_{\mathrm{grav}}(x_i,\tau) \qquad \text{Q model,}$$
(10.107)

where $t_0^{-2} = 6\pi\, G\rho(t_0)$, as can be obtained by Eqs. (10.79) and (10.80). This is the model we refer to as the *quintic (Q) model*.

The interest of this formulation is that, as for the classical static self-gravitating systems in one dimension, E_{grav} is a Lagrangian invariant, proportional to the net mass difference to the right and to the left of a given particle, at a given time. Thus, the evolution of the system is recovered by using a version of the event-driven scheme (Noullez *et al.*, 2001). Nevertheless, within the Q dynamics the crossing times between consecutive particles are computed by solving numerically a *quintic equation*, which in turn explain its name. This observation has non-trivial implications at the level of the numerical procedure adopted, which is thoroughly discussed in Fanelli and Aurell (2002). In Fig. 10.6 we report the results of a numerical simulation: a smooth, sinus-like initial condition is being evolved via the Q dynamics. Figure 10.7 reports instead the evolution of a cosmological initial perturbation.

The density profile as emerging from the non-linear Q evolution of a spatially localized perturbation was studied analytically in Aurell *et al.* (2003), where a general expression was derived and successfully compared to direct simulations. The mass collapse proceeds self-consistently, in different stages of the evolution, the matter being more concentrated in the inner core. Working in the comoving frame, the width of the agglomeration increases exponentially, but at a rate which is slower than the Universe expansion. Hence, physical density as measured in the inertial frame decreases around almost all mass points and the agglomerations spread out over time (albeit more slowly than the expansion of the Universe as a whole). This is an interesting physical conclusion,

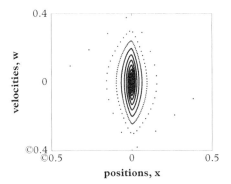

Figure 10.6 *Numerical simulation performed by using the Q model and smooth sinus like initial condition. Here, N = 4096 and $t/t_0 = 6.12 \times 10^4$.*

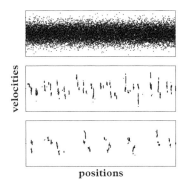

Figure 10.7 *Numerical simulation performed by using the Q model and a 'cosmological' initial condition (upper panel), see Fanelli and Aurell (2002). Here, N = 16384. The second and third plots refer, respectively, to $t/t_0 = 7.0941 \times 10^5$ and $t/t_0 = 8.2167 \times 10^6$.*

so far not fully appreciated, which casts some legitimate doubts on the validity of the one-dimensional approximation to the problem of large-scale structure formation. At variance, within the phenomenological one-dimensional version of the adhesion model, an effective mass collapse takes place, so yielding asymptotic dense objects.

Gabrielli *et al.* (2009) have recently revisited the derivation of one-dimensional models, and questioned their validity for an infinite Universe setting. The force as specified in such models is well defined in the infinite point distribution only if there is a centre of symmetry, a problem which arises because of the naive subtraction of the background (due to the expansion of 'Jeans swindle' in the static case), which leaves an unregulated contribution to the force due to mass surface fluctuations. They resolve the problem by defining the force in infinite point distribution as the limit of an exponentially screened pair interaction. At variance, no modification is requested when the focus is put on a localized perturbation.

10.9   Numerical Simulations in 3D

To address the problem of structure formation in the dark matter component we can alternatively resort to direct numerical simulations. The dynamical formulation is often assumed to be represented by the Vlasov–Poisson equations for a collisionless gas of self-gravitating particles, which defines the mean-field version of the discrete N-body frameworks. Several methods are in principle possible, depending on the chosen scenario. Following Trenti and Hut (2008), we quote for instance:

1. *Direct methods.* These methods do not involve physical approximations and are based on a direct implementation of the governing equations of motion for the N-body system. The computational costs scale as $O(N^2)$ per time step, which limits the number of particles that can possibly define the system under scrutiny. Integration is performed using adaptive (individual) time-steps and often relies on a fourth-order Hermite integrator. Close encounters and bound subsystems are treated exactly in terms of a dedicated transformation of coordinates, the so-called Kustanehimo–Steifel transformation, which builds perturbatively over the analytic two-body solution. A publicly available software is the Aareseth's NBODY6 (Aarseth, 2003). Direct methods were also a boost in the past for the development of special purpose hardware, such as GRAPE (Makino and Taiji, 1998), where the chip architecture has been optimized to serve the gravitational problem. The largest N-body simulation carried out so far is the so-called Millennium Run, which used more than 10 billion particles to trace the evolution of the matter distribution in a cubic region of the Universe over 2 billion light-years on a side. A snapshot of the large-scale dark matter distribution as obtained in the simulation is displayed in Fig. 10.8.

2. *Tree based N-body* (Barnes and Hut, 1986). They provide a fast general integrator for collisionless systems when close encounters can be neglected, and assuming we can deal with a rough approximation for the force contribution coming from distant particles. Small-scale interactions are, in fact, softened and the potential stemming from far away interactions is quantified via a multiple expansion about the group centre of mass. Typical implementations exploit a quadrupole expansion for the potential constructing a tree hierarchy of particles via a recursive binary splitting algorithm. The computational cost can be quantified in $O(N \log N)$.

3. *Particle-mesh codes.* Following this approach the gravitational potential is computed over a grid starting from the density field and solving the associated Poisson equation. Particles do not interact directly, but via the mean-field provided by the underlying density distribution. The complexity is linear with the number of particles and $O(N_g \log N_g)$ with the number of grid cell N_g.

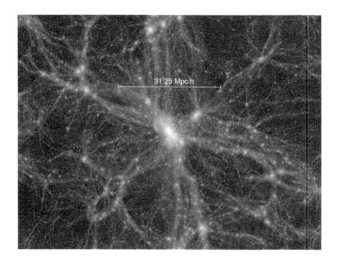

Figure 10.8 *The Millenium Run used the largest supercomputer in Europe at the German Astrophysical Virtual Observatory. The simulation refers to a cube over a billion light years on a side, holding 20 millions galaxies. The Universe is evolved until the present state, where structures like filaments, galaxies, cluster of galaxies are abundant. Reprinted with permission from V. Springel and the Virgo Consortium.*

4. *Grid-based solvers for the collisionless Boltzmann equation, also called Vlasov equation.* Standard computational methods can be employed to solve the associated partial differential equation system. The bottleneck is here represented by the necessary amount of memory.

5. *Fokker–Planck and Monte Carlo methods.* They solve the collisional Boltzmann equation starting from a given distribution function and following the evolution of test particles in the six-dimensional phase space. At each iteration, the particles velocity gets perturbed by means of a collision operator, which depends on the instantaneous particles distribution.

11

Two-Dimensional and Geophysical Fluid Mechanics

Hydrodynamics is another important domain of application in which the statistical mechanics of long-range interactions has important consequences. Although put on firms grounds in the past two decades, it is in general not covered in standard statistical mechanics textbooks. We will present here an introduction to this domain with the aim of describing the main physical idea, keeping apart some mathematical technicalities which might hide the originality of the approach.

Three categories of flows might be encompassed here:

i. Two-dimensional flow freely decaying from a turbulent state are the first examples.

ii. Three-dimensional geophysical flows with a background rotation and/or density stratification inhibiting motions in the third direction. The main example is the quasi-geostrophic model, but several generalizations have been proposed.

iii. More general flows have been studied within this context, with forcing and dissipation which allow these systems to reach an energetic equilibrium.

The paradigmatic image is the Great Red Spot of Jupiter shown in Fig. 11.1. This long-lived giant structure, which has been visible for centuries in the upper Jupiter atmosphere, keeps its coherence despite the intense turbulent environment, the strong rotation of the planet and the important external forcing and dissipation. As briefly discussed later, its existence but also its shape might be explained by resorting to statistical mechanics arguments. Coherent structures have been also observed in pure electron plasmas (Fine, 1995; Kawahara and Nakanishi, 2006; Kiwamoto *et al.*, 2007), which are indeed modelled by the Euler equation in two dimensions.

Although high Reynolds number flows have a very large number of degrees of freedom, we can often identify structures in the flow. This suggests that we could use a much smaller number of effective degrees of freedom to characterize the flow. This remark is particularly valid for two-dimensional flows, where the inverse energy cascade leads to the irreversible formation of large coherent structures (e.g., vortices). A system with a large number of degrees of freedom which can be characterized by a small number

Physics of Long-Range Interacting Systems. First Edition. A. Campa *et al.*
© A. Campa, T. Dauxois, D. Fanelli, and S. Ruffo 2014. Published in 2014 by Oxford University Press.

Figure 11.1 *The stability and the shape of the Great Red Spot of Jupiter can be explained using the statistical mechanics of systems with long-range interactions. Reprinted with permission from NASA/JPL.*

of effective parameters is reminiscent of what happens in thermodynamics (Castaing, 1995), where a few macroscopic variables describe the behaviour of systems composed of many particles. Statistical mechanics for turbulence is nowadays a very active field of research (see Eyink and Sreenivasan, 2006, and Bouchet and Venaille, 2012, for reviews), initiated long ago by Lars Onsager (1949). This chapter is devoted to discussing the long-range character of two-dimensional hydrodynamics.

The theory of fully three-dimensional flows is fundamentally different, as compared to its two-dimensional analogue. It is hence not possible to simply extend the 2D theory to the 3D realm. The dynamics are indeed very different: in three dimensions, there is a cascade of energy transferring the energy from large to small scales where dissipation is dominant. In contrast, in the two-dimensional framework, the introduction of additional constraints is at the origin of an *inverse* energy cascade leading to the formation of a very small number of large structures. This qualitative difference explains why the possibility to apply equilibrium statistical mechanics concepts to 3D fluid mechanics is still debated and questioned. However, as explained in this chapter, some three-dimensional flows of paramount importance from the geophysical point of view can be put in a framework in which statistical mechanics ideas are particularly fruitful.

11.1 Introduction

11.1.1 Elements of fluid dynamics

The governing equations of a two-dimensional incompressible fluid are the Navier–Stokes equation

$$\frac{\partial \mathbf{v}}{\partial t} + (\mathbf{v} \cdot \nabla) \mathbf{v} = -\frac{1}{\rho} \nabla p + \mathbf{F} + \nu \Delta \mathbf{v}, \tag{11.1}$$

together with the incompressibility condition

$$\nabla \cdot \mathbf{v} = 0, \tag{11.2}$$

where $\mathbf{v}$ is the two-dimensional velocity field (u, v) and $\mathbf{r} = (x, y)$ denotes the coordinate on the plane. Finally, ρ is the mass density, p the pressure, ν the kinematic viscosity and $\mathbf{F}$ the external forcing (per unit mass) such as gravity or Coriolis force. Let us consider first the case without external forcing and dissipation: this is the so-called Euler equation.

The incompressibility condition (11.2) is automatically satisfied if we introduce the streamfunction $\psi(\mathbf{r})$

$$u = -\frac{\partial \psi}{\partial y}, \tag{11.3}$$

$$v = +\frac{\partial \psi}{\partial x}, \tag{11.4}$$

as usually done in geophysical fluid mechanics. Note that in the fluid dynamics community, another convention $u = \partial \psi / \partial y$ and $v = -\partial \psi / \partial x$ is generally used. The vorticity ω is related to the velocity field through

$$\omega = \frac{\partial v}{\partial x} - \frac{\partial u}{\partial y} \tag{11.5}$$

and it leads to the Poisson equation for the streamfunction

$$\omega = \Delta \psi. \tag{11.6}$$

Using Green's function $G(\mathbf{r}, \mathbf{r}')$ of the Laplacian operator Δ, we easily find the solution of the Poisson equation in a given domain D

$$\psi(\mathbf{r}) = \int_D d\mathbf{r}' \, \omega(\mathbf{r}') \, G(\mathbf{r}, \mathbf{r}'), \tag{11.7}$$

plus surface terms (Alastuey *et al.*, 2008).

The energy is conserved for the Euler equation and is given by

$$H = \int_D d\mathbf{r} \, \frac{1}{2} (u^2 + v^2) \tag{11.8}$$

$$= \int_D d\mathbf{r} \, \frac{1}{2} (\nabla \psi)^2 \tag{11.9}$$

$$= -\frac{1}{2} \int_D d\mathbf{r} \, \omega(\mathbf{r}) \psi(\mathbf{r}) \tag{11.10}$$

$$= -\frac{1}{2} \int_D \int_D d\mathbf{r} \, d\mathbf{r}' \, \omega(\mathbf{r}') \omega(\mathbf{r}) G(\mathbf{r}, \mathbf{r}'). \tag{11.11}$$

In an infinite domain, the Green function of the Laplacian operator Δ is

$$G(\mathbf{r}, \mathbf{r}') = \frac{1}{2\pi} \ln |\mathbf{r} - \mathbf{r}'|, \tag{11.12}$$

and we can therefore rewrite Eq. (11.11) as

$$H = -\frac{1}{4\pi} \int_D \int_D d\mathbf{r} d\mathbf{r}' \, \omega(\mathbf{r}') \omega(\mathbf{r}) \ln |\mathbf{r} - \mathbf{r}'|. \tag{11.13}$$

This emphasizes that we get a logarithmic interaction between vortices at distant locations, which corresponds to a power-law decay with an effective exponent $\alpha = 0$, well within the case of long-range interactions since $\alpha < d = 2$. For a finite domain D, Green's function contains additional surface terms (Alastuey *et al.*, 2008), which, however, give no contribution to the energy (11.8) if the velocity field is tangent to the boundary of the domain (no outflow or inflow).

In absence of dissipation and of any external drive, the vorticity ω is conserved since its dynamical evolution is governed by

$$\frac{\partial \omega}{\partial t} + (\mathbf{v} \cdot \nabla)\omega = 0. \tag{11.14}$$

The vorticity is, therefore, only advected by the flow, and is consequently a conserved quantity. This implies that any quantity $f(\omega)$, where f is a continuous function, is also conserved. This property is straightforwardly demonstrated by multiplying (11.14) by the derivative $f'(\omega)$ and noting that the obtained equation writes in the equivalent form $df(\omega)/dt = 0$.

Moreover, taking into account the incompressibility condition (11.2), Eq. (11.14) leads to

$$\frac{\partial f(\omega)}{\partial t} + \nabla \cdot (f(\omega) \, \mathbf{v}) = 0. \tag{11.15}$$

Since the surface element $d\mathbf{r}$ is also conserved, we get

$$\frac{d}{dt} \int_D d\mathbf{r} f(\omega) = \int_D d\mathbf{r} \frac{\partial f(\omega)}{\partial t} = - \int_D d\mathbf{r} \, \nabla \cdot (f(\omega)\mathbf{v}) = 0; \tag{11.16}$$

in fact, it is simple to transform the domain integral of the last term in a boundary integral which vanishes due to the impermeability condition. Finally, the above reasoning implies that any functional

$$C_f = \int_D \mathbf{dr} f(\omega),$$ (11.17)

is conserved: this quantity is usually called a Casimir integral.

Considering the two simple examples $f(\omega) = \omega^n$ for $n = 1$ and 2, we get two Casimir integrals of prime interests: the circulation

$$\Gamma = \int_D \mathbf{dr}\, \omega(\mathbf{r}, t)$$ (11.18)

and the enstrophy, defined as

$$\mathcal{A} = \frac{1}{2} \int_D \mathbf{dr}\, [\omega(\mathbf{r}, t)]^2 .$$ (11.19)

Finally, as it will be of later use, it is important to remark that, if we introduce the Jacobian operator $\mathcal{J}(a, b) = a_x b_y - b_x a_y$, Eq. (11.14) can be rewritten as

$$\frac{\partial \omega}{\partial t} + \mathcal{J}(\psi, \omega) = 0,$$ (11.20)

or by taking advantage of the Poisson equation (11.6) as

$$\frac{\partial \Delta \psi}{\partial t} + \mathcal{J}(\psi, \Delta \psi) = 0.$$ (11.21)

It is then clear that any solution $\omega = F(\psi)$ which satisfies the condition $\mathcal{J}(\psi, \omega) = 0$ is a stationary solution of the Euler equation. As we shall see later, these stationary solutions, satisfying $\Delta \psi = F(\psi)$, play an important role in the statistical mechanics of two-dimensional flows.

11.1.2   Illustration of the non-additivity property

In the context of two-dimensional hydrodynamics, a simple example proposed by Venaille (2009) can illustrate the non-additivity property, characteristic of long-range interacting systems. Let us consider an inviscid flow in a canal with frontiers at $y = \pm 1$, as sketched in Fig. 11.2. If the vorticity field is homogeneous ($\omega = 1$) in the canal, with boundary conditions $\psi(x, y = \pm 1) = 0$, it is straightforward to show that $u = -y$, $v = 0$ and

$$\psi(x, y) = (y^2 - 1)/2.$$ (11.22)

Using Eq. (11.10), we get therefore that the total energy per unit length would be

$$E = -\frac{1}{2} \int_D \mathbf{dr}\, \omega(\mathbf{r}) \psi(\mathbf{r}) = -\frac{1}{2} \int_{-1}^{+1} dy\, 1 \cdot \frac{y^2 - 1}{2} = \frac{1}{3}.$$ (11.23)

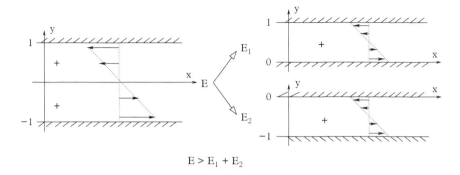

Figure 11.2 *Non-additivity of the energy for two-dimensional hydrodynamics, when one considers the Euler equation with a homogeneous vorticity field $\omega = 1$. The energy of the total system (on the left) is $E = 1/3$, while the energy of each subsystem (on the right) is $E_1 = E_2 = 1/24$. Reprinted from the PhD thesis written by Antoine Venaille, Mélange et circulation océanique une approche par la physique statistique, 1 December 2008, Université de Grenoble, with permission from Antoine Venaille.*

Let us consider now two subsystems, adding for example an additional frontier in $y = 0$, which plays the role of the interface where the streamfunction vanishes. The vorticity being kept equal to 1, we get in the upper canal $u = 1/2 - y$ and $u = -1/2 - y$ in the one below, while in both of them $v = 0$. The streamfunction is $\psi = (y^2 - y)/2$ in the upper canal, while $\psi = (y^2 + y)/2$ in the bottom one. We get now two different subsystems with the same energy per unit length, $E_1 = E_2 = 1/24$. We get, therefore, that the total energy is larger than the sum of the energies of each subsystem: $E > E_1 + E_2$. The additive property cannot be, therefore, invoked to develop the statistical mechanics of two-dimensional hydrodynamics. This latter seems indeed to belong to the class of long-range interacting systems.

11.2 The Onsager Point Vortex Model

11.2.1 The model

The long-range character of the interaction becomes clear if we approximate the vorticity field by N point vortices located at $\mathbf{r}_i = (x_i, y_i)$, with a given circulation Γ_i,

$$\omega(\mathbf{r}) = \sum_{i=1}^{N} \Gamma_i \delta(\mathbf{r} - \mathbf{r}_i). \tag{11.24}$$

The energy of the system reads now

$$H = -\frac{1}{4\pi} \sum_{i \neq j} \Gamma_i \Gamma_j \ln |\mathbf{r}_i - \mathbf{r}_j|, \tag{11.25}$$

where we have dropped the self-energy term because, although singular, it would not induce any motion (Marchioro and Pulvirenti, 1994). The study of point vortices was initiated by Helmholtz in 1858 and the general dynamical equations first derived by Kirchhoff. Considering the two coordinates of the point vortex on the plane, the equations of motion are

$$\Gamma_i \frac{dx_i}{dt} = +\frac{\partial H}{\partial y_i}, \tag{11.26}$$

$$\Gamma_i \frac{dy_i}{dt} = -\frac{\partial H}{\partial x_i}. \tag{11.27}$$

The N point vortices move therefore like N particles and the dynamics of each vortex is usually very complicated, typically chaotic. However, drawing an analogy with a gas of particles in a box, it is not necessary to know the position of each vortex to acquire sufficient information on key global variables to eventually address the thermodynamics of the flow.

This observation allows us to fully appreciate the profound intuition of Onsager (1949), who wrote: 'The formation of large, isolated vortices is an extremely common, yet spectacular phenomenon in unsteady flow. Its ubiquity suggests an explanation on statistical grounds.' This idea was developed during Onsager's talk at the first International Union of Pure and Applied Physics (IUPAP) STATPHYS conference on statistical physics in Florence, Italy (May 1949) and later published in Onsager (1949).

11.2.2   Negative temperatures

This model is two-dimensional, Hamiltonian and with a finite number of degrees of freedom N. Let's try to apply the usual methods of statistical mechanics. It is first interesting to note that both positions (x_i, y_i) of the vortex ith being canonically conjugated as shown by Eqs. (11.26) and (11.27), the phase space coincides with the configurational space.

The phase-space volume contained inside the energy shell $H = E$ can be written as

$$\Phi(E) = \int \prod_{i=1}^{N} d\mathbf{r}_i \, \theta(E - H(\mathbf{r}_1, \ldots, \mathbf{r}_N)), \tag{11.28}$$

where θ stands for the Heaviside step function. The total phase-space volume is $\Phi(\infty) = A^N$, where A is the area of the domain D. We immediately realize that $\Phi(E)$ is a non-negative increasing function of the energy E with limits $\Phi(-\infty) = 0$ and $\Phi(\infty) = A^N$. Its derivative, which is nothing but the microcanonical partition function (1.13), is given by

$$\Omega(E) = \Phi'(E) = \int \prod_{i=1}^{N} d\mathbf{r}_i \, \delta(E - H(\mathbf{r}_1, \ldots, \mathbf{r}_N)), \tag{11.29}$$

and is a non-negative function going to 0 at both extremes $\Omega(\pm\infty) = 0$. Thus the function must achieve at least one maximum at some finite value E_m where $\Omega'(E_m) = 0$.

For energies $E > E_m$, $\Omega'(E)$ will then be negative. Using the entropy $S(E) = \ln \Omega(E)$, we thus get that the inverse temperature dS/dE is negative for $E > E_m$.

Onsager pointed out that negative temperatures could lead to the formation of large-scale vortices by clustering of smaller ones. Although, as anticipated, the canonical solution must be handled with extreme caution for long-range interacting systems, the statistical tendency of vortices of the same circulation sign to cluster in the negative temperature regime can be justified using the canonical distribution $\exp(-\beta H)$. Changing the sign of the inverse temperatures β would be equivalent to reversing the sign of the interaction between vortices, making them to repel (resp. attract) if they have opposite (resp. same) circulation sign.

The above argument for the existence of negative temperatures was proposed by Onsager (1949) two years before the experiment on nuclear spin systems by Purcell and Pound (1951) reported the presence of negative 'spin temperatures'. More generally, any system with a finite phase space will display negative temperatures. However, the vast majority of realistic systems have also a kinetic energy contribution, and this fact forbids the emergence of states characterized by negative temperatures. The point vortex model has an energy strictly speaking purely kinetic, but that can formally be written as a potential energy as shown in (11.25). The vortex gas is, therefore, one of the very few physical systems with negative temperatures.

11.2.3 The statistical mechanics approach

After a long period during which Onsager's statistical theory was not further explored, this domain of research reappeared thanks to the works by Joyce and Montgomery (1973, 1974). The successive generalizations can be presented as follows.

(i) Noninteracting theory

As a first rough approximation, let us neglect the interaction between vortices. Introducing the density field of vortices, $n(\mathbf{r})$, the microcanonical entropy is obtained by counting the microscopic configurations corresponding to the same macroscopic state defined by this density. Distributing the vortices in the available phase space corresponds here to distributing them in the fluid domain. The problem relies on the maximization of the entropy, which as usual can be written as $S = -\sum_i n_i \ln n_i$ or more conveniently in the continuum limit as

$$S = -\int_D d\mathbf{r}\, n \ln n. \tag{11.30}$$

Here, the entropy must be maximized with the constraint of a given total number of vortices $N = \int_D d\mathbf{r}\, n(\mathbf{r})$. Consequently, introducing the associated Lagrange multiplier α, we obtain the most probable density by cancelling the first variation which is

$$\delta S - \alpha \delta N = -\int_D d\mathbf{r}\, \delta n(\mathbf{r})\, (\ln n(\mathbf{r}) + 1 + \alpha) = 0. \tag{11.31}$$

This equality being satisfied for any variation of δn around the most probable state, the term in parenthesis has to vanish and we end up with the uniform density $n_0 = \exp(-(1 + \alpha))$. The entropy maximization leads therefore to a uniform mixing when we neglect the interaction.

(ii) Mean-field theory

The simplest way, again usual in statistical mechanics, to take into account long-range interactions, is to consider that each vortex feels the influence of the mean field due to many others. It is, therefore, natural to simplify the interaction between vortices using the mean-field approximation. Replacing the vorticity ω in Eq. (11.10) by the local density $n\Gamma$, in which all vortices have the same circulation Γ, we get

$$H = -\frac{1}{2} \int_D d\mathbf{r} \, \psi \, n\Gamma, \tag{11.32}$$

while Eq. (11.6) reduces to $\Delta\psi = n\Gamma$.

In this case, the maximization of the entropy is obtained by introducing a second Lagrange multiplier β, associated with the constraint (11.32). Using the Poisson equation (11.6) and an integration by parts, the condition (11.31) on the first-order variation becomes

$$\delta S - \alpha \delta N - \beta \delta H = -\int_D d\mathbf{r} \, \delta n(\mathbf{r}) \, (\ln n(\mathbf{r}) + 1 + \alpha - \beta\Gamma\psi(\mathbf{r})) = 0, \tag{11.33}$$

which leads to the distribution $n = n_0 \exp(+\beta\Gamma\psi)$. The argument of the exponential being analogue to a potential energy, it is interesting to emphasize that we get a canonical-like distribution $\exp(-\beta H)$ although only microcanonical arguments for an isolated system have been invoked. This is indeed only a formal analogy, since no thermal bath with temperature $1/\beta$ has been introduced.

Finally, combining with the Poisson equation (11.6), we end up with the self-consistent mean-field equation

$$\Delta\psi = \Gamma n_0 \exp(+\beta\Gamma\psi), \tag{11.34}$$

an equation identical to the Poisson–Boltzmann equation in the Debye–Hückel theory of electrolytes, Onsager's main domain of research. As recently discovered by Eyink and Sreenivasan (2006) studying Onsager's letters to his peers, Lars Onsager had also found the analogies with the statistics of stars. The Lamé–Emden equation (10.16), which governs the temperature or density profiles for self-gravitating particles, is the exact analogue of the mean-field equation (11.34).

Suppose, for instance, that the circulation Γ is positive. Equation (11.32) shows that this implies that the streamfunction ψ is negative everywhere, since the energy density is positive, as shown by Eq. (11.8). For positive values of the Lagrange multiplier β, Eq. (11.34) shows that the vorticity ω is maximum where the streamfunction vanishes, i.e. close to the boundaries, while it is minimum where ψ is maximum in absolute value: the vortices are, therefore, squeezed against the boundaries. In contrast, negative values of β will induce a maximum of the vorticity at the centre of the domain, which suggests that point vortices will be packed in the centre, as the self-organization of two-dimensional flows. We, therefore, recover the prediction anticipated by Onsager and described in Section 11.2.2.

(iii) Generalization to several species

It is possible to pursue this line of research by considering a system of vortices with different circulations Γ_i, a generalization studied by Onsager but never published (Eyink and Sreenivasan, 2006). The self-consistent equation (11.34) is indeed straightforwardly generalized in the following manner:

$$\Delta \psi = \sum_i \Gamma_i n_i \exp(+\beta \Gamma_i \psi). \tag{11.35}$$

One important particular situation, the one actually studied by Joyce and Montgomery (1973), is the case of zero total circulation, with a symmetric set of positive and negative vortices of unitary circulation. Equation (11.35) reduces to

$$\Delta \psi = n_+ \exp(+\beta \psi) - n_- \exp(-\beta \psi), \tag{11.36}$$

where $n_{\pm}$ are linked to the Lagrange multipliers of both different vortex species. The special case for which both multipliers are equal leads to the sinh-Poisson equation. This equation was extensively studied from the mathematical point of view because of its nice mathematical properties linked to the existence of soliton solutions of its hyperbolic counterpart, the sinh-Gordon equation $\psi_{tt} - \psi_{xx} = \sinh(\beta \psi)$. Moreover, experimental and numerical data showing the functional dependence of vorticity on streamfunction in long-lived structures seemed qualitatively consistent with a hyperbolic sine. However, these studies have been later verified by very long-time, high-resolution numerical studies and more recent work showed some evidence that this result depends on the initial conditions (see Dauxois *et al.*, 1996, and references therein).

Interestingly, in all above three cases, maximizing the entropy at fixed energy, we obtain an equation which links directly the vorticity and streamfunction since $\omega = F(\psi)$. As described at the end of Section 11.1, this condition implies that it corresponds to exact stable stationary solutions of the 2D Euler equation, and is able to describe the macroscopic vortex formation proposed by Onsager for negative temperatures.

11.2.4   Deficiencies of the point vortex model

The point vortex statistical mechanics explains nicely the striking self-organization of two-dimensional flows. However, despite its historical importance and clear pedagogical interests, the point vortex model suffers from a number of deficiencies (Chavanis, 1996; Sommeria, 2002; Turkington, 2008) which limit its applicability as a reference scheme for the Euler system with a continuum velocity field. The four most important limitations are the following:

i. *Each vortex generates a divergent flow at its centre and hence a singular self energy.* As discussed earlier, this difficulty is, however, not too serious.

ii. *The merging of different vortices are not permitted by this theory.* Under continuum fluid mechanics, near collisions result in vortex mergers, in which two vortices coalesce into one vortex and eject a fraction of their total circulation into filament. This is an important process that is not modelled in the classical point vortex dynamics. Associated with this issue, it is important to stress some spurious vorticity concentration.

iii. *Different point vortex discretizations of the same continuum vorticity field lead to different predictions for the equilibrium states.* To give a simple example, let us consider an initial condition with two uniform patches of vorticity $+N$ and $-2N$. It can be described either by N positive vortices $+1$ and the same number of negative vortices -2 or by N vortices $+1$ and $2N$ vortices -1. According to the self-consistent equation (11.35), we get in the first case the equilibrium solution $\omega = n_+ \exp(+\beta\psi) - n_- \exp(-2\beta\psi)$, while we obtain the very different solution $\omega = n'_+ \exp(+\beta\psi) - n'_- \exp(-\beta\psi)$ for the second possibility.

iv. Finally, *this theory cannot be extended to more complicated models* and is restricted to model purely two-dimensional Euler flows.

Consequently, the point vortex model is of prime importance for explaining, at an introductory level, the effect of long-range interactions in two-dimensional hydrodynamics but, from the predictions perspectives, we cannot expect more than a good fit from this theory.

All these remarks motivated the development of a more advanced theory, based on continuum vorticity dynamics and relying directly on the Euler equation, as discussed in the next section.

11.3   The Robert–Sommeria–Miller Theory for the 2D Euler Equation

11.3.1   Introduction

Following an original idea proposed by Kuz'min (1982), Robert (1990) and Miller (1990) have elaborated an equilibrium statistical mechanical theory directly for the continuum 2D Euler equation: nowadays it is usually called RSM theory for Robert–Sommeria–Miller. Nice extensions along these lines can be found in Robert and Sommeria (1991), Michel and Robert (1994) and Chavanis *et al.* (1996), together with applications to the study of the Great Red Spot of Jupiter (Turkington *et al.*, 2001; Bouchet, 2001; Bouchet and Sommeria, 2002; Chavanis and Sommeria, 2002).

As it is well known, the Euler equation develops complex vorticity filaments at finer and finer scales, which makes any deterministic approach almost impossible. We cannot describe the temporal evolution of the fine-grained structure of the flow but, after all, this is often not useful. Indeed, one of the major interests and achievements of equilibrium statistical mechanics is the prediction of the macroscopic state without describing the details of the system configuration at a microscopic level. Guided by this idea, it appeared natural to posit a coarse-grained approximation of the vorticity $\overline{\omega}$, corresponding to a local average. Then, we introduce the probability $\rho(\mathbf{r}, \sigma)$ of finding the vorticity level σ in the neighbourhood of the position $\mathbf{r}$. Appropriately normalized as

$$\int d\sigma \, \rho(\mathbf{r}, \sigma) = 1, \tag{11.37}$$

the coarse-grained vorticity can be readily expressed in terms of this probability density as

$$\overline{\omega}(\mathbf{r}) = \int d\sigma \, \sigma \, \rho(\mathbf{r}, \sigma). \tag{11.38}$$

Having recast the problem in a probabilistic framework and assuming that correlations between vorticity values at different points are negligible (see Bouchet and Corvellec, 2010, for a nice discussion of this subtle point), finding the most probable state relies on maximizing the entropy

$$S = -\int_D d\mathbf{r} \int d\sigma \, \rho(\mathbf{r}, \sigma) \ln \rho(\mathbf{r}, \sigma). \tag{11.39}$$

However, the key point here is to develop a theory which takes into account not only the conservation of energy, as for the point vortex model in Section 11.2.3, but also other conserved quantities. As explained earlier, the vorticity is only advected and is therefore globally preserved during the complicated evolution of the flow controlled by the Euler equation. Consequently, the total areas of the different vorticity levels

$$\gamma(\sigma) = \int_D d\mathbf{r} \, \rho(\mathbf{r}, \sigma) \tag{11.40}$$

are conserved quantities. Similarly, the energy (11.10) is also conserved and can be expressed in terms of the coarse-grained vorticity as

$$H = -\frac{1}{2} \int_D d\mathbf{r} \, \psi \, \overline{\omega}, \tag{11.41}$$

as fluctuations around the average may be neglected (see Bouchet and Corvellec, 2010, and references therein).

The Robert–Sommeria–Miller theory enables us to take into account the constraints (11.40) and (11.41). As mentioned with reference to the point vortex model, the conservation of the energy of interaction does not allow the total mixing, and the competition between mixing and constraints leads to a balance that can be computed using a mixing entropy as for a gas.

To understand the approach in a simple case, let us consider first an initial condition with one patch of uniform vorticity surrounded by an irrotational domain. Afterwards we will generalize easily the calculation to the case of an infinite number of levels, and therefore to more realistic and complicated initial states.

11.3.2   The two levels approximation

Two levels of vorticities, 0 and σ_0, are sufficient to describe an initial state with a single patch of homogeneous vorticity σ_0, but also all successive states during the evolution, since the areas of the different vorticity levels are preserved. The probability density can thus be greatly simplified by introducing the Dirac distribution as

$$\rho(\mathbf{r}, \sigma) = \rho(\mathbf{r}, \sigma_0)\delta(\sigma - \sigma_0) + \rho(\mathbf{r}, 0)\delta(\sigma) \tag{11.42}$$

$$= p(\mathbf{r})\delta(\sigma - \sigma_0) + [1 - p(\mathbf{r})]\,\delta(\sigma), \tag{11.43}$$

where we used the definition $p(\mathbf{r}) = \rho(\mathbf{r}, \sigma_0)$ and the normalization condition (11.37). The entropy (11.39) can thus be simplified as

$$S = -\int d\mathbf{r}\,[p(\mathbf{r})\ln p(\mathbf{r}) + (1 - p(\mathbf{r}))\ln(1 - p(\mathbf{r}))], \tag{11.44}$$

an expression reminiscent of the mixing entropy for a two-component gas.

The most probable distribution is obtained by maximizing entropy (11.44), while taking into account the constraints (11.40) and (11.41); note that in this two levels approximation, the conservation of the vorticity levels (11.40) leads to the conservation of the quantity $I = \int_D p(\mathbf{r})d\mathbf{r}$. Introducing two Lagrange multipliers, β associated with the conservation of the energy and α to the conservation of I, the variational problem $\delta S - \beta \delta H - \alpha \delta I = 0$ leads to

$$-\int_D d\mathbf{r}\,[\ln p(\mathbf{r}) - \ln[1 - p(\mathbf{r})] - \beta \sigma_0 \psi(\mathbf{r}) + \alpha]\,\delta p(\mathbf{r}) = 0. \tag{11.45}$$

This expression being valid for any variation around the optimal state, we get the probability density

$$p(\mathbf{r}) = \frac{e^{-\alpha + \beta \sigma_0 \psi(\mathbf{r})}}{1 + e^{-\alpha + \beta \sigma_0 \psi(\mathbf{r})}}, \tag{11.46}$$

which maximizes the entropy. Introducing this expression in the definition of the coarse-grained vorticity (11.38), the latter can be written as

$$\overline{\omega}(\mathbf{r}) = \sigma_0\,\rho(\mathbf{r}, \sigma_0) + 0\,\rho(\mathbf{r}, 0) = \sigma_0 p(\mathbf{r}) \tag{11.47}$$

$$= \sigma_0 \frac{e^{-\alpha+\beta\sigma_0\psi(\mathbf{r})}}{1 + e^{-\alpha+\beta\sigma_0\psi(\mathbf{r})}} = \Delta\psi(\mathbf{r}). \qquad (11.48)$$

Interestingly, the Boltzmann-like distribution (11.34) obtained by Joyce and Montgomery (1973) for point vortices is modified in the distribution (11.48) reminiscent of the Fermi–Dirac distribution. This formal analogy is after all nothing but a consequence of the local exclusion principle of the two levels of vorticity which comes from the conservation of vorticity. However, as realized by Robert and Sommeria (1991), we recover the Boltzmann distribution (11.34) in the dilute limit in which $\sigma_0 \gg \overline{\omega}$, since we get

$$\overline{\omega}(\mathbf{r}) \simeq \sigma_0 e^{-\alpha}\, e^{+\beta\sigma_0\psi(\mathbf{r})} = \Delta\psi(\mathbf{r}). \qquad (11.49)$$

It is fully transparent here that this approach independently derived by Miller (1990) and Robert and Sommeria (1991) for the Euler equation corresponds exactly to the procedure proposed by Lynden–Bell (1967) for the Vlasov equation in the gravitational framework and reported in Section 9.4 for the two levels initial distribution of the HMF model. The parallel has been drawn by Chavanis (1996).

11.3.3   The generalization to the infinite number of levels

Now that the simplest case of only two levels of vorticities has been presented, let us consider the general case. The expression of the entropy (11.39) must be maximized by introducing a series of Lagrange multipliers: first, the chemical potential of the different levels $\alpha(\sigma)$, and secondly another series $\zeta(\mathbf{r})$ associated with the normalization (11.37) at each point of the physical space. The variational problem (11.45) is therefore generalized in

$$\delta S - \beta\delta H - \int \mathrm{d}\sigma\, \alpha(\sigma)\delta\gamma(\sigma) - \int \mathrm{d}\mathbf{r}\, \zeta(\mathbf{r})\, \delta\left(\int \mathrm{d}\sigma\, \rho(\mathbf{r},\sigma)\right) = 0 \qquad (11.50)$$

$$- \int_D \mathrm{d}\mathbf{r} \int \mathrm{d}\sigma\, [\ln\rho(\mathbf{r},\sigma) + 1 - \beta\sigma\psi + \alpha(\sigma) + \zeta(\mathbf{r})]\, \delta\rho(\mathbf{r},\sigma) = 0. \qquad (11.51)$$

The condition that this expression must be satisfied for any variation $\delta\rho$ leads to the equilibrium distribution

$$\rho(\mathbf{r},\sigma) = \frac{e^{-\alpha(\sigma)+\beta\sigma\psi(\mathbf{r})}}{Z(\mathbf{r})}, \qquad (11.52)$$

where $Z(\mathbf{r}) = \int \mathrm{d}\sigma\, \exp\left[-\alpha(\sigma) + \beta\sigma\psi(\mathbf{r})\right]$ to match Eq. (11.37). Finally, Eq. (11.38) gives the expression for the coarse-grained vorticity $\overline{\omega}(\mathbf{r})$ and therefore the general self-consistent equation

$$\Delta \psi \left(\mathbf{r} \right) = \frac{\displaystyle \int \mathrm{d}\sigma \, \sigma \exp \left[-\alpha(\sigma) + \beta \sigma \psi \left(\mathbf{r} \right) \right]}{\displaystyle \int \mathrm{d}\sigma \, \exp \left[-\alpha(\sigma) + \beta \sigma \psi \left(\mathbf{r} \right) \right]}. \tag{11.53}$$

We find again $\Delta \psi = F(\psi)$. It is therefore important to stress that, as for the point vortex model, we obtain again a steady solution of the Euler equation. However, the relationship between vorticity and streamfunction is different from the one derived for the point vortex model. It is now consistent with the properties of the Euler equation and without singular vorticity.

11.3.4   Ensemble inequivalence

For two-dimensional flows, it turns out that canonical and microcanonical ensembles may be inequivalent (Caglioti *et al.*, 1995; Kiessling and Lebowitz, 1997). In particular for the Euler equation in the region of negative temperature (where vorticity tends to accumulate at the centre of the domain) some hybrid states can develop in the microcanonical ensemble but not in the canonical (Ellis *et al.*, 2000). As far as the point vortex model is concerned, microcanonical stable states, which are unstable in the canonical ensemble, are found for specific geometries (Kiessling and Lebowitz, 1997). In both cases, microcanonical specific heat is negative.

Two-dimensional flows are therefore good candidates for providing experimental evidences of ensemble inequivalence, one of the present challenging issues of the physics of long-range interactions.

11.4   The Quasi-Geostrophic (QG) Model for Geophysical Fluid Dynamics

11.4.1   The quasi-geostrophic model

One of the underlying motivation for studying the statistical mechanics of two-dimensional hydrodynamic flows is the existence of large coherent structures, in the oceans or in the atmosphere of planets as shown in Fig. 11.1. Indeed, with reference to this latter example and as briefly anticipated at the beginning of this chapter, both the strong rotation of the planet and the stratification strongly contribute to the two-dimensionalization of the flow. We will hereafter present a brief introduction to the application of statistical theory to geophysical flows.

In the frame rotating with the planet, the equation describing an incompressible inviscid flow is

$$\frac{\partial \mathbf{v}}{\partial t} + (\mathbf{v}.\nabla)\mathbf{v} = -\frac{1}{\rho}\nabla p - 2\mathbf{\Omega} \times \mathbf{v} + \mathbf{g}, \tag{11.54}$$

where **g** is the gravitational acceleration and **Ω** the angular velocity of the rotating planet. For the sake of simplicity, we consider here the fluid homogeneous and we neglect therefore any stratification. Generalizations of this barotropic model, with an active upper layer above a lower layer at rest, are available for different layers of fluid density. This generalization is termed the baroclinic quasi-geostrophic model.

In addition to the pressure forces, the dynamics of the fluid results, therefore, from a competition between the Coriolis ($2\mathbf{\Omega} \times \mathbf{v}$) and the inertial (($\mathbf{v} \cdot \nabla$) $\mathbf{v}$) terms. Let us introduce the dimensionless Rossby number $\varepsilon = U/(2\Omega L)$, where U and L are, respectively, the characteristic velocity speed and length scale. We will restrict our analysis in the remainder of the section to flows dominated at large scale by the Coriolis force, and therefore in the limit $\varepsilon \to 0$. Taking typical values for the Gulf stream and Kuroshio current, $L = 10^5$ m, $U = 1$ m/s and $\Omega = 0.5 \times 10^{-4}$ s^{-1}, we obtain $\varepsilon \simeq 0.1$.

Focusing in a mid-latitude region around θ_0, one quantity of particular interest is the component of the normal to the planetary surface, namely the Coriolis parameter $f \simeq f_0 + \beta_0 y$, where $f_0 = 2\Omega \sin \theta_0$ and $\beta_0 = 2\Omega \cos \theta_0/r_0$, with r_0 the radius of the planet. This expression is known as the beta-plane approximation, but here, for the sake of simplicity, we will restrict to the f-plane approximation in which one neglects higher order terms: this latter approximation amounts to considering $\beta_0 = 0$.

The rotating shallow water model considers cases in which the horizontal length scales are much larger than the vertical one: this is the typical situation in the dynamics of the oceans if we remember that the averaged depth of the ocean is 4 km, while the radius of the Earth is 6400 km. A careful inspection (Pedlosky, 1987; Vallis, 2006) of the order of the different terms in the third component of Eq. (11.54) shows that the vertical acceleration w can be neglected so that we get the hydrostatic approximation $p_z \simeq -\rho g$, which leads to the expression for the pressure

$$p(x, y, t) = p_0 + \rho g \left(\eta(x, y, t) - z \right), \tag{11.55}$$

where η stands for the free surface altitude. Note that, in doing so, we neglect the horizontal gradients of the atmospheric pressure p_0. Introducing the above expression for the pressure in the first two components of Eq. (11.54), we get, therefore, the two horizontal momentum equations

$$\left(\frac{\partial}{\partial t} + u \frac{\partial}{\partial x} + v \frac{\partial}{\partial y} \right) u - f_0 v = -g \frac{\partial \eta}{\partial x} \tag{11.56}$$

$$\left(\frac{\partial}{\partial t} + u \frac{\partial}{\partial x} + v \frac{\partial}{\partial y} \right) v + f_0 u = -g \frac{\partial \eta}{\partial y}. \tag{11.57}$$

At this stage, an elegant way to pursue the calculation is to rely on an asymptotic expansion in powers of the small parameter ε as

$$u = u_0 + \varepsilon u_1 + \dots \tag{11.58}$$

$$v = v_0 + \varepsilon v_1 + \dots \tag{11.59}$$

$$\eta = \eta_0 + \varepsilon \eta_1 + \dots \quad . \tag{11.60}$$

Introducing these expansions in Eqs. (11.56) and (11.57) gives at the lowest order the geostrophic velocity field $(u_0, v_0) = (g/f_0)(-\partial_y \eta_0, \partial_x \eta_0)$: this is the so-called geostrophic equilibrium. Both velocity components are, therefore, independent of the vertical coordinate z, as stated by the Taylor–Proudman theorem (Pedlosky, 1987; Vallis, 2006). We also realize that the geostrophic streamfunction corresponds to $\psi_0 = g\eta_0(x, y, t)/f_0$, a quantity independent of the altitude z, but proportional to the free surface η. This property is used to infer the surface currents from satellite altimetry.

To get the geostrophic streamfunction, we must consider the next order in the expansion, i.e., by considering ageostrophic perturbations of higher order. It is convenient here to take the curl of the momentum equation (11.54) and project onto the z axis or, equivalently, to subtract the y-derivative of Eq. (11.56) to the x-derivative of (11.57). Following this procedure, we get immediately the dynamical equation for the vorticity ω, defined in Eq. (11.5), which reads

$$\frac{\partial \omega}{\partial t} + \mathbf{v} \cdot \nabla_h \omega + (f + \omega)\nabla_h \cdot \mathbf{v} = 0, \tag{11.61}$$

where ∇_h stands for the horizontal gradient operator $(\partial_x, \partial_y, 0)$.

At the lowest order, this equation gives

$$\frac{\partial \omega_0}{\partial t} + \mathbf{v}_0 \cdot \nabla_h \omega_0 + f\nabla_h \cdot \mathbf{v}_1 = 0, \tag{11.62}$$

since f is of order ε^{-1} and $\nabla_h \cdot \mathbf{v}_0$ identically vanishes according to the expression of the geostrophic velocity field derived earlier.

Introducing D, the depth of fluid, the equation of mass conservation is

$$\frac{\partial \eta}{\partial t} + \nabla_h \cdot [(\eta + D)\, \mathbf{v}] = 0, \tag{11.63}$$

which can be rewritten as

$$\frac{\partial \eta_0}{\partial t} + \mathbf{v}_0 . \nabla_h \eta_0 + D\nabla_h . \mathbf{v}_1 = 0. \tag{11.64}$$

Combining Eqs. (11.62) and (11.64) to get rid of the term $\nabla_h \cdot \mathbf{v}_1$, we obtain

$$\left(\frac{\partial}{\partial t} + \mathbf{v}_0 . \nabla\right)\left(\omega_0 - \frac{f}{D}\eta_0\right) = 0. \tag{11.65}$$

Here, it is interesting to introduce the barotropic Rossby radius of deformation $R = \sqrt{gD}/f$, which characterizes the competition between the gravitational effects which try to flatten the fluid surface and the rotation of the planet which, in contrast, curves it in presence of a flow because of the geostrophic equilibrium (see Pedlosky, 1987, or Vallis, 2006, for a complete discussion). Taking the mean oceanic depth $D = 4$ km, we get $R = 2000$ km at mid-latitudes, a value comparable to the size of the oceanic basins and therefore much larger that the observed coherent structures, since oceanic rings are typically 300 km wide. Consequently, R is often considered as infinite: this is the rigid lid approximation.

Omitting for the sake of convenience the subscript 0, the dynamical equation (11.65) can be rewritten finally as

$$\frac{\partial q}{\partial t} + (\mathbf{v} \cdot \mathbf{\nabla}) q = 0 \qquad (11.66)$$

by defining the generalized vorticity, or the potential vorticity

$$q = \Delta \psi - \frac{\psi}{R^2}, \qquad (11.67)$$

which is, therefore, conserved as it is transported by the fluid parcel.

As expected, the two-dimensional vorticity transport equation (11.14) is recovered when the rotation frequency Ω vanishes since q reduces to ω. In contrast, we can further elaborate on more complicated situations, building on this simple limiting case. In particular, in the beta-plane approximation for which we take into account the variation of the Coriolis parameter with the latitude ($\beta_0 \neq 0$) and/or for a varying bottom bathymetry $b(x, y)$, the expression (11.67) is straightforwardly generalized to

$$q = \Delta \psi - \frac{\psi}{R^2} + \beta_0 y + b(x, y). \qquad (11.68)$$

This quantity is called potential vorticity since in addition to the relative vorticity $\Delta \psi$, there is a contribution of the planetary vorticity and also contributions due to the stretching of the fluid column.

Interestingly, the above expression still applies when considering a two-layer stratification (a layer of density ρ above a layer of density $\rho + \Delta \rho$) provided we replace, in the definition of the barotropic Rossby radius, the gravitational acceleration g by the effective one $g' = \Delta \rho g / \rho$. The internal (or baroclinic) Rossby radius of deformation hence reads $R' = \sqrt{g' D}/f_0$: this is the relevant model for most geophysical flows.

11.4.2   The range of the interaction in the quasi-geostrophic model

For the potential vorticity (11.67), it is possible to generalize the energy (11.8) as

$$H = \int_D \mathbf{dr} \frac{1}{2} \left(|\nabla \psi|^2 + \frac{\psi^2}{R^2} \right), \qquad (11.69)$$

with the usual kinetic term and an additional gravitational potential. As for the Euler equation (see Eq. (11.11)), this expression can be rewritten as

$$H = -\frac{1}{2} \int_D\!\!\int_D \mathrm{d}\mathbf{r}\mathrm{d}\mathbf{r}' \left[q(\mathbf{r}')-\beta_0 y - b(\mathbf{r}')\right] G(\mathbf{r},\mathbf{r}') \left[q(\mathbf{r}')-\beta_0 y' - b(\mathbf{r}')\right], \quad (11.70)$$

where G is the Green function, vanishing on the frontier of the domain and solution of the equation

$$\Delta G(\mathbf{r}-\mathbf{r}') - G(\mathbf{r}-\mathbf{r}')/R^2 = \delta(\mathbf{r}-\mathbf{r}'), \quad (11.71)$$

The solution of such equation is

$$G = -\frac{1}{2\pi} K_0 \left(|\mathbf{r}-\mathbf{r}'|/R\right), \quad (11.72)$$

where K_0 is the zeroth-order modified Bessel function of the third kind. At short distance ($|\mathbf{r}-\mathbf{r}'| \ll R$), we recover the classical logarithmic interaction potential (11.12) so the dynamics is that of the 2D Euler equations. At large distance, in contrast, the interaction potential becomes

$$K_0 \left(\frac{|\mathbf{r}-\mathbf{r}'|}{R}\right) \sim \left(\frac{\pi}{2\,|\mathbf{r}-\mathbf{r}'|}\right)^{1/2} e^{-|\mathbf{r}-\mathbf{r}'|/R}, \quad (11.73)$$

which implies that the Rossby radius R, the scale where kinetic and gravitational potential energies, respectively, balance, plays the role of a screening length. If the screening parameter is much smaller than the typical length scale of the domain where the flow takes place ($R \ll L$), then the system is short-ranged. Otherwise, the system is long-ranged (Venaille and Bouchet, 2011a). The screening length scale, the Rossby radius R, is thus a parameter that allows us to pass from a short-range to an infinite-range interacting system.

11.4.3   The statistical mechanics of the quasi-geostrophic model

Since Eq. (11.66) is analogous to Eq. (11.14), we realize immediately that it is, in principle, possible to directly extend the RSM theory presented earlier for the 2D Euler equation to a three-dimensional geophysical setting.

As previously, in addition to the total energy (11.69), we must take into account the infinity of conserved quantities, the Casimirs $C_f = \int_D \mathrm{d}\mathbf{r} f(q)$, which are conserved for any function f. To keep track of these conservation laws of the transport equation, we introduce the normalized probability density $\rho(\mathbf{r},\sigma)$ of finding the potential vorticity level σ in the neighbourhood of the position $\mathbf{r}$ as described in Section 11.3. The most probable macroscopic states are eventually obtained by maximizing the mixing entropy (11.39).

However, computing the solutions of this variational problem requires the knowledge of the infinity of Casimir constraints, a fact that poses important limitations. First, we generally do not know precisely what is the initial condition, especially if we want to apply this theory for the prediction of oceanic or atmospheric flows. Second, a variational problem with an infinite number of constraints is very difficult to handle. Numerous efforts have been recently devoted to develop alternative approaches to simplify practically and mathematically this problem. In particular, it has been suggested to consider only the conservation of the energy and circulation, see Venaille and Bouchet (2011a) and references therein. Ensemble inequivalence had been numerically observed in the quasi-geostrophic model (Ellis *et al.*, 2000, 2002), but Venaille and Bouchet (2011a) have shown recently that ensemble inequivalence manifests for a large class of two-dimensional and geophysical flows.

Successive attempts to apply the statistical mechanics of the quasi-geostrophic model to real geophysical flows have attracted a lot of attention. Structures visible on the Earth such as oceanic rings and jets in the Gulf stream (Venaille and Bouchet, 2011b) were considered, but also the beautiful pictures (see Fig. 11.1), taken by the Voyager and Galileo spacecraft of the Jupiter Great Red Spot, have led to the search for the most probable state of the Jovian atmosphere. The environments of both physical situations are, of course, highly perturbed and more relevant to out-of-equilibrium statistical mechanics. However, since forcing and dissipation compete to reach an energetic equilibrium, ideas similar to those discussed here have been applied. It appears that in this context, the quasi-geostrophic approximation can be applied with some confidence since the Rossby number ε is smaller than 0.1. It was then possible to show relevant parameter regimes in which coherent vortices embedded in strong zonal shears correspond to the most probable state (Bouchet, 2001; Turkington, 2008).

12

Cold Coulomb Systems

12.1 Introduction

It is useful to begin with a very short summary of the relevant results related to the statistical behaviour of many-body Coulomb systems, underlying the difference from the gravitational interaction, treated in Chapter 10. The similarities are obvious. The Coulomb interaction between charged particles shares the same long-range property with the Newtonian gravitational interaction, i.e. the decay with the inverse of the distance. Concerning the behaviour at short distances, the singularity of the gravitational interaction is also the same as that of the Coulomb interaction between charges of opposite signs.

However, as we know, there is a fundamental difference between the two cases: while the gravitational interaction is always attractive, with Coulomb systems we have both attractive and repulsive forces, and this has profound consequences on the statistical behaviour of the two classes of systems. In fact, we have seen in Chapter 6 that, if we regularize the Coulomb interaction at short distances, the stability property, which requires the energy of the N-body system to be lower bounded by $-CN$ (C being an N independent constant), holds, as shown in Eqs. (6.9) and (6.14). In contrast, this is not true for the gravitational interaction, even if regularized or with the inclusion of a short-range repulsion; e.g., in a uniform distribution, the potential energy would be proportional to $-N^{5/3}$, as we have seen in in Section 6.2 and in the introduction of Chapter 10. On the other hand, the singularity at short distances gives rise to collapse both for the gravitational interaction and for the Coulomb interaction between charges of opposite signs. In the latter case, the collapse is prevented, when we treat the charged system quantum mechanically, by the uncertainty principle, which by itself is, however, not sufficient to avoid instability, since the ground state energy is upper bounded by $-CN^{7/5}$, as shown in Eq. (6.25). The exclusion principle is essential for explaining the stability of matter; in particular, systems where all the particles with charges of a given sign are fermions are stable, regardless of the character of the particles with charges of the other sign. This is obviously the case representing ordinary matter, where the negatively charged particles are the electrons.

Finally, still in Chapter 6, we have considered the existence of the thermodynamic limit; i.e., the conditions for which the free energy per particle tends to a finite value. We have seen that, within the systems where stability holds, the limit exists for globally

Physics of Long-Range Interacting Systems. First Edition. A. Campa *et al.*
© A. Campa, T. Dauxois, D. Fanelli, and S. Ruffo 2014. Published in 2014 by Oxford University Press.

neutral systems or for systems where the charge unbalance is upper bounded by a constant times the power 2/3 of the volume. A system which is not neutral but with a charge unbalance of the latter type can be called quasi-neutral, since the charge per unit volume tends to 0 in the thermodynamic limit. In most Coulomb systems also neutral species are present. For example, if the system is formed by the ionization of atoms in a gas, there will be an equilibrium between the number of electrons, of ions and of neutral atoms, that depends on the temperature.

In the dynamics of a Coulomb system a relevant parameter, as in any many-body system, is the measure of the relative weight of the potential energy with respect to the kinetic energy. In Chapter 8, we have described the general theory for the treatment of the dynamics of long-range systems in the framework of kinetic equations. It has been shown that the Lenard–Balescu equation, which takes into account the collisional effects in a long-range system, going beyond the collisionless mean-field evolution associated with the Vlasov equation, can be derived in a consistent way if we can exhibit a small parameter in these collisional effects. We have seen that the small parameter is related to the ratio between the average interparticle interaction and the average kinetic energy. In this chapter, we will see that in Coulomb systems this ratio is not always small. As a consequence, the kinetic equations studied in Chapter 8 are suited only for the case of weak coupling (defined in more detail later), to which the next chapter is devoted. In this chapter, we will give a short summary of the main properties of the Coulomb systems where the weak coupling condition is not met.

12.2   The Main Parameters in Coulomb Systems

To begin with, we introduce the relevant parameters used to define the characteristics of Coulomb systems (Fortov *et al.*, 2006). On the basis of these parameters, we will make a classification of the systems. For simplicity, we will consider the case of a globally neutral system with only two species of charges, one positive and one negative, equal in magnitude to the absolute value of the electronic charge e. The definitions can be extended to more general cases without the need to invoke new concepts.

We then consider a globally neutral system with an equal number N of ions of charge e and of electrons of charge $-e$, which occupy a volume V at a temperature T. Actually, due to the vastly different masses, it is possible to have a transient in which electrons and ions are at a different temperature. As we pointed out earlier, electrons and ions can recombine to form neutral atoms, and therefore we should take into account also their presence. We defer to later the evaluation of the relative density of neutral atoms compared to that of ions and electrons.

In our neutral system with N ions and N electrons, a measure of the potential energy per particle is $e^2/(4\pi\varepsilon_0 a)$, where a is an average interparticle distance. In general, the latter quantity is the Wigner–Seitz radius, defined as the radius of the sphere that on average contains one particle of one species

$$a = \left(\frac{3}{4\pi n}\right)^{\frac{1}{3}}, \tag{12.1}$$

where $n = N/V$ is the particle density. For a globally non-neutral system, we can take an average density of the species. Therefore, the density dependent measure of the potential energy per particle is

$$\left(\frac{4\pi}{3}\right)^{\frac{1}{3}} \frac{e^2 n^{\frac{1}{3}}}{4\pi\varepsilon_0}. \tag{12.2}$$

To give a measure of the kinetic energy per particle, we should first decide whether the system can be treated classically or if quantum mechanics is necessary. The general criterion for this evaluation is to compare the de Broglie thermal wavelength of a particle of mass M

$$\lambda_M = \left(\frac{2\pi\hbar^2}{Mk_B T}\right)^{\frac{1}{2}}, \tag{12.3}$$

with the average interparticle distance a. Note that in this formula the Boltzmann constant k_B appears. In fact, in contrast to the convention used in the rest of the book, where we have adopted units for which $k_B = 1$, in this and in the next chapter we find more convenient to make this constant appear explicitly.

A classical treatment of the particles of mass M is possible when

$$\frac{\lambda_M}{a} \sim \lambda_M n^{\frac{1}{3}} \ll 1. \tag{12.4}$$

Taking the third power of the last relation gives what can be called the quantum degeneracy criterion

$$n\lambda_M^3 \ll 1, \tag{12.5}$$

i.e., the average number of particles inside a sphere of radius equal to the thermal wavelength is much smaller than 1. In the case of fermions, in particular electrons, this relation can be put in another form that involves the Fermi energy of a system of electrons considered as an ideal Fermi gas,

$$\varepsilon_F = \frac{\hbar^2}{2m}\left(3\pi^2 n\right)^{\frac{2}{3}}, \tag{12.6}$$

where m is the electron mass. We can then define the degeneracy parameter for electrons

$$\xi_e = \frac{\varepsilon_F}{k_B T} = \frac{\hbar^2}{2mk_B T}\left(3\pi^2 n\right)^{\frac{2}{3}} = 4.234 \times 10^{-11} \frac{\left(n\,[\mathrm{cm}^{-3}]\right)^{\frac{2}{3}}}{T\,[\mathrm{K}]}, \tag{12.7}$$

where the right-hand side with the numerical coefficients is useful for making estimates in concrete cases; between square brackets, we have put the dimension in which the quantities must be expressed. When $\xi_e \ll 1$, the system of electrons is non-degenerate and classical mechanics can be applied. Using the thermal wavelength of electrons

$$\lambda_e = \left(\frac{2\pi \hbar^2}{mk_B T} \right)^{\frac{1}{2}}, \tag{12.8}$$

we immediately find the relation

$$n\lambda_e^3 = \frac{8}{3\sqrt{\pi}} \xi_e^{\frac{3}{2}}. \tag{12.9}$$

It is interesting to look at the meaning of the relation $n\lambda_e^3 \sim 1$ under still another point of view. We know that the number $\mathcal{N}$ of quantum states available to a system of electrons in a volume V, restricting to energies up to about $k_B T$, is of the order of

$$\mathcal{N} = V \frac{(2mk_B T)^{\frac{3}{2}}}{8\pi^3 \hbar^3}. \tag{12.10}$$

We immediately see that if the number of electrons is on the same order of $\mathcal{N}$, i.e., if $N \sim \mathcal{N}$, this is equivalent to $n\lambda_e^3 \sim 1$, or in turn to $\lambda_e \sim a$.

A degeneracy parameter can be introduced for ions in a perfectly equivalent way, using their thermal wavelength λ_M given in Eq. (12.3), where now M is the ion mass. We can then introduce the ion degeneracy parameter ξ_i, equivalent to the parameter (12.7) of electrons, defined by

$$\xi_i = \frac{\hbar^2}{2Mk_B T} \left(3\pi^2 n \right)^{\frac{2}{3}} = \frac{m}{M} \xi_e, \tag{12.11}$$

even if this would be related to a Fermi energy, analogously to that in Eq. (12.7), only in the case of fermionic ions. A relation between λ_M and ξ_i equivalent to (12.9) obviously holds.

Let us then begin our analysis of the relative weight of potential and kinetic energies with the classical case, which we obtain when $\xi_e \ll 1$. In this case, the kinetic energy per particle is given, neglecting factors of order 1, by $k_B T$. Then, the coupling constant, i.e., the ratio between the typical potential energy per particle and the typical kinetic energy per particle, usually denoted with Γ, is given by

$$\Gamma = \frac{e^2}{4\pi \varepsilon_0 a k_B T} = \left(\frac{4\pi}{3} \right)^{\frac{1}{3}} \frac{e^2 n^{\frac{1}{3}}}{4\pi \varepsilon_0 k_B T}. \tag{12.12}$$

As before, in order to make estimates in concrete cases, it is useful to write Γ in the following form using the numerical coefficients

$$\Gamma = 2.693 \times 10^{-3} \frac{\left(n \left[cm^{-3} \right] \right)^{\frac{1}{3}}}{T \, [K]} = 2.324 \times 10^{-7} \frac{\left(n \left[cm^{-3} \right] \right)^{\frac{1}{3}}}{(k_B T) \, [eV]}. \tag{12.13}$$

In the case in which the charged particles must be treated quantum mechanically, the estimate of the typical kinetic energy is different. For electrons in the degenerate state, we can simply take the Fermi energy (12.6). Then, the coupling constant is given by

$$\Gamma_e = \frac{e^2}{4\pi\varepsilon_0 a\varepsilon_F} = \frac{2^{\frac{5}{3}} m e^2}{12\pi^2\varepsilon_0\hbar^2}\frac{1}{n^{1/3}} = 6.360 \times 10^7 \frac{1}{\left(n\left[\mathrm{cm}^{-3}\right]\right)^{1/3}}. \tag{12.14}$$

Interestingly, the quantity Γ_e is roughly equal to another ratio, the one between the average interparticle distance a and the Bohr radius

$$a_0 = \frac{4\pi\varepsilon_0\hbar^2}{me^2}. \tag{12.15}$$

In fact, we have that

$$\frac{a}{a_0} = \left(\frac{3}{4\pi}\right)^{\frac{1}{3}}\frac{me^2}{4\pi\varepsilon_0\hbar^2 n^{\frac{1}{3}}} = \left(\frac{81\pi^2}{128}\right)^{\frac{1}{3}}\Gamma_e = 1.842\,\Gamma_e. \tag{12.16}$$

There is also another way to derive the coupling constant Γ_e. We have seen that quantum effects become important when the thermal wavelength of the particle becomes comparable or larger than the average interparticle distance. The relation $\lambda_e/a > 1$ for electrons can be written, using Eq. (12.8), as

$$4\pi\frac{\hbar^2}{2ma^2} > k_{\mathrm{B}}T. \tag{12.17}$$

Neglecting the factor 4π, this relation tells us that the kinetic energy coming from the uncertainty principle and from the localization of the particle in a linear dimension corresponding to the average interparticle distance a becomes larger than the classical kinetic energy when the average interparticle distance becomes smaller than the thermal wavelength. Then, if two particles are localized within a distance a, as is the case if the potential energy is of the order $e^2/(4\pi\varepsilon_0 a)$, it is then necessary to take as the estimate of the typical kinetic energy the quantity

$$\frac{\hbar^2}{2ma^2}. \tag{12.18}$$

The ratio between $e^2/(4\pi\varepsilon_0 a)$ and the last quantity gives again, within numerical factors of order 1, the coupling constant Γ_e. We see that this derivation does not make direct use of the Fermi energy. We can then define the quantum coupling constant for ions, independently from their quantum statistics. It will be given by the ratio between $e^2/(4\pi\varepsilon_0 a)$ and the analogous of Eq. (12.18), with the ion mass M instead of the electron mass m

$$\Gamma_i = \frac{2^{\frac{2}{3}}}{12\pi^2\varepsilon_0}\frac{e^2}{a}\frac{2Ma^2}{\hbar^2} = \frac{2^{\frac{5}{3}}Me^2}{12\pi^2\varepsilon_0\hbar^2}\frac{1}{n^{\frac{1}{3}}} = \frac{M}{m}\Gamma_e \tag{12.19}$$

$$= 6.360 \times 10^7 \frac{M}{m}\frac{1}{\left(n\,[\mathrm{cm}^{-3}]\right)^{\frac{1}{3}}}. \tag{12.20}$$

Before classifying the possible Coulomb systems on the basis of the parameters introduced here, we want to point out, comparing Eq. (12.12) with Eqs. (12.14) and (12.19), the different roles played by the density in the classical and the quantum cases. In the classical regime, we see that at a given temperature the coupling constant increases with the density and therefore, if we want the coupling to be small compared to 1, the density must be lower than a given value. This is easily understood, since $e^2/(4\pi\varepsilon_0 a)$ increases with the density, and therefore at a given temperature (i.e., kinetic energy) we cannot let the density increase too much if we want $\Gamma \ll 1$. In contrast in the case of the degenerate systems an increasing density causes Γ to decrease, since the Fermi energy (12.6) or the quantum kinetic energy (12.18) increases with the density faster than the inverse average interparticle distance; in other words, in contrast to what happens for a classical system, in a degenerate system the relative weight of the potential energy increases if the density decreases. In the degenerate system, the higher the density, the weaker the coupling. From Eq. (12.19), we see that in the degenerate case, ions can be weakly interacting only at extremely high densities, larger than about $10^{33}\,\mathrm{cm}^{-3}$.

12.3   A Classification of Coulomb Systems

When Γ is small compared to 1, the kinetic energy dominates over the potential energy; however, we have seen that the evaluation of Γ requires first an analysis of the quantum degeneracy of the system. Therefore, a classification of Coulomb systems can be done on the basis of two criteria: the value of the coupling constant Γ and the applicability of a classical treatment. We have already pointed out that the classical weak coupling limit, i.e., $\Gamma \ll 1$ with $\xi_e \ll 1$, is the one for which the kinetic equations, like the Landau and the Lenard–Balescu equations, are best suited. Systems with a value of Γ, which is not very small, are often referred to as non-ideal, with an obvious similarity with the terminology used for ordinary gases when the potential energy is not negligible with respect to the kinetic energy.

In the Coulomb systems classification given later, we restrict for simplicity to the case of a fully ionized system with ions of charge e and electrons, although of course there can be systems with more than two charge species, ions more than singly ionized and neutral atoms. In this respect, before showing the classification, it is instructive to see how the degree of ionization can be estimated. We suppose to work in a system in which there are neutral atoms that can undergo a single ionization, losing one electron. Therefore, there will be an equilibrium between neutral atoms, ions and electrons. The equilibrium is governed by the relation

$$\mu_a = \mu_i + \mu_e, \tag{12.21}$$

where μ_a is the chemical potential of neutral atoms, μ_i that of ions and μ_e that of electrons. Exploiting the fact that the chemical potential is the Gibbs free energy per particle, in the hypothesis of a weak interaction the last relation can be transformed in

$$\frac{n_i n_e}{n_a} = \frac{Z_i Z_e}{V Z_a}, \tag{12.22}$$

where V is the volume of the system; n_i, n_e and n_a are the number densities of ions, electrons and neutral atoms, respectively; and Z_i, Z_e and Z_a are the partition functions of the single particles, for ions, electrons and neutral atoms, respectively. The partition function of one electron is simply

$$Z_e = \frac{2V}{\lambda_e^3}, \tag{12.23}$$

where the factor 2 takes into account the spin degeneracy. For the neutral atom and the ion, we must consider both the translational degrees of freedom and the energy levels of the compound. The two contributions factorize in the partition function; in addition, neglecting the mass difference between the neutral atom and the ion (both equal to M), the factor corresponding to the translation is the same for both, and equal to

$$Z_i^{(\mathrm{tr})} = Z_a^{(\mathrm{tr})} = \frac{V}{\lambda_M^3}, \tag{12.24}$$

where the spin degeneracies are to be included in the internal partition functions. Numbering the energy levels of, e.g., the neutral atom with the index s, with $s = 0$ corresponding to the ground state, denoting with $\varepsilon_a^{(s)}$ and $g_a^{(s)}$ the corresponding energies and degeneracies, the internal partition function of the neutral atom is given by

$$Z_a^{(\mathrm{int})} = \sum_s g_a^{(s)} \exp\left(-\frac{\varepsilon_a^{(s)}}{k_{\mathrm{B}} T}\right). \tag{12.25}$$

A completely analogous expression holds for the internal partition function of the ion. Substituting Eqs. (12.23), (12.24) and (12.25) (and the analogous for ions) in Eq. (12.22), we get

$$\frac{n_i n_e}{n_a} = \frac{2}{\lambda_e^3} \frac{Z_i^{(\mathrm{int})}}{Z_a^{(\mathrm{int})}}. \tag{12.26}$$

In the hypothesis that for both the ions and the neutral atoms the energy difference between the first excited state and the ground state is considerably larger than $k_{\mathrm{B}} T$, the former equation simplifies to

$$\frac{n_i n_e}{n_a} = \frac{2}{\lambda_e^3} \frac{g_i^{(0)}}{g_a^{(0)}} \exp\left(-\frac{I}{k_B T}\right),$$ (12.27)

where $I = \varepsilon_i^{(0)} - \varepsilon_a^{(0)}$ is the ionization energy of the neutral atom.

Let us apply this equation to the case of hydrogen. In this case, the ion is simply the proton, and $g_i^{(0)} = 2$ (the spin degeneracy), while the degeneracy of the ground state of the atom is $g_a^{(0)} = 4$ (the spin degeneracies of the electron–proton pair). Supposing in addition that $n_i = n_e$ (there is a free electron for each ionized atom), we thus obtain

$$\frac{n_i^2}{n_a} = \frac{1}{\lambda_e^3} \exp\left(-\frac{I}{k_B T}\right).$$ (12.28)

If we denote by $n = n_i + n_a$ the conserved sum of neutral hydrogen atoms and protons, the last equation can be solved for n_i/n as a function of n and T.

In Fig. 12.1, we plot the fraction of ionized atoms, n_i/n, as a function of the temperature for various values of n. The temperature range goes up to $10^5 K$, which is already too high to trust the approximation embodied in Eq. (12.28); for higher temperatures, the internal partition function of the hydrogen atom must include also the excited levels. We see that at a given temperature the degree of ionization is a decreasing function of the total density n. In fact, at a higher density electrons and protons are closer on average, and the probability of recombination increases. This explains also the apparent inconsistency between the possibility to disregard the excited states of the neutral atom, as in the approximation used in Eq. (12.28), and the relevant role played by

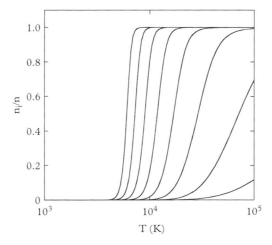

Figure 12.1 *Fraction of ionized hydrogen atoms as a function of the temperature, for different densities. The eight curves, from left to right, correspond to a density n equal to 10^{10}, 10^{12}, 10^{14}, 10^{16}, 10^{18}, 10^{20}, 10^{22} and 10^{24} cm^3, respectively.*

the ionization energy I, which is, of course, higher than the energy difference between an excited state and the ground state: once the ionization has occurred, the probability of recombination is governed not only by $k_B T$, but also by the density.

We now go back to the classification of the Coulomb systems, restricting for simplicity, as anticipated, to the case of a fully ionized systems made of electrons of charge $-e$ and ions of charge e. We first summarize the key parameters encountered so far:

- The coupling parameter Γ, dependent on the density n and on the temperature T, given in Eq. (12.12);

- The electrons degeneracy parameter ξ_e, dependent on the density n and the temperature T, given in Eq. (12.7), and related to the electron thermal wavelength (12.8) or the Fermi energy (12.6);

- The quantum electrons coupling parameter Γ_e, dependent on the density n, given in Eq. (12.14), and related to the Fermi energy or the Bohr radius (12.15);

- The ions degeneracy parameter ξ_i, dependent on the density n and the temperature T, given in Eq. (12.11), and equal to $\xi_e m/M$; and

- The quantum ions coupling parameter Γ_i, dependent on the density n, given in Eq. (12.19), and equal to $\Gamma_e M/m$.

Using these parameters, it is possible to divide the (T, n) plane into regions, in each of which our two-component Coulomb system is in a qualitatively different regime. To this purpose, we draw in Fig. 12.2 the lines corresponding to the unit value of the above

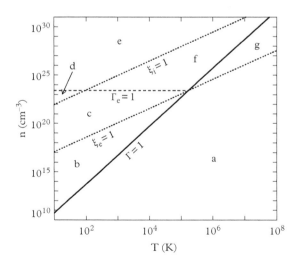

Figure 12.2 *The different regions of the (T, n) plane as obtained by equating to 1 the various parameters characterizing the Coulomb system. The properties of the system in the different regions are described in the text.*

parameters: $\Gamma = 1$, $\xi_e = 1$, $\Gamma_e = 1$ and $\xi_i = 1$. We have not included the line $\Gamma_i = 1$, which occurs at very high density. For the ion mass, we have taken that of the proton. With letters, we have denoted the different regions of the (T, n) plane. It is important to take into account that for each of the parameters Γ, ξ_e and ξ_i, values smaller than 1 are below the corresponding line, while for the parameter Γ_e values smaller than 1 are above the corresponding line. The whole region plotted in the graph is at a density for which $\Gamma_i > 1$.

It happens that the three lines $\Gamma = 1$, $\xi_e = 1$ and $\Gamma_e = 1$ meet at the same point. The line $\Gamma_e = 1$ is not continued after the intersection with the other two, i.e., inside region (a), where it loses its meaning: in that region, the system is classical, and the coupling constant is not defined by Γ_e. At temperatures higher than that of the intersection point quantum effects become important, increasing the density, when the system is still weakly coupled.

According to the values of the parameters, the following classification of the system can be done:

- Region (a), with $\Gamma < 1$ and $\xi_e < 1$. The system is classical, and both electrons and ions are weakly coupled.
- Region (b), with $\Gamma > 1$ and $\xi_e < 1$. The system is classical and both electrons and ions are strongly coupled.
- Region (c), with $\Gamma > 1$, $\xi_e > 1$, $\Gamma_e > 1$ and $\xi_i < 1$. The electrons are degenerate and strongly coupled, while the ions are classical and strongly coupled.
- Region (d), with $\Gamma > 1$, $\xi_e > 1$, $\Gamma_e > 1$ and $\xi_i > 1$. Both electrons and ions are degenerate and strongly coupled.
- Region (e), with $\Gamma > 1$, $\xi_e > 1$, $\Gamma_e < 1$ and $\xi_i > 1$. Both electrons and ions are degenerate, with the electrons weakly coupled and the ions strongly coupled.
- Region (f), with $\Gamma > 1$, $\xi_e > 1$, $\Gamma_e < 1$ and $\xi_i < 1$. The electrons are degenerate and weakly coupled, while the ions are classical and strongly coupled.
- Region (g), with $\Gamma < 1$, $\xi_e > 1$, $\Gamma_e < 1$ and $\xi_i < 1$. The electrons are degenerate and weakly coupled, while the ions are classical and weakly coupled.

In all cases, the negatively charged component of the systems are electrons and therefore, according to the results described in Chapter 6 and summarized at the beginning of this chapter, these systems always satisfy the statistical mechanical property of stability and temperedness, and a proper thermodynamic limit exists.

The first case corresponds to the systems for which usually the term *plasma* is reserved, although the more precise term weakly coupled plasma should be used. Generally they are constituted by a (partially) ionized gas, made of ions, electrons and neutral atoms, in a temperature-dependent dynamical equilibrium between charged species and neutral atoms, as we have shown earlier. From Eq. (12.12) or (12.13), and as clearly visualized in Fig. 12.2, we see that a plasma is characterized by either a sufficiently high temperature for a given density or a sufficiently low density for a given temperature.

The next chapter will be devoted to a review of the main results concerning weakly coupled plasmas.

The picture represented in Fig. 12.2 is somewhat more complicated in systems with multiply ionized atoms. In that case, if Z is the charge of the ions, there are three coupling constants: Γ for the interaction between electrons, $Z\Gamma$ for the interaction between electrons and ions and $Z^2\Gamma$ for the interaction between ions.

Few characteristics of the systems other than weakly coupled plasmas will be described in this chapter. Systems in which the coupling constant is not small are usually called *strongly coupled plasmas*. They are treated in the next section. We will then consider the case of degenerate systems, where the concept of Wigner crystals will be introduced. The title of this chapter can be understood by looking at Fig. 12.2: strongly coupled systems are found at sufficiently low temperatures. We will here restrict to the main equilibrium properties, leaving the treatment of dynamical properties to the case of weakly coupled plasmas, to which the next chapter is entirely devoted.

We will not touch the argument of Coulomb systems that are not globally neutral (or quasi-neutral), and that have been left out of our classification; they are usually called *non-neutral plasmas*. We have seen in Chapter 6 that for these systems a proper thermodynamic limit does not exist. Their confinement requires the use of external fields, which must be taken into account in the study of their statistical behaviour.

12.4   Strongly Coupled Plasmas

Although the majority of the plasmas found in nature are ideal, relevant examples of strongly coupled plasmas exist. One important case is that of the valence electrons in a metal, at room temperature. The electron density is of the order of 10^{22}–10^{23} cm^{-3}, and from the graph in Fig. 12.2 we see that Γ_e is somewhat larger than 1. The system of valence electrons is in region (c) of the (T, n) plane, and therefore it is degenerate and moderately strongly coupled. For comparison, we can consider the electrons system found in the Sun's centre, which has a somewhat higher density, but with a temperature of the order of 10^7 K; this system is in region (a) of the graph, and thus it is non-degenerate and weakly coupled ($\Gamma \ll 1$). As a further example, we can consider the interior of a white dwarf, which is found in region (f) of Fig. 12.2. Therefore, the electrons are degenerate and weakly coupled, while the ions are classical and strongly coupled.

In Chapter 6, we saw that in a Coulomb system a screening effect takes place. In that case, we were interested in estimating the lower energy state of the system, and for that reason we evaluated the screening length as a function of the number N of the particles in the ground state. Now, we would like to estimate this length in the situation where the temperature and the average density are given, i.e., when the system is represented by the canonical ensemble. In the weak coupling limit, e.g., for the plasmas that will be considered in the next chapter, this is done in a consistent way within the framework of the Debye–Hückel theory, in which the Poisson equation of electrostatic is solved using essentially a first-order expansion in Γ. However, the scaling of the screening length λ_D, where the subscript D stands for Debye, with the density n and the temperature T can be obtained by simple considerations, and before obtaining λ_D following the derivation of the Debye–Hückel theory, it is instructive to show how it can be simply estimated. We expect that the electrostatic energy of a given particle with all the particles within a distance λ_D is of the same order of its kinetic energy. The particles within a distance λ_D are of the order $n\lambda_D^3$, and the interaction with each of them is of the order $e^2/(4\pi\varepsilon_0\lambda_D)$. We therefore expect that

$$n\lambda_D^3 \frac{e^2}{4\pi\varepsilon_0\lambda_D} = \frac{ne^2\lambda_D^2}{4\pi\varepsilon_0} \sim k_B T, \tag{12.29}$$

from which we obtain

$$\lambda_D \sim \left(\frac{\varepsilon_0 k_B T}{ne^2}\right)^{\frac{1}{2}}. \tag{12.30}$$

From Eq. (12.12), we know that $\Gamma \sim \left(e^2 n^{\frac{1}{3}}\right)/(\varepsilon_0 k_B T)$; therefore, from Eq. (12.30), we obtain the following estimate of the number of particles inside the screening sphere:

$$n\lambda_D^3 \sim \frac{(\varepsilon_0 k_B T)^{\frac{3}{2}}}{n^{\frac{1}{2}} e^3} \sim \frac{1}{\Gamma^{\frac{3}{2}}}. \tag{12.31}$$

Thus, in the weak coupling limit, the number of particles inside the screening sphere is large; i.e., λ_D is much larger than the interparticle separation a; for strongly coupled plasmas, this number is of order 1 for $\Gamma \sim 1$, or smaller for large values of Γ. It is clear that in the last case the concept itself of screening length loses partly its meaning. The more precise calculation, performed in the framework of the Debye–Hückel theory, yields the same evaluation. However, it is possible to make an amendment to the theory (Nordholm, 1984; Penfold *et al.*, 1990; Tamashiro *et al.*, 1999; Barbosa *et al.*, 2000), which returns a more realistic evaluation. In particular, it shows that the field of a given particle, even in the strong coupling regime, cannot be screened at a distance smaller than the interparticle distance. We now show this derivation. For simplicity, we will perform the calculation with reference to a model which is very often employed for the study of strongly coupled Coulomb systems, i.e., the so-called *one-component plasma* (OCP). In this model, we assume that there are charges of a given sign that are embedded in a uniform neutralizing background, obviously of the opposite sign. Two of the examples of strongly coupled plasmas given above are well represented by this model, the first with negative charges moving in a positive neutralizing background, and the second with the opposite situation: in a metal the positive ions can be seen as a neutralizing background for the valence electrons, while in a white dwarf the weakly coupled degenerate electrons are a neutralizing background for the ions.

The Debye–Hückel theory starts from the representation of the pair distribution function $g(r)$, which is valid in weakly coupled systems, where the kinetic energy is considerably higher than the potential energy

$$g(r) = \exp\left(-\frac{q\phi(r)}{k_\mathrm{B} T}\right), \tag{12.32}$$

where $\phi(r)$ is the electric potential and q the charge of the particle at distance r from the given particle. Then, the spherically symmetric charge distribution around a given particle of our OCP with number density n (we consider the case of positive moving ions of charge e, but the calculations are perfectly identical in the case of negative moving charges) is given by

$$\rho(r) = ne\,[g(r) - 1] = ne\,[\exp\left(-e\phi(r)/k_\mathrm{B} T\right) - 1]. \tag{12.33}$$

The insertion of this equation in the Poisson equation of electrostatics gives an equation for the self-consistent spherically symmetric potential $\phi(r)$

$$\nabla^2 \phi(r) = \frac{1}{r^2}\frac{\mathrm{d}}{\mathrm{d}r}\left(r^2 \frac{\mathrm{d}\phi}{\mathrm{d}r}\right) = -\frac{\rho(r)}{\varepsilon_0}$$

$$= -\frac{ne}{\varepsilon_0}\left[\exp\left(-\frac{e\phi(r)}{k_\mathrm{B} T}\right) - 1\right] - \frac{e}{\varepsilon_0}\delta(\mathbf{r}), \tag{12.34}$$

where the δ function has been added to take into account the given charge itself. Making the substitution $\phi(r) = \psi(r)/r$, and taking into account the boundary condition at $r \to 0$, the last equation is easily solved after expanding up to first order the exponential, obtaining

$$\phi(r) = \frac{e}{4\pi\varepsilon_0 r}\,\mathrm{e}^{-r/\lambda_D}, \tag{12.35}$$

i.e., a Yukawa-type potential with the Debye screening length given by

$$\lambda_\mathrm{D} = \left(\frac{\varepsilon_0 k_\mathrm{B} T}{ne^2}\right)^{\frac{1}{2}}. \tag{12.36}$$

As we have already noted, if Γ becomes larger than 1, as in strongly coupled plasmas, λ_D becomes smaller than the interparticle distance a. This inconsistency can be fixed with the following argument (Nordholm, 1984; Penfold *et al.*, 1990; Tamashiro *et al.*, 1999; Barbosa *et al.*, 2000). We should expect that another ion cannot go closer to the given ion than a distance, to be denoted by h, corresponding to a potential energy equal to the average kinetic energy $k_\mathrm{B} T$; practically, each ion has a 'hole' around it, in which other ions do not penetrate. Then, Eq. (12.34) is substituted by

$$\frac{1}{r^2}\frac{\mathrm{d}}{\mathrm{d}r}\left(r^2\frac{\mathrm{d}\phi}{\mathrm{d}r}\right) = -\frac{\rho(r)}{\varepsilon_0} = \begin{cases} -\dfrac{ne}{\varepsilon_0}\left(\mathrm{e}^{-\frac{e\phi(r)}{k_\mathrm{B} T}} - 1\right) & \text{for } r > h \\[2mm] \dfrac{ne}{\varepsilon_0} - \dfrac{e}{\varepsilon_0}\delta(\mathbf{r}) & \text{for } r < h, \end{cases} \tag{12.37}$$

with the provision that $e\phi(h) = k_B T$. This equation is again solved by expanding the exponential up to first order. This might appear, at first sight, a logical inconsistency, since the computation should in principle apply also when Γ is not very small. However, we should consider that the equation is built in such a way that the exponent is at most equal to 1 in absolute value, and therefore the expansion, although rather approximate, is still applicable. For $r > h$, we again obtain a Yukawa-type potential, although with an effective charge $\tilde{e}$, while for $r < h$ we have the potential corresponding to a point charge plus a uniform negative background

$$\phi(r) = \begin{cases} \dfrac{\tilde{e}}{4\pi\varepsilon_0 r} \exp\left(-\dfrac{r}{\lambda_D}\right) & \text{for } r > h \\[2ex] \dfrac{1}{6\varepsilon_0} ner^2 + \dfrac{e}{4\pi\varepsilon_0 r} + c & \text{for } r < h, \end{cases} \tag{12.38}$$

where the constant c assures continuity at $r = h$. The remaining parameters $\tilde{e}$ and h are obtained by imposing $e\phi(h) = k_B T$ and the continuity of the derivative of $\phi(r)$ at $r = h$. Performing the calculations, we find that the size of the hole h is given by

$$h = \lambda_D \left[f(\Gamma) - 1 \right], \tag{12.39}$$

where $f(\Gamma)$ is

$$f(\Gamma) = \left[1 + (3\Gamma)^{\frac{3}{2}} \right]^{\frac{1}{3}}. \tag{12.40}$$

It can be shown that for Γ large, when $\lambda_D < a$, the value of h tends to a and, at the same time, to the screening length. In fact, by definition, $\frac{4\pi}{3} na^3 = 1$, and therefore, the background charge inside the sphere of radius $h \sim a$ is about $-e$.

This approximate evaluation should give a taste of the type of difficulties that we encounter in the study of the equilibrium properties of a strongly coupled plasma, in contrast to the weak coupling limit, where the correlation between particles is small, i.e. $g(r) \sim 1$, except when r is much smaller than a. For strongly coupled plasmas many results are obtained by the use of Monte Carlo numerical experiments, and we end this section by showing, in Fig. 12.3, the numerical computation of the pair distribution function $g(r)$ for various values of the coupling constant Γ (Ichimaru, 1982). The peak for large values of Γ is clearly reminiscent of what happens in classical dense fluids or solids.

12.5   Wigner Crystals

The one-component plasma can be taken as a model not only for non-degenerate systems, as it has been done in the previous section, but also for the degenerate case. Before considering the latter, we want to show how the model has been used to determine the crystallization in a classical strongly coupled plasma. In Fig. 12.3, the curve

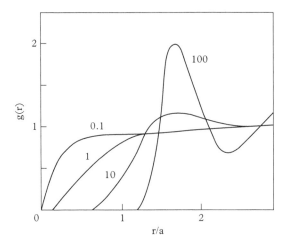

Figure 12.3 *The pair distribution functions for different values of the coupling constant* Γ *(the values are close to the corresponding curve), in a classical OCP. The value of r is normalized to the average interparticle distance a. Adapted from Ichimaru (1982), © 1982 by the American Physical Society.*

corresponding to $\Gamma = 100$ suggests that when the coupling constant is of that order in the classical case, the system crystallizes. The problem of crystallization of the OCP has been studied by Monte Carlo experiments, in which the free energy of the fluid and of the solid phase were computed as a function of the coupling constant Γ; the point of intersection of the two expressions determines the value of Γ at which crystallization occurs. The free energy per particle f can be obtained from the energy per particle u using the thermodynamic relation

$$\varepsilon(\beta) = \frac{\partial}{\partial \beta}(\beta f(\beta)),\tag{12.41}$$

where $\beta = 1/(k_B T)$. Thus, knowing $f(\beta)$ for a particular β_1, we obtain

$$\beta f(\beta) = \beta_1 f(\beta_1) + \int_{\beta_1}^{\beta} d\beta\, u(\beta).\tag{12.42}$$

If $u(\beta)$ is known, e.g., by the Monte Carlo analysis, as a function of Γ, we can exploit the relation $\Gamma = e^2 \beta/(4\pi\varepsilon_0 a)$ (Ichimaru, 1982), to have

$$\beta f(\beta) \equiv \tilde{f}(\Gamma) = \tilde{f}(\Gamma_1) + \frac{4\pi\varepsilon_0 a}{e^2}\int_{\Gamma_1}^{\Gamma} d\Gamma\, u(\Gamma) = \tilde{f}(\Gamma_1) + \int_{\Gamma_1}^{\Gamma} d\Gamma\, \frac{\beta u(\Gamma)}{\Gamma},\tag{12.43}$$

where in the numerator of the integrand on the right-hand side we have written the quantity which is often evaluated in the numerical experiments. Taking into account statistical errors, the comparison between the fluid and the solid phase has given a value of Γ roughly between 160 and 170 for crystallization.

We now turn to the degenerate electrons system. As was pointed out earlier, the density dependence of the coupling constant, for a given temperature, is opposite between the classical and the degenerate systems. To increase the value of Γ in the classical case, we must increase the density, while the degenerate system must be rarefied in order to increase Γ_e: the Fermi energy is proportional to $n^{2/3}$ while the potential energy is proportional to $n^{1/3}$. Therefore, while in a classical system for a given temperature crystallization occurs increasing the density, in a degenerate system we expect it to occur at low density. As a matter of fact, the concept of *Wigner crystal* has come into use after the pioneering work by Wigner (1934) concerning the valence electrons in metals. As we pointed out in the previous section, this system is highly degenerate, since its Fermi energy corresponds to a temperature considerably higher than the room temperature. We know that at the typical densities of metals the electrons system is in the fluid state.

It is not a bad approximation to consider the system as if it were at zero temperature; then the equilibrium state is the ground state. In a classical system, it appears natural that the ground state is the one with the atoms occupying the sites of a crystal, but for a degenerate electrons systems, if the density is too high, the zero point kinetic energy is sufficient to destroy the crystalline structure.

It is not easy to find the density below which the crystallized state is energetically favourable. Here we cite some criteria that have been adopted to make an estimate of the density at which crystallization occurs.

i. We can employ the Lindemann criterion (Lindemann, 1910) according to which, in the classical case, a crystal melts when the average displacement of an atom, or in other words the average amplitude of its thermal vibration, exceeds a certain fraction δ of the average interparticle distance a. Then, the condition for the existence of the crystal is

$$\langle(\Delta\mathbf{r})^2\rangle^{\frac{1}{2}} \leq \delta a, \tag{12.44}$$

where the average thermal vibration $\langle\Delta\mathbf{r}\rangle^{1/2}$ is, apart from numerical factors of order 1, given by the relation

$$M\omega^2\langle(\Delta\mathbf{r})^2\rangle = k_{\mathrm{B}}T, \tag{12.45}$$

where M is the atom mass and ω is the frequency associated with the restoring force that keeps the atom in the crystal site. In a degenerate electrons system, ω is the electron plasma frequency $\omega_e = \left(ne^2/\varepsilon_0 m\right)^{1/2}$ (Tsidil'kovskii, 1987), and the energy associated with the oscillation is taken as the zero point energy at frequency ω_e, namely $\hbar\omega_e/2$. Then, the relation for the displacement becomes

$$m\omega_e\langle(\Delta\mathbf{r})^2\rangle = \frac{1}{2}\hbar. \tag{12.46}$$

Inserting into Eq. (12.44), we obtain, using Eq. (12.1), the following relation to be satisfied by the density in a crystallized electrons system

$$n^{\frac{1}{12}} \leq \delta \left(\frac{3}{4\pi} \right)^{\frac{1}{3}} \left(\frac{4e^2 m}{\varepsilon_0 \hbar^2} \right)^{\frac{1}{4}}.$$

(12.47)

This relation is usually transformed in a relation for the average interparticle distance a normalized to the Bohr radius (12.15), i.e., for the quantity defined in Eq. (12.16), generally denoted by r_s. We easily obtain

$$r_s \geq \frac{1}{12\delta^4}.$$

(12.48)

With the often adopted value of $\delta = 1/4$, we find that an electrons degenerate system is crystallized when

$$r_s \gtrsim 20.$$

(12.49)

Equation (12.48) contains the fourth power of δ, showing that the critical value of r_s is very sensitive to the choice of the value of δ. Therefore, this criterion can give only a rough estimate of the crystallization density.

ii. In the attempt to avoid this large uncertainty, another criterion was proposed (de Wette, 1964). The crystal structure is stable if the potential well in which the electron should reside is sufficiently large and/or deep to have a bound state. However, also this estimate is very sensitive to the details of the computation, in particular to the structure of the potential well. In fact, critical values of r_s between about 50 and 100 have been obtained.

iii. Another criterion is similar to that shown earlier for the crystallization of classical strongly coupled plasmas. According to this procedure, we compare the energies of trial wavefunctions for the liquids and the crystalline states. In this case the critical value of r_s seems to be less sensitive to the details. Various results indicate a critical value of $r_s \approx 70$ (Glyde *et al.*, 1976; Ceperley, 1978).

We hope that the few details presented in the past two sections give a flavour of the theoretical and computational difficulties associated with the study of strongly coupled plasmas, in both the classical and in the degenerate case. The interested reader is urged to consult the literature specialized in this very interesting topic for further insights.

13

Hot Plasma

The plasma state is often referred to as the fourth state of matter (Nicholson, 1983; Chen, 1984). A plasma is an electrified gas where the atoms are dissociated into positive ions (i) and negative electrons (e). The word plasma comes from ancient Greek and was originally devised to indicate something moulded. Tonks and Langmuir (1929) first proposed to make use of it to describe a ionized gas. It is, however, important to realize that any ionized gas *cannot* be called a plasma, as there is always a certain degree of ionization in any gas. A somehow useful definition (Chen, 1984) recites: 'A plasma is a *quasineutral* gas of charged particles and neutral particles which exhibits *collective behaviour*'.

Particularly crucial for the content of this chapter is the second characteristic to which the above definition alludes. Plasmas are, in fact, made of charged particles. As charges move, they generate local concentrations of positive and negative charges, which in turn give rise to *electric and magnetic fields*. These latter induce long-range forces that solicit the motion of other particles, resulting in a rich and complex *self-consistent* scenario. Every selected volume of a plasma exerts a force on any other volume via long-range electromagnetic forces that are responsible for a large repertoire of possible microscopic particles' motions.

Most interesting results concerns the so-called *collisionless* plasmas, in which the long-range couplings dominate over local collisional effects. In this regime, collisions can be neglected all together and the plasma dynamics is approximately described by a set of Vlasov equations (similar to those presented in Chapter 8), one for each species of charged particles, complemented by the Maxwell equations for the fields. Given this scenario, by collective behaviour we mean that the dynamics is not solely sensitive to local conditions but also on the state of the plasma in remote regions. In the following section, we shall elaborate further on the properties of a plasma, so as to eventually return a quantitative definition.

13.1 Temperature, Debye Shielding and Quasi-neutrality

Let us start by considering a gas in equilibrium. As we know from equilibrium statistical mechanics, the particles velocity v obeys the Maxwell distribution, which, in one dimension, we can write as

Physics of Long-Range Interacting Systems. First Edition. A. Campa *et al.*
© A. Campa, T. Dauxois, D. Fanelli, and S. Ruffo 2014. Published in 2014 by Oxford University Press.

$$f(v) = A \exp\left(-\frac{\frac{1}{2}mv^2}{k_B T}\right),$$ (13.1)

where k_B is the Boltzmann constant (we remind that in this chapter, as in the previous one, the constant k_B appears explicitly), A is the normalization constant

$$A = n \left(\frac{m}{2\pi k_B T}\right)^{1/2}$$ (13.2)

and $n = \int dv f(v)$ stands for the particle density. m label the particles' masses, supposed to be identical. The width of the distribution (13.1) is controlled by the constant T, which we call the temperature. It is straightforward to show that the average kinetic energy is equal to $k_B T/2$, the factor 1/2 being replaced by 3/2 if we extend the reasoning to the relevant three-dimensional case. Since temperature and energy are closely related, it is customary in plasma physics to express temperatures in units of energy. To fix ideas, for a 1-eV plasma, we intend a temperature of the charged gas of particles of $T = 1.6 10^{-19}$ [J]/$1.38 10^{-23}$ [J/K] $= 11600$ K (Chen, 1984). Interestingly, a plasma can have several temperatures at the same time. It often happens, indeed, that electrons and ions have separate Maxwell distributions, each characterized by a specific temperature, respectively, T_e and T_i. Collisions drive, in fact, the equilibration process and the thermalization time among ions *or* electrons themselves is definitely faster than the thermalization time between ions *and* electrons. Then each species can be in its own (thermal) equilibrium while the plasma may not last enough to allow for the two temperatures to equalize.

As a further point, we note that in the presence of a magnetic field **B** even a single species, say ions, can have two distinct temperatures. The force acting along the direction parallel to **B** is different from that which is experienced along the perpendicular direction. Velocities can consequently attain different distribution profiles (respectively, along the parallel and perpendicular directions as defined by the vector **B**), being characterized by different temperature values.

A fundamental property of a plasma is its inherent ability to shield out electric potential that are applied to it, a characteristic that is intrinsically connected to the 'collective behaviour' mentioned earlier. To clarify this important aspect, we consider the following conceptual experiment. Take a plasma and insert in it an electric field, which we suppose, for instance, is generated by two charged (macroscopic) balls connected by a battery; see Fig. 13.1.

Each sphere would attract particles of opposite charge: if the plasma were cold, virtually with no thermal motion, there would be as many charges in the cloud as in the ball, yielding what are generally referred to as the perfect shielding conditions. No electric field would be hence present outside the cloud, in the bulk of the plasma. On the other hand, if the temperature is finite, the particles situated at the edge of the cloud would have enough thermal energy to escape the electrostatic potential well. In other words, the edge of the cloud is placed at a radius where the potential energy is approximately equal

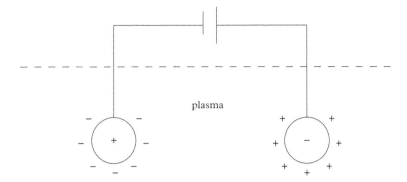

Figure 13.1 *The cartoon visualizes the virtual experiment proposed in the text. Two charged conductive spheres are introduced in a plasma and attract charged particles of opposite signs. Clouds of charged particles materialize around the spheres, so screening the induced electric field.*

to the thermal energy $k_B T$ of the particles and its thickness can be estimated via a simple calculation that we reproduce here.

To have a quantitative estimate of the distance over which a charge is shielded, we consider now that a charge is placed in the plasma at $\mathbf{r} = 0$. We also assume for simplicity that the ions of mass m_i do not move and constitute a uniform background of positive charges. This ansatz is motivated by recalling that $m_i \gg m_e$, the mass of the electrons: the inertia of the ions is so large that it prevents them from moving on the timescale of the considered experiment. The following argument is a simplified version, given here for convenience, of the Debye–Hückel theory, which has been described in Section 12.4. The electric potential field ϕ is given by the Poisson equation

$$\nabla^2 \phi = -\frac{e(n_i - n_e)}{\varepsilon_0}. \tag{13.3}$$

Labelling the density far away as n_∞, we have $n_i = n_\infty$ by virtue of the uniformity condition. In the presence of a potential energy $-e\phi$, the electron velocity distribution gets modified, so as to incorporate the apposite Boltzmann factor, to become

$$f(v) = A' \exp\left(-\frac{\frac{1}{2}mv^2 - e\phi}{k_B T}\right). \tag{13.4}$$

The physical meaning of this formula is transparent: there are more electrons where ϕ is positive, while there are fewer electrons where ϕ is negative: in the latter case this is due to the fact that not all electrons have enough energy to get there.

Integration of Eq. (13.4) over v yields

$$n_e = n_\infty \exp\left(\frac{e\phi}{k_B T}\right). \tag{13.5}$$

Then, inserting (13.5) into Eq. (13.3), we get

$$\nabla^2 \phi = \frac{en_\infty}{\varepsilon_0} \left[\exp\left(e\phi/k_B T\right) - 1 \right]. \tag{13.6}$$

In the region of interest to us, $|e\phi/k_B T| \ll 1$. We can then expand the exponential in Taylor series. Keeping only the linear terms and integrating the obtained differential equation yields the following expression for the electric potential, which obviously depends only on $r = |\mathbf{r}|$

$$\phi(r) = \phi_0(r) \exp\left(-r/\lambda_D\right), \tag{13.7}$$

where $\phi_0(r)$ is the usual potential decaying as $1/r$ that would obtain for the charge in vacuum, and

$$\lambda_D = \sqrt{\frac{\varepsilon_0 k_B T}{e^2 n}} \tag{13.8}$$

is the so called Debye length, a quantitative measure of the shielding distance (the label ∞ has been dropped in the definition of the density n). Note that as n is increased, the Debye length λ_D decreases, as it should since each spherical layer of plasma contains more electrons. Furthermore, λ_D increases with increasing the temperature. This computation, made for a charge placed inside the plasma, can as well be done for each of the particles of the plasma. Then, λ_D describes the shielding distance for the charge of each particle in the plasma.

Given this result, we can now define in rigorous terms the 'quasi-neutrality' concept that was advocated in the definition of a plasma. If the dimensions of a system L are much larger than λ_D, then local concentrations of charges (like the charged balls in the earlier example) get shielded out in a distance that is short when compared to L, thus leaving the bulk of the plasma free from large electric fields. The plasma is hence quasi–neutral, or neutral enough so that it is legitimate to assume $n_i \simeq n_e = n$, this latter quantity being termed the plasma density.

It should be noted, however, that the reasoning developed above relies heavily on the continuum assumption, which certainly holds whenever a large number of particles is present. To translate this observation into a quantitative criterion, we define N_D as the number of particles which are contained in a 'Debye sphere', i.e., a sphere of radius λ_D

$$N_D = n\frac{4}{3}\pi \lambda_D^3 = 1.38 \times 10^3 \; T^{3/2}/n^{1/2} \;, \tag{13.9}$$

with T expressed in K and n in cm^{-3}. A ionized gas behaves as a plasma if $\lambda_D \ll L$ (the quasineutrality condition), while at the same time $N_D \gg 1$ (continuum limit).

A third condition must be further required. Label with ω the frequency of a typical plasma oscillation and τ the mean collision time with neutral atoms. Then the inequality $\omega\tau > 1$ should hold for a gas to behave like a plasma rather than a neutral gas.

Plasma physics touches many domains of applications and fundamental research. Some of the most spectacular examples include space physics (the study of Earth environment in space), astrophysics and the applications to controlled thermonuclear fusion. While describing the specificity of each of the above domains falls beyond the scope of this book, in the remaining part of the chapter we will provide an entry to the basic kinetic theory of a plasma, surfing over a gallery of concepts and techniques of cross-disciplinary interest.

13.2 Klimontovich's Approach for Particles and Waves: Derivation of the Vlasov–Maxwell Equations

The kinetic equation of motion can be derived by resorting to the Klimontovich equation. This latter was already evoked in the general theory described in Chapter 8, and it constitutes the ideal starting point to deduce on a rigorous ground approximate equations that describe the average properties of a discrete system of bodies interacting via electromagnetic forces.

Consider a fluid composed of a mixture of s charged species. Assume that the point particle i of mass m_s occupies position $\mathbf{r}_i$ in the three-dimensional configuration space. Its velocity will be denoted by $\mathbf{v}_i$. The discrete density $f_d^s(\mathbf{r}, \mathbf{v}, t)$ of N_s such particles, belonging to species s, in the six-dimensional phase space $(\mathbf{r}, \mathbf{v})$ reads

$$f_d^s(\mathbf{r}, \mathbf{v}, t) = \frac{1}{N_s} \sum_{i=1}^{N_s} \delta(\mathbf{r} - \mathbf{r}_i(t)) \delta(\mathbf{v} - \mathbf{v}_i(t)), \qquad (13.10)$$

where $\delta(\cdot)$ stands for the Dirac delta function. An exact equation for the evolution of the system of charged particles is obtained by taking the time derivative of the density $f_d^s(\cdot)$. This immediately yields

$$\frac{\partial f_d^s(\mathbf{r}, \mathbf{v}, t)}{\partial t} = -\frac{1}{N_s} \sum_{i=1}^{N_s} \dot{\mathbf{r}}_i \cdot \nabla_\mathbf{r} \delta(\mathbf{r} - \mathbf{r}_i(t)) \delta(\mathbf{v} - \mathbf{v}_i(t))$$

$$-\frac{1}{N_s} \sum_{i=1}^{N_s} \dot{\mathbf{v}}_i \cdot \nabla_\mathbf{v} \delta(\mathbf{r} - \mathbf{r}_i(t)) \delta(\mathbf{v} - \mathbf{v}_i(t)), \qquad (13.11)$$

where $\nabla_\mathbf{r} = (\partial_x, \partial_y, \partial_z)$ and $\nabla_\mathbf{v} = (\partial_{v_x}, \partial_{v_y}, \partial_{v_z})$. The position $\mathbf{r}_i$ satisfies the trivial equation

$$\mathbf{v}_i = \dot{\mathbf{r}}_i \qquad (13.12)$$

and likewise the velocity of particles i obeys the Lorentz force equation

$$m_s \dot{\mathbf{v}}_i = q_s \left[\mathbf{E}(\mathbf{r}_i, t) + \mathbf{v}_i \times \mathbf{B}(\mathbf{r}_i, t) \right], \qquad (13.13)$$

where q_s is the charge of the particle, and $\mathbf{E}(\cdot)$ and $\mathbf{B}(\cdot)$ are the microscopic electric and magnetic field, respectively. These latter are in turn governed by the Maxwell equations (Jackson, 1975), namely

$$\nabla \cdot \mathbf{E}(\mathbf{r}, t) = \frac{\rho(\mathbf{r}, t)}{\epsilon_0} \tag{13.14}$$

$$\nabla \cdot \mathbf{B}(\mathbf{r}, t) = 0 \tag{13.15}$$

$$\nabla \times \mathbf{E}(\mathbf{r}, t) = -\frac{\partial \mathbf{B}(\mathbf{r}, t)}{\partial t} \tag{13.16}$$

$$\nabla \times \mathbf{B}(\mathbf{r}, t) = \mu_0 \mathbf{J}(\mathbf{r}, t) + \mu_0 \varepsilon_0 \frac{\partial \mathbf{E}(\mathbf{r}, t)}{\partial t}. \tag{13.17}$$

The description is completed by the microscopic charge density relation

$$\rho(\mathbf{r}, t) = \sum_s N_s q_s \int d\mathbf{v}\, f_d^s(\mathbf{r}, \mathbf{v}, t), \tag{13.18}$$

while the microscopic current is given by

$$\mathbf{J}(\mathbf{r}, t) = \sum_s N_s q_s \int d\mathbf{v}\, \mathbf{v} f_d^s(\mathbf{r}, \mathbf{v}, t). \tag{13.19}$$

Again, let us recall that, in the above relations, the index s refers to the different species of charged particles that populate the fluid. For plasma-based applications to which our attention is here devoted, ions (i) and electrons (e) will need to be solely considered. Hence, formally $s = i, e$.

By inserting Eq. (13.12) and Eq. (13.13) into Eq. (13.11), we eventually obtain

$$\frac{\partial f_d^s(\mathbf{r}, \mathbf{v}, t)}{\partial t} = -\frac{1}{N_s} \sum_{i=1}^{N_s} \mathbf{v}_i \cdot \nabla_\mathbf{r} \delta(\mathbf{r} - \mathbf{r}_i(t)) \delta(\mathbf{v} - \mathbf{v}_i(t)) \tag{13.20}$$

$$-\frac{1}{N_s} \sum_{i=1}^{N_s} \frac{q_s}{m_s} (\mathbf{E}(\mathbf{r}_i, t) + \mathbf{v}_i \times \mathbf{B}(\mathbf{r}_i, t)) \cdot \nabla_\mathbf{v} \delta(\mathbf{r} - \mathbf{r}_i(t)) \delta(\mathbf{v} - \mathbf{v}_i(t)).$$

Using the properties of the Dirac function yields

$$\frac{\partial f_d^s(\mathbf{r}, \mathbf{v}, t)}{\partial t} = -\frac{1}{N_s} (\mathbf{v} \cdot \nabla_\mathbf{r}) \sum_{i=1}^{N_s} \delta(\mathbf{r} - \mathbf{r}_i(t)) \delta(\mathbf{v} - \mathbf{v}_i(t)) \tag{13.21}$$

$$-\frac{q_s}{m_s} (\mathbf{E}(\mathbf{r}, t) + \mathbf{v} \times \mathbf{B}(\mathbf{r}, t)) \cdot \nabla_\mathbf{v} \sum_{i=1}^{N_s} \delta(\mathbf{r} - \mathbf{r}_i(t)) \delta(\mathbf{v} - \mathbf{v}_i(t))),$$

and recalling the definition of the density function $f_d^s(\cdot)$ the previous equation can be cast in the form

$$\frac{\partial f_d^s(\mathbf{r}, \mathbf{v}, t)}{\partial t} + \mathbf{v} \cdot \nabla_{\mathbf{r}} f_d^s(\mathbf{r}, \mathbf{v}, t) + \frac{q_s}{m_s} (\mathbf{E}(\mathbf{r}, t) + \mathbf{v} \times \mathbf{B}(\mathbf{r}, t)) \cdot \nabla_{\mathbf{v}} f_d^s(\mathbf{r}, \mathbf{v}, t) = 0. \quad (13.22)$$

This is the *Klimontovich equation* (Klimontovich *et al.*, 1967) which, together with the Maxwell equations, provides an exact and general description of any electromagnetic wave–particles system in mutual, self-consistent evolution.

The above equation needs to be complemented by proper initial conditions. By assigning initial particles position and velocity, we can clearly reconstruct the associated densities $f_d^s(\mathbf{r}, \mathbf{v}, t = 0)$. The initial fields are instead chosen to be consistent with the aforementioned Maxwell equations. The system is, hence, completely deterministic and both density and fields can be, in principle, traced as a function of time and in any position of the generalized phase space.

However, the amount of information embedded in the Klimontovich description is enormous. Equation (13.22) explicitly contains, in fact, the orbits of each individual microscopic entity belonging to the system under inspection. The interest of the formulation relies on that it allows for a straightforward simplification to which we will allude in the following and which makes it possible to obtain an approximated description of the relevant, collective degrees of freedom for what it concerns both the particles and the fields. The forthcoming discussion will mainly follow the excellent book by Nicholson (1983).

When looking at average quantities, we are primarily interested in knowing how many particles of type s can be found in a small volume $(\Delta \mathbf{r}, \Delta \mathbf{v})$ of phase space, positioned in $(\mathbf{r}, \mathbf{v})$. In other terms, it is tempting to invoke an appropriate ensemble average $\langle \cdot \rangle$. As explained in Chapter 8, this is an average over realizations of the system prepared according to assigned and reproducible prescriptions. Therefore, we focus attention on the smooth function

$$f_0^s(\mathbf{r}, \mathbf{v}, t) = \langle f_d^s(\mathbf{r}, \mathbf{v}, t) \rangle. \quad (13.23)$$

An equation for the time evolution of the distribution function $f_0^s(\mathbf{r}, \mathbf{v}, t)$ can be recovered from the Klimontovich Eq. (13.22) by this ensemble averaging. To this end we define the quantities δf^s, δE and δB as obeying the relations

$$f_d^s(\mathbf{r}, \mathbf{v}, t) = f_0^s(\mathbf{r}, \mathbf{v}, t)) + \frac{1}{\sqrt{N}} \delta f^s(\mathbf{r}, \mathbf{v}, t), \quad (13.24)$$

$$\mathbf{E}(\mathbf{r}, \mathbf{v}, t) = \mathbf{E}_0(\mathbf{r}, \mathbf{v}, t)) + \frac{1}{\sqrt{N}} \delta \mathbf{E}(\mathbf{r}, \mathbf{v}, t), \quad (13.25)$$

$$\mathbf{B}(\mathbf{r}, \mathbf{v}, t) = \mathbf{B}_0(\mathbf{r}, \mathbf{v}, t)) + \frac{1}{\sqrt{N}} \delta \mathbf{B}(\mathbf{r}, \mathbf{v}, t), \quad (13.26)$$

with $N = \sum_s N_s$. The index 0 labels the averaged quantities, namely $\mathbf{E}_0 = \langle \mathbf{E} \rangle$ and $\mathbf{B}_0 = \langle \mathbf{B} \rangle$, and the factor $1/\sqrt{N}$ takes into account the typical size of relative fluctuations. Inserting this definition in Eq. (13.22) and performing the ensemble averaging, we get

$$\frac{\partial f_0^s(\mathbf{r}, \mathbf{v}, t)}{\partial t} + \mathbf{v} \cdot \nabla_\mathbf{r} f_0^s(\mathbf{r}, \mathbf{v}, t) + \frac{q_s}{m_s} (\mathbf{E}_0(\mathbf{r}_i, t) + \mathbf{v} \times \mathbf{B}_0(\mathbf{r}, t)) \cdot \nabla_\mathbf{v} f_0^s(\mathbf{r}, \mathbf{v}, t)$$
$$= -\frac{1}{N} \frac{q_s}{m_s} \langle (\delta \mathbf{E} + \mathbf{v} \times \delta \mathbf{B}) \cdot \nabla_\mathbf{v} \delta f^s(\mathbf{r}, \mathbf{v}, t) \rangle. \quad (13.27)$$

The right-hand side of Eq. (13.27) is sensitive to the discrete nature of the fluid, while the left-hand side deals with collective variables. In the limit of large systems ($N \rightarrow \infty$), we can formally neglect the right-hand side and consequently obtain the Vlasov equation (Vlasov, 1945)

$$\frac{\partial f_0^s(\mathbf{r}, \mathbf{v}, t)}{\partial t} + \mathbf{v} \cdot \nabla_\mathbf{r} f_0^s(\mathbf{r}, \mathbf{v}, t) + \frac{q_s}{m_s} (\mathbf{E}_0(\mathbf{r}_i, t) + \mathbf{v} \times \mathbf{B}_0(\mathbf{r}, t)) \cdot \nabla_\mathbf{v} f_0^s(\mathbf{r}, \mathbf{v}, t) = 0, \quad (13.28)$$

also referred to as the collisionless Boltzmann equation, as discussed in Chapter 8. The ensemble averaged fields $\mathbf{E}_0$ and $\mathbf{B}_0$ must satisfy the ensemble averaged Maxwell equations, which we give here as

$$\nabla \cdot \mathbf{E}_0(\mathbf{r}, t) = \frac{\rho_0}{\varepsilon_0}, \quad (13.29)$$

$$\nabla \cdot \mathbf{B}_0(\mathbf{r}, t) = 0, \quad (13.30)$$

$$\nabla \times \mathbf{E}_0(\mathbf{r}, t) = -\frac{\partial \mathbf{B}_0(\mathbf{r}, t)}{\partial t}, \quad (13.31)$$

$$\nabla \times \mathbf{B}_0(\mathbf{r}, t) = \mu_0 \mathbf{J}_0(\mathbf{r}, t) + \mu_0 \varepsilon_0 \frac{\partial \mathbf{E}_0(\mathbf{r}, t)}{\partial t}, \quad (13.32)$$

with

$$\rho_0 = \sum_s N_s q_s \int d\mathbf{v} f_0^s(\mathbf{r}, \mathbf{v}, t), \quad (13.33)$$

$$\mathbf{J}_0(\mathbf{r}, t) = \sum_s N_s q_s \int d\mathbf{v} \, \mathbf{v} f_0^s(\mathbf{r}, \mathbf{v}, t). \quad (13.34)$$

The Vlasov equation arises naturally from the Klimontovich equation, as proven above. Alternatively, it can result from the BBGKY hierarchy when the role of collisions is neglected. For a comprehensive account on this derivation, the reader can consult, for example, the book by Nicholson (1983). In the following, we shall analyse in some details the properties of the Vlasov equation, focusing in particular on the case where two species, respectively electrons and ions, are to be considered. When the Vlasov equations, one for each species, are specified together with the Maxwell equations (13.29)–(13.32), we have a complete kinetic description of the behaviour of a plasma.

13.3   The Case of Electrostatic Waves

One of the most interesting applications of the Vlasov theory has to do with the existence of electrostatic waves that have only an electric field component and no magnetic field contribution.

Following Nicholson (1983), we imagine that each species is characterized by an initial equilibrium distribution $f^{s0}(\mathbf{v})$, which is a solution of the time-independent Vlasov equation for $\mathbf{E}_0 = \mathbf{B}_0 = 0$. From hereon, we will drop the (bottom) label 0, so far associated with ensemble averaged quantities. As we shall clarify below, an (upper) 0 will be assumed to indicate unperturbed quantities. We will in fact consider a small perturbation originated by the presence of a small–amplitude wave. In formulae, we require that

$$f^s(\mathbf{r}, \mathbf{v}, t) = f^{s0}(\mathbf{v}) + f^{s1}(\mathbf{r}, \mathbf{v}, t), \tag{13.35}$$

where f^{s1} stems for the aforementioned perturbation. For each species, the following normalization condition holds,

$$\int d\mathbf{v}\, f^{s0}(\mathbf{v}) = n_0, \tag{13.36}$$

where n_0 is the average number of particles per unit configuration space. Let us choose the electric field in the direction of the x axis and focus on waves which are spatially modulated just in the x direction. Then, the Vlasov equation (13.28) reads

$$\frac{\partial f^s}{\partial t} + v_x \partial_x f^s + \frac{q_s}{m_s} E \partial_{v_x} f^s = 0. \tag{13.37}$$

Inserting the ansatz (13.35) and retaining the first-order terms yields

$$\frac{\partial f^{s1}}{\partial t} + v_x \partial_x f^{s1} + \frac{q_s}{m_s} E \partial_{v_x} f^{s0} = 0. \tag{13.38}$$

Looking for plane wave solution proportional to $\exp(ikx - i\omega t)$ returns

$$-i\omega f^{s1} + ik v_x f^{s1} = -\frac{q_s}{m_s} E \partial_{v_x} f^{s0}, \tag{13.39}$$

which, solving for f^{s1}, gives

$$f^{s1}(\mathbf{r}, \mathbf{v}, t) = \frac{-i q_s/m_s}{\omega - k v_x} E \partial_{v_x} f^{s0}(\mathbf{v}). \tag{13.40}$$

The electrostatic field E is determined by the Maxwell equations. However, for electrostatic fields, the only relevant equation is the Poisson's one (13.14), which, in the present case, can be cast in the simple form

$$ikE = \frac{e(n_i - n_e)}{\varepsilon_0}, \tag{13.41}$$

where e is the electron charge; n_i and n_e are, respectively, the ion and electron number density given by

$$n_e = \int d\mathbf{v} f^e, \tag{13.42}$$

$$n_i = \int d\mathbf{v} f^i. \tag{13.43}$$

By manipulating the right-hand side of the Eq. (13.41) upon insertion of (13.42) and (13.43), we are eventually left with the dispersion relation

$$1 + \frac{\omega_e^2}{k^2} \int du \frac{d_u g(u)}{\omega/k - u} = 0, \tag{13.44}$$

where

$$\omega_e = (n_0 e^2 / m_e \varepsilon_0)^{1/2} \tag{13.45}$$

is the electron plasma frequency and the function $g(\cdot)$ is defined as

$$g(v_x) = \frac{m_e}{n_0 m_i} \int dv_y dv_x f^{i0}(\mathbf{v}) + \frac{1}{n_0} \int dv_y dv_x f^{e0}(\mathbf{v}). \tag{13.46}$$

Note that the ion component is damped by a factor m_e/m_i. Ignoring the ion motion implies working in the limit $m_i \to \infty$: This is equivalent to focusing on the dynamics of the electrons and assuming the ions to provide a static background. Under this condition, the expression (13.46) relates the frequency ω and the wave number k for a specific class of high-frequency electron waves, called Langmuir waves. In fact, the condition of sufficiently high frequency assures that the static hypothesis for the ions is satisfied.

Note that the dispersion relation (13.44) includes an integration over an integrand with a singularity in $u = \omega/k$. This latter needs to be handled with extreme care as we shall discuss in the forthcoming section. For the time being, we restrict ourselves to waves such that $\omega/k \gg u$ for all u for which $g(u)$ is appreciably different from 0. The dispersion relation (13.44) can be integrated by parts

$$1 - \frac{\omega_e^2}{k^2} \int_{-\infty}^{+\infty} du \frac{g(u)}{(\omega/k - u)^2} = 0, \tag{13.47}$$

where use has been made of the natural condition $g(u \to \pm\infty) = 0$. Expanding the denominator up to second-order terms in uk/ω we obtain the celebrated Langmuir wave dispersion relation

$$\omega^2 = \omega_e^2 + 3k^2 v_e^2, \qquad (13.48)$$

where $v_e^2 = \int du\, g(u) u^2$, and where we have exploited the normalization of $f^{e0}(\mathbf{v})$ and assumed it to be isotropic, so that $\int du\, g(u) u = 0$. We have further assumed that $k^2 v_e^2 \ll \omega_e^2$. Equation (13.48) is also called the Bohm–Gross dispersion relation (Bohm and Gross, 1949a,b).

In the next section, we shall expand our discussion on the dispersion relation (13.44) and show how the pole in $u = \omega/k$ can be properly dealt with.

13.4 Landau Damping

To carry out the calculation, and following Landau original idea (Landau, 1946), we need to analytically continue the function $g(u)$ to the entire complex plan $u = u_R + iu_I$. Formally, the Vlasov and Poisson equations can be solved in the context of an initial value problem. Taking the Laplace transform and choosing the appropriate contour for the (complex) integration (Nicholson, 1983), we are eventually brought to show that the response to an initial perturbation is characterized by *normal modes* which oscillate at the frequency given implicitly by Eq. (13.47). As discussed in the following, the singularity in the integrand materializes in a imaginary contribution for ω: after a transient, and for a fixed k value, the normal mode with the largest imaginary part dominates the dynamics. More concretely, the dispersion relation (13.47) can be written in the form

$$\varepsilon(k, \omega) = \varepsilon_R + i\varepsilon_I = 0, \qquad (13.49)$$

where $\varepsilon(k, \omega)$ is indeed the dielectric function

$$\varepsilon(k, \omega) = 1 - \frac{\omega_e^2}{k^2} \int_{-\infty}^{+\infty} du\, \frac{d_u g(u)}{\omega/k - u}. \qquad (13.50)$$

Denoting $\omega = \omega_R + i\omega_I$ and Taylor expanding about ω_R yields

$$\varepsilon_R(k, \omega_R) + i\varepsilon_I(k, \omega_R) + i\omega_I \frac{\partial \varepsilon_R(k, \omega)}{\partial \omega}\Big|_{\omega=\omega_R} = 0. \qquad (13.51)$$

In the above development, we have neglected the term proportional to $\omega_I \partial \varepsilon_I / \partial \omega$, which we motivate by assuming ω_I to be small. Under this working hypothesis, the omitted term is indeed a second-order contribution in ω_I since $\varepsilon_I \sim \omega_I$, as we shall see in a moment. Equating to 0 the real and imaginary contributions of (13.51) immediately yields

$$\varepsilon_R(k, \omega_R) = 0 \qquad (13.52)$$

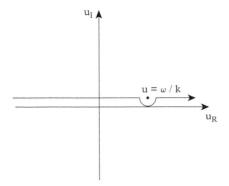

Figure 13.2 *Landau contour of integration used in evaluating the dielectric function when $\omega_I = 0$. The straight line portions of the contour fall exactly on the real axis.*

and

$$\omega_I = -\frac{\varepsilon_I(k, \omega_R)}{\partial \varepsilon_R(k, \omega)/\partial \omega\,|_{\omega=\omega_R}}, \tag{13.53}$$

which shows that $\varepsilon_I \sim \omega_I$. When $\omega_I = 0$, the complex integral is carried out along the contour depicted in Fig. 13.2.

This is generally referred to as the Landau contour. The Plemelj formula applied to the selected contour is

$$\frac{1}{u - a} = \mathcal{P}\left(\frac{1}{u - a}\right) + \pi i \delta(u - a), \tag{13.54}$$

where the symbol $\mathcal{P}$ stands for the principal value. Making use of the above formula in the definition of $\varepsilon(k, \omega_r)$ eventually yields

$$\varepsilon(k, \omega_r) = 1 - \frac{\omega_e^2}{k^2}\mathcal{P}\int_{-\infty}^{+\infty} du \frac{d_u g(u)}{u - \omega_r/k} - i\frac{\omega_e^2}{k^2}d_u g(u)\,|_{u=\omega_r/k}, \tag{13.55}$$

where for any function $f(u)$

$$\mathcal{P}\int_{-\infty}^{+\infty} du \frac{f(u)}{(u - a)} = \lim_{\varepsilon \to 0^+}\left[\int_{-\infty}^{a-\varepsilon} du \frac{f(u)}{(u - a)} + \int_{a+\varepsilon}^{+\infty} du \frac{f(u)}{(u - a)}\right]. \tag{13.56}$$

The complex term on the right-hand side of Eq. (13.55) comes from integrating around the semicircle in Fig. (13.2), which gives one half of $2\pi i$ times the associated residue. Recalling Eq. (13.52) and isolating the real part contribution in (13.55) returns

$$\varepsilon_r = 1 - \frac{\omega_e^2}{k^2}\mathcal{P}\int_{-\infty}^{+\infty} du \frac{d_u g(u)}{u - \omega_r/k} = 0. \tag{13.57}$$

The presence of the principal value allows us to assume, in the computation of this integral, that $d_u g(u) = 0$ at $u = \omega_r/k$. Then, integrating by parts yields

$$\omega_r^2 = \omega_e^2 + 3k^2 v_e^2, \tag{13.58}$$

where use has been made of the additional condition $k^2 v_e^2 \ll \omega_r^2$. We are now in the position to calculate ω_i by evaluating explicitly the derivative $\partial \varepsilon_r/\partial \omega$, via relation (13.57). After some algebraic manipulations and neglecting the second-order terms in $k^2 \lambda_e^2$, where $\lambda_e = v_e/\omega_e$, we obtain

$$\omega_i = \frac{\pi \omega_e^3}{2k^2} \, d_u g(u) \big|_{u=\omega_r/k}. \tag{13.59}$$

In conclusion, combining the above results, the total frequency of the Langmuir wave in the regime with $k\lambda_e \ll 1$ is

$$\omega = \omega_e \left(1 + \frac{3}{2} k^2 \lambda_e^2 \right) + i \frac{\pi \omega_e^3}{2k^2} d_u g(u) \big|_{u=\omega_r/k}, \tag{13.60}$$

and it is clearly affected by the unperturbed (equilibrium) distribution function, via $g(u)$ as defined in Eq. (13.46). In particular, when the slope of the distribution function is negative in correspondence of the phase velocity ω/k, then the Langmuir wave is Landau damped. The particles that exchange energy with the wave are those with velocities close to ω/k. The ones whose velocities are slightly faster lose their energy, transferring it to the wave. Conversely, the particles with speeds slower than the wave phase velocity are speeded up by the wave itself. On average, since the slope of the distribution function in ω/k is assumed to be negative, there are more particles which subtract energy to the wave than those that cede energy to it. This is, for instance, the case when $g(u)$ is a Maxwell distribution centred in 0, and recalling that $\omega/k > 0$ by definition. Following such energy exchange the particle distribution function gets locally flattened in the region where the interaction has occurred and the process comes eventually to a halt.

When the slope of the distribution is instead positive, the wave gets amplified following the interaction with the resonant particles. In this case, it is customary to talk of unstable modes. A paradigmatic example is the so-called 'bump-on-tail' instability, as represented in Fig. 13.3: a Maxwell equilibrium profile is perturbed by injecting a fast electron beam which distorts the right tail of the distribution without affecting the main bulk of it. Waves with their phase velocity in the region of positive slope of the perturbed velocity profile can become unstable and undergo an exponential growth in the linear regime of the evolution for small enough times. Later on, non-linear effects come into play and the wave amplitude saturates to an asymptotic profile. Providing a quantitative characterization of the non-linear evolution of the system is a challenging problem to which we shall return in the forthcoming sections and in the subsequent chapter.

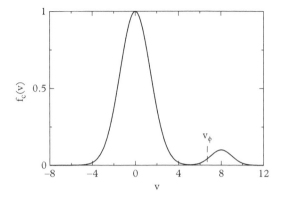

Figure 13.3 *The bump-on tail distribution: a fast electron beam is injected to modify the right tail of the electrons equilibrium distribution. The global distribution is here normalized to its peak value. Waves with phase velocity v_ϕ in the region of positive slope for $f(v)$ get unstable and grow exponentially.*

13.5   Non-linear Landau Damping: An Heuristic Approach

In the previous section, we discussed the linear Landau damping, showing that, under specific condition, the Langmuir waves can be damped (or amplified) due to a self-consistent interaction with the particles. The earlier treatment holds as long as the waves are sufficiently small so that the linear approximation can be invoked. In the following, we will instead shortly discuss the consequences of dealing with finite waves (Nicholson, 1983). For simplicity, let us assume the wave to take the form

$$E(x) = E_0(t)\sin(kx),\tag{13.61}$$

where we chose to operate in a reference frame which is comoving with the wave at the phase velocity $v_\phi = \omega_r/k$. The corresponding electrostatic potential reads

$$\phi(x) = \frac{E_0}{k}\cos(kx).\tag{13.62}$$

In the wave reference frame, ignoring the slow time dependence of $E_0(t)$, i.e. setting $E_0(t) = E_0 = $ const, every single electron feels a time-independent electric field so that its total energy is

$$H = -e\phi(x) + \frac{1}{2}mv^2,\tag{13.63}$$

i.e. a constant of motion when v is measured in the wave reference frame. The corresponding equation of motion takes the form

$$m\ddot{x} = -eE_0\sin(kx).\tag{13.64}$$

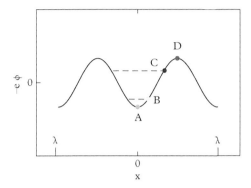

Figure 13.4 *The potential well for the electrons in a Langmuir wave. Depending on their locations particles can display trapped or untrapped dynamics.*

In other words, the particles are moving in the periodic potential well $-e\phi$.

The aim of this section is to shed light on the single-particle dynamics and consequently to learn about the onset of the wave–particles instability and the relevant timescales involved.

Consider at $t = 0$ all particles that have $v = 0$. Those particles have different energies, depending on their specific location on the line. More specifically, $H = -e\phi(x(t = 0))$, as illustrated in Fig. 13.4.

Particles that happen to be in position A, i.e. at the bottom of the well, do not move. Particles in B tends to approach the bottom of the well, while particle in D are marginally stable. In the vicinity of the minimum of the potential, the particles (B in Fig. 13.4) oscillate with a characteristic frequency, the bounce frequency ω_b, which, from Eq. (13.64) can be readily shown to be

$$\omega_b = \left(\frac{eE_0k}{m}\right)^{1/2}. \tag{13.65}$$

Linear Landau damping as derived above relies on the working hypothesis that only small perturbations in the particle motion occur. In light of relation Eq. (13.65), we can therefore expect the linear calculation to be valid for times $t \ll \omega_b^{-1}$. Let us now briefly discuss the associated phase-space dynamics. Each individual particle evolves on a one-dimensional, pendulum–like track, keeping the energy unchanged. Two families of particles are evidenced: those falling within the *cat's eye* (the phase-space region bounded by the two separatrices) are called trapped because their orbits are permanently confined within the original wavelength. Conversely, the particles populating the outer region, beyond the domain where the resonance extends, are *untrapped*. The half-width in velocity v_t of the resonance can be estimated by the condition

$$\frac{1}{2}mv_t^2 = 2\,|\,e\phi_{\text{max}}\,|, \tag{13.66}$$

where $\phi_{\mathrm{max}} = E_0$ is the maximum of the potential. A straightforward calculation yields

$$v_t = 2\omega_b/k. \tag{13.67}$$

Note, however, that this scenario holds as long as we neglect the self-consistent action of the particles on the waves that seed the trapping potential. Indeed, after a substantial fraction of the trapping period, the trapped particles are smeared out around their phase-space orbits and the linear Landau damping regime fades off. The particles interested in the interaction are approximately those that have velocities v in the range $v_\phi - v_t < v < v_\phi + v_t$.

Based on the above, we can draw some conclusive remarks. Label γ_L the Landau damping rate, namely the imaginary component of ω as estimated in the previous section. Assume that $\gamma_L \gg \omega_b$: then the system undergoes the classical linear damping for time smaller than $1/\omega_b$. When $t \simeq \pi/\omega_b$, the trapped particles have gone through half a bounce, as illustrated earlier. While they initiate the second half of the complete rotation, they can return to the wave part of the energy that they initially gained. This reversal of the energy is, however, not complete as the particles are now out of phase with each other. The wave amplitude experiences hence damped oscillations, while the particles' velocity distribution gets locally flattened in the region of the interaction. This picture can be made quantitative by analysing, via combined numerical and analytical means, relevant mathematical models which preserve the self-consistent nature of the Hamiltonian interaction (O'Neil *et al.*, 1971; Elskens and Escande, 2002). The forthcoming chapter is devoted to discussing this important features. For very large times instead, the phases of the electron are completely mixed and the wave amplitude relaxes to a constant value. This regime corresponds to an important class of non-linear waves, the BGK modes, which we will introduce in the next section.

To go beyond this heuristic approach, it is important to mention the recent seminal work by Mouhot and Villani (2011). The essence of their result is the demonstration of asymptotic convergence of an initial distribution to a limiting one. More precisely, considering an initial distribution which is Vlasov stable and adding a *finite* perturbation to it, they prove that the solution of the non-linear Vlasov equation converges to an asymptotic distribution at both $t \to +\infty$ and $-\infty$. The marginal density of this distribution is shown to converge towards the asymptotic density exponentially fast.

13.6   The Asymptotic Evolution: BGK Modes

We will discuss here a special class of non-linear waves, the BGK modes, named after their inventors Bernstein, Greene and Kruskal (Bernstein *et al.*, 1957). To this end we shall need to consider a Vlasov equilibrium distribution in presence of a spatially varying electrostatic potential. Let us start by introducing the convective derivative D/Dt in the 6-dimensional phase space as

$$\frac{D}{Dt} \equiv \frac{\partial}{\partial t} + \mathbf{v} \cdot \nabla_{\mathbf{x}} + \dot{\mathbf{v}} \cdot \nabla_{\mathbf{v}}. \tag{13.68}$$

We can then simply write the Vlasov equation as

$$\frac{Df^s}{Dt} = 0. \tag{13.69}$$

Suppose now we construct a particle distribution function $f^s(\cdot)$ for the species s which can be expressed as a function of selected constants of motions, here termed $C_j(\mathbf{x}, \mathbf{v}, t)$. Then, the following relations apply

$$\frac{D}{Dt} f^s(\cdots, C_j(\mathbf{x}, \mathbf{v}, t), \cdots) = \sum_j \frac{\partial f^s}{\partial C_j} \frac{D}{Dt} C_j = 0, \tag{13.70}$$

which in turn implies that $f^s(\cdots, C_j(\mathbf{x}, \mathbf{v}, t), \cdots)$ is a Vlasov solution. Back to our original problem, imagine that the electrons are subject to an electric field $E(x) = -d\phi(x)/dx$ oriented along the x direction. Then, the particles' constants of motion are the momenta $m_s v_y$ and $m_s v_z$ and the energy $m_s v_x^2/2 + q_s \phi(x)$. Based on the earlier discussion, any function of the type $f(v_x^2 + 2q_s\phi(x)/m_s, v_y, v_z)$ is an equilibrium distribution function. A BGK mode is characterized by such a distribution with the electric potential self-consistently produced by the very same distribution, via the Poisson equation. For the sake of simplicity, following Nicholson (1983), we consider the time-independent setting, writing the Poisson equation as

$$\frac{\partial^2 \phi}{\partial x^2} = \frac{e}{\varepsilon_0} \left[\int_{-\infty}^{+\infty} dv_x dv_y dv_z \, f^e(v_x^2 - 2e\phi(x)/m_e, v_y, v_z) \right.$$
$$\left. - \int_{-\infty}^{+\infty} dv_x dv_y dv_z \, f^i(v_x^2 + 2e\phi(x)/m_i, v_y, v_z) \right]. \tag{13.71}$$

Carrying out the integration over v_y and v_z and labelling $v = v_x$, we get

$$\frac{\partial^2 \phi}{\partial x^2} = \frac{e}{\varepsilon_0} \int_{-\infty}^{+\infty} dv \left[f^e(v^2 - 2e\phi(x)/m_e) - f^i(v^2 + 2e\phi(x)/m_i) \right]. \tag{13.72}$$

Solving this integro-differential equation for $\phi(x)$ subject to proper boundary conditions returns a closed solution of the inspected problem.

To progress with the analytical study, we assume that both ions and electrons are initially distributed in cold beams. In practice, we require that each particle of species s has the same speed in a given spatial position. Mathematically, we impose that the electrons obey

$$f^e(x, v) = 2n_0 v_e \, \delta[v^2 - 2e\phi(x)/m_e - v_e^2]. \tag{13.73}$$

Recall now the general property

$$\delta[f(y)] = \frac{\delta(y - y_0)}{\left|\frac{df}{dy}\right|_{y=y_0}}, \tag{13.74}$$

where y_0 is a solution of $f(y_0) = 0$. Then, when only the positive root inside the delta function is selected, Eq. (13.73) becomes

$$f^e(x, v) = n_0 \frac{v_e}{v} \delta(v - \tilde{v}_e), \tag{13.75}$$

where $\tilde{v}_e = \sqrt{v_e^2 + 2e\phi(x)/m_e}$. Analogously

$$f^i(x, v) = n_0 \frac{v_i}{v} \delta(v - \tilde{v}_i) \tag{13.76}$$

with $\tilde{v}_i = \sqrt{v_i^2 - 2e\phi(x)/m_i}$. Here v_e and v_i are positive constants that we choose large enough so that $\tilde{v}_e$ and $\tilde{v}_i$ are real positive quantities. Note that we have chosen the same normalization constant n_0 in both (13.75) and (13.76), an assumption that implies dealing with an overall neutral plasma.

Once the initial condition is being set, we look for spatially periodic solutions of the Poisson equation (13.72). The latter can be cast in the compact and equivalent form

$$\frac{\partial^2 \phi}{\partial x^2} = -\frac{\partial}{\partial \phi} V(\phi), \tag{13.77}$$

where the *pseudo-potential* $V(\phi)$ is defined by

$$V(\phi) = -\frac{n_0}{\varepsilon_0} \left[m_e v_e^2 \left(1 + \frac{2e\phi(x)}{m_e v_e^2} \right)^{1/2} + m_i v_i^2 \left(1 - \frac{2e\phi(x)}{m_i v_i^2} \right)^{1/2} \right]. \tag{13.78}$$

In obtaining this result, use has been made of Eqs. (13.75) and (13.76): the dependence on $\tilde{v}_e$ (resp. $\tilde{v}_i$) has been removed in favour of v_e (resp. v_i).

Following Nicholson (1983), we assume as a specific example $m_e v_e^2 = m_i v_i^2 \equiv k_B T$. Then, the pseudo-potential becomes

$$V(\phi) = -\frac{n_0}{\varepsilon_0} k_B T \left[\left(1 + \frac{2e\phi(x)}{k_B T} \right)^{1/2} + \left(1 - \frac{2e\phi(x)}{k_B T} \right)^{1/2} \right]. \tag{13.79}$$

The above function is plotted versus ϕ in Fig. 13.5.

With an initial choice of the *energy* between $-(2)^{1/2} n_0 k_B T/\varepsilon_0$ and $-2n_0 k_B T/\varepsilon_0$ the particle oscillates indefinitely in the well, resulting in a spatially periodic potential that oscillates between $-\phi_0$ and $+\phi_0$.

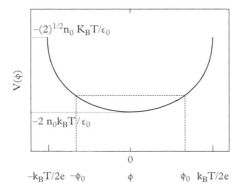

Figure 13.5 *The pseudo–potential (13.79) which arises when studying the BGK modes, under the simplifying hypothesis $m_e v_e^2 = m_i v_i^2 \equiv k_B T$.*

In the limit of very small ϕ_0, we can expand the square roots in Eq. (13.79) to get for Eq. (13.77)

$$\frac{\partial^2 \phi}{\partial x^2} + \frac{2n_0 e^2}{\varepsilon_0 k_B T} \phi = 0, \tag{13.80}$$

whose solution reads

$$\phi(x) = \phi_0 \sin(\sqrt{2}\, x/\lambda_e), \tag{13.81}$$

where $\lambda_e = v_e/\omega_e = \sqrt{\varepsilon_0 k_B T/n_0 e^2}$ is the Debye length. Note that in the linearized version (13.81) the function $\phi(x)$ is sinusoidal, while this is not the case in the general setting considered earlier.

In practice, we have here proved that, within a self-consistent picture where particles influence the fields and vice versa, there is a spatially periodic potential. The ion beam gets accelerated through regions of large negative potential, and thus reduces its density. Conversely the electrons are decelerated when they meet a region of negative potential. The net result is clearly a negative net charge in regions of large negative potential. The opposite holds in regions where the potential is instead positive. The important point is that this process works not only in the linear regime (see Eq. (13.81)) but also in the non-linear one, as governed by Eq. (13.77). Interestingly, it turns out that almost any potential $\phi(x)$ can be constructed by choosing appropriate distributions of (un-)trapped electrons and ions.

There are many practical applications of the BGK modes, including the non-linear stage of the Landau-damped Langmuir waves, theories of shock waves and double layers and many more domains which fall outside the scope of this book. In the following section, we will instead turn to discussing another approach to the study of waves in a plasma, which yields the celebrated Case–Van Kampen modes.

13.7   Case–Van Kampen Modes

In this section, we shall follow a different strategy for investigating the emergence of collective waves in a plasma. As opposed to the procedure previously adopted in Section 13.3, we will here eliminate the electric field E and look for a normal mode solution of the perturbed distribution f^1. Recall that the one-dimensional Vlasov–Poisson system is

$$\partial_t f + v \partial_x f - \frac{eE}{m_e} \partial_v f = 0, \tag{13.82}$$

$$\partial_x E = \frac{e}{\varepsilon_0} \left[n_0 - \int_{-\infty}^{+\infty} dv\, f(v) \right], \tag{13.83}$$

where we now focus for simplicity on a single species (the electrons) and have therefore dropped the redundant label s. In analogy with the preceding discussion, we have also replaced v_x with v. Linearizing the equations and posing $f^0(v) = n_0 g(v)$, we get

$$\partial_t f^1 + v \partial_x f^1 = \frac{n_0 eE}{m_e} \partial_v g, \tag{13.84}$$

$$\partial_x E = \frac{e}{\varepsilon_0} \left[-\int_{-\infty}^{+\infty} dv\, f^1(v) \right]. \tag{13.85}$$

The idea is now to look for normal modes of f^1 proportional to the factor $\exp(-i\omega t + ikx)$. This implies assuming that

$$f^1 = \tilde{f^1}(v) \exp(-i\omega t + ikx), \tag{13.86}$$

$$E = \tilde{E}_0 \exp(-i\omega t + ikx), \tag{13.87}$$

for f^1 and E. Plugging into Eqs. (13.84) and (13.85), we obtain

$$(-i\omega + ikv)\tilde{f^1} = \frac{n_0 eE_0}{m_e} \partial_v g, \tag{13.88}$$

$$E_0 = \frac{-e/\varepsilon_0}{ik} \int_{-\infty}^{+\infty} dv\, \tilde{f^1}. \tag{13.89}$$

By eliminating E_0, we get the integral equation for $\tilde{f^1}$,

$$\left(v - \frac{\omega}{k} \right) \tilde{f^1} = \eta(v) \int_{-\infty}^{+\infty} dv\, \tilde{f^1}, \tag{13.90}$$

where $\eta \equiv (\omega_e/k)^2 \partial_v g$. The solution of Eq. (13.90) can be explicitly written out as

$$\tilde{f^1} = \mathcal{P}\left[\frac{\eta(v)}{v - \omega/k} \right] + \delta(v - \omega/k)\left[1 - \mathcal{P}\int_{-\infty}^{+\infty} dv\, \frac{\eta(v)}{v - \omega/k} \right]. \tag{13.91}$$

Equation (13.91), together with Eq. (13.86), characterizes the Case–Van Kampen modes (Van Kampen, 1955; Case, 1959).

For any value of k, there exists a infinite number of normal modes, one for each value of real ω. This is at variance with the conclusion reached previously where only one normal mode was found for each value of (positive) k. Moreover, as a further distinctive point, the normal modes are not damped, but exist for all times with real frequency ω.

The importance of the modes relies in that they constitute a basis over which one can expand any arbitrary disturbance. Working at fixed k and invoking the linearity of the linearized Vlasov–Poisson system, it is immediately clear that the function

$$f^1(x, v, t) = \exp(ikx) \int_{-\infty}^{+\infty} d\omega \, \exp(-i\omega t) \tilde{f}^1(v, \omega) c(\omega) \qquad (13.92)$$

is a general solution of Eq. (13.84), where ω in $\tilde{f}^1(v, \omega)$ denotes the particular normal mode in Eq. (13.91). Here, $c(\omega)$ is an arbitrary weighting function, which must be chosen so to recover the assigned initial condition at $t = 0$. Using in Eq. (13.92) the solution (13.91) and performing the calculation yields

$$f^1(x, v, t) = e^{ikx} \eta(v) \mathcal{P} \int_{-\infty}^{+\infty} d\omega \frac{e^{-i\omega t} c(\omega)}{v - \omega/k} + e^{ikx} k \, e^{-ikvt} c(\omega = kv)$$
$$- k e^{ikx - ikvt} c(\omega = kv) \mathcal{P} \int_{-\infty}^{+\infty} dv' \frac{\eta(v')}{v' - v}. \qquad (13.93)$$

The electric field can be readily calculated by recalling, from the Poisson equation, that

$$E(x, t) = \frac{-e/\varepsilon_0}{ik} \int_{-\infty}^{+\infty} dv f^1(x, v, t). \qquad (13.94)$$

Inserting (13.92) into (13.94) and carrying out the integration, we are left with

$$E(x, t) = \frac{e}{\varepsilon_0} i \exp(ikx) \int_{-\infty}^{+\infty} dv \, \exp(-ikvt) c(\omega = kv). \qquad (13.95)$$

It can be shown that the latter is in complete agreement with the solution obtained in Sections 13.3 and 13.4, provided the correct choice for the function $c(\omega)$ is being made. On a more fundamental level, it is the self-consistent electric field $E(x, t)$ which selects the slowly Landau-damped normal modes of the system at late times, by picking up the appropriate solution among the family of admissible ones, after initial transient dynamics has died out.

14

Wave–Particles Interaction

As discussed in details in the preceding chapter, the kinetic theory approach to the study of wave–particles interactions relies on the exact Klimontovich equations. Working within this setting, we obtain the Vlasov equation for the time evolution of the single particle distribution function $f(\mathbf{r}, \mathbf{v}, t)$, while the coupled fields obey the Maxwell equations. The same conclusion is reached when operating in the context of a purely Hamiltonian perspective, as pioneered by Bachelard *et al.* (2008b). In the following, we shall review the main steps of the derivation, and so recover once again the self-consistent Vlasov–Maxwell equations. We will then specialize on one-dimensional systems and obtain a particularly important model that finds applications in distinct domains to which we shall allude in the forthcoming discussion.

14.1 Hamiltonian Formulation of Vlasov–Maxwell Equations

Following Bachelard *et al.* (2008b), we recall that the interaction between electromagnetic fields and charged particles (of mass $m = 1$ and charge $e = 1$) is specified by the Hamiltonian (Morrison, 1980; Marsden and Weinstein, 1982; Bianlynicki-Birula *et al.*, 1984)

$$H = \iint d^3q d^3p f(\mathbf{q}, \mathbf{p}) \sqrt{1 + \mathbf{p}^2} + \int d^3q \frac{|\mathbf{E}(\mathbf{q})|^2 + |\mathbf{B}(\mathbf{q})|^2}{2}. \tag{14.1}$$

where for convenience the speed of light c and the vacuum permittivity ε_0 have been put equal to 1. We made this choice not only to stick to the notations of the original paper, but also since it simplifies the analytical derivation.

The two contributions refer to the sum of the kinetic energy of the particles and the energy of the fields. Here, $f(\mathbf{q}, \mathbf{p})$ is the distribution of particles in phase space. Define the Poisson bracket

$$\{F, G\} = \iint d^3q d^3p f \left[\frac{\partial}{\partial \mathbf{p}} \frac{\delta F}{\delta f} \cdot \frac{\partial}{\partial \mathbf{q}} \frac{\delta G}{\delta f} - \frac{\partial}{\partial \mathbf{q}} \frac{\delta F}{\delta f} \cdot \frac{\partial}{\partial \mathbf{p}} \frac{\delta G}{\delta f} \right]$$

Physics of Long-Range Interacting Systems. First Edition. A. Campa *et al.*
© A. Campa, T. Dauxois, D. Fanelli, and S. Ruffo 2014. Published in 2014 by Oxford University Press.

$$-\iint d^3q d^3p\, f\mathbf{B} \cdot \left[\frac{\partial}{\partial \mathbf{p}}\frac{\delta F}{\delta f} \times \frac{\partial}{\partial \mathbf{p}}\frac{\delta G}{\delta f}\right]$$

$$+\iint d^3q d^3p \left[\frac{\delta F}{\delta f}\frac{\partial f}{\partial \mathbf{p}}\cdot\frac{\delta G}{\delta \mathbf{E}} - \frac{\delta G}{\delta f}\frac{\partial f}{\partial \mathbf{p}}\cdot\frac{\delta F}{\delta \mathbf{E}}\right]$$

$$+\int d^3q \left[\left(\nabla \times \frac{\delta F}{\delta \mathbf{B}}\right)\cdot\frac{\delta G}{\delta \mathbf{E}} - \frac{\delta F}{\delta \mathbf{E}}\cdot\left(\nabla \times \frac{\delta G}{\delta \mathbf{B}}\right)\right], \qquad (14.2)$$

where the functionals $F[f(\mathbf{q},\mathbf{p}),\mathbf{E}(\mathbf{q}),\mathbf{B}(\mathbf{q})]$ and $G[f(\mathbf{q},\mathbf{p}),\mathbf{E}(\mathbf{q}),\mathbf{B}(\mathbf{q})]$ play the role of two generic observables. Then by making use of Hamiltonian (14.1) the dynamical evolution of a generic observable F is ruled by

$$\frac{dF}{dt} = \{H,F\}. \qquad (14.3)$$

In particular, one obtains the Maxwell equations for the fields $\mathbf{E}$ and $\mathbf{B}$ and the Vlasov equation for the distribution function f, namely

$$\frac{\partial f}{\partial t} = \{H,f\} = -\mathbf{v}\cdot\nabla f - (\mathbf{E}+\mathbf{v}\times\mathbf{B})\cdot\frac{\partial f}{\partial \mathbf{p}}, \qquad (14.4)$$

$$\frac{\partial \mathbf{E}}{\partial t} = \{H,\mathbf{E}\} = \nabla \times \mathbf{B} - \int d^3p\,\mathbf{v} f, \qquad (14.5)$$

$$\frac{\partial \mathbf{B}}{\partial t} = \{H,\mathbf{B}\} = -\nabla \times \mathbf{E}, \qquad (14.6)$$

where

$$\mathbf{v} = \frac{\mathbf{p}}{\sqrt{1+\mathbf{p}^2}} \qquad (14.7)$$

is the particles' velocity.

14.2 Interaction between a Plane Wave and a Co-propagating Beam of Particles

Starting from the above setting and following the derivation presented in Bachelard *et al.* (2008b), we eventually end up with a compact formulation for the problem at hand. In the following, we will reproduce the main steps of the derivation. For further details, the interested reader can refer to the original paper where these results were discussed for the first time.

14.2.1 Canonical formulation of the Hamiltonian and reduction to a one-dimensional system

It is convenient to obtain the Vlasov–Maxwell equations (14.4)–(14.6) by replacing the fields with the corresponding potentials (Marsden and Weinstein, 1982). The observables are now functionals $F[f_A(\mathbf{q}, \mathbf{p}), \mathbf{A}(\mathbf{q}), \mathbf{Y}(\mathbf{q})]$, and the Hamiltonian and the bracket, respectively, become

$$
H = \iint d^3q \, d^3p \, f_A(\mathbf{q}, \mathbf{p}) \sqrt{1 + (\mathbf{p} - \mathbf{A})^2} + \int d^3q \frac{|\mathbf{Y}|^2 + |\nabla \times \mathbf{A}|^2}{2}, \quad (14.8)
$$

$$
\{F, G\} = \iint d^3q \, d^3p \, f_A \left[\frac{\partial}{\partial \mathbf{p}} \frac{\delta F}{\delta f_A} \cdot \frac{\partial}{\partial \mathbf{q}} \frac{\delta G}{\delta f_A} - \frac{\partial}{\partial \mathbf{q}} \frac{\delta F}{\delta f_A} \cdot \frac{\partial}{\partial \mathbf{p}} \frac{\delta G}{\delta f_A} \right]
$$
$$
+ \int d^3q \left(\frac{\delta F}{\delta \mathbf{Y}} \cdot \frac{\delta G}{\delta \mathbf{A}} - \frac{\delta F}{\delta \mathbf{A}} \cdot \frac{\delta G}{\delta \mathbf{Y}} \right). \quad (14.9)
$$

These latter can be obtained from Eqs. (14.1) and (14.2) by performing the change of coordinates

$$
f(\mathbf{q}, \mathbf{p}) = f_A(\mathbf{q}, \mathbf{p} + \mathbf{A}), \quad (14.10)
$$
$$
\mathbf{E} = -\mathbf{Y}, \quad (14.11)
$$
$$
\mathbf{B} = \nabla \times \mathbf{A}. \quad (14.12)
$$

We note that, this time, the Poisson bracket is canonical but there is no term that couples the particles and the field. However, the coupling is in the Hamiltonian (14.8).

We then shift the potential vector by a static quantity $\mathbf{A}_w(\mathbf{q})$, which we assume to be imposed externally; this is, for instance, the case of an undulator, as we will see later when discussing the free electron laser. This operation is a canonical transformation, and the bracket (14.9) is therefore not changed. The new Hamiltonian reads

$$
H = \iint d^3q \, d^3p \, f_A \sqrt{1 + (\mathbf{p} - \mathbf{A}_w - \mathbf{A})^2}
$$
$$
+ \int d^3q \frac{|\mathbf{Y}|^2 + 2\nabla \times \mathbf{A}_w \cdot \nabla \times \mathbf{A} + |\nabla \times \mathbf{A}|^2}{2}, \quad (14.13)
$$

where we have legitimately dropped the constant quantity $\int d^3q |\nabla \times \mathbf{A}_w|^2/2$.

We now suppose to have a long cavity, elongated in the z direction, where the electrons and the electromagnetic wave interact. For a wave co-propagating with the electrons in the z direction, we can define the k-mode of the wave as

$$
\mathbf{A}_k(\mathbf{q}_\perp) = \frac{1}{L} \int dz \, e^{-ikz} \mathbf{A}(\mathbf{q}), \quad (14.14)
$$

$$
\mathbf{Y}_k(\mathbf{q}_\perp) = \frac{1}{L} \int dz \, e^{-ikz} \mathbf{Y}(\mathbf{q}), \quad (14.15)
$$

in which L is the length of the cavity where the interaction takes place. The above expressions are inverted to give the Fourier expansion in the propagation direction

$$\mathbf{Y}(\mathbf{q}) = \sum_k \mathbf{Y}_k(\mathbf{q}_\perp) e^{ikz} \qquad \text{and} \qquad \mathbf{A}(\mathbf{q}) = \sum_k \mathbf{A}_k(\mathbf{q}_\perp) e^{ikz}, \tag{14.16}$$

where the discrete sum is extended on the allowed wavevectors in the finite cavity, $k = 2n\pi/L$ for $n = 0, \pm1, \pm2, \ldots$. Consider then the *paraxial approximation*. This choice, customarily adopted in the literature, implies neglecting the spatial variations along the transversal x and y directions. The field is, hence, homogeneous in the section S, is transverse to the direction of interaction and vanishes outside the cavity. Moreover, we restrict to a *monochromatic* wave, by solely taking into account one Fourier mode k, in the propagation direction, and consider the case of a *circularly polarized* radiated wave. Other modes can be included in the derivation by proceeding in an analogous fashion. The extension to higher modes will be introduced in the second part of the chapter. Thanks to the above simplifications, we write $\mathbf{A}$ and $\mathbf{Y}$ as

$$\mathbf{A} = -\frac{i}{\sqrt{2}} \left[a e^{ikz} \, \hat{\mathbf{e}} - a^* e^{-ikz} \, \hat{\mathbf{e}}^* \right], \tag{14.17}$$

$$\mathbf{Y} = -\frac{k}{\sqrt{2}} \left[a e^{ikz} \, \hat{\mathbf{e}} + a^* e^{-ikz} \, \hat{\mathbf{e}}^* \right], \tag{14.18}$$

with $\hat{\mathbf{e}} = (\hat{\mathbf{x}} + i\hat{\mathbf{y}})/\sqrt{2}$. Equivalently, the complex field amplitude a can be cast in the form

$$a = \frac{1}{kV} \int d^3q \, e^{-ikz} (-\mathbf{Y} + ik\mathbf{A}) \cdot \hat{\mathbf{e}}^* \tag{14.19}$$

$$= \frac{1}{kS} \int d^2\mathbf{q}_\perp (-\mathbf{Y}_k + ik\mathbf{A}_k) \cdot \hat{\mathbf{e}}^* \tag{14.20}$$

$$= \frac{1}{k} (-\mathbf{Y}_k + ik\mathbf{A}_k) \cdot \hat{\mathbf{e}}^*, \tag{14.21}$$

where $V = LS$ quantifies the volume of the interaction domain.

Assume now that the external field $\mathbf{A}_w$, e.g. created by an undulator as discussed later, only depends on the longitudinal variable z. Then, if the beam of charged particles is injected along the propagation axis with transverse velocity (14.7) equal to $\mathbf{A}_w/\sqrt{1 + |\mathbf{A}_w|^2}$, it keeps this specific value all along the evolution. The motion of the system can be exactly turned into a one-dimensional system, meaning that the distribution function depends on the longitudinal coordinate z and the associated moment p_z. Denoting with $\tilde{f}(z, p_z)$ this distribution function, the bracket reduces to

$$\{F, G\} = \iint dz dp_z \tilde{f}(z, p_z) \left[\frac{\partial}{\partial p_z} \frac{\delta F}{\delta \tilde{f}} \frac{\partial}{\partial z} \frac{\delta G}{\delta \tilde{f}} - \frac{\partial}{\partial z} \frac{\delta F}{\delta \tilde{f}} \frac{\partial}{\partial p_z} \frac{\delta G}{\delta \tilde{f}} \right] + \frac{i}{kV} \left(\frac{\partial F}{\partial a} \frac{\partial G}{\partial a^*} - \frac{\partial F}{\partial a^*} \frac{\partial G}{\partial a} \right), \tag{14.22}$$

and the Hamiltonian becomes

$$H = \iint dz dp_z \, \tilde{f}(z, p_z) \sqrt{1 + p_z^2 + aa^* - i\sqrt{2}(ae^{ikz}\hat{\mathbf{e}} - a^* e^{-ikz}\hat{\mathbf{e}}^*) \cdot \mathbf{A}_w + |\mathbf{A}_w|^2}$$

$$+ k^2 V aa^* - \frac{ikS}{\sqrt{2}} \int dz (ae^{ikz}\hat{\mathbf{e}} - a^* e^{-ikz}\hat{\mathbf{e}}^*) \cdot (\nabla \times \mathbf{A}_w), \tag{14.23}$$

where $k^2 V aa^*$ is the energy associated with the fields $\mathbf{A}$ and $\mathbf{Y}$. Since the motion is now one-dimensional, in the following we drop the label z from the momentum.

14.2.2   Studying the dynamics in the particles–field phase frame

Let us start by assuming a specific form for the $\mathbf{A}_w$ field, which we will cast in the form

$$\mathbf{A}_w = \frac{a_w}{\sqrt{2}} \left(e^{-ik_w z} \, \hat{\mathbf{e}} + e^{ik_w z} \, \hat{\mathbf{e}}^* \right). \tag{14.24}$$

As we will discuss later on, this latter expression is indeed the appropriate one when the interaction is assumed to occur in an undulator, as is the case for a free electron laser.

Then, we neglect the effects of finite sizes, and thus the last term in Eq. (14.23) vanishes since it is a sum of terms proportional to $\int dz \exp[\pm i(k + k_w)z]$. The Hamiltonian therefore becomes

$$H = \iint dz dp \tilde{f} \sqrt{1 + p^2 + aa^* - ia_w(ae^{i(k+k_w)z} - a^* e^{-i((k+k_w)z)}) + a_w^2} + k^2 V aa^*. \tag{14.25}$$

Interestingly, time is not a variable for this model, and we can therefore extend the phase space by adding a new pair of canonically conjugate variables, one being assimilated to time. More precisely, we define the pair of conjugate variables (τ, E), such that the new Hamiltonian and bracket respectively read

$$H_{ext}[\tilde{f}, a, a^*, E, \tau] = H[\tilde{f}, a, a^*] + E, \tag{14.26}$$

and

$$\{F, G\} = \iint dz dp \tilde{f} \left[\frac{\partial}{\partial p} \frac{\delta F}{\delta \tilde{f}} \frac{\partial}{\partial z} \frac{\delta G}{\delta \tilde{f}} - \frac{\partial}{\partial z} \frac{\delta F}{\delta \tilde{f}} \frac{\partial}{\partial p} \frac{\delta G}{\delta \tilde{f}} \right] + \frac{i}{kV} \left(\frac{\partial F}{\partial a} \frac{\partial G}{\partial a^*} - \frac{\partial F}{\partial a^*} \frac{\partial G}{\partial a} \right)$$

$$+ \frac{\partial F}{\partial E} \frac{\partial G}{\partial \tau} - \frac{\partial F}{\partial \tau} \frac{\partial G}{\partial E}, \tag{14.27}$$

so that $\dot{\tau} = \{H_{\text{ext}}, \tau\} = 1$, which means that τ practically is identical to the evolution variable t. We are now in a position to put forward the change of variables

$$\hat{a} = a\,e^{jk\tau}, \tag{14.28}$$

$$\hat{E} = E + k^2 V a a^* \tag{14.29}$$

$$\hat{f}(z, p) = \tilde{f}(z, p), \tag{14.30}$$

$$\hat{\tau} = \tau. \tag{14.31}$$

Under this change of variables, the Poisson brackets are conserved, while the Hamiltonian takes the form

$$\hat{H} = \iint dz dp \hat{f} \sqrt{1 + p^2 + \hat{a}\hat{a}^* - ia_w(\hat{a}e^{i((k+k_w)z - k\tau)} - \hat{a}^* e^{-i((k+k_w)z - k\tau)})} + a_w^2 + \hat{E}. \tag{14.32}$$

Finally, the dynamics can be monitored in the *particles-field phase* frame by imposing $\theta = (k + k_w)z - k\tau$, and considering the (canonical) change of variables

$$\bar{f}(\theta, p) = \hat{f}(z, p)/(k + k_w), \tag{14.33}$$

$$\bar{a} = \hat{a}, \tag{14.34}$$

$$\bar{E} = \hat{E} + \frac{k}{k + k_w} \iint d\theta dp \bar{f} p, \tag{14.35}$$

$$\bar{\tau} = \tau, \tag{14.36}$$

where the factor $1/(k + k_w)$ in the first equation follows from normalization purposes.

Thanks to the above time-dependent change of coordinates, the Hamiltonian has been cast in a time-independent form. The $(\bar{\tau}, \bar{E})$ variables are somehow artificial and formally decoupled from the other variables involved. We can then work in a reduced space by dropping out this additional pair of variables. For notation purposes, the bars labelling the remaining variables are removed.

The Hamiltonian now reads

$$H = \iint d\theta dp f \left[\sqrt{1 + p^2 + a_w^2 - ia_w(ae^{i\theta} - a^* e^{-i\theta}) + aa^*} - \frac{k}{k + k_w} p \right], \tag{14.37}$$

while the resulting Poisson bracket writes

$$\{F, G\} = (k + k_w) \iint d\theta dp f \left[\frac{\partial}{\partial p} \frac{\delta F}{\delta f} \frac{\partial}{\partial \theta} \frac{\delta G}{\delta f} - \frac{\partial}{\partial \theta} \frac{\delta F}{\delta f} \frac{\partial}{\partial p} \frac{\delta G}{\delta f} \right]$$

$$+ \frac{i}{kV} \left(\frac{\partial F}{\partial a} \frac{\partial G}{\partial a^*} - \frac{\partial F}{\partial a^*} \frac{\partial G}{\partial a} \right). \tag{14.38}$$

14.2.3   Resonance condition and high-gain amplification

The final step in the derivation of Bachelard *et al.* (2008b) is to expand Hamiltonian (14.37) around a specific resonant value, which will be evocated later when discussing the free electron laser case study. Following Eq. (14.3), Hamiltonian (14.37) and bracket (14.38) yield the equations of motion

$$\frac{df}{dt} = (k + k_w) \left(\frac{p}{\sqrt{1 + p^2 + a_w^2 - ia_w(ae^{i\theta} - a^*e^{-i\theta}) + aa^*}} - \frac{k}{k + k_w} \right) \frac{\partial f}{\partial \theta}$$

$$-\frac{1}{2} \frac{a_w(k + k_w)(ae^{i\theta} + a^*e^{-i\theta})}{\sqrt{1 + p^2 + a_w^2 - ia_w(ae^{i\theta} - a^*e^{-i\theta}) + aa^*}} \frac{\partial f}{\partial p} \tag{14.39}$$

$$\frac{da}{dt} = -\frac{1}{2kV} \iint d\theta\, dp\, f \frac{ia - a_w e^{-i\theta}}{\sqrt{1 + p^2 + a_w^2 - ia_w(ae^{i\theta} - a^*e^{-i\theta}) + aa^*}}. \tag{14.40}$$

From these equations, we can readily calculate the equilibrium condition. This latter occurs, in fact, for $a = 0$ and $f(\theta, p) = \delta(p - p_r)F(\theta)$, where $F(\theta)$ is a distribution that satisfies the condition $\int d\theta\, e^{-i\theta}F(\theta) = 0$, and p_r is given by

$$\frac{p_r}{\sqrt{1 + a_w^2 + p_r^2}} - \frac{k}{k + k_w} = 0. \tag{14.41}$$

This *resonant momentum* p_r can be associated with a *resonant energy* γ_r for the particles, defined as

$$\gamma_r = \sqrt{1 + a_w^2 + p_r^2} = \sqrt{1 + a_w^2} \frac{k + k_w}{\sqrt{k_w(2k + k_w)}}. \tag{14.42}$$

Note that in the limit $k \gg k_w$, Eq. (14.42) yields $\gamma_r = \sqrt{k(1 + a_w^2)/(2k_w)}$, the usual definition that we shall encounter when discussing the free electron laser application.

This equilibrium can, however, be unstable. When subject to small perturbations, the wave can grow, gaining energy from the particles whose velocities are close to the value p_r. This instability is responsible for the high-gain growth of the wave, which is exploited and well characterized in, e.g., plasma physics. The dynamics of the system can be linearized around the above equilibrium solution: assuming that the momenta of the particles remain sufficiently close to p_r, we can shift p by p_r, by defining $\hat{f}(\theta, \hat{p}) = f(\theta, p)$ with $\hat{p} = p - p_r$. It is also assumed that the amplitude of the radiated field is weak compared to the resonant energy, $|a| \ll \gamma_r$.

After dropping for simplicity the hat over $\hat{p}$ and $\hat{f}$, the following technical steps need then to be performed:

- Expand Hamiltonian (14.37) at the first order in a and at the second order in p.
- Impose the change of variables

$$f'(\theta', p') = \frac{1}{\beta} f(\theta = \theta', p = p'/\beta),\qquad(14.43)$$

$$a' = \varepsilon a.\qquad(14.44)$$

- Rescale time as $t' = \alpha t$, i.e. $H' = H/\alpha$, and consider the new Hamiltonian $\nu H'$ with a new Poisson bracket $\nu^{-1}\{.,.\}$ (which does not alter the dynamics).
- Use

$$\alpha = \frac{1}{\gamma_r}\left(\frac{a_w^2 k_w(k + k_w/2)}{2kV}\right)^{1/3},\qquad(14.45)$$

$$\beta = \frac{2}{k + k_w}\left(\frac{2kVk_w^2(k + k_w/2)^2}{a_w^2}\right)^{1/3},\qquad(14.46)$$

$$\varepsilon = \left(\frac{4k^2V^2k_w(k + k_w/2)}{a_w}\right)^{1/3},\qquad(14.47)$$

$$\nu = 2\left(\frac{2kVk_w^2(k + k_w/2)}{a_w^2}\right)^{1/3}.\qquad(14.48)$$

Then, after some algebraic manipulation, and dropping the primes on the new variables, we eventually obtain the following Hamiltonian and the bracket operator:

$$H = \iint d\theta\,dp f\left[\frac{p^2}{2} - i\left(ae^{i\theta} - a^*e^{-i\theta}\right)\right],\qquad(14.49)$$

$$\{F,G\} = \iint d\theta\,dp f\left[\frac{\partial}{\partial p}\frac{\delta F}{\delta f}\frac{\partial}{\partial \theta}\frac{\delta G}{\delta f} - \frac{\partial}{\partial \theta}\frac{\delta F}{\delta f}\frac{\partial}{\partial p}\frac{\delta G}{\delta f}\right] + i\left(\frac{\partial F}{\partial a}\frac{\partial G}{\partial a^*} - \frac{\partial F}{\partial a^*}\frac{\partial G}{\partial a}\right).\qquad(14.50)$$

Finally, performing the canonical change of variables $(a, a^*) \to (\phi, I)$ such that $a = \sqrt{I}e^{-i\phi}$ (so that $\{\phi, I\} = 1$) makes it possible to retrieve the famous Hamiltonian

$$H[f, I, \phi] = \iint d\theta\,dp f(\theta, p)\left[\frac{p^2}{2} + 2\sqrt{I}\sin(\theta - \phi)\right],\qquad(14.51)$$

which is universally employed in almost all applications where waves and particles are mutually interacting. If f is a Klimontovich discrete distribution,[1] namely

$$f(\theta, p) = \sum_i \delta(\theta - \theta_i(t))\delta(p - p_i(t)),\qquad(14.52)$$

---

[1] We have chosen a distribution with a norm N in contrast to the choice of the unity norm made in (8.3) and (13.10). This alternative choice has no consequences on the generality of the discussion.

we immediately recovers the more familiar discrete Hamiltonian form

$$H = \sum_{i=1}^{N} \left[\frac{p_i^2}{2} + 2\sqrt{I} \sin(\theta_i - \phi) \right], \qquad (14.53)$$

where (θ_i, p_i) and (ϕ, I) are canonical pairs of conjugate variables. This is the Hamiltonian of the Colson–Bonifacio model (Colson, 1976; Bonifacio *et al.*, 1990).

14.3   Alternative Derivation of the Wave–Particles Hamiltonian from the Microscopic Equations

In this section, we will follow an alternative strategy to mathematically characterize the mutual interactions between waves and particles. We will, in particular, start from the discrete microscopic equations of motion and obtain a self-consistent Hamiltonian formulation for the coupled evolution of M modes and N particles which respect the intimate granularity of the scrutinized medium. This is at variance with the above procedure which instead relies on performing the continuum limit, an operation which in turns prevents us from appreciating the crucial role played by finite size corrections. The derivation that we will hereafter sketch is taken from Elskens and Escande (2002); it has been first discussed in Antoni, Elskens and Escande (1998) and can be seen as a sound generalization of the pioneering calculation in O'Neil *et al.* (1971).

Let us focus on the specific example discussed in the preceding chapter devoted to plasma physics and recall that the Langmuir waves are collective motion of the electrons in a plasma. To propose a simple heuristic derivation of the relevant self-consistent Hamiltonian, we imagine that the Langmuir waves are characterized by the following Bohm–Gross dispersion relation, derived in Section 13.3,

$$\omega^2 = \omega_e^2 + 3k^2 v_e^2, \qquad (14.54)$$

where ω_e stands for the plasma frequency (13.45) and v_e refers to the electrons thermal velocity. For a large number of particles N, the retro-action of individual particles on the wave is virtually negligible. Hence, particles can be seen as test particles subjected to the wave influence. Consequently the equation of motion of each individual particle can be assumed of the form

$$\ddot{x} = -A \sin(kx - \omega t - \theta_0), \qquad (14.55)$$

where x denotes the position of the particle, whose associated momentum is labelled with p; θ_0 is the phase of the wave, while A represents instead a positive quantity that measures the intensity of the trapping wave. Equation (14.55) descends from the one-dimensional Hamiltonian

$$H(x, p) = \frac{1}{2}p^2 - A\cos\,(kx - \omega t - \theta_0)\,. \tag{14.56}$$

A straightforward generalization of the Hamiltonian (14.56) to the case where N independent particles of mass m are evolving in the field generated by M propagating longitudinal waves is

$$H_{\text{test}} = \sum_{k=1}^{N} \frac{p^2}{2m} - \sum_{k=1}^{N}\sum_{j=1}^{M} W_j \cos\left(k_j x_k - \omega_{j0} t - \theta_{j0}\right), \tag{14.57}$$

where k_j and ω_{j0} are related through the dispersion relation (14.54) and where W_j represents the amplitude of the potential of the wave j with phase θ_{j0}. The charge of the electrons is implicitly incorporated in the amplitude factor W_j. If we think of Langmuir waves as mechanical objects, it is natural to represent their contributions as harmonic oscillators corresponding to the vibrating bulk electrons. In practice, it sounds natural to generalize the Hamiltonian (14.57) into the self-consistent formulation

$$H = \sum_{k=1}^{N} \frac{p^2}{2m} + \sum_{j=1}^{M} \omega_{j0} I_j - \sum_{k=1}^{N}\sum_{j=1}^{M} c_j \sqrt{I_j} \cos\left(k_j x_k - \theta_j\right), \tag{14.58}$$

where I_j stands for the wave action and is proportional to the energy carried by it. In a self-consistent treatment the phases θ_{j0} of Eq. (14.57) are no more constant, due to the interaction with the electrons. In Eq. (14.58), we have performed the change of variables $\theta_j = \omega_{j0} t + \theta_{j0}$, with the couples $\{I_j, \theta_{j0}\}$ now representing pairs of conjugate variables. Furthermore, the factor W_j has been replaced by $c_j \sqrt{I_j}$. This choice is legitimate if we recall that the electrostatic energy of a wave is proportional to the square of its amplitude; c_j is a constant which embeds the parameters of the model. We remark that the dynamics determined by the Hamiltonian (14.58) conserves the total momentum

$$P = \sum_{k=1}^{N} p_k + \sum_{k=1}^{M} k_j I_j. \tag{14.59}$$

The last Hamiltonian, derived on a heuristic basis, is self-consistent in nature: the particles evolve driven by the waves and in turn influence the evolution of the waves themselves. The derivation can be made rigorous and the details of the calculation can be found in Elskens and Escande (2002). As a final remark, we stress that, upon a trivial change of variables, Hamiltonian (14.58) can be cast in a form equivalent to (14.53) for the case $M = 1$.

In the next section, we turn to discussing a specific application, beyond plasma physics, for which wave–particles interactions are central.

14.4   Free Electron Lasers (FEL)

14.4.1   Introduction

A *free electron laser* (FEL) generates tuneable, coherent, high power radiation, currently spanning wavelengths from millimetre to visible and potentially ultraviolet to X-ray (Colson, 1976). FELs have the optical properties characteristic of conventional lasers such as high spatial coherence. Nevertheless, they differ from conventional lasers in using a relativistic electron beam as its lasing medium, as opposed to bound atomic or molecular states, hence the term free electron. Due to their flexibility, many applications have been made possible in the fields of spectroscopy, solid state physics, biology and medicine.

A FEL is essentially composed of three main elements: an electron beam, a magnetic insertion device (e.g. undulator) and an optical resonator. Two types of FEL setting exist. On the one hand, there is the so-called storage ring free electron laser (SRFEL), where the electrons are stocked in a circular accelerator (Colson, 1976). On the other hand, the currently more popular implementation consists of a linear accelerator. This latter, often referred to as to the single-pass FEL, is generally modelled by resorting to the one-dimensional self-consistent Hamiltonian derived earlier. In the following, we shall briefly discuss the storage ring FEL physics and then move to presenting the single-pass FEL case, more interesting from our perspective. In both cases, the interaction between waves and particles is at the core of the functioning of the devices.

14.4.2   Storage ring FEL

In both single-pass and storage ring FEL, the physical mechanism responsible of the light emission is the interaction between the relativistic electron beam and a magnetostatic periodic field generated in the undulator. Due to the effect of the magnetic field, the electrons are forced to follow curved trajectories, thus emitting synchrotron radiation. In the SRFEL design, presented in the schematic Fig. 14.1, the radiation, which is known as *spontaneous emission*, is stored in the optical cavity and amplified on successive passes through the undulator. The amplification process is made possible by synchronizing consecutive passes of laser and electrons pulses.

An undulator is composed of two magnetic devices, producing a vertical, alternating, magnetic field (Walker, 1996). It is in the undulator that the interaction between the electrons circulating in the ring and the radiated electromagnetic wave stocked in the cavity takes place. For the case of SRFEL, the insertion device usually employed is an optical Klystron (see Fig. 14.2), which is composed of two undulators separated by a dispersive section (i.e., strong magnetic field), favouring the interference between the emission of the undulators. This results in an enhanced amplification gain.

Consider a klystron and assume both the electrons and the electromagnetic wave to proceed along the s axis, which labels the longitudinal direction. If the energy of the electrons $1/\sqrt{1 - v_s^2}$ (in the units in which $m = 1$ and $c = 1$) is equal to the resonant energy γ_r obtained in Section 14.2.3, the electromagnetic field is not amplified. When instead

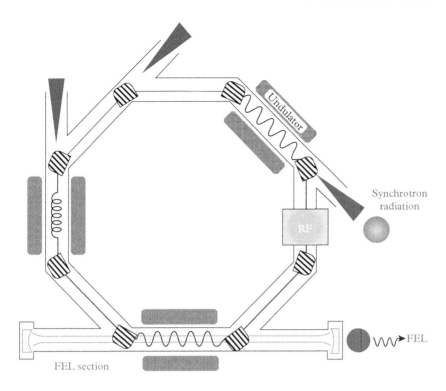

Figure 14.1 *Scheme of a storage ring free electron laser. The electrons are circulating into the storage ring. The synchrotron radiation can be extracted at different locations to serve distinct experiments. A specific section is dedicated to the free electron laser (FEL) facility in which light is amplified and stocked between mirrors.*

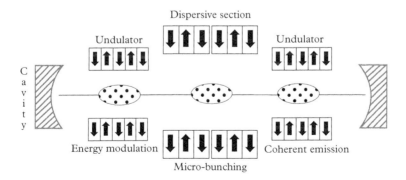

Figure 14.2 *Configuration of an optical klystron.*

the electrons energy is slightly higher than γ_r, the electromagnetic wave receives energy from the electrons. Hence, the laser field gets enhanced. In contrast, if the electrons have slightly lower energy than the resonant one, the energy is transferred from the laser field to the electrons. The effect of the *energy modulation* reflects on the electrons spatial distribution by inducing a fragmentation in *micro-bunches*, separated by one radiation wavelength. This process is responsible for the *coherent emission* of light. This picture is represented in a drastically simplified fashion in Fig. 14.2. The klystron plays the role of the *active medium* for a conventional laser, while the electron beam is equivalent to the *pumping system*.

To gain some insight into the physics of the system, we now consider the trajectory of one individual electron in a planar undulator. The following analysis can be extended to the case of a klystron, albeit with more work, and is not detailed here. Assume the purely transverse periodic magnetic field to be described by

$$\mathbf{B} = B_0 \cos\left(\frac{2\pi s}{\lambda_u}\right) \hat{\mathbf{y}}, \tag{14.60}$$

where B_0 and λ_u are, respectively, the amplitude and the period of the magnetic field (Walker, 1996); s labels the longitudinal position along the undulator axis and $\hat{\mathbf{y}}$ is the vertical unit axis. Suppose that the electron enters in the undulator parallel to the s axis. Therefore

$$\frac{d\mathbf{v}_e}{dt} = -\frac{e}{m\gamma} \mathbf{v}_e \times \mathbf{B}, \tag{14.61}$$

where $\mathbf{r}(t) = (x, y, s)$ and $\mathbf{v}_e$ are the electron position and velocity, respectively. Please note that we have reintroduced, for convenience, the mass and the charge of the electron, m and $-e$, and the velocity of light c. Also note that, due to the form of the Lorentz force, v_e^2, and thus the Lorentz factor, γ, of the electron are constant. Upon integration, assuming that the electrons are injected in the undulator at a velocity very close to c, the trajectory of the electron in the undulator takes the form

$$
\begin{cases}
x = \dfrac{K\lambda_u}{2\pi\gamma}[1 - \cos(\omega_u t)] \\[2ex]
y = 0 \\[2ex]
s = c\left[\left(1 - \dfrac{1}{2\gamma^2} - \dfrac{K^2}{4\gamma^2}\right)t + \dfrac{K^2\lambda_u}{16\pi\gamma^2 c}\sin(2\omega_u t)\right]
\end{cases}, \tag{14.62}
$$

where $\omega_u = 2\pi c/\lambda_u$ and $K = eB_0\lambda_u/(2\pi mc)$ is the so-called deflection parameter.

Consider now a radiation field emitted from a relativistic single electron. The energy received per unit of solid angle $d\Omega$ and per unit of frequency $d\omega$ can be written (Jackson, 1975) at infinity as

$$\frac{d^2I}{d\Omega d\omega} = \frac{e^2\omega^2}{16\pi^3\varepsilon_0 c^2} \left| \int_{-\infty}^{+\infty} dt\, \mathbf{n} \times [\mathbf{n} \times \mathbf{v}_e(t)] \exp\left(i\omega\left[t - \frac{\mathbf{n}\cdot\mathbf{r}(t)}{c}\right]\right) \right|^2, \qquad (14.63)$$

where $\mathbf{n}$ stands for the direction of observation. By inserting Eqs. (14.62) into (14.63) and calculating the total radiation field on the s axis (observation angle $\theta = 0$), under specific approximations not detailed here, we end up with

$$\frac{d^2I}{d\Omega d\omega}(\theta = 0) = \frac{e^2}{4\pi\varepsilon_0 c}\left(\frac{K\mathcal{N}\lambda_u}{\gamma}\right)^2 \sum_{n=1(\text{odd})}^{\infty} \frac{A_n^2}{\lambda^2}\left(\frac{\sin\delta_n}{\delta_n}\right)^2, \qquad (14.64)$$

where $\lambda = 2\pi c/\omega$ and

$$A_n \sim \mathcal{J}_{\frac{n+1}{2}}(n\xi) - \mathcal{J}_{\frac{n-1}{2}}(n\xi), \qquad (14.65)$$

$$\xi = \frac{K^2}{4 + 2K^2}, \qquad (14.66)$$

$$\delta_n = \pi\mathcal{N}(n - \frac{\lambda_r}{\lambda}), \qquad (14.67)$$

$$\lambda_r = \frac{\lambda_u}{2\gamma^2}\left(1 + \frac{K^2}{2}\right). \qquad (14.68)$$

Here, $\mathcal{J}_k$ is the Bessel function of order k, $\mathcal{N}$ is the number of undulator periods and λ_r is the fundamental resonant wavelength of the electromagnetic wave. According to Eq. (14.68), the wavelength λ_r can be tuned by varying either the energy of the ring (namely γ) or, alternatively, the deflection parameter K (by means of B_0). This flexibility is, indeed, one of the most significant characteristics of the SRFEL, of paramount importance in practical applications.

The SRFEL dynamics depends strongly on the longitudinal overlap between the electron bunches and the laser pulses at each pass inside the optical cavity (Billardon *et al.*, 1992). Indeed, dynamical instabilities develop as the SRFEL is moved away from the exact tuning between the period of the electron bunch(es) circulating into the ring and that of the photon pulse stored inside the optical cavity. After the initial amplification, the laser intensity converges to a steady state regime for a perfect electron–photon tuning. A detuning ε different from 0 leads to a cumulative delay between the electrons and the laser pulses: the laser intensity shows a stable pulsed behaviour for intermediate detuning amounts (Billardon *et al.*, 1992), while the so-called 'continuous wave' (cw) regime is recovered for sufficiently large values of ε. These different regimes are clearly represented in Fig. 14.3.

The transition between stable and unstable regimes has been characterized in De Ninno and Fanelli (2004) as a Hopf bifurcation. This result makes it possible to establish a formal bridge with the field of conventional lasers and implement dedicated control strategies to enlarge the region of stable 'cw' signal. The beneficial effects of the closed loop control, as suggested in De Ninno and Fanelli (2004), have been later confirmed

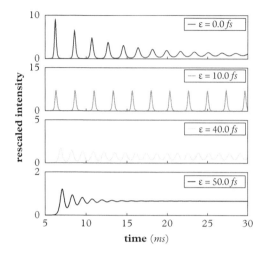

Figure 14.3 *Distinct dynamical regimes exhibited by the FEL model by varying the detuning amount ε. The curves are obtained by integrating a simple model of SRFEL dynamics: see Billardon et al. (1992) and De Ninno and Fanelli (2004) for details.*

in direct experiments performed at Elettra (Trieste, Italy) (De Ninno *et al.*, 2005) and SuperACO (Orsay, France) (Bielawski *et al.*, 2004).

14.4.3   Single-pass FEL

Among different schemes, single-pass high-gain FELs are currently attracting growing interests as promising sources in the UV and X ranges. A schematic picture of the device is given in Fig. 14.4.

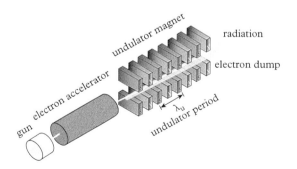

General layout of free-electron laser

Figure 14.4 *Schematic layout of a single-pass FEL.*

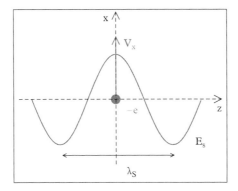

Figure 14.5 *Schematic representation of the interaction between a single electron and the electromagnetic wave.*

As illustrated in Fig. 14.5, the transverse dynamics of the electrons is influenced by the electric field associated with the co-propagating wave.

When the transverse velocity is directed as the electric field, the electrons lose energy, which is, in turn, gained by the wave. Hence, assuming this condition to hold all along the region of interaction, we expect to eventually measure a net amplification of the wave intensity at the end of the undulator. This requirement is met when the delay between the electrons position and the phase of the wave, after one complete transversal oscillation, is equal to the wavelength of the emitted radiation λ_s (see Fig. 14.6).

In formulae, the relativistic energy of the beam γ must take a particular value specified by the relation (Bonifacio *et al.*, 1990)

$$\gamma_r = \sqrt{\frac{\lambda_W (1 + a_W^2)}{2\lambda_s}}, \tag{14.69}$$

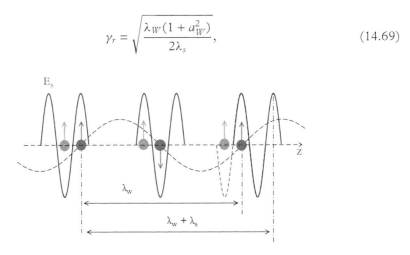

Figure 14.6 *The resonance condition reads as follows: after a period of oscillation of the electron in the transversal space, the delay between the electron and the wave must be equal to the wavelength λ_s.*

where λ_W and a_W relate to the undulator and label, respectively, the spatial period and the magnetic field amplitude. This formula is exactly identical to Eq. (14.42) (where $k \rightarrow k_s$) by performing the limit $\lambda_s \ll \lambda_W$ and noting that $\lambda_s = 1/k_s$ and $\lambda_W = 1/k_W$. We shall hereafter reproduce part of the discussion that can be found in the seminal paper by Bonifacio *et al.* (1990) to obtain a self-consistent formulation of the single-pass FEL dynamics. As already anticipated, we will recover the paradigmatic one-dimensional model (14.53) (or equivalently (14.58)), under realistic approximations. We will here simply outline the main assumptions and list the relevant equations to provide a link with the reference paper (Bonifacio *et al.*, 1990). As it will be clear in the following, the philosophy and the physical hypotheses that inspire the calculation of Bachelard *et al.* (2008b) match those of Bonifacio *et al.* (1990). In fact, it is this latter paper where the model was set up for the first time; it was later revisited by Bachelard *et al.* (2008b) in the framework of a sound and fully consistent canonical formulation.

The evolution of each electron inside the undulator region is described by the Newton–Lorentz equations

$$\frac{d(\gamma m \mathbf{v})}{dt} = -e \left[\mathbf{E} + \mathbf{v} \times (\mathbf{B}_W + \mathbf{B}) \right] \tag{14.70}$$

$$\frac{d(\gamma m c^2)}{dt} = -e \mathbf{E} \cdot \mathbf{v}_\perp, \tag{14.71}$$

where $\mathbf{B}_W$ represents the undulator field, $\mathbf{B}$ and $\mathbf{E}$ describe the electromagnetic wave, m and $\mathbf{v}$ are, respectively, the mass and velocity of the particle. Here, c is the speed of light and e is the electron charge. $\mathbf{B}$ and $\mathbf{E}$ are univocally determined via the vector potential A

$$\mathbf{B} = \nabla \times A, \tag{14.72}$$

$$\mathbf{E} = -\frac{\partial A}{\partial t}. \tag{14.73}$$

Note that these equations are identical to Eqs. (14.11) and (14.12), where use is made of the auxiliary variable $\mathbf{Y} = -\partial A/\partial t$.

Finally, the evolution of A is ultimately governed by

$$\left(\frac{\partial^2}{\partial z^2} - \frac{1}{c^2} \frac{\partial^2}{\partial t^2} \right) A = -\mu_0 \mathbf{J}_\perp (z, t). \tag{14.74}$$

As explained in detail in Bonifacio *et al.* (1990), for wigglers employed in real devices, the emitted radiation contains only odd (higher) harmonics of the fundamental one. Labelling λ_s the fundamental wavelength, the higher contributions are $\lambda_h = \lambda_s/h$. This observation enables us to re-write the vector potential as

$$A = \sum_{h=1, h \text{odd}}^{\infty} \mathbf{A}_h. \tag{14.75}$$

Under particular experimental conditions, when the beam current and emittance are sufficiently small and the waves correctly approximated by planar fields, the motion in the transverse plane can be neglected and the above system reduces to a simpler one-dimensional version (Bonifacio *et al.*, 1990), which accounts for the evolution of the longitudinal coordinate (see Fig. 14.5). Recalling the definition of the plasma frequency

$$\omega_e = \sqrt{\frac{e^2 n}{m\varepsilon_0}}, \tag{14.76}$$

where n stands for the total electron number density, the dimensionless FEL parameter is defined as

$$\rho = \frac{1}{\gamma_r} \left(\frac{a_W}{8\pi} \frac{\omega_e \lambda_W}{c} \right)^{2/3}. \tag{14.77}$$

Making use of such quantities, we can perform an appropriate rescaling of the variables involved, which allows us to recast the equations in a more compact and manageable form. Impose

$$p = \frac{1}{\rho} \frac{\gamma - \langle \gamma \rangle_0}{\langle \gamma \rangle_0}, \tag{14.78}$$

$$A_h = \frac{e |A_h|}{mc^2} \frac{2\pi c}{\lambda_s \omega_e \sqrt{\rho \gamma_r}}, \tag{14.79}$$

$$\bar{z} = 2k_w \rho z, \tag{14.80}$$

where $\langle \gamma \rangle_0$ is the average electron energy at injection. We then introduce the relative phase of the electrons

$$\theta = (k + k_W)z - \frac{2\pi c}{\lambda_s} t. \tag{14.81}$$

In the limit $\rho \ll 1$, which amounts to assuming a low particle density and small energy dispersion, the governing equations (14.70) and (14.74) become

$$\frac{d\theta_j}{d\bar{z}} = p_j, \tag{14.82}$$

$$\frac{dp_j}{d\bar{z}} = -\sum_h F_h (A_h e^{ih\theta_j} + A_h^* e^{-ih\theta_j}), \tag{14.83}$$

$$\frac{dA_h}{d\bar{z}} = \frac{F_h}{N} \sum_j e^{-ih\theta_j} = F_h b_h, \tag{14.84}$$

where N is the number of mutually interacting particles, labelled by the index j. The coupling parameters F_h depend on the experimental setup. As already mentioned, the interested reader can find the detailed derivation of such model in Bonifacio *et al.* (1990). The quantities

$$b_h = \frac{1}{N} \sum_j \exp(-ih\theta_j), \tag{14.85}$$

called *bunching* parameters, are the Fourier coefficients of the single-particle distribution. Remarkably, the above system of equations can be obtained from the Hamiltonian

$$H = \sum_{j=1}^{N} \frac{p_j^2}{2} - i \sum_h \left[\sum_{j=1}^{N} \frac{F_h}{h} (A_h \exp[ih\theta_j] - A_h^* \exp[-ih\theta_j]) \right]. \tag{14.86}$$

Here the canonically conjugated variables are (p_j, θ_j) and $(\sqrt{N}A_h, \sqrt{N}A_h^*)$. Besides the 'energy' H, the total 'momentum' $P = (\sum_j p_j + N \sum_h |A_h|^2)$ is also conserved. Hamiltonian (14.86) can be mapped onto the self-consistent Hamiltonian (14.58) via a change of variables detailed in Antoniazzi *et al.* (2005a). By truncating in (14.58) the sum over the harmonics to retain the dominant contribution ($k = 1$, equivalent to $h = 1$ in (14.86)), and considering the ideal setting with $\omega_{10} = 0$, we hence obtain the Hamiltonian (14.53), which, as we have already stressed, occurs in any domain where waves–particle interaction proves central.

14.4.4   On the dynamical evolution of the single-pass FEL

In the following, we shall present a collection of numerical and analytical facts aimed at unravelling the dynamics of the system derived in the previous section. Despite the fact that we will here develop our reasoning with reference to FEL, the conclusions of the analysis can be transferred almost unchanged to all other settings where the universal Hamiltonian model (14.86) applies. Let us start by focusing on a linear analysis, around an equilibrium solution.

14.4.4.1   *Linear analysis*

In order to estimate the growth rate (gain) associated with the different harmonics, we can perform a linear stability analysis of the governing system (14.82)–(14.84) around a stationary initial condition. As a relevant case study, we consider an initial monokinetic beam and assign a vanishing initial radiation to the field

$$A_{h0} = 0, \qquad \langle \exp[-ih\theta] \rangle_0 = 0, \qquad p_{j0} = p_0. \tag{14.87}$$

Looking for solutions proportional to $\exp(i\alpha z)$ yields a cubic dispersion relation for each harmonics,

$$\alpha^3 - p_0 h\alpha^2 + hF_h^2 = 0. \tag{14.88}$$

Each of these cubic equations admits a complex root with a negative imaginary part, if $p_0 < p_{0c}(h)$. It can be shown (Bonifacio *et al.*, 1990) that $p_{0c}(h)$ is a decreasing function of the harmonic index h. Higher harmonics display hence lower growth rates when compared to the fundamental one.

14.4.4.2 Beyond the linear approximation: numerical results

Numerical simulations based on system (14.82)–(14.84) show that the amplification of the waves occurs in several subsequent steps. Initially, each harmonics experiences an exponential evolution, with a growth rate determined by the solutions of (14.88). Then, the harmonic characterized by the fastest growth rate saturates to an intermediate regime that we identify as a QSS, according to the definition given in Chapter 9. Here the intensity displays oscillations around a well-defined plateau (Bonifacio *et al.*, 1990), as clearly testified in Fig. 14.7. Here and in Fig. 14.8 $I = |A|^2$.

Once the first harmonic has settled down to its corresponding saturated level, the other harmonics gets suddenly stopped. The corresponding intensities display hence a peak but are then rapidly damped to an almost negligible value (see Fig. 14.7). The selected mode, i.e. the dominant harmonic, keeps oscillating around the reference value of the intensity. Similar considerations can be drawn for the evolution of the bunching parameters

$$b_h(t) = \frac{1}{N} \sum_j \exp(ih\theta_j(t)) := \langle \exp(ih\theta(t)) \rangle \qquad (14.89)$$

with $h = 1, 3, 5$; see right panels in Fig. 14.7. The latter quantities provide a quantitative measure of the spatial modulation of the particles distribution.

Asymptotically, a slow time evolution takes the system towards the final equilibrium (see Fig. 14.8) as predicted by the Boltzmann–Gibbs statistics (Firpo, 1999; Firpo and Elskens, 2000) and shown in the next subsection using the large deviations method.

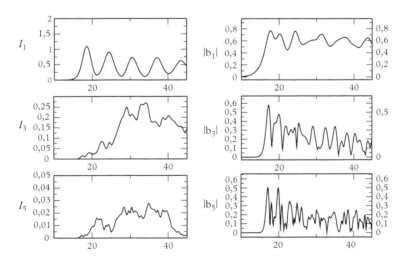

Figure 14.7 *Evolution along the undulator of the intensities (left) and bunching parameters (right) relative to the first three harmonics, for a typical dynamics determined by Eqs. (14.82)–(14.84). A monokinetic beam is considered as an initial condition and the injected wave amplitude is taken to be almost 0.*

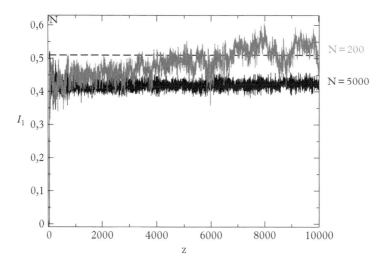

Figure 14.8 *Long time evolution of I_1 for $N = 200$ and $N = 5000$. The dotted line represents the equilibrium value predicted by Boltzmann–Gibbs statistics.*

This process is driven by granularity and the relaxation time diverges with the number of simulated particles N (Barré *et al.*, 2004), in complete analogy with the case of the HMF model (see Chapter 9). This existence of an intermediate QSS phase, which is attained just after the initial violent relaxation, is demonstrated in Fig. 14.8. The final thermal equilibrium, identified in the picture with a dashed line, is reached for unrealistic undulator sizes. The QSS is characterized by an almost constant intensity amount and lasts for a considerable time window before the final transition towards equilibrium takes over.

With reference to the single-particle phase space, when the hth harmonic dominates, a fraction of the particles clusters in phase space and gives birth to h coherent clumps which rotate in the buckets created by the wave, synchronized with the oscillations of the intensity of the selected harmonic. The remaining particles are uniformly spread between two oscillating boundaries, as shown in Fig. 14.9 for a case where the fundamental harmonic prevails (Antoniazzi *et al.*, 2006). The presence of the dense cores, hereafter termed 'macroparticles', gets even more evident when the particles are initialized so to fill a thin, almost one-dimensional, stripe. The relative contrast with the surrounding halo is instead progressively reduced when the initial beam occupies a larger portion of phase space.

This observation opens up the perspective to elaborate a simplified few degrees-of-freedom model to reproduce the dynamical evolution of the principal macroscopic quantities that characterize the system in its QSS phase (Tennyson *et al.*, 1994; Antoniazzi *et al.*, 2006).

Having identified the intermediate regime as a QSS, we can apply the Lynden–Bell maximum entropy technique, described in Chapter 9, to predict the main observables (intensity, bunching, distribution profile) via a self-consistent statistical mechanics

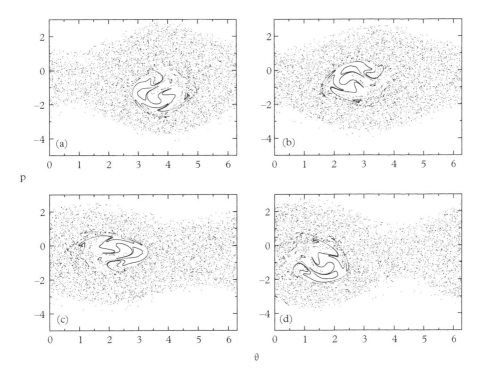

p

θ

Figure 14.9 *Sequence of phase space plots over one bounce period. Figures refer to $\bar{z} = 184$ (a), 185 (b), 187 (c) and 188 (d), with $\bar{z}$ defined in Eq. (14.80). The bulk of particles rotates in the wave separatrices. The simulation is performed for an initial mono-energetic profile ($\gamma = \gamma_r$, 10^4 electrons). Particles are initially uniformly distributed in space.*

approach. The first pre-requisite for the technique to be applicable is to turn the discrete problem specified by Eqs. (14.82)–(14.84) into a continuum analogue, i.e. performing the limit $N \to \infty$. The existence of a continuum picture has been already discussed in the earlier part of the chapter, when revisiting the approach by Bachelard *et al.* (2008b). The next two sections are organized as follows. First, in Section 14.5, we will present an application of the large deviations method to derive the Boltzmann–Gibbs statistical mechanics of the single-pass FEL model. Then, in Section 14.6, and with reference to the same physical problem, we will discuss the out-of-equilibrium Lynden–Bell approach.

14.5 Large Deviations Method Applied to the Colson–Bonifacio Model

14.5.1 Equilibrium solution of the Colson–Bonifacio model

As anticipated, by truncating the sum in Hamiltonian (14.86) over the harmonics and just retaining the dominant contribution, we obtain, upon proper rescaling of the quantity involved (Antoniazzi *et al.*, 2005a) the Hamiltonian (14.53), which occurs in any domain where the interaction between waves and particles is central. As shown in Eq. (14.58), the model can be further modified to account for an additional term which is proportional to the intensity of the field. In the context of FEL physics, the proportionality constant is the so-called detuning δ, a parameter which measures the deviation from the resonance condition. The relevant Hamiltonian can be cast in the form

$$H_N = \sum_{j=1}^{N} \frac{p_j^2}{2} - N\delta A^2 + 2A \sum_{j=1}^{N} \sin(\theta_j - \phi), \tag{14.90}$$

in which A and ϕ define the amplitude and the phase of the dominating mode.

The two conserved quantities are still the 'energy' H_N and the total momentum, which here corresponds to $P = \sum_j p_j + NA^2$. Starting from initial states with a small field A and the electrons uniformly distributed with a small kinetic energy, we observe the growth of the field towards an asymptotic value. As shown by Barré *et al.* (2005), it is indeed possible to calculate the final value of the field analytically by means of the large deviations technique presented in Chapter 3.

The first step in the analysis is the identification of the global variables. Hamiltonian (14.90) can be rewritten easily as

$$H_N = N \left[\frac{u}{2} - \delta A^2 + 2A \left(-b_x \sin \phi + b_y \cos \phi \right) \right], \tag{14.91}$$

where u and v have been defined in Eqs. (4.48) and (4.49), while

$$b_x = \frac{1}{N} \sum_{i}^{N} \cos \theta_i \quad \text{and} \quad b_y = \frac{1}{N} \sum_{i}^{N} \sin \theta_i. \tag{14.92}$$

Relation (14.91) is exact for any N. Defining the phase of the mean field ϕ', i.e. $b_x + ib_y = b \exp(i\phi')$, the global variables are given by $\gamma = (u, v, b, \phi', A, \phi)$. The quantity b measures the degree of bunching of the particles, and it is equivalent to the magnetization m employed earlier when, e.g., discussing the generalized XY-model.

The second step of the calculation is performed by computing the entropy associated with the above global variables. However, as the contribution to the entropy of the two field variables A and ϕ is negligible (precisely of order $1/N$), the generating function is simply

$$\psi(\lambda_u, \lambda_v, \lambda_c, \lambda_s) = \int \left(\prod_i d\theta_i dp_i \right) \tag{14.93}$$

$$\times \exp \left(-\lambda_u \sum_{i=1}^{N} p_i^2 - \lambda_v \sum_{i=1}^{N} p_i - \lambda_c \sum_{i=1}^{N} \cos \theta_i - \lambda_s \sum_{i=1}^{N} \sin \theta_i \right),$$

in which the existence of the integral in (14.93) necessarily implies that $\lambda_u > 0$. We thus get

$$\psi\left(\lambda_u, \lambda_v, \lambda_c, \lambda_s\right) \sim \left[e^{\lambda_v^2/4\lambda_u} \sqrt{\frac{\pi}{\lambda_u}} I_0\left(\sqrt{\lambda_c^2 + \lambda_s^2}\right)\right]^N, \tag{14.94}$$

where I_0 is the modified Bessel function of order 0. In the last expression, we have not reported a constant finite factor inside the square brackets. Besides, the sign $\sim$ indicates that the formula is valid only at leading order; indeed, since the domain of integration Σ is such that $\Sigma \neq \mathbb{R} \times [0, 2\pi[$, the exact expression of $\psi\left(\lambda_u, \lambda_v, \lambda_c, \lambda_s\right)$ would contain neither the Gaussian nor the Bessel functions. However, both of these functions constitute very good approximations to the exact expressions, so that Eq.(14.94) can be used for the following analysis.

Successive steps in the analysis are very similar to what was shown in Section 4.2.1 when deriving the statistical mechanics solution for the generalized XY-model via the large deviations technique. Defining the total momentum density as $\sigma = P/N$, the third step is accomplished by solving the microcanonical variational problem

$$s(\varepsilon, \sigma, \delta) = \sup_\gamma \left[\frac{1}{2} \ln\left(u - v^2\right) + s_{\text{conf}}(b) \,\middle|\, \varepsilon = \frac{u}{2} - \delta A^2 + 2Ab \sin\left(\phi' - \phi\right),\right.$$

$$\left. \sigma = v + A^2 \right], \tag{14.95}$$

where $s_{\text{conf}}(b)$ is given by Eq. (4.57).

Using the constraints of the variational problem, we can express u and v as functions of the other variables, obtaining the following form of the entropy:

$$s(\varepsilon, \sigma, \delta) = \sup_{A, \phi, b, \phi'} \left[\frac{1}{2} \ln\left[2\left(\varepsilon - \frac{\sigma^2}{2}\right) - 4Ab \sin\left(\phi' - \phi\right) + 2(\delta + \sigma)A^2 - A^4\right]\right.$$

$$\left. + s_{\text{conf}}(b)\right]. \tag{14.96}$$

The extremization over the variables ϕ and ϕ' is straightforward, since by direct inspection of formula (14.96), it is clear that the entropy is maximized when $\phi' - \phi = -\pi/2$. We thus get

$$s(\varepsilon, \sigma, \delta) = \sup_{A, b} \left[\frac{1}{2} \ln\left[2\left(\varepsilon - \frac{\sigma^2}{2}\right) + 4Ab + 2(\delta + \sigma)A^2 - A^4\right]\right.$$

$$\left. + s_{\text{conf}}(b)\right] \tag{14.97}$$

$$\equiv \sup_{A, b} s(A, b). \tag{14.98}$$

The non-zero σ case can be reduced to the vanishing σ problem using the identity $s(\varepsilon, \sigma, \delta) = s(\varepsilon - \sigma^2/2, 0, \delta + \sigma)$. Consequently, from now on, we will restrict ourselves to the zero momentum case, changing $\varepsilon - \sigma^2/2 \to \varepsilon$ and $\delta + \sigma \to \delta$. This has also a practical interest, because it is the experimentally relevant initial condition (Barré *et al.*, 2004).

The conditions for having a local stationary point are

$$\frac{\partial s}{\partial A} = \frac{2\left(\delta A - A^3 + b\right)}{2\varepsilon + 2\delta A^2 + 4Ab - A^4} = 0,$$

(14.99)

$$\frac{\partial s}{\partial b} = \frac{2A}{2\varepsilon + 2\delta A^2 + 4Ab - A^4} - B_{\mathrm{inv}}(b) = 0.$$

(14.100)

It is clear that $b = A = 0$ is a solution of conditions (14.99) and (14.100): it exists only for positive ε. We will limit ourselves to study its stability. It must be remarked that this is the typical initial condition studied experimentally in the FEL: it corresponds to having a beat wave with zero amplitude and the electrons uniformly distributed. The lasing phenomenon is revealed by an exponential growth of both A and the electron bunching parameter b.

The second-order derivatives of the entropy $s(A, b)$, computed on this solution, are

$$\frac{\partial^2 s}{\partial A^2}(0,0) = \frac{\delta}{\varepsilon}, \quad \frac{\partial^2 s}{\partial b^2}(0,0) = -2, \quad \frac{\partial^2 s}{\partial A \partial b}(0,0) = \frac{1}{\varepsilon}.$$

(14.101)

The two eigenvalues of the Hessian are the solutions of the equation

$$x^2 - x\left(-2 + \frac{\delta}{\varepsilon}\right) - \frac{2\delta}{\varepsilon} - \frac{1}{\varepsilon^2} = 0.$$

(14.102)

The stationary point is a maximum if the roots of this equation are both negative. This implies that their sum $(-2 + \delta/\varepsilon)$ is negative and their product $-2\delta/\varepsilon - 1/\varepsilon^2$ is positive. Recalling that we restrict to positive ε values, the condition for the sum to be negative is $\varepsilon > \delta/2$ and that for the product to be positive is $\varepsilon > -1/(2\delta)$ with $\delta < 0$. The second condition is more restrictive; hence the only region where the solution $b = A = 0$ exists and is stable is $\varepsilon > -1/(2\delta)$ with $\delta < 0$. When crossing the line $\varepsilon = -1/(2\delta)$ ($\delta < 0$), a non-zero bunching solution ($b \neq 0$) originates continuously from the zero bunching one, producing a second-order phase transition. This analysis fully coincides with that performed in the canonical ensemble in Firpo and Elskens (2000).

For the sake of completeness, it is important to stress that we have not presented the study of the global stability of the different solutions. We will, however, show that it is possible to map this model exactly onto the HMF model, which implies that no surprises are expected to be found in this respect.

14.5.2   Mapping the Colson–Bonifacio model onto HMF

The microcanonical solution of the FEL model (14.90) can be expressed in terms of the HMF Hamiltonian (4.3) using the Laplace representation of the Dirac δ-function and a Gaussian integration. After performing the change of variables $\widetilde{\theta}_i = \theta_i - \phi$, the microcanonical volume of the FEL is given by

$$\Omega(E) = \iiint \prod_i \mathrm{d}p_i \mathrm{d}\widetilde{\theta}_i \mathrm{d}A \, \delta(E - H_N) \tag{14.103}$$

$$= \iiint \prod_i \mathrm{d}p_i \mathrm{d}\widetilde{\theta}_i \mathrm{d}A \, \frac{1}{2i\pi} \int_\Gamma \mathrm{d}\lambda \, \exp\left[\lambda(E - H_N)\right], \tag{14.104}$$

where Γ is a path on the complex λ-plane, going from $-i\infty$ to $+i\infty$, which crosses the real axis at a positive value. Introducing the FEL Hamiltonian (14.90), we get

$$\Omega(E) = \int \prod_i \mathrm{d}p_i \, \exp\left[\lambda\left(E - \sum_{j=1}^N \frac{p_j^2}{2}\right)\right] \iint \prod_i \mathrm{d}\widetilde{\theta}_i \mathrm{d}A$$

$$\times \frac{1}{2i\pi} \int_\Gamma \mathrm{d}\lambda \, \exp\left[\lambda(N\delta A^2 - 2NA\widetilde{m}_y)\right], \tag{14.105}$$

in which $\widetilde{m}_y = (\sum_i \sin\widetilde{\theta}_i)/N$. Performing the Gaussian integral over the field variable A, we get

$$\Omega(E) = \int \prod_i \mathrm{d}p_i \int \prod_i \mathrm{d}\widetilde{\theta}_i \frac{1}{2i\pi} \int_\Gamma \mathrm{d}\lambda \, \frac{\sqrt{\pi}}{2} \, c(\lambda, \widetilde{m}_y)$$

$$\times \exp\left[\lambda\left(E - \sum_{j=1}^N \frac{p_j^2}{2} - N\frac{\widetilde{m}_y^2}{\delta} - \frac{1}{2}\ln\left[\lambda N(-\delta)\right]\right)\right], \tag{14.106}$$

where $1 < |c(\lambda, \widetilde{m}_y)| < 2$ is an unimportant coefficient expressed through the error function of the variable $(-\lambda N\widetilde{m}_y^2/\delta)^{1/2}$. In fact, in the large N-limit, we can safely neglect all constants and the $\ln N$ term, obtaining

$$\Omega(E) \sim \int \prod_i \mathrm{d}p_i \int \prod_i \mathrm{d}\widetilde{\theta}_i \, \delta\left(E - \sum_{j=1}^N \frac{p_j^2}{2} - N\frac{\widetilde{m}_y^2}{\delta}\right). \tag{14.107}$$

We therefore find the microcanonical volume for the Hamiltonian

$$H_N = \sum_{j=1}^N \frac{p_j^2}{2} + N\frac{\widetilde{m}_y^2}{\delta}. \tag{14.108}$$

Hence, apart from an uninteresting constant, for negative values of the parameter δ, solving the microcanonical problem for the FEL Hamiltonian (14.90) is formally equivalent to obtaining the solution of the HMF Hamiltonian (4.3) with $\varepsilon_j = -2/\delta$ and $m_x = 0$.

14.6   Derivation of the Lynden–Bell Solution

In this section, we will present the Lynden–Bell approach to the case of the single-pass FEL. We will, however, start by recalling the relevant continuum Vlasov model. In doing so, we will consider the generalized setting where a discrete set of waves is allowed for. Clearly, by imposing $F_h = 0$ for all $h \neq 1$, we get again the simple case where just one wave is assumed to interact with the particles.

When performing the continuum limit (Firpo and Elskens, 1998) $N \rightarrow \infty$, Eqs. (14.82)–(14.84) converge to the system

$$\frac{\partial f}{\partial z} = -p\frac{\partial f}{\partial \theta} + 2\sum_h F_h (A_h^x \cos h\theta - A_h^y \sin h\theta)\frac{\partial f}{\partial p}, \tag{14.109}$$

$$\frac{dA_h^x}{dz} = +F_h \int d\theta\, dp\, f \cos h\theta, \tag{14.110}$$

$$\frac{dA_h^y}{dz} = -F_h \int d\theta\, dp\, f \sin h\theta, \tag{14.111}$$

where $A_h^x = (A_h + A_h^*)/2$ and $A_h^y = (A_h - A_h^*)/(2i)$. The first is the Vlasov equation, which controls the evolution of the single-particle distribution function f. The other two equations govern the dynamics of the complex field A. The above system conserves the pseudo-energy per particle

$$e(f, A) = \int d\theta\, dp\, \frac{p^2}{2} f + 2\sum_h \frac{F_h}{h} \int d\theta\, dp [A_h^x \sin(h\theta) + A_h^y \cos(h\theta)] f, \tag{14.112}$$

the momentum per particle

$$\sigma(f, A) = \int d\theta\, dp\, pf + \sum_h [(A_h^x)^2 + (A_h^y)^2] \tag{14.113}$$

and the mass

$$m = \int d\theta\, dp\, f. \tag{14.114}$$

Following Antoniazzi *et al.* (2008), we will discuss the case where two waves are simultaneously present ($F_h = 0, h = 5, \ldots, \infty, h$ odd). In principle the analysis can be extended to include other contributions at the price of enhancing the complexity of the treatment. Note that the relevant single wave model can be readily recovered by setting $F_3 = 0$.

For our analysis, we will consider 'water-bag' initial distributions

$$f(\theta, p, 0) = \begin{cases} f_0 = \frac{1}{4\theta_0 p_0} & \text{if } -p_0 < p < p_0 \text{ and } -\theta_0 < \theta < \theta_0, \\ 0 & \text{otherwise.} \end{cases} \tag{14.115}$$

According to the Lynden–Bell principles, we may expect the system to reach an equilibrium that corresponds to the most probable macrostate $\bar{f}$, i.e. the one that maximizes the mixing entropy

$$S(e_0, \sigma_0) = \max_{\bar{f}, A_h^x, A_h^y} \left(s(\bar{f}) \Big| e(\bar{f}, A_h^x, A_h^y) = e_0; \sigma(\bar{f}, A_h^x, A_h^y) = \sigma_0; m(\bar{f}, A_h^x, A_h^y) = 1 \right), \quad (14.116)$$

consistent with all the constraints imposed by the dynamics. The Lynden–Bell entropy $s(\bar{f})$ is defined in expression (9.46). In the expression (14.116), e_0 and σ_0 label, respectively, energy and momentum per particle.

Performing the analytical calculations leads to the equilibrium values

$$\bar{f} = f_0 \frac{e^{-\beta(p^2/2 + \Phi(\theta)) - \lambda p - \mu}}{1 + e^{-\beta(p^2/2 + \Phi(\theta)) - \lambda p - \mu}}, \quad (14.117)$$

$$A_h^x = -\frac{\beta F_h}{h\lambda} \int dp\, d\theta \bar{f} \sin h\theta, \quad (14.118)$$

$$A_h^y = -\frac{\beta F_h}{h\lambda} \int dp\, d\theta \bar{f} \cos h\theta, \quad (14.119)$$

where $\Phi(\theta) = 2F_1(A_1^y \cos \theta + A_1^x \sin \theta) + 2F_3(A_3^y \cos 3\theta + A_3^x \sin 3\theta)/3$, and β/f_0, λ/f_0 and μ/f_0 are the Lagrange multipliers for the energy, momentum and normalization constraints.

Inserting Eqs. (14.117)–(14.119) into the expressions for the constraints, we obtain the system of self-consistent equations

$$\lambda = \beta([A_1^x]^2 + [A_3^x]^2 + [A_3^y]^2), \quad (14.120)$$

$$f_0 \frac{x}{\sqrt{\beta}} \int d\theta\, e^z F_0(x e^z) = 1, \quad (14.121)$$

$$f_0 \frac{x}{2\beta^{3/2}} \int d\theta\, e^z F_2(x e^z) = e_0 + 2(|A_1|^2 + |A_3|^2)^2, \quad (14.122)$$

$$f_0 \frac{x}{\sqrt{\beta}} \int d\theta \sin \theta\, e^z F_0(x e^z) = -(|A_1|^2 + |A_3|^2) \frac{A_1^x}{F_1}, \quad (14.123)$$

$$f_0 \frac{x}{\sqrt{\beta}} \int d\theta \sin 3\theta\, e^z F_0(x e^z) = -3(|A_1|^2 + |A_3|^2) \frac{A_3^x}{F_3}, \quad (14.124)$$

$$f_0 \frac{x}{\sqrt{\beta}} \int d\theta \cos 3\theta\, e^z F_0(x e^z) = -3(|A_1|^2 + |A_3|^2) \frac{A_3^y}{F_3}, \quad (14.125)$$

where we have defined

$$x = \exp(\lambda^2/2\beta - \mu), \quad (14.126)$$

$$z = -\beta \left(2F_1 A_1^x \sin \theta + \frac{2F_3}{3} \left[A_3^y \cos(3\theta) + A_3^x \sin(3\theta) \right] \right), \quad (14.127)$$

$$F_0(y) = \int dv \, \frac{e^{-v^2/2}}{1 + y\,e^{-v^2/2}}, \tag{14.128}$$

$$F_2(y) = \int dv \, \frac{v^2 e^{-v^2/2}}{1 + y\,e^{-v^2/2}}. \tag{14.129}$$

Note that in Eq. (14.120) we have assumed that the total momentum σ_0 is 0. This system (14.120)–(14.125) for the six variables $(\beta, \lambda, \mu, A_1, A_3^x, A_3^y)$ can be solved numerically, via a Newton–Raphson method to determine the values of the amplitudes of the two harmonics and the Lagrange multipliers. Note that the system is invariant under a phase translation $\theta \to \theta + \psi$; thus we can set $\int d\theta \bar{f} \cos\theta = 0$, which implies that $A_1^y = 0$, without losing generality.

14.7   Comparison with Numerical Results

14.7.1   The single-wave model

As a first test of the theoretical predictions we here consider the single-wave case, which implies setting $F_1 \neq 0$ and $F_h = 0, h = 3, \ldots, \infty$ in Eqs. (14.82)–(14.84). Assume a water-bag initial distribution as specified by Eq. (14.115), where $\theta_0 \leq \pi$. Further, label with $|b_0|$ the initial bunching, i.e. the quantity given by $|b_0| = |\langle \exp(i\theta(0)) \rangle|$. The case $b_0 = 0$ corresponds to homogeneous initial conditions and has been the object of investigations in Barré *et al.* (2004), where a remarkably good agreement between theory and numerics was first reported. Such a comparison is here documented in Fig. 14.10, where the average intensity $I = (A_1^x)^2 + (A_1^y)^2$ is plotted as a function of the initial energy dispersion $\langle p^2/2 \rangle$.

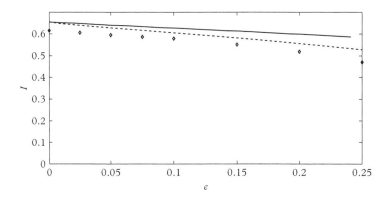

Figure 14.10 *Average laser intensity as a function of the initial energy dispersion. The diamonds represents the numerical solutions ($N = 1600$, averaged over 20 different realizations) and the dashed line refers to the Lynden–Bell solution described in Section 14.6. The solid line is the theoretical solution, resulting from the Boltzmann statistical mechanics derived using the large deviations method (see Section 14.5).*

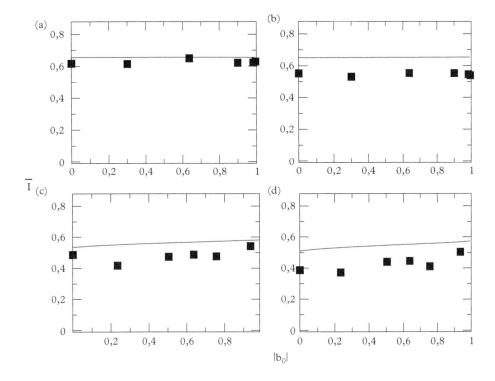

Figure 14.11 *Average intensity $\bar{I}$ as a function of the initial bunching $|b_0|$ for different values of the initial energy, respectively: (a) $e_0 = 0.01$, (b) $e_0 = 0.16$, (c) $e_0 = 0.21$ and (d) $e_0 = 0.315$. The continuous lines correspond to the theoretical predictions (14.118) and (14.119), while symbols represent numerical results obtained by averaging intensity fluctuations in the saturated regime over ten different realizations of initial conditions giving the same θ_0 and p_0. Reprinted from Curbis et al. (2007), © 2007, with kind permission of the* European Physical Journal *(EPJ).*

The analysis can be extended to bunched initial conditions $b_0 \neq 0$, a configuration of paramount importance in several experimental applications. In Fig. 14.11, the average intensity at saturation is reported as a function of $|b_0|$, for different values of the initial kinetic energy. Results are compared with the analytical estimates obtained following the procedure outlined in the previous section. Analogous plots for the average value of the bunching parameter at saturation are reported in Fig. 14.12. A direct inspection of the figures confirms the adequacy of the proposed theoretical framework: predictions based on the Lynden–Bell theory correlate well with numerical curves.

The theory is sufficiently accurate for all the cases displayed in Figs. 14.11 and 14.12, that is, for any $|b_0|$ and for all the energies below a critical energy threshold $e_{0c} = 0.315$, where the system experiences a dynamical transition (see Section 14.4.4.1). Above e_{0c}, in fact, the initial state is stable, for $\theta_0 = \pi$ (i.e. $|b_0| = 0$), and the wave oscillates indefinitely without getting amplified. The system is hence prevented from eventually reaching the state predicted by the theory and, as a consequence of this purely dynamical effect, the

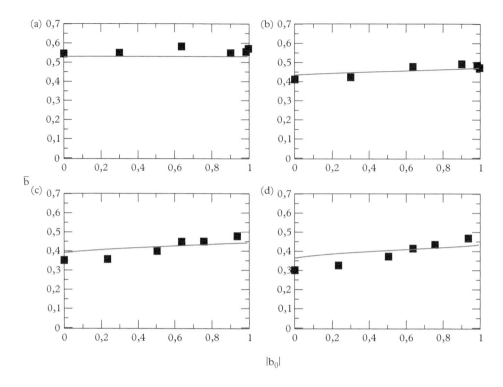

Figure 14.12 *Average values of the bunching parameter $|b|$ as a function of $|b_0|$. Same choice of parameters as in Fig. (14.11). Reprinted from Curbis et al. (2007), © 2007, with kind permission of the European Physical Journal (EPJ).*

agreement is punctually lost. When increasing the value of the nominal energy e_0, the instability mechanism takes place only at larger value of b_0, and the agreement between theory and numerics progressively deteriorates.

As a further check of the theory, we can focus on the velocity distribution in the QSS. To this aim numerical results are compared to the prediction (14.117). Two curves corresponding to $e_0 = 0.16$ and different values of the water-bag spatial support θ_0 are displayed in Fig. 14.13. For large θ_0 (right plots in the figure), the agreement is excellent, and the solid line (theory) interpolates correctly the numerical histogram. Conversely, when the initial distribution becomes narrower, which in turn corresponds to smaller of θ_0 (i.e. to larger bunching), the discrepancy between theory and numerics is enhanced.

The source of this disagreement is explained by looking at particles dynamics in the phase space. As discussed in Section 14.4.4.2, when starting with a relatively large bunching parameter, the particle distribution gets more filamented during the initial violent relaxation and the cluster which asymptotically persists in the saturated regime is

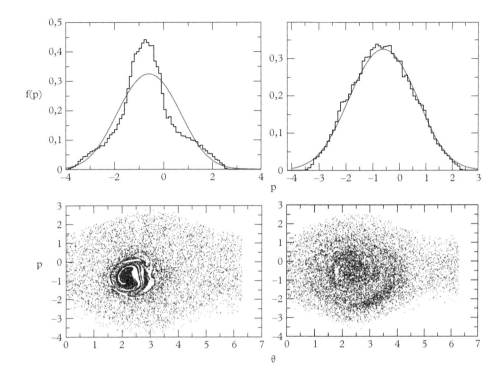

Figure 14.13 *(Top) Velocity distributions of the particles for $e_0 = 0.1667$ and $\theta_0 = 0.785$ (left), 2.355 (right). The histograms are obtained from simulations with $\bar{z} = 100$ and result from an average over ten different realizations. The continuous lines represent the analytical result predicted by Eq. (14.117). (Bottom) Phase-space portraits corresponding to the two cases mentioned above. As before, $e_0 = 0.1667$ and $\theta_0 = 0.785$ (left), 2.355 (right). Reprinted from Curbis et al. (2007), © 2007, with kind permission of the* European Physical Journal *(EPJ).*

definitely more pronounced (see Fig. 14.13). This dense core of particles results in the central peak which appears in the velocity profile for smaller θ_0, as shown in Fig. 14.13.

However, the presence of such localized structure, which apparently contradicts the mixing hypothesis assumed within the Lynden–Bell scenario, does not affect the quality of the predictions of the average intensity and bunching parameter. Indeed, the presence of the macroparticle mainly influences the oscillations of such quantities, their mean value being correctly reproduced by the theory.

14.7.2 The case of two harmonics

The case where the third harmonic ($F_3 \neq 0$) is also present in the system (14.82)–(14.84) was explicitly considered in Antoniazzi *et al.* (2008). In the following, we will review the main conclusions and results of this analysis. The quantities F_1 and F_3 are

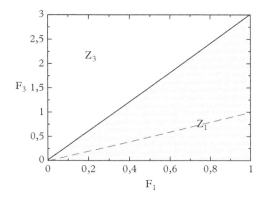

Figure 14.14 *The two zones Z_1 (grey) and Z_3 (white) in the plane (F_1, F_3) predicted by Lynden–Bell theory. In Z_1 (resp. Z_3) the maximum entropy state is S_1 (resp. S_3). Above the dashed line, a linear analysis (Bonifacio et al., 1990) shows that the growth rate of A_3 is faster than that of A_1. Hence, in the region between the dashed and the full line the system tends to relax towards S_3 and the convergence to the entropy maximum S_1 state is hindered. Reprinted from Antoniazzi et al. (2008), © 2008, with permission from Elsevier.*

treated as free parameters and allowed to scan a finite portion of the plane $[0, 1] \times [0, 3]$, beyond the constraints imposed by a real experimental setup.

Consider a homogeneous initial condition, of the form given by Eq. (14.115), with $\theta_0 = \pi$ and assume $p_0 = 0.1$. This choice corresponds to a small energy amount, namely $H/N = 1.6 \times 10^{-3}$. Within the explored region in the plane (F_1, F_3), Lynden–Bell's theory predicts two stationary equilibria, one with $A_1 \neq 0$ and $A_3 = 0$ (denoted hereafter as S_1), and the other with $A_3 \neq 0$ and $A_1 = 0$ (labelled S_3). They are associated with two distinct zones, as illustrated in Fig. 14.14:

1. Z_1: S_1 is the entropy maximum: thus A_1 should dominate and A_3 turns out to be negligible.

2. Z_3: S_3 is the entropy maximum: thus solely A_3 should be amplified.

The system undergoes a first-order phase transition when passing from Z_1 to Z_3, a conclusion that is fully confirmed by direct numerical investigations. Inside each of the regions in which the parameters space is partitioned, the average intensity and bunching $|b_h|$ match Lynden–Bell's predictions (see Fig. 14.15).

A third, intermediate region (enclosed between the two lines in Fig. 14.14), included in Z_1, can be identified, where the system initially relaxes to S_3 and only successively converges towards S_1, which is associated with the global maximum of the entropy (see Fig. 14.16 for the time evolution). The simulations also show that the time duration of the transient regime diverges with the number of particles N, and hence in the mean-field limit ($N \to \infty$) the system gets trapped in a state that does not correspond to

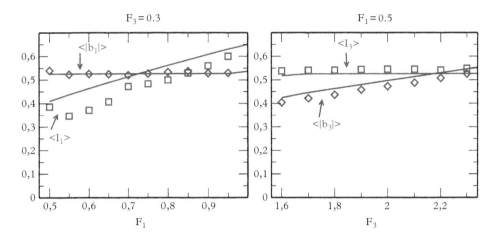

Figure 14.15 *(Left) Zone 1. Time average intensity $\langle I_1 \rangle$ and bunching $\langle |b_1| \rangle$ as a function of F_1 for $F_3 = 0.3$. (Right) Zone 3. Time average intensity $\langle I_3 \rangle$ and bunching $\langle |b_3| \rangle$ as a function of F_3 for $F_1 = 0.5$. The simulations are performed for a homogeneous initial condition with $N = 10^4$ and $H/N = 1.6 \cdot 10^{-3}$. Reprinted from Antoniazzi et al. (2008), © 2008, with permission from Elsevier.*

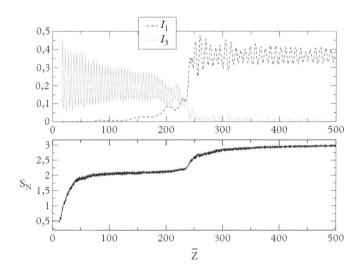

Figure 14.16 *(Top) Evolution of the intensities I_1 and I_3 in the intermediate region. (Bottom) The time evolution of the (coarse-grained) entropy (numerically estimated) indicates that the first metastable state corresponds to a saddle point. Reprinted from Antoniazzi et al. (2008), © 2008, with permission from Elsevier.*

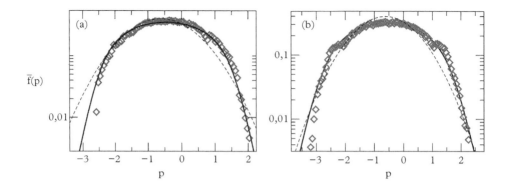

Figure 14.17 *Final equilibrium distributions in Z_1(a) and in Z_3(b). The diamonds refer to the numerical results, while the continuous lines represent the analytical result predicted in Section 14.6. The dashed lines are the curves predicted by the Boltzmann-Gibbs statistical mechanics in the microcanonical ensemble. Reprinted from Antoniazzi et al. (2008), © 2008, with permission from Elsevier.*

the maximum entropy one. This peculiar behaviour can be explained by analysing the initial exponential regime, which occurs on short timescales ($z = O(1)$). In fact, in the intermediate region delimited by the two lines in Fig. 14.14, A_3 displays a faster linear growth rate that A_1 (Bonifacio *et al.*, 1990). The scenario illustrated implies that S_3 is a saddle point for the Lynden–Bell entropy in Z_1. Unfortunately, performing a stability analysis of stationary points proves cumbersome. The time evolution of the entropy as seen in N-body simulations, presented in Fig. 14.16, provides, however, an indirect support to the above claim.

Concerning the velocity profile, a disagreement is systematically observed in the inner core, for small energy values. This discrepancy is the signature of the *macroparticle*, the localized particles agglomeration that spontaneously form in phase space. As opposed to this regime, for sensibly higher values of the energy a good agreement between theory and simulations is observed. This can be appreciated in Fig. 14.17, where the profiles relative to the Z_1 and Z_3 configurations are represented, for $p_0 = 1$, which corresponds to setting $H/N = 0.16$. The agreement is satisfactory both in the core and in the tails of the distributions. It is also important to emphasize the net improvement of Lynden–Bell prediction (solid line) over the equilibrium Boltzmann–Gibbs solution (dashed line).

14.8   Analogies with the Traveling Wave Tube

In the single-pass free electron laser, relativistic electron beams are involved, a complication that prevents the possibility of investigating experimentally the particles distribution function. Nevertheless, the model introduced to describe the FEL dynamics is quite general and applies to many different systems characterized by wave–particles interaction. Among the other apparatus that exploit the wave–particles interaction, it is worth recalling the traveling wave tubes (TWT), where an electron beam co-propagates with electrostatic longitudinal waves. This device is flexible and has been often invoked as a testbed for theory validation (Dimonte and Malmberg, 1978; Tsunoda *et al.*, 1987; Doveil and Macor, 2006). It is, in fact, relatively easy to vary the initial conditions and perform fast repetitions so to build up consistent statistics. Moreover, in a TWT the electron velocity is much less that the speed of light, and this makes it possible to easily obtain the velocity distribution functions.

A traveling wave tube (Pierce, 1950; Gilmour, 1994) consists of two main elements: an electron gun, and a slow wave structure along which waves propagate with a phase velocity much lower than the speed of light. Short TWT are commonly found in communication satellites, as a powerful signal amplifier.

The long TWT sketched in Fig. 14.18 was constructed in the laboratories of the *Equipe de Turbulence et Plasma* in Marseille (France) to mimic beam–plasma interaction (Dimonte and Malmberg, 1978; Tsunoda *et al.*, 1987; Doveil and Macor, 2006), i.e. the interaction of an electron beam inserted in a thermalized plasma. The slow wave structure is formed by a wire helix with axially movable antennas. The electron gun produces an electron beam which propagates along the axis of the slow wave structure and is confined by a strong axial magnetic field, so to prevent direct collisions between the single electrons. The central core of the TWT consists in the grid–cathode subassembly of a ceramic microwave triode and the anode is replaced by a copper plate with an on-axis hole whose aperture defines the beam diameter.

Two correction coils (not shown in Fig. 14.18) provide perpendicular magnetic fields to control the tilt of the electron beam with respect to the axis of the helix. The helix is rigidly held together by three threaded alumina rods and is enclosed by a glass vacuum tube. Four antennas can insert or detect the wave in the frequency range from 5 to 95 MHz. The wave propagates along the helix, causing the electric field to remain parallel to the beam direction.

The cumulative changes of the electron beam distribution are measured with a trochoidal velocity analyser at the end of the interaction region. A small fraction (0.5%) of the electrons passes through a hole in the centre of the front collector, and is slowed down by three retarding electrodes. By operating a selection of electrons through the use of the drift velocity caused by an electric field perpendicular to the magnetic field, the direct measurement of the current collected behind a tiny off-axis hole gives the time-averaged beam axial energy distribution (Guyomarc'h and Doveil, 2000). Retarding potential and measured current are computer-controlled, allowing an easy acquisition and treatment with an energy resolution lower than 0.5 eV.

If we put forward the hypothesis of a one-dimensional beam, the equations of motion for N electrons can be derived from the paradigmatic Hamiltonian

$$H = \sum_{i=1}^{N} \frac{p_i^2}{2} + \sum_{i=1}^{N} \sum_{k} \frac{1}{k} \left(A_k \exp[ikq_i] + c.c. \right), \tag{14.130}$$

where (p_i, q_i) represents the ith particle coordinate in the phase space, while A_k correspond to the complex amplitude of the kth harmonic of the field.

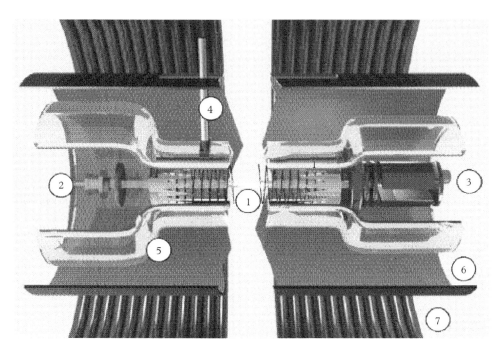

Figure 14.18 *Travelling wave tube rendering: (1) helix, (2) electron gun, (3) trochoidal analyser, (4) antenna, (5) glass vacuum tube, (6) slotted rf ground cylinder, (7) magnetic coil. Reprinted from the PhD thesis written by Alessandro Macor, D'un faisceau test à l'auto-cohérence dans l'interaction onde-particule, 17 April 2007, Université de Provence, with permission from Fabrice Doveil.*

As already discussed in the preceding sections, numerical simulations performed for the model (14.130) show a violent relaxation phase characterized by a fast exponential growth of the intensity. After this initial fast evolution, the system attains a QSS, where the intensity oscillates around a plateau. The same behaviour is also found in real experiments. The TWT could therefore materialize in an interesting and viable possibility to allow benchmarking theory and experiments.

14.9   Collective Atomic Recoil Laser (CARL)

Another interesting experimental application, which builds on the crucial interplay between particles and fields, is the collective atomic recoil laser (CARL). A CARL consists of a collection of cold two-level atoms driven by a far-detuned laser pump of frequency ω_p which radiates at the frequency $\omega \sim \omega_p$ in the direction opposite to the pump (Bonifacio and De Salvo, 1994). As for the FEL, in a CARL device the radiation process is seeded by a collective instability which originates from a symmetry breaking in the spatial distribution, i.e. a self-bunching of particles which group together in regions smaller than the wavelength. In the limit of negligible radiation pressure as due to the pump laser, the CARL is described by the very same dimensionless FEL equations

$$\frac{d\theta_j}{d\bar{t}} = p_j, \tag{14.131}$$

$$\frac{dp_j}{d\bar{t}} = -\left(Ae^{i\theta_j} + A^*e^{-i\theta_j}\right), \tag{14.132}$$

$$\frac{dA}{d\bar{t}} = \frac{1}{N}\sum_{i=1}^{N} e^{-i\theta_j} + i\delta A - \kappa A, \tag{14.133}$$

with the additional presence of a damping term $-\kappa A$ in the field equation. This latter contribution accounts for the radiation losses from a ring cavity surrounding the atoms. Although CARLs and FELs evolve with a similar dynamics, the dimensionless variables involved are very different, and so are the typical timescales of the respective processes. In a CARL, the phase and the normalized momentum of the atoms j are $\theta_j = 2k(z_j(t) - \langle v_z\rangle_0 t)$ and $p_j = m(v_{zj}(t) - \langle v_z\rangle_0)/(2\hbar k\rho_C)$ (where $z_j(t)$ and $v_{zj}(t)$ are the position and velocity of the jth atom along the direction of the scattered field and $\langle v_z\rangle_0$ is the average initial velocity), whereas A stands for the normalized complex amplitude of the radiation field, $A = (\varepsilon_0/\hbar\omega n_a\rho_C)^{1/2}E_0$, where n_a is the atomic density. The scaled time is $\bar{t} = (8\omega_{\rm rec}\rho_C)t$, where $\omega_{\rm rec} = \hbar k^2/2m$ is the recoil frequency, $\delta = (\omega - \omega_p - 2k\langle v_z\rangle_0)/(8\omega_{\rm rec}\rho_C)$ is the pump-probe detuning, $\kappa = \kappa_c/(8\omega_{\rm rec}\rho_C)$ is the scaled loss of a ring cavity with length L_{cav}, transmission T and $\kappa_c = cT/L_{cav}$, and finally $\rho_C = (\Gamma/8)(c\sigma_0 n_a/\Delta^2\omega_{\rm rec}^2)^{1/3}(I/I_{sat})^{1/3}$, where Γ is the natural decay rate of the excited state, $\sigma_0 = 3\lambda^2/2\pi$ is the scattering cross-section, $\Delta = \omega_0 - \omega_p$ is the pump-atom detuning, I is the pump intensity and $I_{\rm sat} = \hbar\omega\Gamma/2\sigma_0^2$ is the saturation intensity.

There have been different experiments that have observed the CARL effect in room temperature gases (Hemmer *et al.*, 1996), cold atomic samples from magneto-optical traps (MOT) (Kruse *et al.*, 2003; von Cube *et al.*, 2004) or Bose–Einstein condensates (BEC) (Inouye *et al.*, 1999; Bonifacio *et al.*, 2004; Slama *et al.*, 2007). CARL experiments were performed either in high-finesse optical cavities (Kruse *et al.*, 2003) or in free space (Bonifacio *et al.*, 2004), where the effect was originally interpreted as super-radiant Rayleigh scattering (Inouye *et al.*, 1999). However, it was later emphasized that these experiments in free space can be seen as a CARL process in the superradiant regime (Slama *et al.*, 2008). We refer to super-radiance when the radiation intensities scale as N^2 from a self-bunched system.

In Fallani *et al.* (2005), the CARL experiment is, for example, realized with a cigar-shaped BEC of ^{87}Rb produced in a Ioffe–Pritchard magnetic trap by means of RF-induced evaporative cooling. After 2 ms of free expansion, when the magnetic trap field is completely switched off and the atomic cloud still has an elongated shape (at this time the radial and axial sizes of the condensate are typically 10 and 70 μm, respectively), a square pulse of light is applied along the z axis (see Fig. 14.19). The size of the laser beams is larger than 0.5 mm, far larger than the condensate free fall during the interaction with light. In this geometry, the CARL process causes the pump light to be

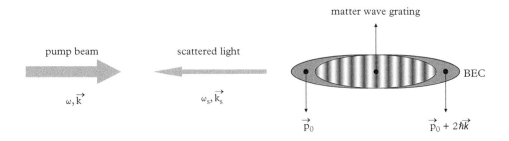

Figure 14.19 *Schematics of superradiant light scattering from a Bose–Einstein condensate (BEC). An elongated BEC is illuminated by a far off-resonant laser beam (pump beam) with frequency ω and wavevector $\vec{k}$ directed along its axial direction. After backscattering of photons with $\vec{k}_s \simeq -\vec{k}$ and the subsequent recoil of atoms, a matter wave grating forms, due to the quantum interference between the two momentum components of the wavefunction of the condensate. The effect of this grating is to further scatter the incident light in a self-amplifying process. From Bachelard* et al. *(2010), © SISSA Medialab Srl. Reproduced by permission of IOP Publishing. All rights reserved.*

backscattered and the self-amplified matter wave propagates in the same direction as the incident light.

This experiment allows for a great flexibility in the preparation of the initial state of the system: for example, it is possible to prepare the atoms in an initially bunched state by imposing an electromagnetic standing wave before the pump laser is activated. Regarding the momenta spread, it can be varied by cooling only partially the atoms. Finally, a non-zero pump-probe detuning δ can be induced by giving the atoms an initial momentum, which can be up to $1000\hbar k$: indeed, to change the initial detuning of the probe, we can either change the initial atomic momentum or change the cavity frequency, since conservation of energy imposes the frequency of the backscattered photon (Bonifacio *et al.*, 2004).

Back to the equations of motion (14.131)–(14.133), it is evident that when the dissipation term κ is small (a value $\kappa \ll 1$ means a good cavity regime), the CARL is governed by exactly the same equations as those which rule the FEL dynamics. On the other hand, the CARL experiment can be realized either without cavity or within a low-quality one, a choice which requires invoking a $\kappa > 0$ term in Eq. (14.133). In this case, if the dissipation is large enough ($\kappa \sim 1$), an adiabatic treatment of the wave dynamics can be performed (Fallani *et al.*, 2005), which corresponds to setting $dA/dt = 0$. Hence,

$$A = \frac{1}{\kappa - i\delta} \frac{1}{N} \sum_j e^{-i\theta_j}. \tag{14.134}$$

Inserting this expression into the equations of motion (14.132), we end up with the following equation for the momentum of particle j (the equation for its position is unchanged)

$$\dot{p}_j = \frac{2}{\delta^2 + \kappa^2} \frac{1}{N} \sum_m \left(\delta \sin \left(\theta_j - \theta_m \right) - \kappa \cos \left(\theta_j - \theta_m \right) \right). \tag{14.135}$$

If we now invoke the additional limit where the detuning δ is large with respect to κ ($|\delta| \gg \kappa$), Eq. (14.135) simply turns into

$$\dot{p}_j = \frac{2}{\delta} \frac{1}{N} \sum_m \sin \left(\theta_j - \theta_m \right). \tag{14.136}$$

Then, using the normalization

$$\tilde{\theta}_j = \theta_j, \tag{14.137}$$

$$\tilde{p}_j = p_j \sqrt{|\delta|/2}, \tag{14.138}$$

$$\tilde{t} = \bar{t} \sqrt{|\delta|/2}, \tag{14.139}$$

$$\tilde{H} = |\delta| H/2, \tag{14.140}$$

the CARL system is therefore mapped into the HMF model (4.1). The coupling factor $\varepsilon_J = -sign(\delta)$ can be made positive or negative, thus making it possible to probe ferromagnetic or antiferromagnetic interactions. Numerical simulations confirm that in the above-mentioned limit, despite small differences in the dynamics, the CARL and HMF yield similar QSS regimes (see Fig. 14.20).

The existence of a formal link between CARL and HMF suggests that CARL systems could eventually display the rich zoology of out-of-equilibrium features as seen in the framework of the HMF model. Consider, for instance, the ferromagnetic case: Fig. 14.21 depicts the phase diagram of the HMF model in the (m, ε) plane as predicted by the Lynden–Bell prescription, and those obtained by direct N-body simulations of both the CARL dynamics and the HMF model. An excellent agreement between the

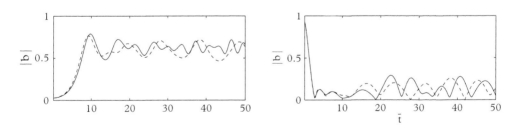

Figure 14.20 *Comparison between the dynamics of the CARL (continuous lines) and HMF (dashed lines) dynamics, in magnetized regime (left: $b_0 = 0$, $\Delta \tilde{p} = 0.1$) and unmagnetized regime (right: $b_0 = 0.94$, $\Delta \tilde{p} = 1.5$). Simulations performed with $N = 10000$ particles; CARL parameters: $\kappa = 1$ and $\delta = -4$. From Bachelard et al. (2010), © SISSA Medialab Srl. Reproduced by permission of IOP Publishing. All rights reserved.*

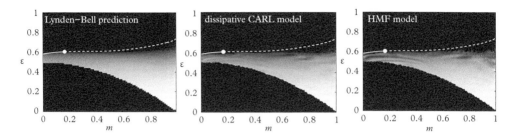

Figure 14.21 *Bunching factor as a function of the initial bunching and energy of the system, as predicted by Lynden–Bell's violent relaxation approach (left), and by N-body simulations of the CARL (centre) and HMF (right) dynamics at finite length. The white line stands for the transition as predicted by Lynden–Bell, of the first-order type below $m_c \approx 0.17$, and of the second-order kind above. N-body simulations realized with $N = 10000$ until $\bar{t} = 40$; CARL simulations performed with $\kappa = 0.5$ and $\delta = -5$. Note that the energy ε here refers to the normalized energy $\varepsilon = \Delta \tilde{p}^2/6 + (1-m^2)/2$. Note that the magnetization m in the HMF model corresponds to the bunching parameter b_0 in the CARL. From Bachelard et al. (2010), © SISSA Medialab Srl. Reproduced by permission of IOP Publishing. All rights reserved.*

CARL phase diagram and its reduced counterpart, the HMF, is observed, as well as with the Lynden–Bell theoretical prediction. This allows us to conclude on the presence of an out-of-equilibrium phase transition in the CARL device, of the same type of that predicted for the HMF model (Antoniazzi *et al.*, 2007b). CARL is therefore a particularly attractive experimental platform that could eventually enable us to explore and characterize a wide range of peculiar phenomena as displayed by long-range interacting systems.

15

Dipolar Systems

15.1 Introduction

Mean-field magnetic models were extensively treated in Chapter 4, while Chapter 5 has considered magnetic systems with slowly decaying (i.e. long-range) ferromagnetic exchange interactions. This chapter is devoted to the study of systems with magnetic dipolar interactions, which in 3D are only marginally long range, as already noted. From the statistical mechanics point of view, electric and magnetic dipolar systems do not differ, since the structure of the dipole–dipole interaction between electric dipoles and between magnetic dipoles is the same. However, since magnetic dipolar systems are more easily realized in real systems, we have chosen here to focus on magnetic interactions.

The fundamental physical field describing magnetic interactions, $\mathbf{B}(\mathbf{r})$, is usually denoted as the magnetic induction field, keeping the term magnetic intensity field for the $\mathbf{H}(\mathbf{r})$, introduced in the presence of macroscopic bodies, as shown later. Both are fields and both are due to magnetic properties; anyway, $\mathbf{H}(\mathbf{r})$ is often called magnetic field, without further specification, and $\mathbf{B}(\mathbf{r})$ magnetic induction. In this chapter we will conform to this use.

We know that the magnetic induction $\mathbf{B}(\mathbf{r})$ created by a magnetic dipole $\mathbf{m}_1$ placed in $\mathbf{r}'$ is given by

$$\mathbf{B}(\mathbf{r}) = \frac{\mu_0}{4\pi} \frac{3\,(\mathbf{r}-\mathbf{r}')\left[(\mathbf{r}-\mathbf{r}')\cdot\mathbf{m}_1\right]-\mathbf{m}_1\,|\mathbf{r}-\mathbf{r}'|^2}{|\mathbf{r}-\mathbf{r}'|^5}. \tag{15.1}$$

To derive the interaction energy between the magnetic dipole $\mathbf{m}_1$ and another magnetic dipole $\mathbf{m}_2$, we use the fact that the energy of the magnetic dipole $\mathbf{m}_2$, once placed in a magnetic induction $\mathbf{B}$, is $-\mathbf{m}_2 \cdot \mathbf{B}$. We stress that this is the energy of a *permanent* dipole $\mathbf{m}_2$, namely we are considering the case in which, when $\mathbf{m}_2$ moves in the magnetic induction $\mathbf{B}$, no work must be done to maintain the modulus m_2 of $\mathbf{m}_2$ fixed. This is the setting that applies to the cases of interest here, with the dipoles resulting from the intrinsic spin of the particles or from molecular orbitals. Therefore, the dipole–dipole interaction between the dipole $\mathbf{m}_1$ placed in $\mathbf{r}'$ and the dipole $\mathbf{m}_2$ placed in $\mathbf{r}$ is

Physics of Long-Range Interacting Systems. First Edition. A. Campa *et al.*
© A. Campa, T. Dauxois, D. Fanelli, and S. Ruffo 2014. Published in 2014 by Oxford University Press.

$$\frac{\mu_0}{4\pi} \frac{\mathbf{m}_1 \cdot \mathbf{m}_2 |\mathbf{r} - \mathbf{r}'|^2 - 3\left[(\mathbf{r} - \mathbf{r}') \cdot \mathbf{m}_1\right]\left[(\mathbf{r} - \mathbf{r}') \cdot \mathbf{m}_2\right]}{|\mathbf{r} - \mathbf{r}'|^5}. \tag{15.2}$$

The first thing to note is the anisotropy of this interaction. For example, for two parallel dipoles of modulus m_1 and m_2, at distance r, the interaction energy is equal to $-2m_1 m_2/r^3$ if the distance vector $\mathbf{r}$ is parallel to the dipoles, while it is equal to $m_1 m_2/r^3$ if the distance vector is orthogonal to the dipoles. This rather complex angular dependence determines, as we will see, the properties of magnetized bodies.

At the microscopic level, the magnetic dipoles (or magnetic moments) stem from the angular momentum and from the intrinsic spin of the electrons. Indicating the operators with a circumflex over the symbol, for an atom we have

$$\hat{\mathbf{m}} = -\mu_B \left(\hat{\mathbf{L}} + g\hat{\mathbf{S}}\right), \tag{15.3}$$

where $\hat{\mathbf{L}}$ and $\hat{\mathbf{S}}$ are the total electronic orbital angular momentum and the total electronic spin, respectively (in units of $\hbar$), $\mu_B = e\hbar/(2m_e)$ is the Bohr magneton (with m_e the electron mass) and g is the electron spin g-factor. Therefore the order of magnitude of the dipolar interactions between two atoms at a distance r is μ_B^2/r^3. It turns out that this quantity, computed for a typical interatomic distance, is generally $10^2 - 10^3$ times smaller than the Heisenberg exchange interactions between nearby spins. The exchange interaction in magnetic systems is usually restricted to nearest neighbour spins or, at most, next to nearest neighbour spins. Therefore, magnetic systems are generally characterized by the presence of both a strong short-range exchange interaction and a weak dipole–dipole interaction between the electronic dipole moments shown in Eq. (15.3). Then, the interaction between nearby dipoles will be dominated by the exchange interaction.

Although the dipole–dipole interaction is pairwise much weaker than the exchange interaction, the long-range character of the former prevails in macroscopic systems in determining the large scale properties of the spontaneous magnetization of magnetic systems. It could be argued that this is to be expected, since a non-integrable interaction (although only marginally because of the $1/r^3$ decay), even if pairwise much weaker than a short-range interaction, should become dominant for sufficiently large systems. However, the anisotropy of the dipole–dipole interaction makes the last argument not obvious; as a matter of fact, we will see that, due to the anisotropy, the limit of the energy per dipole in the thermodynamic limit is finite (in other words the energy is extensive), without the use of any such normalization as the Kac's prescription. We can say that the dipole–dipole interaction is conditionally integrable.

In this chapter, only the most relevant properties of dipolar systems can be treated. The reader interested in going deeper into the subject can consult textbooks or reviews on the theory of magnetism and of the properties of magnetic materials, e.g. Kittel (1949), Morrish (1965) and White (1970). In the spirit of this book, our purpose here is to present the characteristics of these systems, mainly in comparison to those of other long-range systems. To this end, we will be therefore interested in showing the proof of the existence of the thermodynamic limit, i.e. of the limit of the free energy per particle. The proof is due to Griffiths (1968). We will then see the consequences of this fact on the structure of the magnetization of ferromagnetic materials. Finally, we will give an overview of some experimental studies of systems where the dipolar interactions play the preeminent role.

We should note that the origin of the magnetic dipoles from intrinsic spins and from molecular orbitals would in turn imply a quantum mechanical treatment of the problem, with the magnetic dipoles represented, as in Eq. (15.3), by operators expressed in terms of the spins and the angular momenta. However, in the proof of the existence of the thermodynamic limit, we will adopt a classical approach (but in Section 15.3.5 we will show the necessary further steps for the proof of the subadditivity of the free energy in the quantum case). In fact, the statistical properties of the dipole–dipole interaction, as determined by its long-range character, do not change between the classical and the quantum case, and therefore we can avoid the technical difficulties that arise from a fully quantum treatment. This does not mean that, in the application to a specific real system, the quantitative determination of its properties can avoid the use of quantum mechanics.

We will see that the marginal long rangedness of the dipolar interaction, together with its rather complicated dependence on the dipole orientation, induces a behaviour quite different from that resulting from long-range ferromagnetic exchange interactions. Therefore, we find it useful at first to remind some of the results obtained for systems with exchange type long-range ferromagnetic interactions, and to deduce some further properties of these systems, not explicitly considered previously.

It was shown that the thermodynamic limit in both the canonical and microcanonical ensembles can be solved exactly for periodic boundary conditions. In this case, the equilibrium state of the system is either a demagnetized state or a uniformly magnetized state. In the more realistic case of free boundary conditions, the calculations cannot, in general, be completed analytically. The α-Ising model in a one-dimensional lattice, normalized with Kac's prescription, was studied in detail in Section 5.2, showing how the large deviations theory makes it possible to arrive at a self-consistent equation in the functional space of the magnetization profile; this equation must be solved numerically. This procedure can be extended to lattices of larger dimension and of any structure (e.g. in 3D simple cubic or fcc), and also to rotators (e.g. XY models).

The key point that we want to stress here is the size and shape dependence of the magnetization profile that we should expect in long-range magnetic systems, a fact that has already been pointed out. The just recalled treatment of the one-dimensional α-Ising model offers a simple way of understanding the origin of this dependence. In fact, if we consider the α-Ising model in a larger dimension (again normalized with Kac's prescription), it is clear that in the thermodynamic limit we arrive at a self-consistent equation similar to that for the one-dimensional case, but now in a different functional space, namely in the space of the functions of as many variables as the number of dimensions. Although the solution will be obtained only numerically, we can infer that it will have the following properties:

(i) Above a certain critical temperature in the canonical ensemble, or above a certain critical energy in the microcanonical ensemble, the equilibrium state will be demagnetized (i.e. the magnetization profile will be constant, equal to zero), while below the critical values the magnetization profile will be nonhomogeneous;

(ii) The nonhomogeneous magnetized states will be characterized by larger magnetization values near the centre of the lattice, and smaller values near the edges of the lattice;

(iii) The up–down symmetry of the Ising Hamiltonian (or more generally, the rotational invariance for the rotators Hamiltonian) leaves the direction of the magnetization undetermined; however this direction will be the same throughout the lattice;

(iv) At zero temperature, the system will be in its uniformly and fully magnetized ground state;

(v) the self-consistent equation is obtained after normalizing the sample size to a given region of $\mathbb{R}^3$; therefore for a given lattice structure and sample shape the magnetization profile of any (large) system will be the same after normalization to the size;

(vi) However, for a given lattice structure, this profile will be shape-dependent;

(vii) As a consequence, the energy per site in the microcanonical ensemble or the free energy per site in the canonical ensemble will be shape-dependent;

(viii) Also the energy per site in the ground state will be shape-dependent, despite the uniform full magnetization.

If we do not introduce Kac's prescription, the previous points maintain their validity despite the fact that thermodynamic quantities like energy and free energy are no more extensive, with the only difference that the critical values of the thermodynamic variables will be size-dependent.

We want, in particular, to call attention to points (iii), (vii) and (viii), where the most significant differences in the dipolar interaction case will be found. As far as the last two points are concerned, we will show that a rather general theorem by Griffiths proves that the free energy per site at any temperature is shape independent in the thermodynamic limit; therefore, in particular, also the energy per site in the ground state (i.e. the free energy per site at zero temperature) is shape independent. From this, we will be able to deduce, although not on the basis of a rigorous mathematical proof, that point (iii) does not apply in dipolar systems.

In the second section, we start with the macroscopic magnetostatic equations. We will show the origin of the so-called demagnetizing field, which arises in the interior of magnetized bodies. Many of the expressions derived in that section will prove to be useful also in the treatment of the microscopic dipolar interaction, studied in the third section. The main purpose there will be the proof of the shape independence of the free energy in the thermodynamic limit. The fourth section will be devoted to studying the physical consequences of the theorem for the equilibrium configuration of magnetic systems. In the last section, we will try to give a flavour of some recent experimental studies dedicated to the observation of equilibrium and dynamical properties in dipolar systems.

15.2   The Demagnetizing Field

Macroscopic magnetostatics is described by the equations

$$\nabla \times \mathbf{H} = \mathbf{J}, \tag{15.4}$$

$$\nabla \cdot \mathbf{B} = 0, \tag{15.5}$$

where $\mathbf{J}$ is the macroscopic current density. The magnetic field $\mathbf{H}$ is defined by the relation

$$\mathbf{B} = \mu_0 \left(\mathbf{H} + \mathbf{M} \right), \tag{15.6}$$

where $\mathbf{M}$ is the magnetization density (or magnetic moment density) of the macroscopic bodies. To have a closed set of equations, we need a constitutive relation between $\mathbf{B}$ and $\mathbf{H}$. In non-ferromagnetic bodies, these two quantities are linearly related (for isotropic bodies the two vectors are proportional), but we are treating ferromagnetic systems, where the relation is more complicated.

However, there is a simplifying feature that comes from the fact that we are interested in the case in which there are no macroscopic currents, i.e. $\mathbf{J} = 0$. Then, we consider the equations

$$\nabla \times \mathbf{H} = 0, \tag{15.7}$$

$$\nabla \cdot \left(\mathbf{H} + \mathbf{M} \right) = 0. \tag{15.8}$$

A constitutive relation is still necessary to have a closed set of equations, but now we are going to study these equations assuming the magnetic moment density field $\mathbf{M}$ as given, thus bypassing this necessity. This is a physically justified procedure for hard ferromagnets, which have a magnetization density independent from the applied external magnetic fields, at least for moderate field intensities. Furthermore, this procedure is useful for our purpose here, where we want to show the dependence of the energy of the system on the magnetization profile.

Therefore, we are going to find the solution of Eqs. (15.7) and (15.8) in the case in which there is a region of space occupied by a magnetized body with a given magnetization density $\mathbf{M}(\mathbf{r})$. The equations imply that at the surface of the body the tangential component of the vector $\mathbf{H}$ and the normal component of the vector $\mathbf{H} + \mathbf{M}$ are continuous. The first equation also implies that the magnetic field $\mathbf{H}$ is the gradient of a scalar field, i.e. the scalar potential Φ

$$\mathbf{H} = -\nabla \Phi. \tag{15.9}$$

Therefore, we must solve the equation

$$\Delta \Phi = \nabla \cdot \mathbf{M}, \tag{15.10}$$

subject to the mentioned boundary conditions at the surface of the magnetized body. The solution is given by

$$\Phi(\mathbf{r}) = -\frac{1}{4\pi} \int_V d\mathbf{r}' \frac{\nabla' \cdot \mathbf{M}(\mathbf{r}')}{|\mathbf{r}-\mathbf{r}'|} + \frac{1}{4\pi} \int_S dS \frac{\mathbf{M}(\mathbf{r}') \cdot \mathbf{n}}{|\mathbf{r}-\mathbf{r}'|} \tag{15.11}$$

$$= \frac{1}{4\pi} \int_V d\mathbf{r}' \, \mathbf{M}(\mathbf{r}') \cdot \nabla' \frac{1}{|\mathbf{r}-\mathbf{r}'|}, \tag{15.12}$$

where the subscripts V and S denote integrals extended to the volume and the surface of the magnetized body, respectively, and where the prime on the symbol ∇ means that it acts on the variable $\mathbf{r}'$; $\mathbf{n}$ is the outward normal to the surface S.

15.2.1   Uniformly magnetized bodies

The expression for $\Phi(\mathbf{r})$ in Eq. (15.11) shows clearly that in a uniformly magnetized body the scalar potential Φ and thus the magnetic field $\mathbf{H}$ depend only on the magnetization density at the surface of the body. With $\mathbf{M}$ the constant magnetization density, the use of Eq. (15.9) and of Eq. (15.12) allows us to write

$$\mathbf{H}(\mathbf{r}) = \frac{1}{4\pi} \nabla \left(\mathbf{M} \cdot \nabla \int_V d\mathbf{r}' \frac{1}{|\mathbf{r}-\mathbf{r}'|} \right). \tag{15.13}$$

Denoting the symmetric tensor (also called demagnetization tensor)

$$\overline{D}(\mathbf{r}) = -\frac{1}{4\pi} \nabla\nabla \left(\int_V d\mathbf{r}' \frac{1}{|\mathbf{r}-\mathbf{r}'|} \right), \tag{15.14}$$

we can also write

$$\mathbf{H}(\mathbf{r}) = -\overline{D}(\mathbf{r}) \cdot \mathbf{M}. \tag{15.15}$$

We now treat in detail the simplest case, i.e. that in which V is a sphere; we denote by $\mathbf{r}_0$ and by a the position of the centre of the sphere and its radius, respectively. The relations

$$\int_{V_{\mathbf{r}_0,a}} d\mathbf{r}' \frac{1}{|\mathbf{r}-\mathbf{r}'|} = \begin{cases} \frac{4\pi a^3}{3} \frac{1}{|\mathbf{r}-\mathbf{r}_0|} & \text{for } |\mathbf{r}-\mathbf{r}_0| \geq a \\[2mm] 2\pi a^2 - \frac{2\pi}{3} |\mathbf{r}-\mathbf{r}_0|^2 & \text{for } |\mathbf{r}-\mathbf{r}_0| \leq a, \end{cases} \tag{15.16}$$

which can be derived via elementary calculations, are a consequence of the Gauss theorem; $V_{\mathbf{r}_0,a}$ indicates the volume occupied by the sphere. We now consider separately the cases in which $|\mathbf{r}-\mathbf{r}_0| \geq a$ and $|\mathbf{r}-\mathbf{r}_0| \leq a$, i.e. points outside and inside the sphere, respectively.

In the first case, we have

$$\overline{D}(\mathbf{r}) = -\frac{a^3}{3} \left[3 \frac{(\mathbf{r} - \mathbf{r}_0)(\mathbf{r} - \mathbf{r}_0)}{|\mathbf{r} - \mathbf{r}_0|^5} - \frac{I}{|\mathbf{r} - \mathbf{r}_0|^3} \right],$$ (15.17)

where I is the unit tensor. Therefore, from Eq. (15.15), the magnetic field is

$$\mathbf{H}(\mathbf{r}) = \frac{a^3}{3} \frac{3(\mathbf{r} - \mathbf{r}_0)[(\mathbf{r} - \mathbf{r}_0) \cdot \mathbf{M}] - \mathbf{M}|\mathbf{r} - \mathbf{r}_0|^2}{|\mathbf{r} - \mathbf{r}_0|^5}.$$ (15.18)

From Eq. (15.6), we see that the magnetic induction field $\mathbf{B}(\mathbf{r})$ is equal to $\mu_0 \mathbf{H}(\mathbf{r})$, since outside the sphere $\mathbf{M} = 0$. By comparison with Eq. (15.1), we then obtain that the magnetic induction is the same as that generated by a dipole equal to $(4\pi a^3/3)\,\mathbf{M}$, i.e. equal to the total magnetization of the sphere, placed at the centre of the sphere. Only for a magnetized body of spherical shape, the property that the magnetic induction outside the body is identical to that generated by a dipole holds.

On the other hand, inside the sphere, we have

$$\overline{D}(\mathbf{r}) = \frac{1}{3} I,$$ (15.19)

i.e. a constant diagonal tensor. Then, inside the sphere the magnetic field is uniform and equal to

$$\mathbf{H}(\mathbf{r}) = -\frac{1}{3}\mathbf{M}.$$ (15.20)

The magnetic induction is therefore uniform, and equal to

$$\mathbf{B}(\mathbf{r}) = \mu_0 \frac{2}{3}\mathbf{M}.$$ (15.21)

In Fig. 15.1, we see a schematic representation of the three constant vectors $\mathbf{M}$, $\mathbf{H}$ and $\mathbf{B}$ inside a spherical uniformly magnetized body.

From Eq. (15.16), we can derive also the scalar potential Φ outside and inside the sphere. We have

$$\Phi(\mathbf{r}) = \begin{cases} \dfrac{a^3}{3} \dfrac{\mathbf{M} \cdot (\mathbf{r} - \mathbf{r}_0)}{|\mathbf{r} - \mathbf{r}_0|^3} & \text{for } |\mathbf{r} - \mathbf{r}_0| \geq a, \\[2ex] \dfrac{1}{3} \mathbf{M} \cdot (\mathbf{r} - \mathbf{r}_0) & \text{for } |\mathbf{r} - \mathbf{r}_0| \leq a; \end{cases}$$ (15.22)

two expressions which will be useful later.

We note that because of the linearity of the magnetostatic equations, if we place the sphere in a pre-existing external magnetic induction $\mathbf{B}_0$ constant throughout all space,

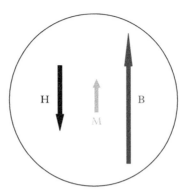

Figure 15.1 *Schematic representation of the fields inside a uniformly magnetized sphere. The magnetic induction field is parallel to the magnetization density, while the magnetic intensity field in antiparallel. The length of the arrows for $\mathbf{H}$ and $\mathbf{B}$ are proportional to the magnitudes of $\mathbf{H}$ and $\mathbf{B}/\mu_0$, while for clarity the length of the arrow for $\mathbf{M}$ has been decreased.*

then inside the sphere the magnetic field $\mathbf{H}$ will be given by $-\frac{1}{3}\mathbf{M} + \frac{1}{\mu_0}\mathbf{B}_0$, and the magnetic induction $\mathbf{B}$ by $\frac{2\mu_0}{3}\mathbf{M} + \mathbf{B}_0$. When there is no external magnetic induction $\mathbf{B}_0$, we see that in the case of a spherical body the magnetic field $\mathbf{H}$ is antiparallel to the magnetization density. For bodies of arbitrary shape, the magnetic field inside the body will not be uniform but, in general, it will tend to have a negative projection on the uniform magnetization density, as in the case of the sphere; this is the origin of the term demagnetizing field for $\mathbf{H}(\mathbf{r})$ inside the magnetized body.

It turns out that the uniformity of the magnetic field inside a uniformly magnetized body holds also for any ellipsoidal shape, characterized by the three semi-axes a, b and c. To prove this fact we must choose the coordinate axes parallel to the ellipsoidal axes, and must solve the Laplace equation in elliptical coordinates (Landau and Lifshitz, 1984). For $\mathbf{r}$ inside the ellipsoid, the off-diagonal components of the tensor $\overline{D}(\mathbf{r})$ vanish, while the diagonal components are constant, with values (Akhiezer *et al.*, 1968; Landau and Lifshitz, 1984) given by

$$\overline{D}_1 = \frac{abc}{2} \int_0^\infty ds \frac{1}{(a^2 + s)\, R_s}, \tag{15.23}$$

$$\overline{D}_2 = \frac{abc}{2} \int_0^\infty ds \frac{1}{(b^2 + s)\, R_s}, \tag{15.24}$$

$$\overline{D}_3 = \frac{abc}{2} \int_0^\infty ds \frac{1}{(c^2 + s)\, R_s}, \tag{15.25}$$

where

$$R_s = \left[(a^2 + s)\, (b^2 + s)\, (c^2 + s) \right]^{\frac{1}{2}}. \tag{15.26}$$

The three quantities $\overline{D}_1$, $\overline{D}_2$ and $\overline{D}_3$ are called the demagnetization coefficients. Being a tensor, $\overline{D}(\mathbf{r})$ for an ellipsoidal shape will be constant also in a generic coordinate frame, although the off-diagonal components in general will not vanish. Since $\overline{D}$ is not proportional to the unit tensor, the uniform magnetic field inside the body is not antiparallel to the magnetization density but, in general, it will have, as already noted, a negative projection on the uniform magnetization density. Outside the ellipsoid, the magnetic field will not have the simple dipolar form as for the sphere; its exact form, obtained again with the use of elliptical coordinates, is not of interest here. We only note that far from the body it will tend to assume the dipolar form.

It can be easily verified that the three demagnetization coefficients depend only on the ratios of the axes, e.g. on a/b and a/c, and that they sum up to 1; therefore, for a given ellipsoidal shape they do not depend on the volume of the body. From the result for the sphere, we see that for that geometry we have $\overline{D}_1 = \overline{D}_2 = \overline{D}_3 = 1/3$; the equality of the three quantities is an obvious consequence of the rotational invariance of the sphere. In general, if $a > b > c$, then we will have $\overline{D}_1 < \overline{D}_2 < \overline{D}_3$. It is interesting to see the expressions of the demagnetization coefficients for ellipsoids of revolution.

In the case of a prolate ellipsoid of revolution, i.e. $a \geq b = c$, it is not difficult to derive that

$$\overline{D}_1 = \frac{1-v^2}{2v^3}\left(\log\frac{1+v}{1-v} - 2v\right) \qquad \text{with} \qquad v \equiv \left(1 - \frac{b^2}{a^2}\right)^{1/2}, \qquad (15.27)$$

$$\overline{D}_2 = \overline{D}_3 = \frac{1}{2}\left(1 - \overline{D}_1\right). \qquad (15.28)$$

It can be shown that for a varying between b and infinite, i.e. for $0 \leq v \leq 1$, $\overline{D}_1$ decreases monotonically from 1/3 to 0. While for $v = 0$ we recover the case of the sphere, $v = 1$ represents a cylinder. We thus see that a uniform magnetization density parallel to the axis of a cylindrical body does not produce a magnetic field inside it. For a very elongated ellipsoid, we have a situation very close to that of a cylinder (see Fig. 15.2). The oblate ellipsoid of revolution is given by axes such that $a \leq b = c$. In this case, we find that

$$\overline{D}_1 = \frac{1+w^2}{w^3}(w - \arctan w) \qquad \text{with} \qquad w \equiv \left(\frac{b^2}{a^2} - 1\right)^{\frac{1}{2}}, \qquad (15.29)$$

$$\overline{D}_2 = \overline{D}_3 = \frac{1}{2}\left(1 - \overline{D}_1\right). \qquad (15.30)$$

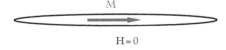

M

H ≈ 0

Figure 15.2 *The magnetic field inside a very elongated ellipsoid of revolution is practically 0, if the uniform magnetization density is parallel to the long axis of the ellipsoid.*

For a varying between b and 0, i.e. for $0 \leq w \leq \infty$, $\overline{D}_1$ increases monotonically from 1/3 to 1. The upper bound represents a disk-shaped body. We see that for this shape a uniform magnetization density without a component perpendicular to the disk plane does not produce a magnetic field inside the body.

For a general shape, the mentioned dependence of the demagnetization coefficients only on the ratio of the axes allows us to conclude that the constant value of the field inside the body, for a given $\mathbf{M}$, does not depend on the size of the ellipsoid. Actually, this property of size independence can be extended to any shape, just by looking at the expression for the scalar field $\Phi(\mathbf{r})$ in Eq. (15.11). For a constant $\mathbf{M}$, only the surface integral is present and, by dimensional analysis, it is clear that, when the body size at constant shape is increased, the magnetic field profile inside the body will not change.

The last property suggests that the magnetostatic energy of a uniformly magnetized body will be an extensive quantity, for a given body shape. We emphasize the importance of this fact, in view of the (marginal) long-range character of the dipolar interaction. This important property will be studied later in this chapter, in general terms, when treating the statistical mechanics of a system of dipoles. For the moment, we shall discuss it with reference to the case of uniformly magnetized macroscopic bodies.

15.2.2   The magnetostatic energy

The energy of a magnetic dipole $\mathbf{m}$ placed in a magnetic induction $\mathbf{B}$, as noted in the introduction of this chapter, is $-\mathbf{m} \cdot \mathbf{B}$. Since the components of the magnetic induction created by a dipole (see Eq. (15.1)) depend linearly on the components of the dipole, it is a simple matter to show that the magnetostatic energy of a system of N magnetic dipoles is

$$-\frac{1}{2} \sum_{i=1}^{N} \mathbf{m}_i \cdot \mathbf{B}_i, \tag{15.31}$$

where $\mathbf{B}_i$ is the magnetic induction created at the position of the magnetic dipole $\mathbf{m}_i$ by all the other magnetic dipoles. We want now to extend this expression to the case in which the dipoles $\mathbf{m}_i$ are substituted by a continuous magnetization density $\mathbf{M}(\mathbf{r})$. Some care must be adopted since, as in Eq. (15.31), we must consider at any point the induction created by the magnetization density present outside that point. Previously, we noted the magnetic induction inside a uniformly magnetized sphere, with magnetization density $\mathbf{M}$, placed in a pre-existing uniform external magnetic induction $\mathbf{B}_0$, is $\frac{2\mu_0}{3}\mathbf{M} + \mathbf{B}_0$. If the external induction is general, but the magnetized sphere is infinitesimal, we infer that the magnetic induction inside the sphere will be $\frac{2\mu_0}{3}\mathbf{M} + \mathbf{B}$, where now $\mathbf{B}$ is the induction created at the position of the infinitesimal sphere by all the other sources. Therefore, we can say that the magnetic induction created inside a uniformly polarized infinitesimal sphere by all sources outside it will be equal to the total magnetic induction inside the infinitesimal sphere minus $\frac{2\mu_0}{3}\mathbf{M}$. Thus, we arrive at the conclusion that the magnetostatic energy of a magnetized body is

$$E = -\frac{1}{2} \int_V dr \left(\mathbf{M}(\mathbf{r}) \cdot \mathbf{B}(\mathbf{r}) - \frac{2\mu_0}{3} M^2(\mathbf{r}) \right),$$
(15.32)

$$= -\frac{\mu_0}{2} \int_V dr \left(\mathbf{M}(\mathbf{r}) \cdot \mathbf{H}(\mathbf{r}) + \frac{1}{3} M^2(\mathbf{r}) \right),$$
(15.33)

where $M^2 = \mathbf{M} \cdot \mathbf{M}$. The integral of the first term is called the demagnetizing energy. If we apply Eq. (15.33) to the case of a uniformly magnetized ellipsoid, with the help of Eq. (15.15), we have

$$E = \frac{\mu_0 V}{2} \left(\mathbf{M} \cdot \overline{D} \cdot \mathbf{M} - \frac{1}{3} M^2 \right).$$
(15.34)

We clearly see the extensivity of this expression but, at the same time, we also see its shape dependence, since the values of the elements of the tensor $\overline{D}$ depend on the shape of the ellipsoid. We note that for the particular case of a sphere, using Eq. (15.19), we have $E = 0$.

To conclude this subsection, we show another expression for the magnetostatic energy, useful for obtaining a lower bound for a generic magnetization profile. The integrals in Eq. (15.33) can be clearly extended to all space since outside the volume V, where $\mathbf{M} = 0$, they give no contribution. The first term inside brackets can be transformed expressing $\mathbf{M}$ through $\mathbf{B}$ and $\mathbf{H}$. We then obtain

$$-\frac{1}{2} \int dr \, \mathbf{M}(\mathbf{r}) \cdot \mathbf{H}(\mathbf{r}) = -\frac{1}{2\mu_0} \int dr \, \mathbf{B}(\mathbf{r}) \cdot \mathbf{H}(\mathbf{r}) + \frac{1}{2} \int dr \, H^2(\mathbf{r}).$$
(15.35)

But

$$\mathbf{B} \cdot \mathbf{H} = -\mathbf{B} \cdot \nabla \Phi = -\nabla \cdot (\Phi \mathbf{B}) + \Phi \nabla \cdot \mathbf{B} = -\nabla \cdot (\Phi \mathbf{B}).$$
(15.36)

The integral of the right-hand side on all space becomes an integral on the surface at infinite, and then it vanishes. We therefore have

$$E = \frac{\mu_0}{2} \int dr \, H^2(\mathbf{r}) - \frac{\mu_0}{6} \int_V dr \, M^2(\mathbf{r}).$$
(15.37)

The first term on the right-hand side, i.e. the demagnetizing energy, is positive definite; therefore, the second term gives an extensive lower bound to the magnetostatic energy.

15.3   The Thermodynamic Limit for Dipolar Media

We now turn to the statistical mechanics of dipolar media, and therefore to a microscopic treatment. Then, we will study a system of N magnetic dipoles that reside on the sites of a given three-dimensional lattice; the ith dipole $\mathbf{m}_i$ is placed at position $\mathbf{r}_i$. We suppose for the moment that the dipole–dipole interaction is the only one; the presence of *short-range* exchange interactions will be considered later. Using Eqs. (15.1) and (15.31), we can write the total energy of the systems as

$$E = \frac{\mu_0}{8\pi} \sum_{i=1}^{N} \sum_{j=1}^{N}{}' \frac{\mathbf{m}_i \cdot \mathbf{m}_j \, |\mathbf{r}_i - \mathbf{r}_j|^2 - 3 \left[(\mathbf{r}_i - \mathbf{r}_j) \cdot \mathbf{m}_i\right]\left[(\mathbf{r}_i - \mathbf{r}_j) \cdot \mathbf{m}_j\right]}{|\mathbf{r}_i - \mathbf{r}_j|^5} \tag{15.38}$$

$$= \frac{\mu_0}{8\pi} \sum_{i=1}^{N} \sum_{j=1}^{N}{}' \frac{1}{|\mathbf{r}_i - \mathbf{r}_j|^3} \left\{ \mathbf{m}_i \cdot \mathbf{m}_j - 3 \frac{\left[(\mathbf{r}_i - \mathbf{r}_j) \cdot \mathbf{m}_i\right]\left[(\mathbf{r}_i - \mathbf{r}_j) \cdot \mathbf{m}_j\right]}{|\mathbf{r}_i - \mathbf{r}_j|^2} \right\}, \tag{15.39}$$

where the prime on the summation symbol means that the terms with $i = j$ are absent. We consider the general case in which the moduli of the dipoles need not be equal. For later use, we denote by m_0 the largest dipole module.

In the thermodynamic limit, we could expect that the energy per site diverges as the logarithm of the size of the lattice, because of the $1/r^3$ behaviour of the dipole–dipole interaction. However, we have shown that there is an extensive lower bound for the magnetostatic energy of macroscopic bodies; we expect this to hold also here and to find that, in the ground state, the energy per site remains finite. We will now show that actually in any configuration the energy per site remains finite in the thermodynamic limit.

It could be argued that the extensivity of the lower bound is sufficient to have a proper thermodynamic behaviour. In fact, we will show that the extensivity of the free energy obtained from the canonical ensemble is proven using only the lower bound. Nevertheless, we think that finding also an extensive upper bound for the energy is a useful exercise.

15.3.1   Some useful relations

Our starting point will be a relation like that reported in Eq. (15.16). We imagine each lattice site $\mathbf{r}_i$ to be the centre of a sphere V_i; the common radius a of all the spheres is less than one half of the minimum distance between the lattice sites (Griffiths, 1968). Therefore there is no overlap between the spheres. We will now derive several relations, whose physical meaning will become clear shortly, with the computation of the bounds for the interaction energy.

Considering a particular sphere V_j of radius a, we have

$$\frac{1}{|\mathbf{r} - \mathbf{r}_j|} = \frac{3}{4\pi a^3} \int_{V_j} d\mathbf{r}' \frac{1}{|\mathbf{r} - \mathbf{r}'|} \qquad \text{for} \qquad |\mathbf{r} - \mathbf{r}_j| \geq a. \tag{15.40}$$

With a change of integration variable this becomes

$$\frac{1}{|\mathbf{r} - \mathbf{r}_j|} = \frac{3}{4\pi a^3} \int_{V_a} d\mathbf{r}' \frac{1}{|\mathbf{r} - \mathbf{r}_j - \mathbf{r}'|} \qquad \text{for} \qquad |\mathbf{r} - \mathbf{r}_j| \geq a, \tag{15.41}$$

where now the integration is over the sphere V_a of radius a centred in the origin. From this relation, we can derive a vectorial and a tensorial equality. In the following expressions, it is understood that $|\mathbf{r}-\mathbf{r}_j| \geq a$, unless otherwise specified. Applying the negative of the gradient operator to Eq. (15.41), we have

$$\frac{\mathbf{r}-\mathbf{r}_j}{|\mathbf{r}-\mathbf{r}_j|^3} = -\frac{3}{4\pi a^3} \int_{V_a} d\mathbf{r}'\, \boldsymbol{\nabla} \frac{1}{|\mathbf{r}-\mathbf{r}_j-\mathbf{r}'|} \tag{15.42}$$

$$= +\frac{3}{4\pi a^3} \int_{V_a} d\mathbf{r}'\, \boldsymbol{\nabla}' \frac{1}{|\mathbf{r}-\mathbf{r}_j-\mathbf{r}'|}. \tag{15.43}$$

Now, we use the following relation, valid for any volume V bounded by the surface S

$$\int_V d\mathbf{r}'\, \boldsymbol{\nabla}' f(\mathbf{r}') = \int_S dS'\, \mathbf{n} f(\mathbf{r}'), \tag{15.44}$$

with $\mathbf{n}$ the outward normal to the surface. For the sphere of radius a centred in the origin, we have

$$\mathbf{n}\, dS' = a\mathbf{r}'\, d\Omega', \tag{15.45}$$

where $d\Omega'$ is the surface element of the sphere with unit radius (this slight abuse of notation will not create confusion). We therefore obtain the vectorial relation

$$\frac{\mathbf{r}-\mathbf{r}_j}{|\mathbf{r}-\mathbf{r}_j|^3} = \frac{3}{4\pi a^2} \int_{S_a} d\Omega' \frac{\mathbf{r}'}{|\mathbf{r}-\mathbf{r}_j-\mathbf{r}'|}. \tag{15.46}$$

Applying again the gradient operator, we derive the tensorial relation that we need

$$\frac{\mathbf{I}}{|\mathbf{r}-\mathbf{r}_j|^3} - 3\frac{(\mathbf{r}-\mathbf{r}_j)(\mathbf{r}-\mathbf{r}_j)}{|\mathbf{r}-\mathbf{r}_j|^5} = -\frac{3}{4\pi a^2} \int_{S_a} d\Omega' \frac{(\mathbf{r}-\mathbf{r}_j-\mathbf{r}')\mathbf{r}'}{|\mathbf{r}-\mathbf{r}_j-\mathbf{r}'|^3} \tag{15.47}$$

$$= -\frac{3}{4\pi a^2} \int_{S_a} d\Omega' \frac{\mathbf{r}'(\mathbf{r}-\mathbf{r}_j-\mathbf{r}')}{|\mathbf{r}-\mathbf{r}_j-\mathbf{r}'|^3}. \tag{15.48}$$

The tensors inside the two integrals in (15.47) and (15.48) are the transpose of each other, but their integrals give the same result for any $\mathbf{r}-\mathbf{r}_j$, as can be shown for a general $\mathbf{r}-\mathbf{r}_j$, with a change of integration variables (i.e. a rotation of the axes with respect to which $\mathbf{r}'$ is defined). This is also obvious from the fact that on the left-hand side there is a symmetric tensor.

Equations (15.46)–(15.48) will be the basis for the computation of the lower and upper bounds of the total energy (15.39) of our dipole system.

15.3.2   The lower bound

We imagine that each dipole $\mathbf{m}_i$ is substituted by the uniform magnetization density of the sphere V_i, equal to $3\mathbf{m}_i/(4\pi a^3)$. From the properties of the uniformly magnetized spheres, studied in the previous section, we can deduce that the magnetic induction $\mathbf{B}$ and the magnetic field $\mathbf{H}$ outside the spheres are exactly the same as those generated by the system of dipoles. This is the physical basis of the computation of the lower bound for the energy.

If we contract, with $\mathbf{m}_i$ on the left and with $\mathbf{m}_j$ on the right, the tensors on both sides of Eq. (15.47), and we put $\mathbf{r} = \mathbf{r}_i$, we obtain

$$\frac{\mathbf{m}_i \cdot \mathbf{m}_j}{|\mathbf{r}_i - \mathbf{r}_j|^3} - 3\frac{\left[(\mathbf{r}_i - \mathbf{r}_j) \cdot \mathbf{m}_i\right]\left[(\mathbf{r}_i - \mathbf{r}_j) \cdot \mathbf{m}_j\right]}{|\mathbf{r}_i - \mathbf{r}_j|^5}$$

$$= -\frac{3}{4\pi a^2} \int_{S_a} d\Omega' \frac{\left[\mathbf{m}_i \cdot (\mathbf{r}_i - \mathbf{r}_j - \mathbf{r}')\right]\left[\mathbf{m}_j \cdot \mathbf{r}'\right]}{|\mathbf{r}_i - \mathbf{r}_j - \mathbf{r}'|^3}. \tag{15.49}$$

On the left-hand side, we have obtained, apart from the factor $\mu_0/4\pi$, the interaction energy between dipoles $\mathbf{m}_i$ and $\mathbf{m}_j$. We will now see that on the right-hand side there is a quantity related to the magnetic fields generated by our imaginary magnetized spheres V_i and V_j.

From Eq. (15.22), giving the scalar potential due to a uniformly magnetized sphere, we can easily derive the following relation for the scalar potential $\Phi_i(\mathbf{r})$ generated by the sphere V_i

$$\Delta\Phi_i(\mathbf{r}) = -\frac{3}{4\pi a^4} \left[\mathbf{m}_i \cdot (\mathbf{r} - \mathbf{r}_i)\right] \delta\left(|\mathbf{r} - \mathbf{r}_i| - a\right). \tag{15.50}$$

The gradient of $\Phi_i(\mathbf{r})$ is the opposite of the magnetic field $\mathbf{H}_i(\mathbf{r})$ generated by the sphere V_i. Therefore, we can write

$$\int d\mathbf{r}\, \mathbf{H}_i(\mathbf{r}) \cdot \mathbf{H}_j(\mathbf{r}) = \int d\mathbf{r}\, \nabla\Phi_i(\mathbf{r}) \cdot \nabla\Phi_j(\mathbf{r}) \tag{15.51}$$

$$= -\int d\mathbf{r}\, \Phi_i(\mathbf{r}) \cdot \Delta\Phi_j(\mathbf{r}) \tag{15.52}$$

$$= \frac{1}{(4\pi)^2} \int d\mathbf{r}\, \frac{\mathbf{m}_i \cdot (\mathbf{r} - \mathbf{r}_i)}{|\mathbf{r} - \mathbf{r}_i|^3} \frac{3}{a^4}\left[(\mathbf{r} - \mathbf{r}_j) \cdot \mathbf{m}_j\right] \delta\left(|\mathbf{r} - \mathbf{r}_j| - a\right) \tag{15.53}$$

$$= -\frac{3}{(4\pi a)^2} \int_{S_a} d\Omega' \frac{\mathbf{m}_i \cdot (\mathbf{r}_i - \mathbf{r}_j - \mathbf{r}')}{|\mathbf{r}_i - \mathbf{r}_j - \mathbf{r}'|^3} \mathbf{m}_j \cdot \mathbf{r}', \tag{15.54}$$

where the integrals in Eqs. (15.51)–(15.53) are extended over all space. In the expression appearing in Eq. (15.53), we could use the expression of $\Phi_i(\mathbf{r})$ valid outside the sphere V_i since the spheres do not overlap. The integral in Eq. (15.54) is equal to the right-hand side of Eq. (15.49) up to a factor 4π. Therefore, the total energy E in Eq. (15.39) can be written as

$$E = \frac{\mu_0}{2} \sum_{i=1}^{N} \sum_{j=1}^{N'} \int d\mathbf{r}\, \mathbf{H}_i(\mathbf{r}) \cdot \mathbf{H}_j(\mathbf{r}). \tag{15.55}$$

The 'self-energy' of each magnetized sphere is obtained by Eq. (15.54) by putting $i = j$ (also in this case it is possible to use the expression of $\Phi_i(\mathbf{r})$ valid outside and on the surface of the sphere V_i). We easily obtain

$$\frac{\mu_0}{2} \int d\mathbf{r}\, \mathbf{H}_i(\mathbf{r}) \cdot \mathbf{H}_i(\mathbf{r}) = \mu_0 \frac{m_i^2}{8\pi a^3} \tag{15.56}$$

and we then finally get

$$E = \frac{\mu_0}{2} \int d\mathbf{r} \left| \sum_{i=1}^{N} \mathbf{H}_i(\mathbf{r}) \right|^2 - \frac{\mu_0}{8\pi a^3} \sum_{i=1}^{N} m_i^2. \tag{15.57}$$

The first term on the right-hand side being clearly positive definite, we obtain the following lower bound for the energy of the system

$$E \geq -\frac{\mu_0}{8\pi a^3} \sum_{i=1}^{N} m_i^2 \geq -N\mu_0 \frac{m_0^2}{8\pi a^3}, \tag{15.58}$$

where m_0 is the largest dipole module.

Therefore the dipole–dipole interaction is a stable interaction, despite the $1/r^3$ decay. This statement is obviously conditioned on the existence of a minimum distance between dipoles, uniformly on N.

15.3.3 The upper bound

The physical basis of the computation of the upper bound of the interaction energy is different. Now we imagine that each of the spheres V_i has a certain electric charge density on its surface, and that it rotates with a uniform angular velocity around an axis parallel to the dipole $\mathbf{m}_i$, in the anti-clockwise direction. This will generate a current density and therefore a magnetic induction. In the following, we see how this picture, together with the relation previously derived, is used to obtain the bound.

As a first thing, we need the vector potential due to a dipole $\mathbf{m}_j$ placed in $\mathbf{r}_j$. We know from elementary magnetostatic (Jackson, 1975) that this vector potential is

$$\mathbf{A}(\mathbf{r}) = \mathbf{A}_j(\mathbf{r}) = \frac{\mu_0}{4\pi} \frac{\mathbf{m}_j \times (\mathbf{r} - \mathbf{r}_j)}{|\mathbf{r} - \mathbf{r}_j|^3}. \tag{15.59}$$

It is a straightforward calculation to derive the magnetic induction $\mathbf{B}(\mathbf{r})$ of the dipole computing the curl of $\mathbf{A}(\mathbf{r})$.

If we now make the vector product of the two sides of Eq. (15.46) with $\mathbf{m}_j$, we obtain

$$\frac{\mathbf{m}_j \times (\mathbf{r} - \mathbf{r}_j)}{|\mathbf{r} - \mathbf{r}_j|^3} = \frac{3}{4\pi a^2} \int_{S_a} d\Omega' \, \frac{\mathbf{m}_j \times \mathbf{r}'}{|\mathbf{r} - \mathbf{r}_j - \mathbf{r}'|}. \tag{15.60}$$

On the other hand, we know that the vector potential due to a given current density $\mathbf{J}(\mathbf{r})$ is

$$\mathbf{A}(\mathbf{r}) = \frac{\mu_0}{4\pi} \int d\mathbf{r}' \, \frac{\mathbf{J}(\mathbf{r}')}{|\mathbf{r} - \mathbf{r}'|}. \tag{15.61}$$

We therefore deduce that the vector potential due to the dipole $\mathbf{m}_j$ is the same as that due to the current density

$$\mathbf{J}_j(\mathbf{r}) = \frac{3}{4\pi a^4} \left[\mathbf{m}_j \times (\mathbf{r} - \mathbf{r}_j) \right] \delta \left(|\mathbf{r} - \mathbf{r}_j| - a \right). \tag{15.62}$$

Obviously this is valid only for $|\mathbf{r} - \mathbf{r}_j| \geq a$, the range of validity of Eq. (15.46). A bit of thought confirms the picture just described: in fact, this is the current density resulting from a surface charge density σ on the sphere of radius a, rotating with an angular velocity ω in the anti-clockwise direction around an axis parallel to $\mathbf{m}_j$, such that $\sigma\omega = (3\mathbf{m}_j)/(4\pi a^4)$.

Equations (15.47) and (15.48) will help us in the computation of the upper bound by using another type of tensor contraction. Given a generic tensor $\overline{T}$, we define a contraction with the two vectors $\mathbf{v}$ and $\mathbf{w}$ by

$$\mathbf{v} : \overline{T} : \mathbf{w} \equiv \sum_{k=1}^{3} \sum_{l=1}^{3} \left[v_k \overline{T}_{kl} w_l \right] - (\mathbf{v} \cdot \mathbf{w}) \sum_{k=1}^{3} \overline{T}_{kk} \tag{15.63}$$

$$= \sum_{k=1}^{3} \sum_{l=1}^{3} \left[v_k \overline{T}_{kl} w_l \right] - (\mathbf{v} \cdot \mathbf{w}) \, \mathrm{Tr} \, \overline{T}. \tag{15.64}$$

If we apply this definition to the tensor product of two vectors, i.e. $\overline{T} = \mathbf{xy}$, we have

$$\mathbf{v} : \mathbf{xy} : \mathbf{w} = \sum_{k=1}^{3} \sum_{l=1}^{3} [v_k x_k y_l w_l] - (\mathbf{v} \cdot \mathbf{w}) \sum_{k=1}^{3} x_k y_k \tag{15.65}$$

$$= (\mathbf{v} \cdot \mathbf{x})(\mathbf{y} \cdot \mathbf{w}) - (\mathbf{v} \cdot \mathbf{w})(\mathbf{x} \cdot \mathbf{y}) \tag{15.66}$$

$$= (\mathbf{v} \times \mathbf{y}) \cdot (\mathbf{x} \times \mathbf{w}), \tag{15.67}$$

where the last equality comes from a known vector identity. On the other hand, applying (15.63) to the identity tensor, we find that

$$\mathbf{v} : \mathbf{I} : \mathbf{w} = \sum_{k=1}^{3} \sum_{l=1}^{3} [v_k \delta_{kl} w_l] - (\mathbf{v} \cdot \mathbf{w}) \sum_{k=1}^{3} \delta_{kk} \qquad (15.68)$$

$$= (\mathbf{v} \cdot \mathbf{w}) - 3 (\mathbf{v} \cdot \mathbf{w}) \qquad (15.69)$$

$$= -2 (\mathbf{v} \cdot \mathbf{w}) . \qquad (15.70)$$

Using the expression with the scalar product of Eq. (15.66) and Eq. (15.69), we obtain the result of the contraction with the vectors $\mathbf{m}_i$ and $\mathbf{m}_j$ applied to the left-hand side of Eq. (15.47) with $\mathbf{r} = \mathbf{r}_i$

$$\mathbf{m}_i : \frac{\mathbf{I}}{|\mathbf{r}_i - \mathbf{r}_j|^3} - 3 \frac{(\mathbf{r}_i - \mathbf{r}_j)(\mathbf{r}_i - \mathbf{r}_j)}{|\mathbf{r}_i - \mathbf{r}_j|^5} : \mathbf{m}_j = \frac{\mathbf{m}_i \cdot \mathbf{m}_j}{|\mathbf{r}_i - \mathbf{r}_j|^3} - 3 \frac{[(\mathbf{r}_i - \mathbf{r}_j) \cdot \mathbf{m}_i][(\mathbf{r}_i - \mathbf{r}_j) \cdot \mathbf{m}_j]}{|\mathbf{r}_i - \mathbf{r}_j|^5},$$
$$(15.71)$$

i.e. the interaction energy between dipoles $\mathbf{m}_i$ and $\mathbf{m}_j$, apart from the factor $\mu_0/4\pi$. The application to Eq. (15.48) gives

$$\frac{\mu_0}{4\pi} \left[\mathbf{m}_i : -\frac{3}{4\pi a^2} \int_{S_a} d\Omega' \, \frac{\mathbf{r}'(\mathbf{r}_i - \mathbf{r}_j - \mathbf{r}')}{|\mathbf{r}_i - \mathbf{r}_j - \mathbf{r}'|^3} : \mathbf{m}_j \right]$$

$$= \frac{\mu_0}{4\pi} \left[-\frac{3}{4\pi a^2} \int_{S_a} d\Omega' \, \frac{[\mathbf{m}_i \times (\mathbf{r}' + \mathbf{r}_j - \mathbf{r}_i)] \cdot [\mathbf{m}_j \times \mathbf{r}']}{|\mathbf{r}' + \mathbf{r}_j - \mathbf{r}_i|^3} \right], \qquad (15.72)$$

using the expression with the vector product in Eq. (15.67). Equations (15.59) and (15.62), and then the use of Eq. (15.61), allow us to write the right-hand side as

$$- \int d\mathbf{r} \, \mathbf{A}_i(\mathbf{r}) \cdot \mathbf{J}_j(\mathbf{r}) = -\frac{\mu_0}{4\pi} \int d\mathbf{r} \int d\mathbf{r}' \, \frac{\mathbf{J}_i(\mathbf{r}') \cdot \mathbf{J}_j(\mathbf{r})}{|\mathbf{r} - \mathbf{r}'|} . \qquad (15.73)$$

Some readers, remembering the expression of the energy of a magnetostatic configuration (Jackson, 1975), might wonder about a 'wrong' sign in the last expressions. The same argument mentioned at the beginning of the chapter applies here: this is the interaction energy of two *permanent* dipoles, i.e. without the contribution of the work that would be necessary to maintain the currents generating the dipoles.

The total energy E in Eq. (15.38) can then be written as

$$E = -\frac{\mu_0}{8\pi} \sum_{i=1}^{N} \sum_{j=1}^{N} \int d\mathbf{r} \int d\mathbf{r}' \, \frac{\mathbf{J}_i(\mathbf{r}') \cdot \mathbf{J}_j(\mathbf{r})}{|\mathbf{r} - \mathbf{r}'|} . \qquad (15.74)$$

As for the lower bound, we must compute the 'self-energy'. This can be done by putting $i = j$ in Eq. (15.72). A straightforward computation gives

$$-\frac{\mu_0}{8\pi} \int d\mathbf{r} \int d\mathbf{r}' \, \frac{\mathbf{J}_i(\mathbf{r}') \cdot \mathbf{J}_i(\mathbf{r})}{|\mathbf{r} - \mathbf{r}'|} = -\frac{\mu_0}{4\pi} \frac{\mathbf{m}_i^2}{a^3} . \qquad (15.75)$$

We then finally have

$$E = -\frac{\mu_0}{8\pi} \int d\mathbf{r} \int d\mathbf{r}' \frac{\mathbf{J}_T(\mathbf{r}') \cdot \mathbf{J}_T(\mathbf{r})}{|\mathbf{r}-\mathbf{r}'|} + \frac{\mu_0}{4\pi a^3} \sum_{i=1}^{N} m_i^2, \quad (15.76)$$

with

$$\mathbf{J}_T(\mathbf{r}) = \sum_{i=1}^{N} \mathbf{J}_i(\mathbf{r}). \quad (15.77)$$

We can show that the double integral on the right-hand side is positive definite. In fact, given any function $f(\mathbf{r})$, we have

$$\int d\mathbf{r} \int d\mathbf{r}' \frac{f(\mathbf{r}')f(\mathbf{r})}{|\mathbf{r}-\mathbf{r}'|} > 0. \quad (15.78)$$

This can be proven remembering from electrostatics that if

$$\Psi(\mathbf{r}) = \int d\mathbf{r}' \frac{f(\mathbf{r}')}{|\mathbf{r}-\mathbf{r}'|}, \quad (15.79)$$

then

$$\Delta\Psi(\mathbf{r}) = -4\pi f(\mathbf{r}). \quad (15.80)$$

Therefore, the left-hand side of Eq. (15.78) can be written as

$$-\frac{1}{4\pi} \int d\mathbf{r}\, \Psi(\mathbf{r})\Delta\Psi(\mathbf{r}) = \frac{1}{4\pi} \int d\mathbf{r}\, |\nabla\Psi(\mathbf{r})|^2 > 0. \quad (15.81)$$

Applying this relation for each component of $\mathbf{J}_T(\mathbf{r})$, we reach the desired result. Then, we finally obtain the upper bound for E

$$E \le \frac{\mu_0}{4\pi a^3} \sum_{i=1}^{N} m_i^2 \le N\mu_0 \frac{m_0^2}{4\pi a^3}. \quad (15.82)$$

15.3.4   The thermodynamic limit

Introducing the explicit representation of the dependence of the interaction energy E on the dipoles configuration, $E = E(\{\mathbf{m}_i\})$, the canonical partition function is

$$Z_N(\beta) = \sum_{\{\mathbf{m}_i\}} \exp\{-\beta E(\{\mathbf{m}_i\})\}, \quad (15.83)$$

where the subscript denotes the dependence on the number of dipoles. We use a sum notation, but obviously an integral must be understood in case of dipole components assuming continuous values. The associated total free energy is

$$F_N(\beta) = -\frac{1}{\beta}\ln Z_N(\beta). \tag{15.84}$$

The existence of the thermodynamic limit

$$f(\beta) = \lim_{N\to\infty}\frac{1}{N}F_N(\beta) \tag{15.85}$$

is proven on the basis of a stability condition, and on the temperedness of the interaction (Fisher, 1964). The last condition is not satisfied by long-range interactions as discussed in Chapter 1. Nevertheless, it can be easily shown that for dipole–dipole interactions, the following bound holds. If we divide the lattice with N sites into two non-overlapping lattices with N_1 and N_2 sites, so that $N_1 + N_2 = N$, then we have

$$F_N(\beta) \le F_{N_1}(\beta) + F_{N_2}(\beta). \tag{15.86}$$

This is the situation that occurs in strongly tempered interactions (i.e. when the two-body potential strictly vanishes beyond a given distance), and it is called subadditivity of the free energy. The two bounds in Eq. (15.86) and in Eq. (15.58), taken together, assure the existence of the thermodynamic limit for $f(\beta)$.

The proof of (15.86) is quite simple. Denoting with $\{\mathbf{m}_{1,i}\}$ and $\{\mathbf{m}_{2,i}\}$ the dipoles of the two subsystems with N_1 and N_2 dipoles, respectively, we have

$$Z_N(\beta) = \sum_{\{\mathbf{m}_{1,i}\},\{\mathbf{m}_{2,i}\}} \exp\left\{-\beta\left[E_1(\{\mathbf{m}_{1,i}\}) + E_2(\{\mathbf{m}_{2,i}\}) + E_{12}(\{\mathbf{m}_{1,i}\},\{\mathbf{m}_{2,i}\})\right]\right\}, \tag{15.87}$$

where E_1 (resp. E_2) is composed of all terms with two dipoles of the first (resp. the second) subsystem, while the interaction between the two subsystems, E_{12}, is composed of all terms in which there is one dipole of system 1 and one dipole of system 2. We now make the assumption that for each configuration of the dipoles of the whole system, there is the configuration in which anyone of the dipoles is reversed, e.g. $\mathbf{m}_i \to -\mathbf{m}_i$. This quite natural assumption is based on the use of the time-reversal operator (Griffiths, 1968). From the bilinearity of the dipole–dipole interaction, it follows that under the transformation $\{\mathbf{m}_{1,i}\},\{\mathbf{m}_{2,i}\} \to \{\mathbf{m}_{1,i}\},-\{\mathbf{m}_{2,i}\}$, E_1 and E_2 remain invariant, while E_{12} changes sign. This property and the above assumption allow us to write

$$Z_N(\beta) = \sum_{\{\mathbf{m}_{1,i}\},\{\mathbf{m}_{2,i}\}} \exp\left\{-\beta\left[E_1(\{\mathbf{m}_{1,i}\}) + E_2(\{\mathbf{m}_{2,i}\}) - E_{12}(\{\mathbf{m}_{1,i}\},\{\mathbf{m}_{2,i}\})\right]\right\}. \tag{15.88}$$

Summing Eqs. (15.87) and (15.88) and dividing by two, we have

$$Z_N(\beta) = \sum_{\{\mathbf{m}_{1,i}\},\{\mathbf{m}_{2,i}\}} \exp\left\{-\beta\left[E_1(\{\mathbf{m}_{1,i}\}) + E_2(\{\mathbf{m}_{2,i}\})\right]\right\} \cosh\left\{\beta\left[E_{12}(\{\mathbf{m}_{1,i}\},\{\mathbf{m}_{2,i}\})\right]\right\}$$

$$\ge \sum_{\{\mathbf{m}_{1,i}\},\{\mathbf{m}_{2,i}\}} \exp\left\{-\beta\left[E_1(\{\mathbf{m}_{1,i}\}) + E_2(\{\mathbf{m}_{2,i}\})\right]\right\} = Z_{N_1}(\beta) Z_{N_2}(\beta). \tag{15.89}$$

The bound (15.86) is an immediate consequence of the last inequality.

15.3.5   Further remarks

In the introduction, we mentioned that magnetic systems have also, generally, a short-range exchange interaction that, pairwise, is much stronger than the dipole–dipole interaction. We expect that the addition of a short-range interaction does not spoil the proof of the existence of the thermodynamic limit. It is easy to see this.

The inclusion of a short-range exchange interaction means that the interaction energy of the system of N dipoles contains, in addition to the dipolar energy shown in Eq. (15.39), also a term of the form

$$E_{\text{exc}} = -\frac{1}{2} \sum_{i=1}^{N} \sum_{j=1}^{N} \mathcal{J}_{ij}\, \mathbf{m}_i \cdot \mathbf{m}_j, \tag{15.90}$$

where the coupling constants $\mathcal{J}_{ij}$ have the properties that $\mathcal{J}_{ij} = \mathcal{J}_{ji}$, $\mathcal{J}_{ii} = 0$; their dependence on i and j reflects the translational symmetry of the underlying lattice, and they vanish if dipoles $\mathbf{m}_i$ and $\mathbf{m}_j$ are further apart than a given distance. Actually the proof can be extended also to the case in which the coupling constants decay with this distance faster than $1/r^3$. Positive (resp. negative) coupling constants describe ferromagnetic (resp. antiferromagnetic) exchange interactions.

The key for the extension of the proof is that the fundamental bounds for the existence of the thermodynamic limit, i.e. Eqs. (15.58) and (15.86), hold also in this case. Concerning the stability bound, the coefficient of N in Eq. (15.58) will be different, since it will take into account also the bound on the exchange interaction between nearby dipoles. The subadditivity will hold unaffected, since the argument made to obtain it depends only on the bilinearity of the interaction.

Finally, we show the very simple extension of the proof that is required in the quantum case. It can be shown (Griffiths, 1968) that, although the dipoles $\mathbf{m}_i$ must be replaced by appropriate operators, the lower bound on the interaction energy (in this case to be understood as a lower bound on the eigenvalues of the Hamiltonian) is obtained in much the same way as in the classical case. To obtain the subadditivity, we need the time-reversal operator and Peierls' theorem (Peierls, 1938). In fact, for our system divided into two subsystems, the energies E_1, E_2 and E_{12} must be replaced by the corresponding operators $\hat{U}_1$, $\hat{U}_2$ and $\hat{U}_{12}$. The partition function will be

$$Z_N(\beta) = \text{Tr}\left(\exp\left\{ -\beta \left[\hat{U}_1 + \hat{U}_2 + \hat{U}_{12} \right] \right\} \right). \tag{15.91}$$

We now suppose that $\{\phi_i\}$ and $\{\psi_j\}$ are the complete sets of orthonormal eigenfunctions of $\hat{U}_1$ and $\hat{U}_2$, respectively, with eigenvalues μ_i and ν_j. The states $\{\phi_i \psi_j\}$ are a complete orthonormal set for the whole system, but they will not be eigenfunctions of $\hat{U} = \hat{U}_1 + \hat{U}_2 + \hat{U}_{12}$. The expectation value of $\hat{U}$ in each of these states can be written as

$$\left(\phi_i \psi_j, \hat{U} \phi_i \psi_j \right) = \mu_i + \nu_j + \xi_{ij}, \tag{15.92}$$

where

$$\xi_{ij} = \left(\phi_i \psi_j, \hat{U}_{12} \phi_i \psi_j \right). \tag{15.93}$$

The Peierls's theorem (Peierls, 1938) assures that

$$Z_N(\beta) \geq \sum_{ij} \exp \left\{ -\beta \left[\mu_i + \nu_j + \xi_{ij} \right] \right\}. \tag{15.94}$$

Applying the time reversal operator to the functions $\{\phi_i\}$ we obtain another complete set of orthonormal eigenfunctions of $\hat{U}_1$, which can be denoted by $\{\phi_i'\}$, with the same eigenvalues. However, now we have

$$\xi_{ij}' = \left(\phi_i' \psi_j, \hat{U}_{12} \phi_i' \psi_j \right) = -\xi_{ij}. \tag{15.95}$$

Therefore, the inequality coming from Peierls's theorem can also be written as

$$Z_N(\beta) = \text{Tr} \left(\exp \left\{ -\beta \left[\hat{U}_1 + \hat{U}_2 + \hat{U}_{12} \right] \right\} \right) \geq \sum_{ij} \exp \left\{ -\beta \left[\mu_i + \nu_j - \xi_{ij} \right] \right\}. \tag{15.96}$$

Summing Eqs. (15.94) and (15.96) and dividing by two, we obtain

$$Z_N(\beta) \geq \sum_{ij} \exp \left\{ -\beta \left[\mu_i + \nu_j \right] \right\} \cosh \left(\beta \xi_{ij} \right) \tag{15.97}$$

$$\geq \sum_{ij} \exp \left\{ -\beta \left[\mu_i + \nu_j \right] \right\} = Z_{N_1}(\beta) Z_{N_2}(\beta), \tag{15.98}$$

an inequality similar to Eq. (15.89).

15.4 The Physical Consequences of the Existence of the Thermodynamic Limit

The proof of the existence of the thermodynamic limit in a system of dipoles interacting through the dipole–dipole interaction, and possibly also through a short-range exchange interaction, does not depend on the global shape of the underlying crystal. Therefore, the proof implies that the limit is independent of the sample shape, with the only weak condition that all three linear dimensions of the sample increase to infinity. We therefore have the result, anticipated in the introduction of this chapter, that the properties (vii) and (viii), valid for long-range exchange interactions, do not hold in dipolar systems: the free energy per site and the energy per site in the ground state are shape independent in the thermodynamic limit.

This section is dedicated to the consequences of Griffiths theorem on the existence of the thermodynamic limit for dipolar systems. To begin with, we must find an expression for the magnetostatic energy of a system of dipoles. We will find that the expression that was derived in macroscopic magnetostatic is useful also here, although it will be necessary to introduce corrections in order to take into account more precisely the interaction energy between nearby dipoles.

15.4.1   The computation of the dipolar energy

At the end of Section 15.2, we computed the magnetostatic energy of a macroscopic magnetized body, obtaining the two equivalent expressions in Eqs. (15.33) and (15.37). On the other hand, the magnetostatic energy of a system of N microscopic dipoles is given by Eq. (15.31) that, after substitution for the magnetic induction B_i at the position of the dipole m_i, becomes the expression given in Eq. (15.38) or (15.39).

From the argument that has been used to obtain the energy of a macroscopic magnetized body, it is clear that the same expressions found for that case are also an approximation for the energy of an ensemble of dipoles. In fact, the sum over the dipoles in Eq. (15.31) can be approximated by an integration over the volume occupied by the dipoles. To this purpose, if v is the volume of the unit cell of the lattice under consideration, we make the substitution

$$\mathbf{M}(\mathbf{r}_k) = \frac{1}{v} \sum_{i \in k} \mathbf{m}_i, \qquad (15.99)$$

where the sum is extended over the dipoles $\mathbf{m}_i$ included in the kth cell (therefore in general the sum will include only few terms); the position of this cell is $\mathbf{r}_k$. We expect that the magnetization density $\mathbf{M}(\mathbf{r}_k)$ so defined will vary appreciably only on a length scale much larger than the distance between nearest dipoles (or unit cells). This will be especially true for ferromagnetic samples, where the exchange interaction will align nearby dipoles (but we are computing here only the dipole–dipole interaction contribution to the magnetostatic energy). Then, $\mathbf{M}(\mathbf{r}_k)$ can be extended to a magnetization density $\mathbf{M}(\mathbf{r})$ defined on all space, and the sum in Eq. (15.31) (where it must be reminded that the magnetic induction associated with each dipole is the one created by all the other dipoles) to an integral. Therefore, we would obtain the expressions in Eq. (15.33). However, due to the $1/r^3$ divergence, at small distances, of the expression of the magnetic induction due to a dipole, this approximation is not good for the representation of the magnetostatic energy between nearby dipoles. Then, we need to treat separately this contribution. This can be done in the following way.

We denote by b a typical distance between nearest dipoles. Indicating by R the typical (macroscopic) distance over which the magnetization density $\mathbf{M}(\mathbf{r})$ varies appreciably, we suppose that it is possible to define a length ℓ such that $b \ll \ell \ll R$. We then imagine to surround each dipole by a sphere of radius ℓ, the Lorentz cavity. The magnetic induction created at the position of the dipole can then be separated into that created by the dipoles outside the Lorentz cavity and that created by the dipoles inside. Therefore, the magnetic induction at the position of the ith dipole will be given by

$$B_i(r_i) = \sum_{j,|r_i-r_j|\le\ell} B_{ij} + \sum_{j,|r_i-r_j|>\ell} B_{ij}, \tag{15.100}$$

with an obvious meaning for B_{ij}. For the contribution of the dipoles outside the cavity, we use the continuum approximation, while for those inside we keep the exact expression. In turn, the continuum approximation of the magnetic induction created by the dipoles outside the Lorentz sphere will be equal to the continuum approximation of the magnetic induction created by the whole sample minus that created by the dipoles in the sphere. Remembering, as explained at the end of Section 15.2, that to obtain the field experienced by a dipole we must subtract $\frac{2\mu_0}{3}M$, we obtain

$$\sum_{j,|r_i-r_j|>\ell} B_{ij} \rightarrow \left(B_{\text{sample}}(r_i) - \frac{2\mu_0}{3}M(r_i)\right) - \left(B_{\text{sphere}}(r_i) - \frac{2\mu_0}{3}M(r_i)\right)$$

$$= B_{\text{sample}}(r_i) - B_{\text{sphere}}(r_i). \tag{15.101}$$

Since $\ell \ll R$, we can assume that $B_{\text{sphere}}(r_i)$ is that created by a sphere with a uniform magnetization density $M(r_i)$. Then

$$\sum_{j,|r_i-r_j|>\ell} B_{ij} \rightarrow B_{\text{sample}}(r_i) - \frac{2\mu_0}{3}M(r_i). \tag{15.102}$$

To compute the first term on the right-hand side of (15.100), we can use the same uniformity assumption for the magnetization in the length scale $\ell \ll R$. We first define the tensor

$$\overline{B}(r_i) = \frac{1}{4\pi}\frac{v}{p^2}\sum_{j\in C_i}\sum_{k,|r_k-r_j|\le\ell}\left[3\frac{(r_k-r_j)(r_k-r_j)}{|r_k-r_j|^5} - \frac{I}{|r_k-r_j|^3}\right], \tag{15.103}$$

where C_i denotes the unit cell containing the ith dipole, and p is the number of dipoles in a unit cell. The average over the sites j belonging to the same cell and the fact that $\ell \gg b$ assure that $\overline{B}$ is a constant tensor. We can then finally write

$$\sum_{j,|r_i-r_j|\le\ell} B_{ij} \rightarrow \mu_0\overline{B}\cdot M(r_i). \tag{15.104}$$

Collecting the expressions given in Eqs. (15.102) and (15.104), we can finally obtain the expression for the dipole–dipole magnetostatic energy. Normalizing to the volume of the systems, we have

$$\frac{E}{V} = -\frac{1}{2V}\int_V dr\left(\mu_0 M(r)\cdot\overline{B}\cdot M(r) + M(r)\cdot B(r) - \frac{2\mu_0}{3}M^2(r)\right) \tag{15.105}$$

$$= -\frac{\mu_0}{2V}\int_V dr\left(M(r)\cdot\overline{B}\cdot M(r) + M(r)\cdot H(r) + \frac{1}{3}M^2(r)\right). \tag{15.106}$$

The difference with respect to the expressions in Eq. (15.33), which we have derived in macroscopic magnetostatic, is the presence of the first term. We could have chosen a different shape for the Lorentz cavity, obtaining the same result: the tensor $\overline{B}$ would have been different, but the sum of the first and the third terms in the integrand of the last expressions would have been the same. The third term is called the Lorentz energy term. The sum of these two terms depends on the crystal structure, but not on the sample shape (and so the integral of the sum does not depend on the sample shape). In contrast, the integral of the second term, i.e. the demagnetizing energy, does not depend on the crystal structure, but we have already seen that it depends on the sample shape.

If we specialize to a uniformly magnetized system, we have

$$\frac{E}{V} = \frac{\mu_0}{2}\left(\mathbf{M}\cdot\overline{B}\cdot\mathbf{M} + \mathbf{M}\cdot\overline{D}\cdot\mathbf{M} - \frac{1}{3}\mathbf{M}^2 \right) \tag{15.107}$$

$$= \varepsilon_0 + \frac{\mu_0}{2}\mathbf{M}\cdot\overline{D}\cdot\mathbf{M} + \varepsilon_{\mathrm{L}}, \tag{15.108}$$

where ε_0 and ε_{L} are the short-range contributions, coming from the discrete summation in Eq. (15.104) and from the spherical Lorentz cavity, respectively. Once the shape of the Lorentz cavity has been chosen, the dependence on the crystal structure remains in ε_0. We remind that the shape-dependent demagnetizing energy, the second term on the right-hand side of Eq. (15.107), is positive definite. This occurs, as we have seen, not only for uniform magnetization, but for a general magnetization profile.

If also a short-range exchange interaction is present, we must add another term to expression (15.106)

$$\frac{E}{V} = \varepsilon_{\mathrm{exc}} - \frac{\mu_0}{2V}\int_V \mathrm{d}\mathbf{r}\,\left(\mathbf{M}(\mathbf{r})\cdot\overline{B}\cdot\mathbf{M}(\mathbf{r}) + \mathbf{M}(\mathbf{r})\cdot\mathbf{H}(\mathbf{r}) + \frac{1}{3}\mathbf{M}^2(\mathbf{r}) \right), \tag{15.109}$$

where $\varepsilon_{\mathrm{exc}} = E_{\mathrm{exc}}/V$, with E_{exc} given by Eq. (15.90). Specializing again to a uniformly magnetized system, we have

$$\frac{E}{V} = \varepsilon_{\mathrm{exc}} + \varepsilon_0 + \frac{\mu_0}{2}\mathbf{M}\cdot\overline{D}\cdot\mathbf{M} + \varepsilon_{\mathrm{L}}. \tag{15.110}$$

Like ε_0, $\varepsilon_{\mathrm{exc}}$ depends on the crystal structure, but not on the sample shape.

The form of the demagnetizing energy is worth emphasizing. An energy term proportional to the square of the magnetization density of the system, as clearly shown in Eqs. (15.107) and (15.110), is the familiar one that we encountered in the study of mean-field exchange interactions in Chapter 4. In the thermodynamic limit, in order to obtain a finite value of the energy per particle for a given magnetization, we had to introduce Kac's prescription. We note that here we obtain the same kind of mean-field expression. In fact, since the energy of a magnetic dipole in a magnetic induction is proportional to the product of the dipole and the induction, a term like $\mathbf{M}^2$, means that the effective magnetic induction on any given dipole is proportional to the average of the magnetization over all system. However, in this case, there is no Kac's prescription, despite the long-range character of the dipole–dipole interaction. The complex angular dependence of this interaction is the ultimate reason for this.

15.4.2   The large-scale structure of the magnetization profile: domains and curling

We must reconcile two seemingly contrasting results. On the one hand, we have seen that the free energy per site and the energy per site in the ground state are shape-independent in the thermodynamic limit. On the other hand, we have found that a uniformly magnetized system of dipoles has a shape-dependent magnetostatic energy. The first consequence is that the equilibrium state of a ferromagnet, in the thermo-dynamic limit, cannot be a uniformly magnetized state for general sample shapes, no matter how strong the short-range exchange interaction is. If the magnetization density is not uniform, but it has a preferred direction, i.e. at any point the scalar product of the magnetization density with a given unit vector has the same sign, we can argue that the magnetostatic energy will be shape-dependent also in this case. In fact, if, e.g., the preferred direction is one of the principal axes of an ellipsoidal shape, the energy will depend on the value of the eigenvalue of the demagnetization tensor associated with that axis. We thus conclude that property (iii), mentioned in Section 15.1, is also not valid in dipolar systems.

We therefore arrive to the conclusion that, even in the presence of a strong ferromag-netic short-range exchange interaction, in the thermodynamic limit the magnetization density in the equilibrium state must assume a rather complex profile, such that the magnetostatic energy (or the free energy) does not depend on the sample shape. The system will not be simply overall magnetized along a direction.

For the sake of simplicity, we concentrate in the following on the ground state of ferromagnetic systems. Mathematically, this means that we consider systems at zero tem-perature; physically, this is a plausible procedure even at finite (e.g., room) temperatures if the short-range exchange interaction is strong enough to produce, at least for moderate temperatures, a temperature-independent magnetization density modulus. This assump-tion, valid for hard ferromagnets, is equivalent to that made in the study of macroscopic magnetostatic in Section 15.2. In other words, we consider the case in which the energy contribution to the free energy is much more important than the entropic contribution.

Therefore the equilibrium configuration is the one that minimizes the right-hand side of Eq. (15.109) or Eq. (15.110). We see that in those expressions there are contrast-ing terms. In fact, the positive definite demagnetizing energy is clearly minimized by a vanishing magnetization density. On the other hand, the Lorentz energy term ε_L favours magnetization. In addition, there are terms coming from the discrete summation, i.e. the contribution of the dipolar energy from nearby dipoles, ε_0, and the exchange energy term ε_{exc}. The last two cannot be further specified without knowing the details of the underlying crystal structure.

These two terms ε_0 and ε_{exc} are the contributions to the interaction energy coming from pairs of dipoles which are close. The exchange part ε_{exc}, as already emphasized, is generally restricted to nearest neighbour or next to nearest neighbour dipoles, and since it is much stronger than the dipole–dipole interaction, it dominates the short-range structure of ferromagnets. It strongly depends on the crystal structure, and obviously on which atoms or molecules give rise to the magnetic dipoles. Very often this interaction is anisotropic, and this introduces hard and easy axes in the ferromagnet.

The anisotropy of ε_0 was the first thing to be mentioned in this chapter. We emphasize here that the competition between the first and the second terms in Eq. (15.2) is frustration: for example, if we have a small 2D system of dipoles, like 4 dipoles arranged in a square or 3 dipoles arranged in an equilateral triangle, it is not possible to minimize at the same time the dipole–dipole interaction between all pairs of dipoles. It can be seen that a square geometry favours an antiferromagnetic configuration, while a triangular geometry favours a ferromagnetic configuration (Bramwell, 2009). In general, like in these examples, the most favourable magnetic configuration will depend on the point symmetry of the crystal structure. Since ε_{exc} is dominant over ε_0 at short distances, the isotropy properties of the magnetic systems will tend to be determined by the former.

15.4.3   Isotropic and anisotropic ferromagnets

Let us suppose that, as a consequence of ε_{exc}, the modulus M of the magnetization density is given. This is the situation that we have indicated earlier as occurring generally in hard ferromagnets. There is one case in which the magnetization profile of the all system is easily determined. This is when the shape of the system is that of a very elongated ellipsoid. We have seen that the demagnetization coefficient corresponding to the long axis of cylinder is 0, and that as a consequence (see Fig. 15.2) the magnetic field inside a very elongated ellipsoid, with a uniform magnetization density parallel to the long axis, is practically 0. Therefore, such configuration minimizes the positive definite contribution of the demagnetizing energy to the magnetostatic energy. In conclusion, in this case, we have an exception to the fact mentioned earlier, and the system is overall magnetized along a direction.

At the same time, we also have a method for computing the value of the specific energy in the ground state: we take a sample in the form of a very elongated ellipsoid with the long axis parallel to the hard axis, as determined by the point symmetry of the underlying crystal structure (Banerjee *et al.*, 2004; Bramwell, 2009). According to the theorem proved earlier, macroscopic samples of generic shapes will have the same energy density, and thus the magnetization will be such to self-consistently yield this value.

It should be clear by now that the determination of the magnetization profile for general shapes is not an easy task. The problem was treated already some time ago (see, e.g., Sauer (1940), Luttiger and Tisza (1946)). Here, we will restrict ourselves just to rather qualitative arguments.

It has been known, since a long time, that magnetic domains are a common property of magnetic systems. For theoretical works on this, see Kittel (1949) and (1951), and Van Vleck (1937). We can give a simple argument that helps to understand why the formation of domains decreases the demagnetizing energy. From Eq. (15.106), we see that, even if the magnetization density modulus M is given throughout the sample, we

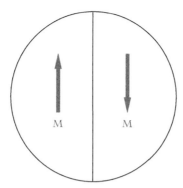

Figure 15.3 *A spherical sample in which the magnetization density is constant and has a given direction in one half of the sphere, and the opposite direction in the other half. There will be a gain in demagnetizing energy, with respect to the uniformly magnetized sphere (with the same magnetization modulus M), since the magnetic field will be smaller. On the other hand, there will be an energy cost in the exchange energy near the boundary between the two regions. For sufficiently large samples, the demagnetizing energy gain will be larger than the exchange energy cost, and the configuration with two domains will be favourable.*

can hope to decrease as much as possible the demagnetizing energy if we find a profile for the orientation of the magnetization density such that the magnetic field is small. If the magnetic field is decreased throughout the sample, it is clear that the gain in specific energy is of order $O(1)$. Let us then suppose that the magnetization profile in, e.g. a spherical sample, is like that in Fig. 15.3. The gain in demagnetizing energy must be contrasted with the cost in exchange energy. For a ferromagnetic exchange interaction, the boundary between the two regions will cost a term proportional to the area of the boundary. Therefore the cost in specific energy will be of the order $O(1/R)$, where R is the radius of the sphere. If R is sufficiently large, the gain will dominate over the cost, even if the exchange coupling constant is large.

We can expect that the formation of more domains decreases further the magnetic field. For a given sample size, the formation of domains will stop when the gain in demagnetizing energy will not be any more larger than the cost in exchange energy.

The formation of domains with sharp boundaries is not the only way in which the magnetic field inside a magnetized body can be decreased. Another mechanism is the curling of the magnetization profile, i.e. a behaviour like that in Fig. 15.4. It can be shown (Banerjee *et al.*, 2004) that, in this case, the cost in specific exchange energy is $O(\ln R/R)$. Again, the formation of such structure is favoured with respect to the uniform magnetization. The curling is found more easily in soft ferromagnets, which do not have hard axes, since the exchange energy is either isotropic or slightly anisotropic (Bramwell, 2009).

A more detailed treatment of the magnetization profiles in ferromagnets is outside the scope of this book. The interested reader can consult the cited literature.

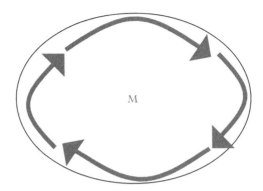

Figure 15.4 *A spherical sample in which the magnetization density profile has assumed a vortex-like structure. As for the sphere with two domains, the demagnetizing energy is smaller than for the uniformly magnetized sphere.*

15.5   Experimental Studies of Dipolar Interactions

The formation of domains, or more generally of complex magnetization profiles, is a consequence of the long-range character of the dipole–dipole interaction, as we have seen. We can consider this as the only possible 'revenge' that this interaction can have, in magnetic macroscopic samples, over the exchange interaction, which is much stronger pairwise. It would be interesting to have real many-body systems where the dipolar interaction is not overwhelmed by another interaction, in the hope of observing equilibrium configurations and dynamical effects directly ascribable to it.

In recent years some very interesting steps in this direction have been made. In this last section, we will try to describe the main characteristics of a few of such experimental systems. Actually, in the second subsection, we will not describe an actual experiment, but rather the computations that have been performed for a concrete real system, and that have a high probability to be experimentally confirmed in the near future.

Dipolar systems are probably one of the best candidates, among laboratory systems, to find clear signatures of the many properties that characterize long-range interacting systems.

15.5.1   Spin ice systems

It turns out that in some particular crystals the dipole–dipole interaction between nearby atoms is stronger than the exchange interaction. One such case is that of the holmiun titanate, $Ho_2Ti_2O_7$, in which the crystal lattice is face-centred cubic, with a tetrahedral basis (Bramwell and Gingras, 2001). The unusual strength of the dipole–dipole magnetic interaction stems from the large magnetic moment in the ground state, which is about 10.65 μ_B, i.e. about 10 times larger than a typical magnetic moment. The further particular property of this material is that the configuration of the magnetic moments can be mapped exactly on the configuration of the hydrogens in water ice, which must lie on the line of contact between two nearby oxygen atoms. For this reason, this material has been termed spin ice. We then expect to find the same properties as in water ice, among which the large zero point entropy. This observation has been actually achieved (Bramwell and Gingras, 2001).

15.5.2   2D optical lattices

Another system where the signature of the dipole–dipole interaction could be observed in the near future is an optical lattice of dipolar bosonic gases. An optical lattice is a system in which cold atoms are held on the sites of a lattice by means of traps created by the intersection of laser fields. In the case of non-polar atoms, the system is described by the Bose–Hubbard Hamiltonian

$$H = -\mathcal{J} \sum_{<ij>} \hat{a}_i^\dagger \hat{a}_j + \frac{\mathcal{U}}{2} \sum_i \left[\hat{a}_i^\dagger \hat{a}_i \left(\hat{a}_i^\dagger \hat{a}_i - 1 \right) \right] - \mu \sum_i \hat{a}_i^\dagger \hat{a}_i, \qquad (15.111)$$

where $\hat{a}_i$ is the annihilation operator for one boson at site i, with wavefunction equal to the Wannier function localized at the bottom of that site, the parameter $\mathcal{U}$ is related to the scattering length describing the contact interaction between the bosons, and the interaction parameter $\mathcal{J}$, connecting only nearest neighbours $< ij >$, is related to the strength of the trap. The chemical potential μ fixes the total number of particles. This Hamiltonian is known to give rise to a transition between an insulating phase, the so-called Mott phase, with each site occupied by a given number of atoms, and a superfluid phase, when the interaction $\mathcal{J}$ is large enough to make hopping energetically favourable. The parameters of the Hamiltonian can be controlled, and the insulating-superfluid transition has recently been observed (Jaksch *et al.*, 1998; Greiner *et al.*, 2002).

In the case of dipolar bosons, the Hamiltonian must be augmented with a term that takes into account the dipole–dipole interaction. In the literature, the case of a 2D lattice, with dipoles uniformly oriented, has been considered (Góral *et al.*, 2002; Menotti *et al.*, 2007, 2008). The extra term is given by

$$H_{dd} = \frac{A}{2} \sum_{i,j} \frac{1}{r_{ij}^3} \left[1 - 3 \cos^2 \theta_{ij} \right], \qquad (15.112)$$

where $A = \mu_0 m^2 / (4\pi)$, with m the common strength of all dipoles, while r_{ij} is the modulus of the vector distance $\mathbf{r}_{ij}$ between the sites i and j, and θ_{ij} is the angle between $\mathbf{r}_{ij}$ and the direction of the dipoles. Actually a term with a Dirac delta function, describing the on-site dipole–dipole interaction (a sort of self-energy), has not been explicitly written on the right-hand side of Eq. (15.112); it modifies the value of the parameter $\mathcal{U}$ of Eq. (15.111). It has been shown that the dipole–dipole interaction causes the presence of other phases, like the so-called supersolid phase, in which the superfluid phase coexists with a periodically modulated order parameter (Góral *et al.*, 2002). Also a checkerboard configuration, in which a site occupied exactly by one atom is followed by an empty site, has been reported (Góral *et al.*, 2002).

But also another very interesting feature has been found in dipolar optical lattices, the presence of a large number of metastable states (Menotti *et al.*, 2007, 2008). Basically, the same phases found in the equilibrium ground state, Mott insulating, superfluid, supersolid, checkerboard, are found also at higher energies, locally stable and therefore metastable. The reader will remember that metastable states are an ubiquitous property of long-range interactions, as shown in great detail in previous chapters.

Hopefully, the very interesting properties that have been briefly shown to occur in optical lattices of dipolar atoms will soon find an experimental confirmation.

15.5.3   Bose–Einstein condensates

Recently a Bose–Einstein condensate of chromium (^{52}Cr), at a temperature of the order of $10^{-5}K$, has been obtained (Menotti *et al.*, 2007; Griesmaier, 2007). ^{52}Cr has a large magnetic moment, and this makes it a good candidate for the observation of the effects of dipolar interactions.

In these ultra-cold degenerate gases, the interaction between atoms is generally described by a contact interaction

$$U_c(\mathbf{r}) = \frac{4\pi\hbar^2 b}{M}\delta(\mathbf{r}),$$

(15.113)

characterized by a scattering length b, and where M is the atomic mass. If the atoms have a magnetic moment m, the relative strength of the dipole–dipole interaction with respect to the contact interaction is usually described (Menotti *et al.*, 2008) by the dimensionless parameter

$$\varepsilon_{dd} = \frac{AM}{12\pi\hbar^2 b}$$

(15.114)

(where A is the quantity introduced in Eq. (15.112)), which is generally very small for Bose–Einstein condensates, of the order 10^{-3}, but for the chromium condensate it is about 0.16, also thanks to a large magnetic moment of $6\mu_B$. This value can be increased considerably by using a Feshback resonance (Werner *et al.*, 2005), since by varying opportunely the applied magnetic field the scattering length b can approach 0.

After the release of the trap, the atomic cloud begins to expand, and it is possible to show that the aspect ratio of the cloud does not change, during the expansion, if there is only a pure contact interaction. On the other hand, with a dipolar interaction the aspect ratio changes, since it is energetically favourable to have particles close to the magnetization axis. Therefore a change of the aspect ratio during expansion is a clear signature of the dipole–dipole interaction. Increasing the value of ε_{dd} makes this effect very strong. Spectacular results of this phenomenon have recently been obtained by Lahaye *et al.* (2007).

In conclusion, we emphasize again that we have here offered only a very brief excursus of the recent experimental results on dipolar systems. We hope that this rapid description has been sufficient to show the strong promises that reside in this kind of systems, concerning the observation of the equilibrium and of the dynamical properties of long-range interactions.

Appendix A
Features of the Main Models Studied throughout the Book

The table here summarizes the main features of the models studied throughout this book, the variable (Var.) being either continuous (C) or discrete (D). Particular attention is made on ensemble inequivalence, negative microcanonical specific heat c_V, ergodicity breaking and whether the entropy is computable in the microcanonical ensemble by one or more of the methods studied (direct evaluation of the microcanonical partition function, large deviations method, min–max procedure).

Cells of the table are filled by Y for Yes or N for No. In few cases an asterisk denotes a result not proven but nevertheless considered very probable.

We recall that BEG refers to the Blume–Emery–Griffiths model, HMF to the Hamiltonian Mean-Field model, SGR to the self-gravitating ring model, while L+S means the model with long- plus short-range interactions.

Name of the model	Var.	Ens. Inequiv.	$c_V < 0$	Ergod. Break.	Comput. Entropy	Section
BEG	D	Y	Y	Y	Y	2.3
3 states Potts	D	Y	Y	N	Y	3.3
Ising L+S	D	Y	Y	Y	Y	5.1
α-Ising	D	Y	N	N*	Y	5.2
HMF	C	N	N	N	Y	4.1
XY L+S	C	Y	Y	Y	Y	5.3
α-HMF	C	N	N	N*	N	5.4
Generalized XY	C	Y	Y	Y	Y	4.2
Mean-field ϕ^4	C	Y	N	N*	Y	4.3
Colson–Bonifacio	C	N	N	N	Y	14.5
Point vortex	C	Y	Y	Y	Y	11.2
Quasi-geostrophic	C	Y	Y	Y	Y	11.4
SGR	C	Y	Y	Y	Y	4.4

Appendix B
Evaluation of the Laplace Integral Outside the Analyticity Strip

The microcanonical partition function in formula (2.76) can be expressed using the Laplace representation of the Dirac δ function as

$$\Omega(\varepsilon, N) = \frac{1}{2\pi i} \int_{\beta - i\infty}^{\beta + i\infty} d\lambda \, e^{N\lambda\varepsilon} Z(\lambda, N), \tag{B.1}$$

with $\beta > 0$; this is the expression given in (2.78), with $E = N\varepsilon$. As explained in Section 2.5, we divide the integral in (B.1) into three intervals, defined by $\lambda_I < -\delta$, $-\delta < \lambda_I < \delta$ and $\lambda_I > \delta$, respectively, with $0 < \delta < \Delta$. Here we show that the contribution to the integral in λ coming from values of λ_I outside the strip, i.e. for values of λ_I with $|\lambda_I| > \Delta$, is exponentially small in N.

Let us then consider first the value of $Z(\lambda, N)$ in the two external intervals, i.e. for $|\lambda_I| > \Delta$. We have

$$Z(\beta + i\lambda_I, N) = \sum_{\{S_1, \dots, S_N\}} \exp\{-\beta H(\{S_i\})\} \exp\{-i\lambda_I H(\{S_i\})\}. \tag{B.2}$$

We see that this expression is proportional to $\langle \exp(-i\lambda_I H) \rangle$, the canonical expectation value of $\exp(-i\lambda_I H)$, which we expect to be exponentially small for large N. We confirm this expectation rewriting the last expression as

$$Z(\beta + i\lambda_I, N) = N \int d\varepsilon \, \exp\left(-N\left[\beta\varepsilon + i\lambda_I\varepsilon - s(\varepsilon)\right]\right), \tag{B.3}$$

where we have used the expression of the canonical partition in terms of the microcanonical entropy as in Eq. (1.64). This is an integral with a large phase. For N going to infinity, its value will be determined by the value of the integrand for ε equal to the integration extremes and to the values of the possible nonanalyticities of $s(\varepsilon)$, all denoted by ε_k (see, e.g. Bender and Orszag (1978)). We then have

$$Z(\beta + i\lambda_I, N) \sim N \sum_k c_k \exp\left(-N\left[\beta\varepsilon_k + i\lambda_I\varepsilon_k - s(\varepsilon_k)\right]\right), \tag{B.4}$$

where c_k are coefficients that could in principle be determined. It is clear that the successive integration over λ_I in any one of the two external intervals of integration will then give a vanishing contribution, due to the very large oscillations. We are then left with

$$\Omega(\varepsilon, N) \overset{N \to +\infty}{\sim} \frac{1}{2\pi i} \int_{\beta - i\delta}^{\beta + i\delta} d\lambda \, e^{N\lambda\varepsilon} Z(\lambda, N), \qquad (B.5)$$

which is the first equality given in Eq. (2.79).

Appendix C
The Equilibrium Form of the One-Particle Distribution Function in Short-Range Interacting Systems

We here present the derivation of the equilibrium form, Eq. (7.21), of the one-particle reduced distribution function $f_1(\mathbf{q}_1, \mathbf{p}_1)$ in short-range systems. We start from the equilibrium form of the N-body distribution function $f_N(x_1, \ldots, x_N)$, with $x_i \equiv (\mathbf{q}_i, \mathbf{p}_i)$, and the definition given in Eq. (7.6), which for $s = 1$ reads

$$f_1(x_1) = N \int dx_2 \ldots dx_N \, f_N(x_1, \ldots, x_N). \tag{C.1}$$

We will perform the computation of $f_1(x_1)$ for two cases: in the microcanonical and in the canonical ensemble. We begin with the latter, which is easier.

In the canonical ensemble, the equilibrium form of f_N is given by Eq. (1.30), which we rewrite here by making use of a slightly different notation,

$$f_N(x_1, \ldots, x_N) = \frac{\exp\left[-\beta H\left(\{x_i\}\right)\right]}{\int dx_1' \ldots dx_N' \, \exp\left[-\beta H\left(\{x_i'\}\right)\right]}, \tag{C.2}$$

where β is the inverse temperature characterizing the canonical distribution function, and where $H(\{x_i\})$ is the Hamiltonian (7.1). In this case f_N is clearly the product of a function of the coordinates times a function of the momenta, and the integration on the two sets of variables can be performed independently. Hence, we immediately derive

$$f_1(\mathbf{q}_1, \mathbf{p}_1) = N \left(\frac{\beta}{2\pi m}\right)^{\frac{3}{2}} e^{-\beta \frac{p_1^2}{2m}} \frac{\int d\mathbf{q}_2 \ldots d\mathbf{q}_N \, \exp\left[-\beta \sum_{1 \le i < j \le N} V_{ij}\right]}{\int d\mathbf{q}_1' \ldots d\mathbf{q}_N' \, \exp\left[-\beta \sum_{1 \le i < j \le N} V_{ij}\right]}. \tag{C.3}$$

Since the total potential is a sum of two-body functions of the relative coordinate between pairs of particles, in both the numerator and denominator of the last equation, we can operate a change of variables: $\mathbf{q}_i \to \mathbf{q}_i - \mathbf{q}_1$, for any $i = 2, \ldots, N$. The error that we make in keeping the same limits of integrations (i.e. each new variable integrated in the macroscopic volume V) is negligible because of the short-rangedness of the interaction. The total potential in the newvariables does not depend on $\mathbf{q}_1$, and the ratio of integrals in Eq. (C.3) gives simply $1/V$. Thus, we obtain Eq. (7.21).

The microcanonical form of the equilibrium f_N is given by Eq. (1.15); i.e. in Eq. (C.2) the function $\exp\left[-\beta H\left(\{x_i\}\right)\right]$ is replaced by the function $\delta\left[E - H\left(\{x_i\}\right)\right]$. Now f_N does not have a product form, and the integration is somewhat more difficult. Let us denote by K and U the kinetic and potential energy of the Hamiltonian, i.e. the first and second term on the right-hand side of Eq. (7.1). By using the identity

$$\delta\left[E - H\left(\{x_i\}\right)\right] = \int dE_k\, \delta\left[K\left(\{p_i\}\right) - E_k\right]\delta\left[E - E_k - U\left(\{q_i\}\right)\right], \tag{C.4}$$

we can write

$$f_1(\mathbf{q}_1,\mathbf{p}_1) = N\,\frac{\displaystyle\int dE_k d\mathbf{q}_2 \ldots d\mathbf{q}_N d\mathbf{p}_2 \ldots d\mathbf{p}_N\,\delta\left[K\left(\{p_i\}\right) - E_k\right]\delta\left[E - E_k - U\left(\{q_i\}\right)\right]}{\displaystyle\int dE_k d\mathbf{q}_1' \ldots d\mathbf{q}_N' d\mathbf{p}_1' \ldots d\mathbf{p}_N'\,\delta\left[K\left(\{p_i'\}\right) - E_k\right]\delta\left[E - E_k - U\left(\{q_i'\}\right)\right]}. \tag{C.5}$$

In the thermodynamic limit, the integrals in the phase-space variables $\{x_i\}$ is strongly peaked at a particular value of E_k, so that we make an error that in the thermodynamic limit is negligible, if we substitute the integrals in E_k by the expression obtained for that particular value, denoted hereafter E_k^*. Afterwards, by making the same change of variables in the coordinates as before, we arrive at

$$f_1(\mathbf{q}_1,\mathbf{p}_1) = n\,\frac{\displaystyle\int d\mathbf{p}_2 \ldots d\mathbf{p}_N\,\delta\left[K\left(\{p_i\}\right) - E_k^*\right]}{\displaystyle\int d\mathbf{p}_1' \ldots d\mathbf{p}_N'\,\delta\left[K\left(\{p_i'\}\right) - E_k^*\right]}, \tag{C.6}$$

where $n = N/V$. Both integrals in this expression are simply related to the surface of a hypersphere of radius R in a given number of dimensions. We remind that in dimension d such surface S_d is given by

$$S_d = \frac{2\pi^{d/2}}{\Gamma\left(d/2\right)}R^{d-1}. \tag{C.7}$$

Since each momentum $\mathbf{p}_i$ has three components, the integral in the numerator of (C.6) corresponds to a hypersphere in $3(N-1)$ dimensions and radius $(2mE_k^* - \mathbf{p}_1^2)^{1/2}$, while the one in the denominator has $3N$ dimensions and a radius $(2mE_k^*)^{1/2}$. We obtain therefore

$$f_1(\mathbf{q}_1,\mathbf{p}_1) = \frac{n}{\pi^{\frac{3}{2}}}\,\frac{\Gamma\left(\frac{3N}{2}\right)}{\Gamma\left(\frac{3N-3}{2}\right)}\,\frac{1}{(2mE_k^*)^{3/2}}\left[1 - \frac{p_1^2}{2mE_k^*}\right]^{\frac{3N-4}{2}}. \tag{C.8}$$

At this point, we substitute the relation between the temperature and the total kinetic energy, $E_k^* = 3N/(2\beta)$, and we perform the limit $N \to \infty$ in the last equation, to obtain

$$f_1(\mathbf{q}_1, \mathbf{p}_1) = n \left(\frac{\beta}{2\pi m} \right)^{\frac{3}{2}} e^{-\beta \frac{p_1^2}{2m}}, \tag{C.9}$$

which is exactly Eq. (7.21), the one-particle reduced distribution function in short-range interacting systems.

Appendix D
The Differential Cross-Section of a Binary Collision

In the derivation of the Boltzmann equation in Chapter 7, we introduced the cross-section related to the binary collisions to transform Eq. (7.46) into Eq. (7.48). As underlined in the derivation, the collisions between two particles are computed neglecting the interaction with the other particles in the system, i.e. as if the two particles were isolated. In this Appendix, we explain in more details the binary collision process and the meaning of the relation in Eq. (7.47).

Let us start from the Hamiltonian of a system of two identical particles

$$H = \frac{p_1^2}{2m} + \frac{p_2^2}{2m} + V\left(|\mathbf{q}_1 - \mathbf{q}_2|\right), \tag{D.1}$$

where $p_i = |\mathbf{p}_i|$. Making the canonical transformation of coordinates defined in Eqs. (7.35) and (7.36), here rewritten as

$$\mathbf{Q} = \frac{1}{2}\left(\mathbf{q}_1 + \mathbf{q}_2\right) \tag{D.2}$$

$$\mathbf{q} = \mathbf{q}_1 - \mathbf{q}_2, \tag{D.3}$$

the Hamiltonian becomes

$$H = \frac{P^2}{4m} + \frac{p^2}{m} + V\left(|\mathbf{q}|\right), \tag{D.4}$$

where $\mathbf{P}$ and $\mathbf{p}$ are the momenta canonically conjugated to $\mathbf{Q}$ and $\mathbf{q}$, respectively. We therefore have the free motion of a particle of mass $2m$, the centre of mass, plus the dynamics of a particle of mass $m/2$ (the reduced mass) subject to the potential V. Let us then concentrate on the motion of a particle of mass $\mu = m/2$ in a central potential. We know that such a motion develops in a plane. Using polar coordinates (r, θ) in this plane, the energy of this mechanical problem is

$$E = \frac{\mu}{2}\left(\dot{r}^2 + r^2\dot{\theta}^2\right) + V(r). \tag{D.5}$$

The angular momentum $\ell = \mu r\dot{\theta}$ is a constant of the motion. Substituting in the expression of the energy, we get

$$E = \frac{\mu}{2}\dot{r}^2 + V(r) + \frac{\ell^2}{2\mu r^2},$$ (D.6)

showing that the problem is equivalent to a one-dimensional motion in the effective potential $V(r) + \ell^2/(2\mu r^2)$. We are interested in the case of open orbits, which occur when $E > 0$. Assuming, without loss of generality, that the particle approaches the origin from $r = \infty$ at the angle $\theta = 0$, the general solution, expressing θ as a function of r during the approach, is given by (Goldstein, 1980)

$$\theta(r) = \int_r^{+\infty} dr' \frac{1}{r'^2} \frac{\ell}{\sqrt{2\mu\,[E - V(r')] - \frac{\ell^2}{r'^2}}}.$$ (D.7)

The distance of closest approach, r_m, is the largest positive root of the expression under square root in the denominator in Eq. (D.7); if we denote by θ_m the corresponding value of θ, the angle of deflection of the particle, i.e. the angle between the direction of approach and the direction along which the particle goes back to $r = \infty$, is simply related to θ_m. Again, without loss of generality, we can assume that ℓ is positive; then χ is given by $\pi - 2\theta_m$, i.e.

$$\chi = \pi - 2\int_{r_m}^{\infty} dr \frac{1}{r^2} \frac{\ell}{\sqrt{2\mu\,[E - V(r)] - \frac{\ell^2}{r^2}}}.$$ (D.8)

Both the energy E and the angular momentum ℓ are a simple function of the impact parameter b and of the modulus v of the velocity of the incoming particle at infinity. We have

$$E = \frac{1}{2}\mu v^2 \quad \text{and} \quad \ell = \mu b v.$$ (D.9)

Substituting in Eq. (D.8), we get

$$\chi = \pi - 2\int_{r_m}^{\infty} dr \frac{1}{r^2} \frac{b}{\sqrt{1 - 2\frac{V(r)}{\mu v^2} - \frac{b^2}{r^2}}}.$$ (D.10)

This expression determines χ as a function of b and v; for most of the potentials $V(r)$ of interest, this function, for given v, is a monotonic decreasing function of b.

Let us now recall the definition of the differential cross-section $d\sigma/d\Omega$ in a scattering problem from a centre of force. If there is a flux I of impinging monoenergetic particles (of energy ε determined by their velocity of incidence v) per unit area and per unit second, they will be scattered to different angles according to their impact parameter. The number $\mathcal{N}(\Omega)d\Omega$ of incident particles scattered per unit second into the solid angle element $d\Omega$ about the direction Ω defines the differential cross-section by

$$\mathcal{N}(\Omega)d\Omega = I\frac{d\sigma\,(\Omega,\varepsilon)}{d\Omega}d\Omega. \tag{D.11}$$

On the other hand, we have just seen that the angle of deflection is uniquely determined by the impact parameter b and by v. Considering the cylindrical symmetry of the problem, we then have

$$\mathcal{N}(\Omega)d\Omega = Ibdbd\phi, \tag{D.12}$$

where ϕ is the azimuthal angle defining, together with r and θ, the cylindrical coordinates that describe the scattering. We obtain therefore Eq. (7.47):

$$\frac{d\sigma}{d\Omega}d\Omega = bdbd\phi. \tag{D.13}$$

Appendix E
Autocorrelation of the Fluctuations of the One-Particle Density

Using the definition of the Fourier transform and formula (8.13), we get

$$\langle \delta f\, (k, p, 0)\, \delta f\, (k', p', 0) \rangle = \int_0^{2\pi} \frac{d\theta}{2\pi} \int_0^{2\pi} \frac{d\theta'}{2\pi}\, e^{-i(k\theta + k'\theta')}\, \langle \delta f(\theta, p, 0)\, \delta f(\theta', p', 0) \rangle \quad \text{(E.1)}$$

$$= \int_0^{2\pi} \frac{d\theta}{2\pi} \int_0^{2\pi} \frac{d\theta'}{2\pi}\, e^{-i(k\theta + k'\theta')}$$
$$\times N \left[\langle f_d\, (\theta, p, 0)\, f_d\, (\theta', p', 0) \rangle - f_0\, (p)\, f_0\, (p') \right]. \quad \text{(E.2)}$$

The expression of the discrete density function (8.3) leads then to

$$\langle f_d\, (\theta, p, 0)\, f_d\, (\theta', p', 0) \rangle = \frac{1}{N^2} \left\langle \sum_{j=1}^{N} \delta\, (\theta - \theta_j)\, \delta\, (p - p_j)\, \delta\, (\theta - \theta')\, \delta\, (p - p') \right.$$

$$\left. + \sum_{i \neq j} \delta\, (\theta - \theta_j)\, \delta\, (p - p_j)\, \delta\, (\theta' - \theta_i)\, \delta\, (p' - p_i) \right\rangle \quad \text{(E.3)}$$

$$= \frac{1}{N^2} \left[N\, \langle f_d\, (\theta, p, 0) \rangle\, \delta\, (\theta - \theta')\, \delta\, (p - p') \right.$$

$$\left. + N(N-1) f_2\, (0, \theta, p, \theta', p') \right] \quad \text{(E.4)}$$

$$= \frac{1}{N} f_0\, (p)\, \delta\, (\theta - \theta')\, \delta\, (p - p') + f_0\, (p)\, f_0\, (p')$$
$$+ h_2\, (\theta, p, \theta', p', 0), \quad \text{(E.5)}$$

where we used the following definition of the correlation function h_2:

$$f_2(\theta, p, \theta', p', 0) = \langle \delta\, (\theta - \theta_j)\, \delta\, (p - p_j)\, \delta\, (\theta' - \theta_i)\, \delta\, (p' - p_i) \rangle \quad \text{(E.6)}$$

$$= \frac{N}{N-1} \left[f_0\, (p)\, f_0\, (p') + h_2(\theta, p, \theta', p') \right]. \quad \text{(E.7)}$$

Substituting expression (E.5) in Eq. (E.2), we find that

$$
\langle \delta f\,(k,p,0)\,\delta f\left(k',p',0\right)\rangle = \int_0^{2\pi} \frac{d\theta}{2\pi}\,\frac{f_0\,(p)}{2\pi}\,e^{-i(k+k')\theta}\,\delta\left(p-p'\right)
$$

$$
+ \int_0^{2\pi} \frac{d\theta}{2\pi} \int_0^{2\pi} \frac{d\theta'}{2\pi}\,N e^{-i(k\theta+k'\theta')}\,h_2\left(\theta,p,\theta',p'\right) \qquad \text{(E.8)}
$$

$$
= \frac{f_0\,(p)}{2\pi}\,\delta_{k,-k'}\,\delta\left(p-p'\right) + \frac{1}{2\pi}\,\delta_{k,-k'}\,\mu\left(k,p,p'\right) \qquad \text{(E.9)}
$$

$$
= \frac{\delta_{k,-k'}}{2\pi}\left[f_0\,(p)\delta\,(p-p') + \mu\,(k,p,p')\right]. \qquad \text{(E.10)}
$$

In the passage from (E.8) to (E.9), we have used the fact that h_2 depends only on the difference $\theta-\theta'$. Besides, it decays rapidly to 0 in a range $(\theta-\theta') \sim 1/N$, so that $\mu\,(k,p,p')$ is of order 1.

Appendix F
Derivation of the Fokker–Planck Coefficients

As mentioned in Section 9.2, the derivation of the Fokker–Planck equation (9.9) is performed in the time range $1 \ll t \ll N$. Within this approximation, expressions (9.10) and (9.11) are thus replaced by

$$A(p, t) = \frac{1}{t} \langle (p(t) - p(0)) \rangle_{p(0)=p} \tag{F.1}$$

$$B(p, t) = \frac{1}{t} \langle (p(t) - p(0))^2 \rangle_{p(0)=p}. \tag{F.2}$$

Looking at Eq. (9.8), we see that, at the order $1/N$, we need to compute the averages

$$\frac{1}{\sqrt{N}} \left\langle \frac{\partial \delta v}{\partial \theta} \Big|_t \right\rangle = \frac{1}{\sqrt{N}} \left\langle \frac{\partial \delta v}{\partial \theta} (\theta(0) + p(0)t, t) \right\rangle \tag{F.3}$$

and

$$\frac{1}{N} \left\langle \frac{\partial^2 \delta v}{\partial \theta^2} \Big|_{t_1} \frac{\partial \delta v}{\partial \theta} \Big|_{t_2} \right\rangle = \frac{1}{N} \left\langle \frac{\partial^2 \delta v}{\partial \theta^2} (\theta(0) + p(0)t_1, t_1) \frac{\partial \delta v}{\partial \theta} (\theta(0) + p(0)t_2, t_2) \right\rangle \tag{F.4}$$

for the first moment, and

$$\frac{1}{N} \left\langle \frac{\partial \delta v}{\partial \theta} \Big|_{t_1} \frac{\partial \delta v}{\partial \theta} \Big|_{t_2} \right\rangle = \frac{1}{N} \left\langle \frac{\partial \delta v}{\partial \theta} (\theta(0) + p(0)t_1, t_1) \frac{\partial \delta v}{\partial \theta} (\theta(0) + p(0)t_2, t_2) \right\rangle \tag{F.5}$$

for the second one. The fact that the position and the momentum of the test particle at time 0 are given is specified inside the dependence on θ of the derivatives of δv. The condition of given initial angle and momentum of the test particle produces corrections of the averages in Eqs. (F.3)–(F.5), with respect to their unconditioned values. These corrections being of order $1/\sqrt{N}$, we get that at order $1/N$ they have to be taken into account only for Eq. (F.3). Note that the unconditioned value of that equation vanishes.

Let us begin with Eq. (F.5). To compute it, we start from Eq. (9.3) that we rewrite here

$$\langle \delta v(k,\omega)\delta v(k',\omega')\rangle$$

$$= 2\pi^3 \delta_{k,-k'} \left(4\delta_{k,0} + \delta_{k,1} + \delta_{k,-1}\right) \frac{\delta(\omega+\omega')}{\left|\tilde{D}(\omega,k)\right|^2} \int dp' \, f_0(p')\delta(\omega-p'k). \tag{F.6}$$

By the inverse Fourier-Laplace transform, we then have

$$\left\langle \frac{\partial\delta v}{\partial\theta}\Big|_{t_1} \frac{\partial\delta v}{\partial\theta}\Big|_{t_2} \right\rangle = \left\langle \frac{\partial\delta v}{\partial\theta}(\theta(0)+pt_1,t_1)\frac{\partial\delta v}{\partial\theta}(\theta(0)+pt_2,t_2) \right\rangle \tag{F.7}$$

$$= \frac{1}{(2\pi)^2} \sum_{k=-\infty}^{+\infty} \sum_{k'=-\infty}^{+\infty} \int_{-\infty}^{+\infty} d\omega \int_{-\infty}^{+\infty} d\omega' \, (-kk')$$

$$\times e^{ik(\theta(0)+pt_1)-i\omega t_1} e^{ik'(\theta(0)+pt_2)-i\omega' t_2} \langle \delta v(k,\omega)\delta v(k',\omega')\rangle \tag{F.8}$$

$$= \frac{\pi}{2} \sum_{k=-\infty}^{+\infty} \int_{-\infty}^{+\infty} d\omega \, k^2 e^{i(kp-\omega)(t_1-t_2)} \frac{\delta_{k,1}+\delta_{k,-1}}{\left|\tilde{D}(\omega,k)\right|^2} \int_{-\infty}^{+\infty} dp' \, f_0(p')\delta(\omega-p'k) \tag{F.9}$$

$$= \frac{\pi}{2} \int_{-\infty}^{+\infty} d\omega \, \frac{f_0(\omega)}{\left|\tilde{D}(\omega,1)\right|^2} \left[e^{i(p-\omega)(t_1-t_2)} + e^{-i(p-\omega)(t_1-t_2)} \right]. \tag{F.10}$$

Equation (9.8) shows that the second moment appearing on the right-hand side of Eq. (F.2) is determined by the integral of the last expression in t_1 and t_2 from 0 to t. More precisely

$$B(p,t) = \frac{1}{t}\frac{1}{N}\frac{\pi}{2}\int_0^t dt_1 \int_0^t dt_2 \int_{-\infty}^{+\infty} d\omega \, \frac{f_0(\omega)}{\left|\tilde{D}(\omega,1)\right|^2}\left[e^{i(p-\omega)(t_1-t_2)}+e^{-i(p-\omega)(t_1-t_2)}\right] \tag{F.11}$$

$$= \frac{1}{t}\frac{1}{N}\pi \int_{-\infty}^{+\infty} d\omega \, \frac{f_0(\omega)}{\left|\tilde{D}(\omega,1)\right|^2} \int_0^t ds \int_0^{t-s} dt_2 \left[e^{i(p-\omega)s}+e^{-i(p-\omega)s}\right] \tag{F.12}$$

$$= \frac{1}{t}\frac{1}{N}\pi \int_{-\infty}^{+\infty} d\omega \, \frac{f_0(\omega)}{\left|\tilde{D}(\omega,1)\right|^2} \int_0^t ds(t-s)\left[e^{i(p-\omega)s}+e^{-i(p-\omega)s}\right] \tag{F.13}$$

$$\overset{t\to+\infty}{\sim} \frac{1}{N} 2\pi \int_{-\infty}^{+\infty} d\omega \, \frac{f_0(\omega)}{\left|\tilde{D}(\omega,1)\right|^2} \pi\delta(p-\omega) \tag{F.14}$$

$$= \frac{1}{N} 2\pi^2 \frac{f_0(p)}{\left|\tilde{D}(p,1)\right|^2} \tag{F.15}$$

$$= \frac{2}{N}D(p), \tag{F.16}$$

which is expression (9.13).

For what concerns Eq. (F.4), starting again from Eq. (F.6), we obtain

$$\left\langle \frac{\partial^2 \delta v}{\partial \theta^2} \bigg|_{t_1} \frac{\partial \delta v}{\partial \theta} \bigg|_{t_2} \right\rangle = \left\langle \frac{\partial^2 \delta v}{\partial \theta^2} (\theta(0) + pt_1, t_1) \frac{\partial \delta v}{\partial \theta} (\theta(0) + pt_2, t_2) \right\rangle \tag{F.17}$$

$$= \frac{1}{(2\pi)^2} \sum_{k=-\infty}^{+\infty} \sum_{k'=-\infty}^{+\infty} \int_{-\infty}^{+\infty} d\omega \int_{-\infty}^{+\infty} d\omega' \, (-ik^2 k')$$

$$\times e^{ik(\theta(0)+pt_1)-i\omega t_1} \, e^{ik'(\theta(0)+pt_2)-i\omega' t_2} \, \langle \delta v(k,\omega) \delta v(k',\omega') \rangle \tag{F.18}$$

$$= i\frac{\pi}{2} \sum_{k=-\infty}^{+\infty} \int_{-\infty}^{+\infty} d\omega \, k^3 \, e^{i(kp-\omega)(t_1-t_2)} \frac{\delta_{k,1} + \delta_{k,-1}}{\left| \tilde{D}(\omega,k) \right|^2}$$

$$\int_{-\infty}^{+\infty} dp' \, f_0(p') \delta(\omega - p'k) \tag{F.19}$$

$$= i\frac{\pi}{2} \int_{-\infty}^{+\infty} d\omega \frac{f_0(\omega)}{\left| \tilde{D}(\omega,1) \right|^2} \left[e^{i(p-\omega)(t_1-t_2)} - e^{-i(p-\omega)(t_1-t_2)} \right]. \tag{F.20}$$

Equation (9.8) shows that the contribution $A_1(p, t)$ to the first moment appearing on the right-hand side of Eq. (F.1) is given by

$$A_1(p, t) = \frac{1}{t} \frac{1}{N} \int_0^t du \int_0^u du_1 \int_0^{u_1} du_2 \left\langle \frac{\partial^2 \delta v}{\partial \theta^2} (\theta(0) + pu, u) \frac{\partial \delta v}{\partial \theta} (\theta(0) + pu_2, u_2) \right\rangle \tag{F.21}$$

$$= \frac{1}{t} \frac{1}{N} \frac{i\pi}{2} \int_0^t du \int_0^u du_1 \int_0^{u_1} du_2 \int_{-\infty}^{+\infty} d\omega \frac{f_0(\omega)}{\left| \tilde{D}(\omega,1) \right|^2}$$

$$\times \left[e^{i(p-\omega)(u-u_2)} - e^{-i(p-\omega)(u-u_2)} \right] \tag{F.22}$$

$$= -\frac{1}{t} \frac{1}{N} \pi \, \text{Im} \int_{-\infty}^{+\infty} d\omega \frac{f_0(\omega)}{\left| \tilde{D}(\omega,1) \right|^2} \int_0^t du \int_0^u du_2 (u - u_2) \, e^{i(p-\omega)(u-u_2)} \tag{F.23}$$

$$= -\frac{1}{t} \frac{1}{N} \pi \, \text{Re} \int_{-\infty}^{+\infty} d\omega \frac{f_0(\omega)}{\left| \tilde{D}(\omega,1) \right|^2} \frac{\partial}{\partial \omega} \int_0^t du \int_0^u du_2 \, e^{i(p-\omega)(u-u_2)} \tag{F.24}$$

$$= \frac{1}{t} \frac{1}{N} \pi \, \text{Re} \int_{-\infty}^{+\infty} d\omega \left(\frac{\partial}{\partial \omega} \frac{f_0(\omega)}{\left| \tilde{D}(\omega,1) \right|^2} \right) \int_0^t du \frac{i}{p - \omega} \left[1 - e^{i(p-\omega)u} \right] \tag{F.25}$$

$$= \frac{1}{t} \frac{1}{N} \pi \int_{-\infty}^{+\infty} d\omega \left(\frac{\partial}{\partial \omega} \frac{f_0(\omega)}{\left| \tilde{D}(\omega,1) \right|^2} \right) \int_0^t du \frac{\sin(p-\omega)u}{p - \omega} \tag{F.26}$$

$$= \frac{1}{t}\frac{1}{N}2\pi \int_{-\infty}^{+\infty} d\omega \left(\frac{\partial}{\partial \omega} \frac{f_0(\omega)}{\left|\tilde{D}(\omega,1)\right|^2} \right) \frac{1}{(p-\omega)^2} \sin^2 \frac{(p-\omega)t}{2} \tag{F.27}$$

$$\underset{\sim}{\overset{t\to +\infty}{\sim}} \frac{1}{t}\frac{1}{N}2\pi \left(\frac{\partial}{\partial p} \frac{f_0(p)}{\left|\tilde{D}(p,1)\right|^2} \right) \int_{-\infty}^{+\infty} d\omega \frac{1}{(p-\omega)^2} \sin^2 \frac{(p-\omega)t}{2} \tag{F.28}$$

$$= \frac{1}{N}\pi^2 \left(\frac{\partial}{\partial p} \frac{f_0(p)}{\left|\tilde{D}(p,1)\right|^2} \right) \tag{F.29}$$

$$= \frac{1}{N}\frac{d}{dp}D(p) . \tag{F.30}$$

For what concerns Eq. (F.3), let us rewrite Eq. (8.34) (where the Fourier transform of the potential for the HMF model is given by $\widehat{V}(k) = \left(2\delta_{k,0} - \delta_{k,1} - \delta_{k,-1}\right)/2$) as

$$\widetilde{\delta v}(k,\omega) = -\frac{\pi \left(\delta_{k,1} + \delta_{k,-1} \right)}{\tilde{D}(\omega,k)} \int_{-\infty}^{+\infty} dp' \frac{\widehat{\delta f}(k,p',0)}{i(p'k-\omega)}, \tag{F.31}$$

where we have neglected the $k = 0$, which will not contribute. Since the initial angle and momentum of the test particle are given, we get

$$\left\langle \widehat{\delta f}(k,p',0) \right\rangle = \frac{1}{\sqrt{N}} \frac{1}{2\pi} e^{-ik\theta(0)} \delta(p-p'). \tag{F.32}$$

We therefore have

$$\left\langle \frac{\partial \delta v}{\partial \theta}(\theta(0) + pu, u) \right\rangle = -\frac{1}{\sqrt{N}} \frac{1}{(2\pi)^2} \sum_{k=-\infty}^{+\infty} \int_C d\omega \, (ik) \, e^{ik(\theta(0)+pu)-i\omega u}$$

$$\times \frac{\pi \left(\delta_{k,1} + \delta_{k,-1} \right)}{\tilde{D}(\omega,k)} \frac{e^{-ik\theta(0)}}{i(pk-\omega)} \tag{F.33}$$

$$= -\frac{1}{\sqrt{N}} \frac{1}{4\pi} \int_C d\omega \left[\frac{e^{i(p-\omega)u}}{(p-\omega)\tilde{D}(\omega,1)} + \frac{e^{-i(p+\omega)u}}{(p+\omega)\tilde{D}(\omega,-1)} \right] \tag{F.34}$$

$$= -\frac{1}{\sqrt{N}} \frac{1}{4\pi} \left(\mathcal{P} \int_{-\infty}^{+\infty} d\omega \, \frac{1}{p-\omega} \left[\frac{e^{i(p-\omega)u}}{\tilde{D}(\omega,1)} + \frac{e^{-i(p-\omega)u}}{\tilde{D}(-\omega,-1)} \right] \right.$$

$$\left. + i\pi \left[\frac{1}{\tilde{D}(p,1)} - \frac{1}{\tilde{D}(-p,-1)} \right] \right) \tag{F.35}$$

$$= -\frac{1}{\sqrt{N}}\frac{1}{4\pi}\left(2\operatorname{Re}\mathcal{P}\int_{-\infty}^{+\infty}d\omega\,\frac{e^{i(p-\omega)u}}{p-\omega}\,\frac{\tilde{D}^*(\omega,1)}{\left|\tilde{D}(\omega,1)\right|^2}\right.$$

$$\left.+2\pi^3\frac{f_0'(p)}{\left|\tilde{D}(p,1)\right|^2}\right),\tag{F.36}$$

where $\mathcal{P}$ stands for the principal value.

From Eq. (9.8), we see that the contribution $A_2(p,t)$ of this term to the first moment appearing on the right-hand side of Eq. (F.1) is given by

$$A_2(p,t) = \frac{1}{t}\frac{1}{N}\frac{1}{4\pi}\int_0^t du\left(2\pi^3\frac{f_0'(p)}{\left|\tilde{D}(p,1)\right|^2}\right.$$

$$\left.+2\operatorname{Re}\mathcal{P}\int_{-\infty}^{+\infty}d\omega\,\frac{1}{p-\omega}\frac{\tilde{D}^*(\omega,1)}{\left|\tilde{D}(\omega,1)\right|^2}e^{i(p-\omega)u}\right)\tag{F.37}$$

$$= \frac{1}{N}\frac{\pi^2}{2}\frac{f_0'(p)}{\left|\tilde{D}(p,1)\right|^2}+\frac{1}{t}\frac{1}{N}\frac{1}{4\pi}$$

$$\times 2\operatorname{Re}\mathcal{P}\int_{-\infty}^{+\infty}d\omega\,\frac{1}{(p-\omega)^2}\frac{\tilde{D}^*(\omega,1)}{\left|\tilde{D}(\omega,1)\right|^2}\left[2i\sin^2\frac{(p-\omega)t}{2}+\sin(p-\omega)t\right]$$

$$\tag{F.38}$$

$$\overset{t\to+\infty}{\sim}\frac{1}{N}\frac{1}{2}\pi^2\frac{f_0'(p)}{\left|\tilde{D}(p,1)\right|^2}+\frac{1}{N}\frac{1}{2}\operatorname{Re}\frac{i\tilde{D}^*(p,1)}{\left|\tilde{D}(p,1)\right|^2}\tag{F.39}$$

$$= \frac{1}{N}\frac{1}{2}\pi^2\frac{f_0'(p)}{\left|\tilde{D}(p,1)\right|^2}+\frac{1}{N}\frac{1}{2}\pi^2\frac{f_0'(p)}{\left|\tilde{D}(p,1)\right|^2}\tag{F.40}$$

$$= \frac{1}{N}\pi^2\frac{f_0'(p)}{\left|\tilde{D}(p,1)\right|^2}\tag{F.41}$$

$$= \frac{1}{N}\frac{1}{f_0}\frac{\partial f_0}{\partial p}D(p)\,.\tag{F.42}$$

Adding Eqs. (F.30) for $A_1(p,t)$ and (F.42) for $A_2(p,t)$, we obtain Eq. (9.12) for $A(p,t)$.

References

Aarseth, S. J. (2003). *Gravitational N-Body Simulations: Tools and Algorithms*. Cambridge University Press, Cambridge.

Akhiezer, A. I., Baryakhtar, A. I. and Peletminskij, S. V. (1968). *Spin Waves*. North Holland, Amsterdam.

Alastuey, A., Magro, M. and Pujol, P. (2008). *Physique et Outils Mathématiques: Méthodes et Exemples*. CNRS Editions and EDP Sciences.

Angelani, L., Casetti, L., Pettini, M., Ruocco, G. and Zamponi, F. (2003). *Europhysics Letters*, **62**, 775.

Anteneodo, C. and Tsallis, C. (1998). *Physical Review Letters*, **80**, 5313.

Antoni, M., Hinrichsen, H. and Ruffo, S. (2002). *Chaos, Solitons and Fractals*, **13**, 393.

Antoni, M. and Ruffo, S. (1995). *Physical Review E*, **52**, 2361.

Antoni, M., Ruffo, S. and Torcini, A. (2002). *Physical Review E*, **66**, 025103.

Antoni, M., Ruffo, S. and Torcini, A. (2004). *Europhysics Letters*, **66**, 645.

Antoni, M. and Torcini, A. (1998). *Physical Review E*, **57**, R6233.

Antoni, M., Elskens, Y. and Escande, D. (1998). *Physics of Plasmas*, **5**, 841.

Antoniazzi, A., Califano, F., Fanelli, D. and Ruffo, S. (2007a). *Physical Review Letters*, **98**, 150602.

Antoniazzi, A., De Ninno, G., Fanelli, D., Guarino, A. and Ruffo, S. (2005). *Journal of Physics: Conference Series*, **7**, 143.

Antoniazzi, A., Elskens, Y., Fanelli, D. and Ruffo, S. (2006). *European Physical Journal B*, **50**, 603.

Antoniazzi, A., Fanelli, D., Barré, J., Chavanis, P. H., Dauxois, T. and Ruffo, S. (2007b). *Physical Review E*, **75**, 011112.

Antoniazzi, A., Fanelli, D., Ruffo, S. and Yamaguchi, Y. Y. (2007c). *Physical Review Letters*, **99**, 040601.

Antoniazzi, A., Johal, Ramandeep, S., Fanelli, D. and Ruffo, S. (2008). *Communications in Nonlinear Science and Numerical Simulation*, **13**, 2.

Antonov, V. A. (1962). *Vestnik Leningradskogo Universiteta*, **7**, 135.

Aronson, E. B. and Hansen, C. J. (1972). *Astrophysical Journal*, **177**, 145.

Asslani, M., Fanelli, D., Turchi, A., Carletti, T. and Leoncini, X. (2012). *Physical Review E*, **85**, 021148.

Aurell, E., Fanelli, D. and Muratore-Ginanneschi, P. (2001). *Physica D*, **148**, 272.

Aurell, E., Fanelli, D., Gurbatov, S. N. and Moshkov, A. Yu. (2003). *Physica D*, **186**, 171.

Ayuela, A. and March, N. H. (2008). *Physics Letters A*, **372**, 5617.

Bachelard, R., Chandre, C., Fanelli, D., Leoncini, X. and Ruffo, S. (2008a). *Physical Review Letters*, **101**, 260603.

Bachelard, R., Chandre, C. and Vittot, M. (2008b). *Physical Review E*, **78**, 036407.

Bachelard, R., Manos, T., de Buyl, P., Staniscia, F., Cataliotti, F. S., De Ninno, G., Fanelli, D. and Piovella, N. (2010). *Journal of Statistical Mechanics: Theory and Experiment*, P06009.

Baldovin, F. and Orlandini, E. (2006a). *Physical Review Letters*, **96**, 240602.

Baldovin, F. and Orlandini, E. (2006b). *Physical Review Letters*, **97**, 100601.

Balescu, R. (1960). *Physics of Fluids*, **3**, 52.

Balescu, R. (1963). *Statistical Mechanics of Charged Particles*. Interscience, New York.

Balescu, R. (1975). *Equilibrium and Nonequilibrium Statistical Mechanics*. Wiley, New York.

Balian, R. (1992). *From Microphysics to Macrophysics: Methods and Applications of Statistical Physics*. Springer-Verlag, Berlin.

Banerjee, S., Griffiths, R. B. and Widom, M. (2004). *Journal of Statistical Physics*, **93**, 109.

Barbosa, M. C., Deserno, M. and Holm, C. (2000). *Europhysics Letters*, **52**, 80.

Barnes, J. E. and Hut, P. (1986). *Nature*, **324**, 466.

Barré, J. (2002). *Physica A*, **305**, 172.

Barré, J., Bouchet, F., Dauxois, T. and Ruffo, S. (2005). *Journal of Statistical Physics*, **119**, 677.

Barré, J., Bouchet, F., Dauxois, T., Ruffo, S. and Yamaguchi, Y. Y. (2006). *Physica A*, **365**, 177.

Barré, J., Dauxois, T., De Ninno, G., Fanelli, D. and Ruffo, S. (2004). *Physical Review E*, **69**, 045501(R).

Barré, J., Mukamel, D. and Ruffo, S. (2001). *Physical Review Letters*, **87**, 030601.

Barré, J., Mukamel, D. and Ruffo, S. (2002). Ensemble inequivalence in mean-field models of magnetisms. In (Dauxois *et al.*, 2002a).

Bender, C. M. and Orszag, S. A. (1978). *Advanced Mathematical Methods for Scientists and Engineers*. McGrawHill, New York, Chapter 6.

Bernstein, I. B., Greene, J. M. and Kruskal, M. D. (1949). *Physical Review*, **108**, 546.

Bielawski, S., Bruni, C., Orlandi, L. G., Garzella, D. and Couprie, M. E. (2004). *Physical Review E*, **69**, 0465502(R).

Billardon, M., Garzella, D. and Couprie, M. E. (1992). *Physical Review Letters*, **69**, 2368.

Bialynicki-Birula, I., Hubbard, J. C. and Turski, L. A. (1984). *Physica A*, **128**, 509.

Binney, J. and Tremaine, S. (1987). *Galactic Dynamics*. Princeton Series in Astrophysics. Princeton University Press, Princeton.

Blume, M. (1966). *Physical Review*, **141**, 517.

Blume, J., Emery, V. J. and Griffiths, R. B. (1971). *Physical Review A*, **4**, 1071.

Bogoliubov, N. N. (1946). *Journal of Physics (USSR)* **10**, 265.

Bohm, D. and Gross, E. P. (1949a). *Physical Review*, **75**, 1851.

Bohm, D. and Gross, E. P. (1949b). *Physical Review*, **75**, 1864.

Boltzmann, L. (1872). *Wiener Berichte*, **66**, 275.

Bonifacio, R., Casagrande, F., Cerchioni, G., De Salvo Souza, L., Pierini, P. and Piovella, N. (1990). *Rivista del Nuovo Cimento*, **13**, 1.

Bonifacio, R. and De Salvo, L. (1994). *Nuclear Instruments and Methods A*, **341**, 360.

Bonifacio, R., Cataliotti, F. S., Cola, M., Fallani, L., Fort, C., Piovella, N. and Inguscio, M. (2004). *Optical Communications*, **233**, 155.

Borgonovi, F., Celardo, G. L., Maianti, M. and Pedersoli, E. (2004). *Journal of Statistical Physics*, **116**, 1435.

Borgonovi, F., Celardo, G. L., Musesti, A., Trasarti-Battistoni, R. and Vachal, P. (2006). *Physical Review E*, **73**, 039903.

Born, M. and Green, H. S. (1949). *A General Kinetic Theory of Liquids*. Cambridge University Press, Cambridge.

Boucher, C., Ellis, R. S. and Turkington, B. (1999). *Annals of Probability*, **27**, 297. Erratum, *Annals of Probability*, **30**, 2113 (2002).

Bouchet, F. (2001). *Mécanique Statistique pour des Écoulements Géophysiques*. PhD thesis, Université Joseph Fourier, Grenoble.

Bouchet, F. (2004). *Physical Review E*, **70**, 036113.

Bouchet, F. and Barré, J. (2005). *Journal of Statistical Physics*, **118**, 1073.

Bouchet, F. and Corvellec, M. (2010). *Journal of Statistical Mechanics*, P08021.

Bouchet, F. and Dauxois, T. (2005). *Physical Review E*, **72**, 045103(R).

Bouchet, F., Dauxois, T., Mukamel, D. and Ruffo, S. (2008). *Physical Review E*, **77**, 011125.

Bouchet, F., Gupta, S. and Mukamel, D. (2010). *Physica A*, **389**, 4389.

Bouchet, F. and Sommeria, J. (2002). *Journal of Fluid Mechanics*, **464**, 165.

Bouchet, F. and Venaille, A. (2012). *Physics Reports*, **515**, 227.

Bramwell, S. T. (2009). *Dipolar Effects in Condensed Matter*, in Dauxois *et al.* (2009).

Bramwell, S. T. and Gingras, M. J. P. (2001). *Science*, **294**, 1495.

Braun, W. and Hepp, K. (1977). *Communications in Mathematical Physics*, **56**, 101.

Brydges, D. C. and Martin, P. A. (1999). *Journal of Statistical Physics*, **96**, 1163.

Buchert, T., Domínguez, A. and Perez-Mércader, J. (1999). *Astronomy and Astrophysics*, **349**, 343.

Caglioti, E., Lions, P. L., Marchioro, C. and Pulvirenti, M. (1992). *Communications in Mathematical Physics*, **143**, 501.

Caglioti, E., Lions, P. L., Marchioro, C. and Pulvirenti, M. (1995). *Communications in Mathematical Physics*, **174**, 229.

Caglioti, E. and Rousset, F. (2008). *Archive for Rational Mechanics and Analysis*, **190**, 517.

Caldwell, R. R., Dave, R. and Steinhardt, P. J. (2001). *Physical Review Letters*, **80**, 1582.

Camm, G. L. (1950). *Monthly Notices of the Royal Astronomical Society*, **110**, 305.

Campa, A., Chavanis, P. H., Giansanti, A. and Morelli, G. (2008a). *Physical Review E*, **78**, 040102(R).

Campa, A., Dauxois, T. and Ruffo, S. (2009). *Physics Reports*, **480**, 57.

Campa, A. and Giansanti, A. (2004). *Physica A*, **340**, 170.

Campa, A., Giansanti, A., Morigi, G. and Sylos Labini, F. (Eds) (2008b). *Dynamics and Thermodynamics of Systems with Long-Range Interactions: Theory and Experiment*. AIP Conference Proceedings, 970.

Campa, A., Giansanti, A. and Morelli, G. (2007a). *Physical Review E*, **76**, 041117.

Campa, A., Giansanti, A. and Moroni, D. (2003). *Journal of Physics A*, **36**, 6897.

Campa, A., Giansanti, A., Mukamel, D. and Ruffo, S. (2006). *Physica A*, **365**, 120.

Campa, A., Khomeriki, R., Mukamel, D. and Ruffo, S. (2007b). *Physical Review B*, **76**, 064415.

Campa, A. and Ruffo, S. (2007). *Physica A*, **385**, 233.

Campa, A., Ruffo, S. and Touchette, H. (2007c). *Physica A*, **369**, 517.

Capel, H. W. (1968). *Physica*, **32**, 966.

Case, K. M. (1959). *Annals of Physics (N. Y.)*, **7**, 349.

Casetti, L. and Kastner, M. (2007). *Physica A*, **384**, 318.

Castaing, B. (1995). An Introduction to hydrodynamics, in *Hydrodynamics and Nonlinear Instabilities*, ed. C. Godrèche, P. Manneville. Cambridge University Press, Cambridge.

Celardo, G. L., Barré, J., Borgonovi, F. and Ruffo S. (2006). *Physical Review E*, **73**, 011108.

Ceperley, D. M. (1978). *Physical Review B*, **18**, 3126.

Chabanol, L.-L., Corson, F. and Pomeau, Y. (2000). *Europhysics Letters*, **50**, 148.

Chandrasekhar, S. (1942). *Principles of Stellar Dynamics*. University of Chicago Press, Chicago.

Chandrasekhar, S. (1943). *Review of Modern Physics*, **15**, 1.

Chapuis, G., Brunisholz, G., Javet, C. and Roulet, R. (1983). *Inorganic Chemistry*, **22**, 455.

Chavanis, P. H. (1996). *Contribution à la mécanique statistique des tourbillons bidimensionnels. Analogie avec la relaxation violente des systèmes stellaires*. PhD thesis, École Normale Supérieure de Lyon.

Chavanis, P. H. (2001). *Physical Review E*, **64**, 026309.

Chavanis, P. H. (2002a). *Physical Review E*, **65**, 056123.

Chavanis, P. H. (2002b). Statistical mechanics of two-dimensional vortices and stellar systems. In Dauxois *et al.* (2002a).

Chavanis, P. H. (2005). *Astronomy and Astrophysics*, **432**, 117.

Chavanis, P. H. (2006a). *International Journal of Modern Physics B*, **20**, 3113.

Chavanis, P. H. (2006b). *Physica A*, **361**, 55.

Chavanis, P. H. (2006c). *Physica A*, **361**, 81.

Chavanis, P. H. (2006d). *European Physical Journal B*, **52**, 61.

Chavanis, P. H. (2008a). Dynamics and thermodynamics of systems with long-range interactions: interpretation of the different functionals. In Campa *et al.* (2008b).

Chavanis, P. H. (2008b). *Physica A*, **387**, 787.

Chavanis, P. H. (2008c). *Physica A*, **387**, 1504.

Chavanis, P. H. and Lemou, M. (2005). *Physical Review E*, **72**, 061106.

Chavanis, P. H. and Lemou, M. (2007). *European Physical Journal B*, **59**, 217.

Chavanis, P. H. and Sommeria, J. (2002). *Physical Review E*, **65**, 026302.

Chavanis, P. H., Sommeria, J. and Robert, R. (1996). *Astrophysical Journal*, **471**, 385.

Chavanis, P. H., Vatteville, J. and Bouchet, F. (2005). *European Physical Journal B*, **46**, 61.

Chen, F. F. (1984). *Introduction to Plasma Physics and Controlled Fusion*. Plenum, New York.

Chomaz, P. (2008). *Phase Transitions in Finite Systems Using Information Theory*. In Campa *et al.* (2008b).

Chomaz, P. and Gulminelli, F. (2002). Phase transitions in finite systems. in Dauxois *et al.* (2002a).

Colson, W. B. (1976). *Physics Letters A*, **59**, 187.

Compagner, A., Bruin, C. and Roelse, A. (1989). *Physical Review A*, **39**, 5989.

Creutz, M. (1983). *Physical Review Letters*, **50**, 1411.

Curbis, F., Antoniazzi, A., De Ninno, G. and Fanelli, D. (2007). *European Physical Journal B*, **59**, 527.

Dauxois, T., Fauve, S. and Tuckerman, L. (1996). *Physics of Fluids*, **8**, 487.

Dauxois, T., Latora, V., Rapisarda, A., Ruffo, S. and Torcini, A. (2002b). The Hamiltonian mean field model: from dynamics to statistical mechanics and back. In Dauxois *et al.* (2002a).

Dauxois, T., Lepri, S. and Ruffo, S. (2003). *Communications in Nonlinear Science and Numerical Simulations*, **8**, 375.

Dauxois, T., Ruffo, S., Arimondo, E. and Wilkens, M. (Eds) (2002a). *Dynamics and Thermodynamics of Systems with Long-Range Interactions*, Lecture Notes in Physics **602**. Springer, Berlin.

Dauxois, T., Ruffo, S. and Cugliandolo, L. F. (Eds) (2009). *Long-Range Interacting Systems*. Oxford University Press, Oxford.

de Buyl, P., De Ninno, G., Fanelli, D., Nardini, C., Patelli, A., Piazza, F. and Yamaguchi, Y. Y. (2013). *Physical Review E*, **87**, 042110.

de Buyl, P., Mukamel, D. and Ruffo, S. (2005). *AIP Conferences Proceedings*, **800**, 533.

Del Castillo-Negrete, D. (1998a). *Physics Letters A*, **241**, 99.

Del Castillo-Negrete, D. (1998b). *Physics of Plasmas*, **5**, 3886.

Dembo, A. and Zeitouni, O. (1998). *Large Deviations Techniques and their Applications*. Springer-Verlag, New York.

De Ninno, G. and Fanelli, D. (2004). *Physical Review Letters*, **92**, 094801.

De Ninno, G., Antoniazzi, A., Diviacco, B., Fanelli, D., Giannessi, L., Meucci, R. and Trovó, M. (2005). *Physical Review E*, **71**, 066505.

Desai, R. and Zwanzig, R. (1978). *Journal of Statistical Physics*, **19**, 1.

Dimonte, G. and Malmberg, J. H. (1978). *Physics Fluids*, **21**, 1188.

Doveil, F. and Macor, A. (2006). *Physics of Plasmas*, **13**, 055704.

Dubin, D. H. and O'Neil, T. M. (1999). *Review of Modern Physics*, **71**, 87.

Dubin, D. H. and Jin, D. Z. (2001). *Physics Letters A*, **284**, 112.

Dubin, D. H. (2003). *Physics of Plasmas*, **10**, 1338.

Dupas, A., Le Dang, K., Renard, J.-P. and Veillet, P. (1977). *Journal of Physics C: Solid State Physics*, **10**, 3399.

Dyson, F. J. (1967). *Journal of Mathematical Physics*, **8**, 1538.

Dyson, F. J. (1969). *Communications in Mathematical Physics*, **12**, 91.

Eddington, A. S. (1926). *The Internal Constitution of Stars*. Cambridge University Press, Cambridge.

Eldridge, O. C. and Feix, M. (1962). *Physics of Fluids*, **5**, 1076.

Ellis, R. S. (1985). *Entropy, Large Deviations, and Statistical Mechanics*. Springer-Verlag, New York.

Ellis, R. S. (1999). *Physica D*, **133**, 106.

Ellis, R. S., Haven, K. and Turkington, B. (2000). *Journal of Statistical Physics*, **101**, 999.

Ellis, R. S., Haven, K. and Turkington, B. (2002). *Nonlinearity*, **15**, 239.

Ellis, R. S., Touchette, H. and Turkington, B. (2004). *Physica A*, **335**, 518.

Elskens, Y. and Escande, D. (2002). *Microscopic Dynamics of Plasmas and Chaos*. IOP Publishing, Bristol.

Emden, R. (1907). *Gaskugeln*. Teubner, Leipzig.

English, L. Q., Sato, M. and Sievers, A. J. (2003). *Physical Review B*, **67**, 024403.

Eyink, G. L. and Sreenivasan, K. R. (2006). *Review of Modern Physics*, **78**, 87.

Fallani, L., Fort, C., Piovella, N., Cola, M., Cataliotti, F. S., Inguscio, M. and Bonifacio, R. (2005). *Physical Review A*, **71**, 033612.

Fanelli, D. and Aurell, E. (2002). *Astronomy and Astrophysics*, **395**, 399.

Farago, J. (2000). *Europhysics Letters*, **52**, 379.

Feix, M. R. and Bertrand, P. (2005). *Transport Theory and Statistical Physics*, **34**, 7.

Fine, K. S., Cass, A. C., Flynn, W. G. and Driscoll, C. F. (1995). *Physical Review Letters*, **75**, 3277.

Firpo, M.-C. (1999). *Etude dynamique et statistique de l'interaction onde-particule*. PhD thesis, Université de Provence.

Firpo, M.-C. and Elskens, Y. (1998). *Journal of Statistical Physics*, **93**, 193.

Firpo, M.-C. and Elskens, Y. (2000). *Physical Review Letters*, **84**, 3318.

Fisher, M. E. (1964). *Archive for Rational Mechanics and Analysis*, **17**, 377.

Fisher, M. E. and Ruelle, D. (1966). *Journal of Mathematical Physics*, **7**, 260.

Fortov, V. E., Iakubov, I. T. and Khrapak, A. G. (2006). *Physics of Strongly Coupled Plasma*. Oxford University Press, Oxford.

Gabrielli, A., Joyce, M. and Sicard, F. (2009). *Physical Review E*, **80**, 041108.

Gallavotti, G. (1999). *Statistical Mechanics: A Short Treatise*. Springer, Berlin.

Gilmour, A. S. (1994) *Principles of Travelling Wave Tubes*. Artech House, London.

Glyde, H. R., Keech, G. H., Mazighi, R. and Hansen, J. P. (1976). *Physics Letters A*, **58**, 226.

Goldstein, H. (1980). *Classical Mechanics*. Addison-Wesley, San Francisco.

Góral, K., Santos, L. and Lewenstein, M. (2002). *Physical Review Letters*, **88**, 170406.

Greiner, M., Mandel, O., Esslinger, T., Hänsch, T. W. and Bloch, I. (2002). *Nature*, **415**, 39.

Griesmaier, A. (2007). *Journal of Physics B*, **40**, 91.

Griffiths, R. B. (1968). *Physical Review*, **176**, 655.

Griesmaier, A., Werner, J., Hensler, S., Stuhler, J. and Pfau, T. (2005). *Physical Review Letters*, **94**, 160401.

Gross, D. H. E. (2001). *Microcanonical Thermodynamics*. World Scientific, Singapore.

Gross, D. H. E. (2002). Thermo-statistics or topology of the microcanonical entropy surface. In Dauxois *et al.* (2002a).

Gross, D. H. E. and Votyakov, E. V. (2000). *European Physical Journal B*, **15**, 115.

Gupta, S. and Mukamel, D. (2010). *Physical Review Letters*, **105**, 040602.

Gurbatov, S. N., Saichev, A. I. and Shandarin, S. F. (1989). *Monthly Notices of the Royal Astronomical Society*, **236**, 385.

Guyomarc'h, D. and Doveil, F. (2000). *Review of Scientific Instruments*, **71**, 4087.

Hahn, I. and Kastner, M. (2005). *Physical Review E*, **72**, 056134.

Hahn, I. and Kastner, M. (2006). *European Physical Journal B*, **50**, 311.

Hauray, M. and Jabin, P. E. (2007). *Archive for Rational Mechanics and Analysis*, **183**, 489.

Heggie, D. and Hut, P. (2003). *The Gravitational Million-Body Problem*. Cambridge University Press, Cambridge.

Hemmer, P. R., Bigelow, N. P., Katz, D. P., Shahriar, M. S., DeSalvo, L. and Bonifacio, R. (1996). *Physical Review Letters*, **77**, 1468.

Hénon, M. (1964). *Annales d'Astrophysique*, **27**, 83.

Hertel, P. and Thirring, W. (1971). *Annals of Physics*, **63**, 520.

Hohl, F. and Feix, M. R. (1967). *Astrophysical Journal*, **147**, 1164.

Huang, K. (1987). *Statistical Mechanics*. Wiley, New York.

Ichimaru, S. (1973). *Basic Principles of Plasma Physics*. W. A. Benjamin, Reading, MA.

Ichimaru, S. (1982). *Reviews of Modern Physics*, **54**, 1017.

Inagaki, S. (1993). *Progress in Theoretical Physics*, **90**, 577.

Inagaki, S. and Konishi, T. (1993). *Publications of the Astronomical Society of Japan*, **45**, 733.

Inouye, S., Chikkatur, A. P., Stamper-Kurn, D. M., Stenger, J., Pritchard, D. E. and Ketterle, W. (1999). *Science*, **285**, 571.

Ispolatov, I. and Cohen, E. G. D. (2001). *Physica A*, **295**, 475.

Jackson, J. D. (1975). *Classical Electrodynamics*. Wiley, New York.

Jaksch, D., Bruder, C., Cirac, I., Gardiner, C. W. and Zoller, P. (1998). *Physical Review Letters*, **81**, 3108.

Jellinek, J., Beck, T. L. and Berry, A. S. (1986). *Journal of Chemical Physics*, **84**, 2783.

Joyce, G. and Montgomery, D. (1973). *Journal of Plasma Physics*, **10**, 107.

Kadomtsev, B. B. and Pogutse, O. P. (1970). *Physical Review Letters*, **25**, 1155.

Kandrup, H. E. (1981). *Astrophysical Journal*, **244**, 316.

Kardar, M. (1983a). *Physical Review B*, **28**, 244.

Kardar, M. (1983b). *Physical Review Letters*, **51**, 523.

Katz, J. (1978). *Monthly Notices of the Royal Astronomical Society*, **183**, 765.

Kawahara, R. and Nakanishi, H. (2006). *Journal of Physical Society of Japan*, **75**, 054001.

Kiessling, M. K. H. (1989). *Journal of Statistical Physics*, **55**, 203.

Kiessling, M. K. H. and Lebowitz, J. L. (1997). *Letters in Mathematical Physics*, **42**, 43.

Kiessling, M. K. H. and Neukirch, T. (2003). *Proceedings of the National Academy of Sciences (USA)*, **100**, 1510.

King, I. (1962). *Astronomical Journal*, **67**, 471.

King, I. (1965). *Astronomical Journal*, **70**, 376.

King, I. (1966). *Astronomical Journal*, **71**, 471.

Kirkwood, J. G. (1946). *Journal of Chemical Physics*, **14**, 180.

Kittel, C. (1949). *Review of Modern Physics*, **21**, 541.

Kittel, C. (1951). *Physical Review*, **82**, 965.

Kiwamoto, Y., Hashizume, N., Soga, Y., Aoki, J. and Kawai, Y. (2007). *Physical Review Letters*, **99**, 115002.

Klimontovich, Yu. L. (1967). *The Statistical Theory of Non-equilibrium Processes in a Plasma*. MIT Press, Cambridge, MA.

Klinko, P. J. and Miller, B. N. (2000). *Physical Review E*, **62**, 5783.

Konishi, T. and Kaneko, K. (1992). *Journal of Physics A*, **25**, 6283.

Kramers, H. A. and Wannier, G. H. (1941). *Physical Review*, **60**, 252.

Kruse, D., von Cube, Ch., Zimmermann, C. and Courteille, Ph.W. (2003). *Physical Review Letters*, **91**, 183601.

Kuz'min, G. A. (1982). Statistical mechanics of the organization into two-dimensional coherent structures. In *Structural Turbulence*, Acad. Naouk CCCP Novosibirsk, Institute of thermophysics, ed. M. A. Goldshtik, 103–14.

Labastie, P. and Whetten, R. L. (1990). *Physical Review Letters*, **65**, 1567.

Lahaye, T., Koch, T., Frölich, B., Fattori, M., Metz, J., Griesmaier, A., Giovanazzi, S. and Pfau, T. (2007). *Nature*, **448**, 672.

Landau, L. D. (1936). *Physikalische Zeitschrift der Sowjetunion*, **10**, 154.

Landau, L. D. (1946). *Journal of Physics (USSR)*, **10**, 25.

Landau, L. D. and Lifshitz, E. M. (1980). *The Classical Theory of Fields*. Butterworth-Heinemann, London.

Landau, L. D. and Lifshitz, E. M. (1984). *Electrodynamics of Continuous Media*, Course of Theoretical Physics Vol. 8. Pergamon Press, Oxford.

Landau, L. D. and Lifshitz, E. M. (1991). *Quantum Mechanics: Non-relativistic Theory*. Pergamon Press, Oxford.

Latora, V., Rapisarda, A. and Ruffo, S. (1999). *Physical Review Letters*, **83**, 2104.

Latora, V., Rapisarda, A. and Tsallis, C. (2001). *Physical Review E*, **64**, 056134.

Lebowitz, J. L. and Lieb, E. H. (1969). *Physical Review Letters*, **22**, 631.

Lebowitz, J. L. and Penrose, O. (1966). *Journal of Mathematical Physics* 7, 98.

Lebwohl, P. A. and Lasher, G. (1972). *Physical Review A*, **6**, 426.

Lenard, A. (1960). *Annals of Physics*, **10**, 390.

Lenard, A. (1961). *Journal of Mathematical Physics*, **2**, 682.

Lenard, A. and Dyson, F. J. (1967). *Journal of Mathematical Physics*, **8**, 423.

Lenard, A. and Dyson, F. J. (1968). *Journal of Mathematical Physics*, **9**, 698.

Levin, Y., Pakter, R. and Teles, T. (2008a). *Physical Review Letters*, **100**, 040604.

Levin, Y., Pakter, R. and Rizzato, F. B. (2008b). *Physical Review E*, **78**, 021130.

Levin, Y., Pakter, R., Rizzato, F. B., Teles, T. N. and Benetti, F. P. C. (2014). *Physics Reports*, **535**, 1.

Leyvraz, F. and Ruffo, S. (2002). *Journal of Physics A: Mathematical and General*, **35**, 285.

Lieb, E. H. and Lebowitz, J. L. (1972). *Advances in Mathematics*, **9**, 316.

Lifshitz, E. M. and Pitaevskij, L. P. (1981). *Physical Kinetics*. Pergamon Press, Oxford.

Lindemann, F. A. (1910). *Zeitschrift fur Physik*, **11**, 609.

Lori, L. (2008). *Modelli con interazione a lungo e corto raggio con variabili microscopiche continue*. Tesi di Laurea, Università di Firenze.

Luttiger, J. M. and Tisza, M. (1946). *Physical Review*, **70**, 954.

Luwel, M., Severne, G. and Rousseeuw, P. J. (1984). *Astrophysics and Space Science*, **100**, 261.

Lynden–Bell, D. (1967). *Monthly Notices of the Royal Astronomical Society*, **136**, 101.

Lynden–Bell, R. M. (1995). *Molecular Physics*, **86**, 1353.

Lynden–Bell, R. M. (1996). In *Gravitational Dynamics* ed. O. Lahav, E. Terlevich, R. J. Terlevich. Cambridge University Presss, Cambridge.

Lynden–Bell, D. (1999). *Physica A*, **263**, 29.

Lynden–Bell, D. and Lynden–Bell, R. M. (2008). *Europhysics Letters*, **42**, 83001.

Lynden–Bell, D. and Wood, R. (1968). *Monthly Notices of the Royal Astronomical Society*, **138**, 495.

MacDowell, L. G., Virnau, P., Müller, M. and Binder, K. (2004). *Journal of Chemical Physics*, **120**, 5293.

Macor, A. (2007). *D'un faisceau test à l'auto-cohérence dans l'interaction onde-particule*. PhD thesis, Université de Provence, Marseille.

Maday, J. M. J. (1971). *Journal of Applied Physics*, **42**, 1906.

Makino, J. and Taiji, M. (1998). *Scientific Simulations with Special Purpose Computer: The GRAPE System*. Wiley.

Mallbor, A., Fanelli, D., Turchi, A., Carletti, T. and Leoncini, X. (2012). *Physical Review E*, **85**, 021148.

Marchioro, C. and Pulvirenti, M. (1994). *Mathematical Theory of Incompressible Nonviscous Fluids*. Springer-Verlag, New York Heidelberg.

Marksteiner, S., Ellinger, K. and Zoller, P. (1996). *Physical Review A*, **53**, 3409.

Marsden, J. E. and Weinstein, A. (1982). *Physica D*, **4**, 394.

Maxwell, J. C. (1876). *Cambridge Philosophical Society's Transactions* **XII**, 90.

Menotti, C., Trefzger, C. and Lewenstein, M. (2007). *Physical Review Letters*, **98**, 235301.

Menotti, C., Lewenstein, M., Lahaye, T. and Pfau, T. (2008). Dipolar interaction in ultra-cold atomic gases. In Campa *et al.* (2008b).

Messer, J. and Spohn, H. (1982). *Journal of Statistical Physics*, **29**, 561.

Meylan, G. and Heggie, D. (1997). *The Astronomy and Astrophysics Review*, **8**, 1.

Miccichè, S. (2008). *Physical Review E*, **79**, 031116.

Michel, J. and Robert, R. (1994). *Communications in Mathematical Physics*, **159**, 195.

Miller, J. (1990). *Physical Review Letters*, **65**, 2137.

Miloshevich, G., Dauxois, T., Khomeriki, R. and Ruffo, S. (2013). *Europhysics Letters*, **104**, 17011.

Mori, T. (2013). *Journal of Statistical Mechanics: Theory and Experiment*, P10003.

Morrish, A. H. (1965). *The Physical Principles of Magnetism*. Wiley, New York.

Morrison, P. J. (1980). *Physics Letters A*, **80**, 383.

Mouhot, C. and Villani, C. (2011). *Acta Mathematica*, **207**, 29.

Moyano, L. G. and Anteneodo, C. (2006). *Physical Review E*, **74**, 021118.

Mukamel, D., Ruffo, S. and Schreiber, N. (2005). *Physical Review Letters*, **95**, 240604.

Nagle, J. F. (1970). *Physical Review A*, **2**, 2124.

Neunzert, H. (1975). Neuere qualitative und numerische Methoden in der Plasmaphysik, Skriptum einer Gastvorlesung an der Gesamthochschule Paderborn.

Neunzert, H. (1978). *Fluid Dynamics Transactions*, **8**.

Nicholson, D. R. (1983). *Introduction to Plasma Theory*. Wiley, New York.

Nordholm, S. (1984). *Chemical Physics Letters*, **105**, 302.

Noullez, A., Fanelli, D. and Aurell, E. (2001). Heap base algorithm. cond-mat/0101336.

Oja, A. S. and Lounasmaa, V. (1997). *Review of Modern Physics*, **69**, 1.

O'Neil, T. M., Winfrey, J. H. and Malmberg, J. H. (1971). *Physics of Fluids*, **14**, 1204.

Onsager, L. (1949). *Nuovo Cimento Supplement*, **6**, 279.

Padmanabhan, T. (1990). *Physics Reports*, **188**, 285.

Padmanabhan, T. (2009). Statistical Mechanics of gravitating systems in static and expanding backgrounds. In Dauxois *et al.* (2009).

Pakter, R. and Levin, Y. (2011). *Physical Review Letters*, **106**, 200603.

Pedlosky, J. (1987). *Geophysical Fluid Dynamics*, Springer-Verlag.

Peebles, P. J. (1980). *The Large-scale Structure of the Universe*. Princeton University Press, Princeton.

Peierls, R. (1938). *Physical Review*, **54**, 918.

Penfold, R., Nordholm, S., Jönsson, B. and Woodward, C. E. (1990). *Journal of Chemical Physics*, **92**, 1915.

Pierce, J. R. (1950). *Traveling Wave Tubes*. Van Nostrand, New York.

Posch, H. A., Narnhofer, H. and Thirring, W. (1990). *Physical Review A*, **42**, 1880.

Posch, H. A. and Thirring, W. (2006). *Physical Review E*, **74**, 051103.

Potapenko, I. F., Bobylev, A. V., de Azevedo, C. A. and de Assis, A. S. (1997). *Physical Review E*, **56**, 7159.

Prager, S. (1961). *Advances in Chemical Physics*, **4**, 201.

Purcell, E. M. and Pound, R. V. (1951). *Physical Review*, **81**, 279.

Ramirez-Hernandez, A., Larralde, H. and Leyvraz, F. (2008a). *Physical Review Letters*, **100**, 120601.

Ramirez-Hernandez, A., Larralde, H. and Leyvraz, F. (2008b). *Physical Review E*, **78**, 061133.

Reidl, C. J. and Miller, B. N. (1991). *Astrophysical Journal*, **371**, 260.

Reidl, C. J. and Miller, B. N. (1992). *Physical Review A*, **46**, 837.

Robert, R. (1990). *Comptes Rendus de l'Académie des Sciences: Série I Maths*, **311**, 575.

Robert, R. (1991). *Journal of Statistical Physics*, **65**, 531.

Robert, R. and Sommeria, J. (1991). *Journal of Fluid Mechanics*, **229**, 291.

Rouet, J. L., Feix, M. R. and Navet, M. (1990). *Vistas in Astronomy*, **33**, 357.

Rouet, J. L., Jamin, E. and Feix, M. R. (1991). Fractal properties in the simulation of a one-dimensional spherically expanding universe. *Lecture Notes in Physics: Applying Fractals in Astronomy*, 161.

Ruelle, D. (1969). *Statistical Mechanics: Rigorous Results*. Benjamin, New York.

Ruffo, S. (1994). Hamiltonian dynamics and phase transitions. In *Transport and Plasma Physics*, ed. Benkadda, S., Elskens, Y. and Doveil, F. World Scientific, Singapore.

Rybicki, G. B. (1971). *Astrophysical and Space Sciences*, **14**, 56.

Salzburg, A. M. (1965). *Journal of Mathematical Physics*, **6**, 158.

Sato, M. and Sievers, A. J. (2004). *Nature*, **432**, 486.

Sato, M. and Sievers, A. J. (2005). *Physical Review B*, **71**, 214306.

Sauer, J. A. (1940). *Physical Review*, **57**, 142.

Severne, G. and Luwel, M. (1986). *Astrophysics and Space Science*, **122**, 299.

Shandarin, S. F. and Zeldovich, Ya. B. (1989). *Review of Modern Physics*, **61**, 185.

Slama, S., Krenz, G., Bux, S., Zimmermann, C. and Courteille, P. W. (2008). Scattering in a high-Q ring cavity. In Campa *et al.* (2008b).

Slama, S., Bux, S., Krenz, G., Zimmermann, C. and Courteille, P. W. (2007). *Physical Review Letters*, **98**, 053603.

Sommeria, J. (2002). Two dimensional turbulence. In *New Trends in Turbulence*, Les Houches Session LXXIV, ed. M. Lesieur, A. Yaglom, F. David, Springer-Verlag, EDP Sciences, Berlin.

Sota, Y., Iguchi, O., Morikawa, M., Tatekawa, T. and Maeda, K. (2001). *Physical Review E*, **64**, 056133.

Spohn, H. (1991). *Large Scale Dynamics of Interacting Particles*. Springer, Berlin.

Stahl, B., Kiessling, M. K. H. and Schindler, K. (1995). *Planetary and Space Sciences*, **43**, 271.

Tamarit, F. and Anteneodo, C. (2000). *Physical Review Letters*, **84**, 208.

Tamashiro, M. N., Levin, Y. and Barbosa, M. C. (1999). *Physica A*, **268**, 24.

Tatekawa, T., Bouchet, F., Dauxois, T. and Ruffo, S. (2005). *Physical Review E*, **71**, 056111.

Tennyson, J. L., Meyss, J. D. and Morrison, P. J. (1994). *Physical Review E*, **71**, 1.

Thirring, W. (1970). *Zeitschrift für Physik*, **235**, 339.

Tonks, L. and Langmuir, I. (1929). *Physical Review*, **33**, 195.

Torcini, A. and Antoni, M. (1999). *Physical Review E*, **59**, 2746.

Touchette, H. (2009). *Physics Reports*, **78**, 1.

Trenti, M. and Hut, P. (2008). *Scholarpedia*, **3**(5):3930.

Tsidil'kovskii, I. M. (1987). *Soviet Physics Uspekhi*, **30**, 676.

Tsuchiya, T. and Gouda, N. (2000). *Physical Review E*, **61**, 948.

Tsuchiya, T., Gouda, N. and Konishi, T. (1996). *Physical Review A*, **53**, 2210.

Tsuchiya, T., Konishi, T. and Gouda, N. (1994). *Physical Review E*, **50**, 2607.

Tsunoda, S. I., Doveil, F. and Malmberg, J. H. (1987). *Physical Review Letters*, **58**, 1112.

Turkington, B., Majda, A., Haven, K. and DiBattista, M. (2001). *Proceedings of the National Academy of Sciences (USA)*, **98**, 12346.

Turkington, B. (2009). Statistical mechanics of two-dimensional and quasi-geostrophic flows. In Dauxois *et al.* (2009).

Valageas, P. (2006a). *Astronomy and Astrophysics*, **450**, 445.

Vallis, G. K. (2006). *Atmospheric and Oceanic Fluid Dynamics*. Cambridge University Press, Cambridge.

Van Hove, L. (1949). *Physica*, **15**, 951.

Van Kampen, N. G. (1955). *Physica*, **21**, 949.

Van Kampen, N. G. (1992). *Stochastic Processes in Physics and Chemistry*. North-Holland, Elsevier.

Van Vleck, J. H. (1937). *Physical Review*, **52**, 1178.

Velazquez, L. and Curilef, S. (2009). *Journal of Physics A*, **42**, 095006.

Venaille, A. (2008). *Mélange et circulation océanique: une approche par la physique statistique*. PhD thesis, Université de Grenoble.

Venaille, A. and Bouchet, F. (2009). *Physical Review Letters*, **102**, 104501.

Venaille, A. and Bouchet, F. (2011a). *Journal of Statistical Physics*, **143**, 346.

Venaille, A. and Bouchet, F. (2011b). *Journal of Physical Oceanography*, **41**, 1860.

Vergassola, M., Dubrulle, B., Frisch, U. and Noullez, A. (1994). *Astronomy and Astrophysics*, **289**, 325.

Villain, J. (2008). *Reflets de la Physique*, **7**, 10.

Vlasov, A. A. (1945). *Journal of Physics (USSR)*, **9**, 25.

von Cube, C., Slama, S., Kruse, D., Zimmermann, C., Courteille, Ph. W., Robb, G. R. M., Piovella, N. and Bonifacio, R. (2003). *Physical Review Letters*, **93**, 083601.

Walker, R. P. (1996). Insertion devices: undulators and wigglers. In *Grenoble 1996, Synchroton Radiation and Free Election Lasers*, ed. Turner, S. Ed.

Weinberg, S. (1972). *Gravitation and Cosmology*. Wiley, New York.

Werner, J., Griesmaier, A., Hensler, S., Stuhler, J., Pfau, T., Simoni, A. and Tiesinga, E. (2005). *Physical Review Letters*, **94**, 183201.

de Wette, F. W. (1964). *Physical Review*, **135**, A287.

White, R. M. (1970). *Quantum Theory of Magnetism*. McGraw-Hill, New York.

Wigner, E. P. (1934). *Physical Review*, **46**, 1002.

Willis, C. R. and Picard, R. H. (1974). *Physical Review A*, **9**, 1343.

Wrubel, J. P., Sato, M. and Sievers, A. J. (2005). *Physical Review Letters*, **95**, 264101.

Yamaguchi, Y. Y. (2003). *Physical Review E*, **68**, 066210.

Yamaguchi, Y. Y. (2008). *Physical Review E*, **78**, 041114.

Yamaguchi, Y. Y., Barré, J., Bouchet, F., Dauxois, T. and Ruffo, S. (2004). *Physica A*, **337**, 36.

Yamaguchi, Y. Y., Bouchet, F. and Dauxois, T. (2007). *Journal of Statistical Mechanics: Theory and Experiment*, P01020.

Yano, T. and Gouda, N. (1998). *Astrophysical Journal Supplement Series*, **118**, 267.

Youngkins, V. P. and Miller, B. N. (2000). *Physical Review E*, **62**, 4583.

Yvon, J. (1935). *La Théorie des Fluides et l'Équation d'État*, Hermann.

Zeldovich, Ya. B. and Novikov, I. D. (1974). *The Structure and the Evolution of the Universe*. Nauka, Moscow.

Index